T0328212

Fundamentals of Molecular Structural Biology

Fundamentals of Molecular Structural Biology

Subrata Pal

Academic Press is an imprint of Elsevier
125 London Wall, London EC2Y 5AS, United Kingdom
525 B Street, Suite 1650, San Diego, CA 92101, United States
50 Hampshire Street, 5th Floor, Cambridge, MA 02139, United States
The Boulevard, Langford Lane, Kidlington, Oxford OX5 1GB, United Kingdom

Notices

Knowledge and best practice in this field are constantly changing. As new research and experience broaden our understanding, changes in research methods, professional practices, or medical treatment may become necessary.

Practitioners and researchers must always rely on their own experience and knowledge in evaluating and using any information, methods, compounds, or experiments described herein. In using such information or methods they should be mindful of their own safety and the safety of others, including parties for whom they have a professional responsibility.

To the fullest extent of the law, neither the Publisher nor the authors, contributors, or editors, assume any liability for any injury and/or damage to persons or property as a matter of products liability, negligence or otherwise, or from any use or operation of any methods, products, instructions, or ideas contained in the material herein.

Library of Congress Cataloging-in-Publication Data
A catalog record for this book is available from the Library of Congress

British Library Cataloguing-in-Publication Data
A catalogue record for this book is available from the British Library

ISBN: 978-0-12-814855-6

For information on all Academic Press publications visit our website at https://www.elsevier.com/books-and-journals

Publisher: Andre Gerhard Wolff
Acquisition Editor: Mary Preap
Editorial Project Manager: Megan Ashdown
Production Project Manager: Punithavathy Govindaradjane
Cover Designer: Mark Rogers

Typeset by SPi Global, India

Dedication

To my mother

Contents

About the Author

Subrata Pal

Bachelor of Science in Physics, Calcutta University, Kolkata, India, 1972.
Master of Science in Physics, Calcutta University, Kolkata, India, 1974.
Ph.D. in Molecular Biology, Calcutta University, Kolkata, India, 1982.
Asst. Professor, Physics, in a Calcutta University-affiliated college, 1979–82.
Visiting Fellow, National Institutes of Health, National Cancer Institute, Bethesda, MD, United States, 1984–87.
Fellow/Claudia Adams Barr Investigator, Harvard Medical School, Dana-Farber Cancer Institute, Boston, MA, United States, 1988–92.
Assoc. Professor, Molecular Biology, Jadavpur University, Kolkata, India, 1993–2001.
Professor, Molecular Biology, Genomics and Proteomics, Jadavpur University, Kolkata, India, 2001–15.
Claudia Adams Barr special investigator award for basic contribution to cancer research at the Dana-Farber Cancer Institute (Harvard Medical School), Boston, MA, United States, 1991.

Preface

Physics is gaining a leading position in biology. This is not to be understood that physics is conquering biology and shoving biologists to the background. Simply, the science we used to call biology is evolving into physics, thereby confirming the view that natural science is passing through a period of reconstruction on a unified foundation.

Alexander Kitaigorodsky
I am a physicist (1971)

The contribution of physics-based techniques to biology had begun with the development of the microscope. It marked the beginning of the studies of cell structure—structural cell biology. Subsequently, with the discovery of the two most important macromolecules of the living cell—DNA and protein—biology had turned molecular. In order to understand how these molecules function, it became necessary to elucidate their structures.

Visible light-dependent microscopy has its limitations. Visualization of molecular structures at atomic resolution by such microscopes is not even a theoretical possibility. The problem has been addressed, in the last century, by the development of three powerful physical methodologies—X-ray crystallography, nuclear magnetic resonance (NMR), and electron crystallography. Within a span of less than ten years, the double helical structure of DNA and high-resolution structures of two proteins, myoglobin and hemoglobin, were published and, with these discoveries, the age of molecular structural biology had begun.

Physics has provided not only the techniques and tools for elucidation of the structure of biomolecules, but also concepts and laws for investigation of their dynamics. Based on physical laws it is possible to compute biomolecular interactions—as for example, how two molecules would bind, bend, and rotate each other—leading to mechanistic explanations for different biological processes.

Using the techniques mentioned above, more than 150,000 macromolecular structures have already been solved and deposited in the Worldwide Protein Data Bank (wwPDB). The number continues to grow at a rapid pace. These structures have been the basis for molecular biological investigations of various activities within and involving the cell.

Keeping pace with the scientific developments, institutional biology curricula have also been evolving. In 1965, the first edition of *Molecular Biology of the Gene* was published as "The field of molecular biology was ripe for its first textbook, defining for the first time the curriculum for undergraduate courses in this topic" (in Preface to the 7th edition of the book). Presently, there are molecular biology and related courses in universities across the world. These courses have been aptly supported by a number of popular textbooks written by renowned authors.

"Molecular biology" named courses, either introductory or advanced, mostly center around the "basic genetic mechanisms," namely, maintenance and expression of the genome and their regulation. Such courses are usually catered at the undergraduate level with due emphasis on biological and experimental aspects. However, the subject is transcending from its descriptive nature to a more mechanistic format. Even in the traditional molecular biology courses, explanation and understanding of the basic genetic mechanisms are becoming ever more dependent on macromolecular structure-function correlation. This is proving to be true not only for the graduate and upper-division courses, but also for the lower-division courses.

Looking at the graduate-level courses, we find titles such as structural biology, molecular interactions, molecular systems biology, biophysical methods, structural biochemistry, structure and function of biological macromolecules, macromolecular machines, techniques in structural biology, protein folding, molecular recognition and interactions, structure-function relationship of proteins, structural biology and proteomics, biophysical methods in life sciences, etc. Clearly and expectedly, these courses are interdisciplinary in nature with a judicious blend of physical and biological sciences. As such, it appears that no single textbook does or can cover all aspects of the above-named graduate level courses and no individual course can depend exclusively on one text.

During my tryst with molecular biology, first as a student and researcher and subsequently as a teacher, I have experienced how wonderfully the subject has evolved in the last five decades. Keeping pace with the developments, several landmark textbooks, written by world-famous authors, appeared one after another. My learning as well as my teaching have been enormously benefitted by these textbooks. Yet, particularly in my later years of teaching, I felt the need for a textbook going deeper into the structural aspects of molecular biology, at least as a supplement and not a substitute to the already available excellent textbooks.

It may be reasonable to consider that a textbook containing the words "molecular," "structural," and "biology" in its title should embrace macromolecular structures and interactions with the arms of conventional molecular biology featured by the "basic genetic mechanisms." Such a textbook would be useful for the interdisciplinary courses as well as the conventional molecular biology courses upgraded with the inclusion of macromolecular structures.

With all such considerations as outlined above, the present book is an attempt to cater to the fundamentals of molecular structural biology in a concise, yet comprehensive manner. The layout of the text has been stepwise. The historical perspective in the *Introduction* forms the philosophical basis of the entire book, for it is my personal belief that science bereft of philosophy produces technicians and not scientists. Chapters 1–4 provide the mathematical and physical concepts required for the subject. They are likely to be useful in a varying degree to the reader depending on his/her background. Based on these concepts, Chapters 5–7 deal with macromolecular structures—formation, determination, and representation. In Chapters 8–13, these structures are utilized to explain the basic genetic mechanisms. The chapters that follow discuss some advanced issues surrounding macromolecular structures, namely, structural genomic/proteomics, cell signaling and systems biology, macromolecular assemblies, and computational molecular biology. Covering a wide spectrum, the book can be used as a reference as well as a text partially fulfilling the requirements of the related courses.

Needless to say, in science, every theory is preceded by experiments and observations. Accordingly, a number of published text books have elegantly described the experiments that have laid the foundation of different concepts and theories in molecular biology. The present book, though alluding to the results of many of these experiments, refrains from duplicating such efforts.

Instead, in order to do justice to the title, the book emphasizes the basic concepts and takes care not to overburden the reader with excessive information. Redundancy is avoided as much as possible. For example, it is mentioned in Chapter 11 that the enzyme aminoacyl-tRNA synthetase links an amino acid to a specific tRNA. Although, based on their active site architecture and consensus sequence elements, the family of aminoacyl-tRNA synthetases is divided into two classes, I and II, all these enzymes are united by a common aminoacylation reaction that occurs in two steps. The chapter highlights only one enzyme belonging to class I to understand the mechanism of the aminoacylation reaction.

In the same chapter, out of three release factors, bacterial RF1 and RF2 and eukaryotic eRF1, that are responsible for translation termination, only RF2 is selected for detailed discussion. The underlying expectation is that if the basic concepts are aptly conveyed, the interested reader is competent enough to explore the literature for mechanisms involving the other factors and identify the subtle differences.

It will be absolutely unwise to claim that the book is flawless. Critical comments and suggestions are most welcome to prompt the author to make necessary amends and improvements in subsequent editions, if and when opportunity arrives.

Subrata Pal

Acknowledgments

The book is dedicated to the memory of my mother who passed away while the manuscript was in preparation.

I respectfully express my indebtedness to my mentors—late Professors Binayak Dutta-Roy and Ramendra Kumar Poddar—from whom I learnt quantum mechanics and molecular biology, respectively. I will sorrowfully miss them when the book is published.

I gathered the initial confidence for undertaking this daunting task from Professor Geoffrey M. Cooper, author of the widely circulated book *The Cell: A Molecular Approach* and my mentor at Dana-Farber Cancer Institute/Harvard Medical School, Professor Utpal C. Chaudhuri, my teacher in Calcutta University, and Professor Chanchal Das Gupta, my academic senior.

My sincere thanks to Professor Al Burlingame of University of California in San Francisco and Dr. Samuel Wilson of National Institute of Environmental Health Sciences, Durham, North Carolina, for all their encouragement.

My friend and colleague Professor Amitava Das, himself a prolific writer, has provided plenty of valuable guidance in my endeavor and my Ph.D. student Dr. Sutripta Sarkar has always been very enthusiastic about the work.

Special acknowledgments

Most of the figures/structures in the book have been drawn with UCSF Chimera, developed by the Resource for Biocomputing, Visualization, and Informatics at the University of California, San Francisco. My deep gratitude to Professor Tom E. Ferrin, the Director of the facility, for giving me not only permission, but also encouragement to use the program.

Enormous thanks to Mica H. Haley, Megan Ashdown, and the entire Elsevier team for their unhesitating help and cooperation in addressing my queries and difficulties right from the inception of the project to the publication of the book.

And last but not the least, my daughter Sharmishtha and my wife Subhra have been extremely supportive of this effort in all possible ways.

Subrata Pal

CHAPTER

Introduction—A historical perspective

1

Man began to take interest in nature, which includes plants and animals, not in any aesthetic or philosophic sense, but only in relation to his immediate need primarily for food. The observational knowledge of nature, particularly the habits of animals and properties of plants, formed the basis of present day biological sciences.

1.1 Biology begins as natural history

Biology, as we all know, is the study of living organisms and their vital processes. It began as natural history which aimed to understand the whole organism in context. Natural history, which is essentially the study of plants and animals, was based on observational methods. It can be found from the earliest recorded history of biology that the Babylonians, Egyptians, Chinese, and Indians made innumerable observations in the course of their agricultural and medical practices. Yet, the accumulated biological information did not immediately and automatically lead to any rational concept or hypothesis. Rapid progress started to be made with the advent of Greek civilization.

Greek civilization produced legendary personalities who, by virtue of their astounding insight, examined the phenomena of the natural world and made seminal contributions to natural philosophy. The most distinguished of them was Aristotle (384–322 BC) whose interests spanned all branches of knowledge including biology. He is recognized as the originator of the scientific study of life.

Aristotle was the first to undertake a systematic classification of animals based on specific principles, some of which remain valid even today. However, he made little contribution to the study of plants. This was left to his student, Theophrastus, who is said to have done for botany what Aristotle did for zoology.

1.2 Nature of matter

The ancient Greek philosophers did not keep themselves confined to classification and categorization. They speculated about the nature of matter and formulated hypotheses. Democritus (470–380 BC) proposed that all matter consisted of tiny particles which were further indivisible. These particles were called *atamos*

Fundamentals of Molecular Structural Biology. https://doi.org/10.1016/B978-0-12-814855-6.00001-8

(Greek meaning: indivisible). The early "atomists" thought that the infinite universe consisted of *atamos* (or atoms) and void space.

One of the earliest Greek philosophers, Thales of Miletus, held the view that the universe contained a creative "force" which he called *physis* (precursor of the term physics). Thales thought that the basic element of matter was water; on the other hand, Anaximenes believed that it was air, while Heraclitus maintained that it was fire. Empedocles combined the ideas of Thales, Anaximenes, and Heraclitus and added one of his own—earth. Accordingly, the notion emerged that all matter was made of differing amounts of fire, air, water, and earth held together by "forces" of attraction and repulsion.

Physis as the creative force was accepted also by Hippocrates, the eminent Greek physician. However, members of the Hippocratic school gave little importance to the roles thought to be played by fire, air, water, and earth. Instead, they believed that all living bodies were made up of four humors (Latin meaning: liquid or fluid)—blood, phelgm, choler (yellow bile), and melancholy (black bile).

Needless to say, for the ancient Greeks, it was not possible to verify all such speculations and hypotheses through experiments. The civilization had no dearth of creative minds, but lacked proper tools to conduct scientific investigations. The world had to wait hundreds of years before the microscope revealed the basic structure of both plants and animals.

1.3 Microscope reveals internal structure of living organisms

The magnifying power of segments of a glass sphere was known for nearly 2000 years. During the first century AD, glass was invented and the Romans were testing different shapes of clear glass to see if they could magnify an object. Sometime towards the end of 16th century, two Dutch spectacle-makers, Hans Jensen and his son Zacharias, invented the compound microscope by putting several lenses in a tube. However, their instruments were of little practical utility since the magnification was only around 9 × and the images were blurred.

In the late 17th century, another Dutchman, Antonie van Leeuwenhoek, became the first man to make and use a real microscope. Using single lenses instead of their combinations, Leeuwenhoek observed a number of biological specimens including protozoans (which he called "animalcules") and bacteria. His microscopes achieved remarkable magnifications up to 270 ×.

1.4 Cell theory

Improvements in microscopy facilitated the introduction of a new concept in the internal structure of living organisms—the cell. The credit for the first description of the cell goes to Robert Hooke, an English physicist and microscopist. Examining a slice of cork under the microscope, Hooke found air-filled compartments and introduced the term "cells" (cell in Latin means a small room) or "pores" to refer to these units. Hooke's observations were published in 1665 in Micrographia. Nevertheless, the discovery had to wait nearly 200 years for its significance to be appreciated.

In the meantime, microcopy continued to undergo technical improvement. One problem with the earlier microscopes was that of "chromatic aberration," as a result of which the resolving power of the instruments was compromised. The problem was satisfactorily addressed by the introduction of achromatic microscopes in the 1830s.

In 1838, German botanist Mathias Jacob Schleiden postulated that every structural element of plants is composed of cells or their products. Subsequently, in the following year, German zoologist Theodor Schwann extended the proposition to include animals. Biological science saw a rapprochement between botany and zoology. The conclusions of Schleiden and Schwann together formed the cell theory—a gigantic advance in the study of living organisms. Added to this in the 1850s was Rudolf Virchow's aphorism *omnis cellula a cellula* (every cell from a preexisting cell). Now, the attention of the scientific world shifted to the "living" processes inside the cell.

Together with the cell theory, two other landmark developments during the second half of the 19th century made the study of intracellular components all-the-more compelling. These developments were associated with two legendary individuals—Charles Darwin and Gregor Mendel.

1.5 Theory of natural selection and laws of heredity

In 1858, Charles Darwin published his theory of natural selection in "On the Origin of Species by Means of Natural Selection". The crux of the theory is as follows: In the randomly varying nature, some variations are more advantageous than others. There is always a struggle for existence and those organisms which are better adjusted to their environment, even slightly, will most likely survive and transmit their advantageous traits to the next generation.

On the other hand, Gregor Mendel, universally acclaimed as the father of genetics, carried out extensive fertilization experiments with garden peas and formulated a set of laws, known as the laws of heredity, for the transmission of trait units (later came to be known as "genes") from one generation to another through reproductive mechanism. The first principle, the law of segregation, states that the hereditary units are paired in the parent and segregate during the formation of gametes. Secondly, the law of independent assortment states that each pair of units is inherited independently of all other pairs. The third tenet, the law of dominance, maintains that the trait units act as pairs. In the case of a pair with contrasting traits, the 'dominant' one appears in the hybrid offspring, although the 'recessive' one is also present.

1.6 Gene and genetics

Ironically, Mendel's findings were not recognized during his lifetime. It was only at the turn of the century that his work was rediscovered and a spurt of activities in relation to the laws of heredity ensued thereafter. In 1909, Danish botanist Wilhelm Johanssen coined the term "gene" as a physical and functional unit of heredity. Earlier, in 1905, British geneticist William Bateson had introduced the term "genetics."

Now, the question was where in the cell the genes are located. The cell nucleus and the chromosome it contained were already known by then. Between 1825 and 1838, the nucleus was reported by three investigators including Robert Brown, who is credited for coining the term "nucleus." Subsequently, German anatomist Walther Flemming, who is said to be the founder of the science of cytogenetics, was the first to observe and systematically describe the movement of chromosomes in the cell nucleus during normal cell division. In 1915, American geneticist Thomas Morgan and his students asserted that genes are the fundamental units of heredity. Their research confirmed that specific genes are found on specific chromosomes and that genes are indeed physical objects. The chromosome theory of inheritance emerged.

1.7 Nature of "physical objects" in the cell

Once it was established that genes are physical objects, the next step, as expected, was to investigate the physical (and chemical) nature of these objects. Major advances were already made in the investigation of another kind of important "physical objects" of the cell—the proteins.

In 1789, French chemist Antoine Fourcroy had recognized several distinct varieties of proteins (though the term was not used then) from animal sources. These were albumin, fibrin, gelatin, and gluten. Several years later, in 1837, Dutch chemist Geradus Johannes Mulder determined the elemental composition of many of these molecules. The term "protein" was subsequently used by Mulder's associate Jacob Berzelius to describe these molecules. The name was derived from the Greek word πρωτειοζ which means "primary," "in the lead," or "standing in front."

One class of proteins are the enzymes which catalyze the biological processes in living organisms. The first enzyme to be discovered was diastase (now called amylase). It was extracted from malt solution at a French sugar factory by Anselme Payne in 1833. The term "enzyme" was coined in 1878 by German physiologist Wilhelm Kühne. In 1897, Eduard Buchner, a German chemist and zymologist, fermented sugar with yeast extracts in the absence of live organisms.

Nucleic acid, on the other hand, was discovered by a Swiss physician Friedrich Mischer. Working in the laboratory of biochemist Felix Hoppe-Seyer at Tübingen, Mischer initially intended to study proteins in leucocytes (blood cells containing nuclei). However, in his experiments he noticed a precipitate of an unknown substance. On further examination, the substance was found to be neither a protein nor a lipid. Unlike proteins, it contained a large amount of phosphorous. Since the substance was from the cells' nuclei, it was named "nuclein."

Later, Albrecht Kossel, another scientist in Hoppe-Seyer's laboratory, found that nuclein consisted of four bases and sugar molecules. Kossel provided the present chemical name "nucleic acid." In 1909, a Russian-born American scientist, Phoebus Levine, isolated nucleotides, the basic building blocks of ribonucleic acid (RNA).

1.8 DNA as the genetic material

All these advances notwithstanding, till the late 1920s, the question regarding the nature of the genetic material remained unanswered. The scenario started changing in 1928 when British bacteriologist Fred Griffith carried out an experiment on the pathogenicity of *Streptococcus pneumonia*. Though not conclusive, the experiment did lay the foundation for later discovery that DNA is the genetic material. The results showed that apparently something in the cell debris of a virulent strain of *Streptococcus* had "transformed" an avirulent strain to become virulent. This something was called the "transforming principle"; it remained unclear what the transforming principle was—RNA, DNA, protein, or a polysaccharide.

The conclusive evidence was provided 16 years later, in 1944, by American microbiologists Oswald Avery, Colin MacLeod, and Maclyn McCarty. They established that the active genetic principle was DNA since its transforming activity could be destroyed by deoxyribonuclease, an enzyme that specifically degrades DNA.

A confirmation of the conclusion made by Avery and his colleagues came in 1952 from two scientists, Alfred Hershey and Martha Chase, who were working at Cold Spring Harbor Laboratory. The protein coat and DNA core of a bacteriophage were labeled with ^{35}S and ^{32}P, respectively. By infecting a bacterial culture with the radiolabeled phage, they showed that the parental DNA, and not the parental protein, was present in the progeny phage.

1.9 Biology turns molecular—natural science becomes unified

Undoubtedly, with the discovery of the two most important macromolecules of the living cell—DNA, that carries the blueprint of life, and protein, that executes the plan—biology had turned molecular. In fact, as early as in 1938, Warren Weaver, who was the director of the Natural Sciences section of the Rockefeller Foundation at that time, introduced the term molecular biology. The cellular processes were now required to be explained in terms of molecular interactions. It became ever more evident that the living system conforms to the laws of physics and chemistry.

Eminent geneticist Hermann Mueller recognized the similarity between the contemporary developments in physics and genetics. In 1936, he even made a fervent appeal to the physicists and chemists to join forces with him and his colleagues in unraveling the fundamental properties of genes and their actions. Eight years later, in a book entitled "What is Life," Erwin Schrödinger, one of the founders of quantum mechanics, made a similar plea and expressed his thoughts, many of which were similar to those of Mueller.

Physicists indeed joined forces with biologists with the goal of understanding how the two disciplines complemented each other. As Alexander Kitaigorodsky, a renowned physicist and passionate popularizer of science, wrote (1971)—"Physics is gaining a leading position in biology. This is not to be understood that physics is conquering biology and shoving biologists to the background. Simply the science we used to call biology is evolving into physics, thereby confirming the view that natural science is passing through a period of reconstruction on a unified foundation." Several X-ray crystallographers as well as structural chemists, such as Linus Pauling, dedicated their knowledge to the investigation of macromolecular structure.

1.10 Deeper into the structure of matter

We have already seen that the contribution of physics-based techniques to biology had begun with the development of the microscope. It marked the beginning of the studies of cell structure—structural cell biology. Thereafter, improvements in microscopy have facilitated ever more revelations of the cellular and subcellular structures.

Nevertheless, visible light-dependent microscopy had its limitations. As the resolution was inversely proportional to the wavelength of the observing light, the microscope had reached the theoretical limit of its resolving power. The problem was overcome by the introduction of the electron microscope. Here, visible light waves are replaced by electron waves which are bent and focused by electromagnetic lenses. The first electron microscopes were manufactured in the 1930s. However, traditional electron microscopes were not able to determine the precise structures of macromolecules such as proteins and nucleic acids.

Molecules, including biomolecules, are made up of atoms and bonds with dimensions around 0.1 nm. Three different methods can provide structural information at this resolution—X-ray crystallography, nuclear magnetic resonance (NMR) spectroscopy, and electron crystallography.

X-rays were discovered by Wilhelm Conrad Röntgen in Germany in the year 1895. Seventeen years later, Max von Laue, speculating that the wavelength of X-rays might be comparable to the interatomic distances, observed the diffraction pattern from a blue crystal of copper sulfate. Then, in 1913, the father-and-son team of Henry Bragg and Lawrence Bragg determined the structure of crystalline common salt and initiated the field of X-ray crystallography. As we shall see below, X-ray crystallography has become the most widely used technique in the structural analysis of biological molecules.

The phenomenon of nuclear magnetic resonance was first observed in 1946 by the physicists Felix Bloch of Stanford University and Edward Purcell of Harvard

University independent of each other. In 1950, Warren Proctor and Fu Chun Yu discovered that the two nitrogen atoms in NH_4NO_3 produced two different NMR signals—a phenomenon that became known as "chemical shift." This discovery, together with the contributions of physicist Albert Overhauser and many others, has made NMR spectroscopy the second most useful tool in the field of structural biology.

Electron crystallography is based on electron diffraction phenomenon. In 1968, David DeRosier and Aaron Klug laid the foundation of electron crystallography by demonstrating that electron microscopic images of a two-dimensional crystal generated at different angles to the electron beam could be combined with electron diffraction data to produce a three-dimensional structure. Since then, electron crystallography has been used to determine the structures of a relatively few but important proteins.

Understandably, with each of the above technical advances, view of the cell became more distinct and enlarged than before. From the whole cell and its organelles, structural investigation could go down to the molecular level.

1.11 Molecular biology endowed with structures

By the early 1950s, DNA had become a prominent target for structural studies. Several investigators engaged themselves in obtaining X-ray diffraction photographs of DNA fibers. The most prominent of them were those by Rosalind Franklin at King's College, London. The photographs she obtained were, according to John Desmond Bernal (affectionately called "Sage"), "among the most beautiful X-ray photographs of any substance ever taken."

At the same time, Francis Crick and James Watson were working together at the Cavendish Laboratory, University of Cambridge, to build a model of DNA. Based on the X-ray diffraction data of Rosalind Franklin and Maurice Wilkins (also at King's College) and profound theoretical insight of Crick, they published the double helical structure of DNA in the journal Nature in 1953.

It is interesting to note here that the proposal of the double helical structure of DNA, though helping to explain two vital dynamic processes of the living cell—DNA replication and DNA recombination—was verified at atomic detail by single-crystal diffraction much later, in the late 1970s. The crystal structures of all forms of the DNA double helix—A, B, and Z—were published during this period. On the other hand, the structure of a complete A-RNA turn was visualized only in the late 1980s.

Around the same time at the Cavendish Laboratory, Max Perutz and John Kendrew were pursuing X-ray crystallographic studies of proteins since the 1940s. Perutz was working on horse hemoglobin and Kendrew on a protein four times smaller, sperm whale myoglobin. It was Kendrew who, in 1957, met with the first success which was, nevertheless, aided by Perutz' solution to the phase problem. The structure of myoglobin was initially determined at 6 Å resolution, soon improved to 2 Å.

In 1960, Perutz published the structure of hemoglobin determined at 5.5 Å and later improved to 2.8 Å. It will not be inappropriate to affirm here that with the discovery of the structures of these crucial biological macromolecules, first the DNA and then the two proteins, the age of molecular structural biology had begun.

Following the successes of Perutz and Kendrew, the structure of seven additional proteins were published in the 1960s; hen egg white lysozyme was the first enzyme whose structure was known in 1965. Nevertheless, all the proteins including enzymes, whose crystal structures had been determined till the late 1970s, were water-soluble. In contrast, membrane-located proteins, being insoluble, were more difficult to crystallize. The first membrane protein to be crystallized in 1980 was *Halobacterium halobium* rhodopsin, whereas the photosynthetic reaction center (PSRC) from *Rhodoppseudomonas viridis* became the first transmembrane protein to have its three-dimensional structure determined by X-ray crystallography in 1984.

1.12 Structural complex(ity) disentangled

From single polypeptide chains, X-ray crystallography progressed towards multi-subunit structures. In 1989, the three-dimensional structure of *Escherichia coli* RNA polymerase core, a five-subunit enzyme, was determined by electron crystallography. This was closely followed by the determination of the three-dimensional structure of yeast RNA polymerase II, an enzyme more complex than the bacterial polymerase, at 16 Å resolution. Eventually, in 2001, Roger Kornberg and his colleagues at Stanford University published the three-dimensional structure of a 10-subunit variant of the yeast enzyme (out of the complete set of 12 distinct polypeptides), determined by X-ray crystallography at 2.8 Å resolution. The publication enormously facilitated a better understanding of the multiple steps of RNA polymerase II-mediated transcription in eukaryotes.

Another remarkable feat in macromolecular structure analysis was the determination of the structure of the ribosome—a huge molecular complex that translates genetic messages in the living cell. Ribosomes, prokaryotic or eukaryotic, consists of two subunits—small and large. Each subunit contains one to three RNAs and several proteins. Initially, the crystals of complete ribosomes were refractory to X-ray diffraction studies. Therefore, individual subunits were crystallized and their structures determined separately. The structure of a bacterial ribosomal subunit was first known in 1987. The eukaryotic subunits, which were more complex than their bacterial counterparts, had to wait till the turn of the century to have their structures resolved.

1.13 Molecular structural biology confronts human disease

In the last few decades, molecular structural biology has made a profound impact on the understanding of human disease and discovery of its remedy. Preliminary efforts in this direction had started with the publication of hemoglobin structure. Later in the

1980s, a large number of pharmaceutical companies became interested in utilizing structural data to design therapeutic molecules that would target specific proteins or nucleic acids. Among others were Agouron Pharmaceuticals in California, USA, and Molecular Discovery in UK.

A shining example of the rational drug design effort in the 1990s has been the development of drugs for the treatment of human immunodeficiency virus (HIV) infection. Work on the inhibitors of HIV aspartic protease, whose structure was already available, led to Food and Drug Administration's (FDA) endorsement of four successful drugs. Additional protease inhibitors were developed later and, eventually, a fatal disease became a manageable infection.

1.14 From gene to genome

It is no surprise that genetics, which studies individual genes and their roles in inheritance, has its limitations. The study of individual genes, one or only a few at a time, gives partial information on most of the metabolic processes and interaction networks within and across the cells of an organism. So, it became both necessary and important to focus on the entire genome, which is the complete set of DNA (including genes) of an organism. The science of genomics emerged towards the end of last century.

As the focus shifted (or rather expanded) from gene to genome, a global effort to determine the genome sequences of various organisms—from bacteria to human—was underway. In 1995, the genome sequence of the bacterium *Haemophilus influenzae* was published and, by 2007, sequence of the entire human genome was declared to be finished.

Concurrently, the number of protein sequences, mostly derived from genome sequences by bioinformatic means, kept growing with unimaginable rapidity. Advances in structure determination were relatively slower. As a result, the gap between genomic and structural information was widening. To address the issue, several structural genomics (SG) initiatives were created in America, Europe, and Asia at the beginning of the 21st century. The aim of structural genomics is to create a representative set of three-dimensional structures of all macromolecules found in nature. With this objective, the SG centers have put special efforts into the development of high-throughput methodologies for faster and more accurate determination of protein structures.

1.15 In lieu of a conclusion

Thus, it appears that the genomic and postgenomic projects are rapidly moving towards providing sequence as well as structural information on the entire set of molecular components present in an organism. As a welcome consequence, molecular biology is no longer restricted to the studies of single (or even a few) macromolecules. It is becoming ever more competent to understand and predict the behavior of biological systems based on a set of molecules involved and the relationships between them.

References and Further Reading

Bernal, J.D., 2010. Science in History. Faber & Faber, London.

Brooks-Bartlett, J.C., Garman, E.F., 2015. The Nobel science: one hundred years of crystallography. Interdiscip. Sci. Rev. 40 (3), 244–264.

Campbell, I.D., 2008. The Croonian lecture 2006 structure of the living cell. Philos. Trans. R. Soc. B 363, 2379–2391.

Dahm, R., 2005. Friedrich Miescher and the discovery of DNA. Dev. Biol. 278, 274–288.

Grabowski, M., et al., 2016. The impact of structural genomics: the first quindecinnial. J. Struct. Funct. Genom. 17 (1), 1–16.

Harrison, S.C., 2004. Whither structural biology. Nat. Struct. Mol. Biol. 11 (1), 12–15.

Jaskolski, M., Dauter, Z., Wlodawer, A., 2014. A brief history of macromolecular crystallography, illustrated by a family tree and its Nobel fruits. FEBS J. 281, 3985–4009.

Kitaigorodsky, A., 1971. I Am a Physicist. MIR Publishers, Moscow.

Masters, B.R., 2008. History of the optical microscope in cell biology and medicine. In: Encyclopedia of Life Sciences (ELS). John Wiley & Sons, Chichester https://doi.org/10.1002/9780470015902.a0003082.

Mazzarello, P., 1999. A unifying concept: the history of cell theory. Nat. Cell Biol. 1, E13–E15.

Schrödinger, E., 1967. What is Life. Cambridge University Press, Cambridge, UK.

Shi, Y., 2014. A glimpse of structural biology through X-ray crystallography. Cell 159, 995–1014.

CHAPTER

Mathematical tools

2

Scientific investigation of the structure and dynamics of an object, be it inanimate or living, requires (a) the measurement of different physical quantities and (b) appropriate mathematical tools. Here, in this chapter, we present a brief description of the standards of measurement followed by an overview of the mathematical concepts that are important for quantitative analyses of molecular structure and dynamics.

2.1 Measurements: Standards and units

Some examples of the physical quantities we may like to measure are mass, volume, density, charge, temperature, and so on. The result of each measurement is a "value" that consists of a number and a unit of measurement. For communication and comparison of the results from different laboratories, it is imperative to have a universally accepted set of standards for units of measurement. The accepted standards are required to be precise and reproducible, remaining invariable with the passage of time or changes in the environment.

The General Conference on Weights and Measures (Conférence Générale des Poids et Measures, CGPM) has selected the internationally accepted standards of units known as Le Système International d'Unités (abbreviated as SI). SI is based on seven primary units—those of time (second), length (meter), mass (kilogram), amount of substance (mole), thermodynamic temperature (kelvin), electric current (ampere), and luminous intensity (candela) (Table 2.1).

The CGPM, at different times, has adopted specific definitions for each of the seven primary or base units. The latest definitions are given in Table 2.2.

Besides the physical quantities that are expressed in terms of the primary units, all other physical quantities are expressed in units which are combinations of the base units. Although some of the secondary units have their own names, they are related to the SI units. As for example, the SI units of force and energy, called newton (N) and joule (J), respectively, are defined in terms of the base units as

$$1\,\text{N} = 1\,\text{kg m s}^{-2}$$

$$1\,\text{J} = 1\,\text{kg m}^2\,\text{s}^{-2}$$

Fundamentals of Molecular Structural Biology. https://doi.org/10.1016/B978-0-12-814855-6.00002-X

Table 2.1 SI primary units

Quantity	Name	Symbol
Time	Second	s
Length	Meter	m
Mass	kilogram	kg
Amount of substance	Mole	mol
Thermodynamic temperature	Kelvin	K
Electric current	ampere	A
Luminous intensity	candela	cd

Table 2.2 CGPA definitions of primary units of measurement

	Unit	Definition
1.	Second	The second is the duration of 9,192,631,770 periods of the radiation emitted due to the transition between two hyperfine levels of the ground state of cesium-133 atom.
2.	Meter	The meter is the length of the path traveled by light in vacuum during a time interval of 1/299,792,458 of a second.
3.	Kilogram	The kilogram is the unit of mass—it is equal to the mass of the international prototype of the kilogram made of platinum-iridium and kept at the International Bureau of Weights and Measures.
4.	Mole	The mole is the amount of substance in a system which contains as many elementary entities as there are atoms in 0.012 kg of carbon-12. The elementary entities may be atoms, molecules, ions, electrons, other particles, or specified groups of such particles.
5.	Kelvin	The kelvin is the fraction 1/273.16 of the thermodynamic temperature of the triple point of water.
6.	Ampere	The ampere is the constant current which, if maintained in two straight parallel conductors of infinite length and negligible cross section placed 1 m apart in vacuum, would produce between the conductors a force equal to 2×10^{-7} N per meter of length.
7.	Candela	The candela is the luminous intensity, in a given direction, of a source that emits monochromatic radiation of frequency 540×10^{12} hertz and that has a radiant intensity in that direction of 1/683 watt per steradian (steradian is the unit of solid-angle measure in the SI).

Both primary and secondary units can be scaled up or down by powers of 10 and appropriate prefixes be used with the changed values. Here are some examples:

$$\text{femtosecond (fs)} = 10^{-15}\ \text{s}$$
$$\text{picomole} = 10^{-12}\ \text{mol}$$
$$\text{nanometer} = 10^{-9}\ \text{m}$$
$$\text{kilojoules} = 10^{3}\ \text{J}$$
$$\text{megahertz} = 10^{6}\ \text{Hz}$$

In physicochemical measurements, some non-SI units are also conventionally used. However, those units have defined relations with corresponding SI units. For example, degree Celsius (°C) is related to kelvin by °C = K − 273.15 and liter (L), which is often used as the unit of volume, is $10^{-3}\,m^3$.

As we go down to the atomic or molecular level, some numbers and units (which may or may not belong to the SI) become important. One such unit relates to the atomic mass—atomic mass unit. The mass of carbon-12 has been defined to be 12 atomic mass units (12 amu). The unit, which is not an SI unit, is also called Dalton. The relation between amu and the primary standard is given by

$$1\,\text{amu} = 1.661 \times 10^{-27}\,\text{kg}$$

Related to the primary unit mole is a number that is ubiquitous in physicochemical literature—the Avogadro's number, N_A (also called Avogadro's constant). One mole of carbon-12, which has a mass of 0.012 kg, contains N_A number of atoms. N_A has been experimentally determined to be $6.02214199 \times 10^{23}\,\text{mol}^{-1}$ with an uncertainty of about one part in a million. One mole of any other substance contains the same number of elementary entities. Thus, 1 mol of helium gas contains N_A atoms of He, while 1 mol of water contains N_A molecules of H_2O.

A unit that has been extensively used in the description of atomic and molecular structures is the ångström (symbolized as Å) which is not an SI unit. In 1907, the International Astronomical Union had defined ångström by declaring the wavelength of the red line of cadmium in air to be 6438.46963 Å. The latest CGPM definition of meter has made ångström exactly equal to 0.1 nm.

$$1\,\text{Å} = 10^{-10}\,\text{m} = 0.1\,\text{nm}$$

2.2 Algebraic functions

In science, most often we may want to represent a physical object by a point and a process by the movement of the point. Let us consider that we are studying the movement of a particle denoted geometrically by a point P. We can measure the distance of P with respect to a reference point O at a certain instant of time measured with respect to an initial time. The result of each of the two measurements will produce a number with a unit assigned to it. However, if the particle keeps changing its position, say along a straight line (one dimension), we would rather denote the distance by a "variable," say x. Similarly, the time would be denoted by another variable, t.

To generalize, physical systems (including biological systems) and their dynamics are quantitatively described in terms of "observables" or entities which can be measured. In a specific physical problem, we may have an entity whose value is denoted by a variable x. Now, it may happen that there is another entity denoted by a variable y which is related to x. The relation between the two variables may be expressed as

$$y = f(x) \tag{2.1}$$

Formally, it is said that if there is a unique value of y for each value of x, then y is a function of x. The set of all permitted values of x is called a domain and that of all permitted values of y, a range.

On the other hand, if there be a unique value of x for each value of y, we can write

$$x = g(y) \tag{2.2}$$

which is defined as the reverse function.

Each of the functions $f(x)$ and $g(y)$ is a function of a single variable, x or y. However, we may consider a function $f(x, y)$ which depends on two variables x and y. In this case, it is required that for any pair of values (x, y), $f(x, y)$ has a well-defined value. The notion can be extended to a function $f(x_1, x_2, \ldots, x_n)$ that depends on n number of variables $x_1, x_2, \ldots, x_n$. Functions of two variables may be represented by a surface in a three-dimensional space; functions with higher number of variables are usually difficult to visualize.

In certain physical problems, a function can be expressed as a polynomial in x.

$$f(x) = a_0 + a_1x + a_2x^2 + a_3x^3 + \ldots\ldots\ldots + a_{n-1}x^{n-1} + a_nx^n \tag{2.3}$$

When $f(x)$ is set equal to zero, the polynomial equation

$$a_0 + a_1x + a_2x^2 + a_3x^3 + \ldots\ldots\ldots + a_{n-1}x^{n-1} + a_nx^n = 0 \tag{2.4}$$

is satisfied by specific values of x known as the roots of the equation. n, an integer >0, is the degree of the polynomial and the equation. The coefficients $a_0, a_1, \ldots, a_n$, which are real quantities ($n \neq 0$), are determined by the physical properties of the system under study.

$f(x)$ is a univariate function which can be used to describe and analyze an one-dimensional system. Graphically, it is represented by a line. Examples of two univariate functions are shown in Fig. 2.1.

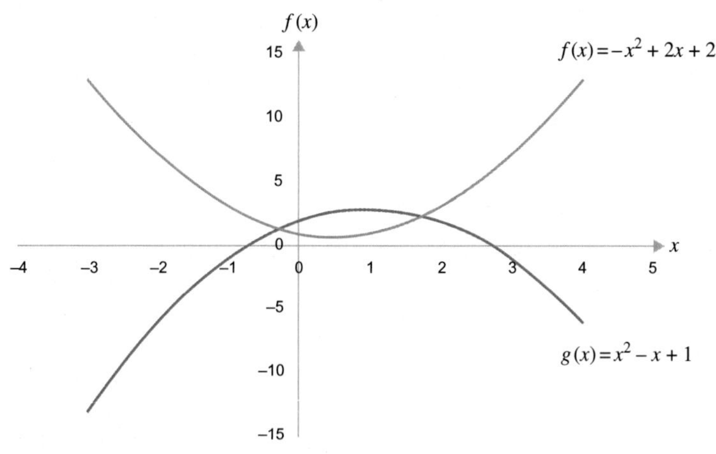

FIG. 2.1

Graphical representation of univariate algebraic functions.

Functions involving more than one variable are called multivariate. A function $f(x, y)$ of two variables describes a two-dimensional system and graphically it is represented by a surface.

In Eq. (2.4), if $n=1$, the equation takes the form of a linear equation

$$a_0 + a_1 x = 0 \tag{2.5}$$

whose solution (root) is given by $\alpha_1 = -a_0/a_1$.

For $n=2$, Eq. (2.4) becomes a quadratic equation

$$a_0 + a_1 x + a_2 x^2 = 0 \tag{2.6}$$

the roots of which are

$$\alpha_{1,2} = \frac{-a_1 \pm \sqrt{a_1{}^2 - 4a_0 a_2}}{2a_2} \tag{2.7}$$

$n=3$ gives a cubic equation.

2.3 Trigonometric functions

Physical systems which involve the periodic motion of a point can be represented by trigonometric functions which are also periodic in nature. Here the point in question is considered as the projection of another point moving along a circle. To illustrate, let us consider a circle in the xy-plane (Fig. 2.2). The radius of the circle is r and its center is at the origin of xy-coordinate system. P' is a point on the circle whose coordinates are given by (x, y) and makes an angle ϕ with the x-axis. P is the projection of P' on the x-axis.

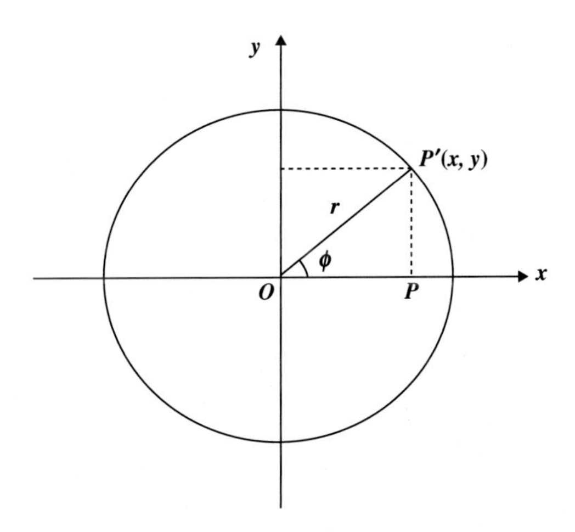

FIG. 2.2

Position of point P' indicated by Cartesian coordinates (x, y) and polar coordinates (r, ϕ).

As P' moves counterclockwise along the circle starting from the x-axis, ϕ increases from 0 to 2π radians or 360°. With $P'(x, y)$ associated with angle ϕ, we have the following definitions:

$$\begin{aligned}\cos\phi &= \frac{OP}{OP'} = \frac{x}{r}\\ \sin\phi &= \frac{PP'}{OP'} = \frac{y}{r}\\ \tan\phi &= \frac{PP'}{OP} = \frac{y}{x}\end{aligned} \tag{2.8}$$

and the reciprocal relations

$$\begin{aligned}\sec\phi &= 1/\cos\phi = r/x\\ \operatorname{cosec}\phi &= 1/\sin\phi = r/y\\ \cot\phi &= 1/\tan\phi = x/y\end{aligned} \tag{2.9}$$

If the circle be of unit radius, that is, $r=1$

$$x = \cos\phi,\;\; y = \sin\phi,\;\; y/x = \tan\phi$$
$$1/x = \sec\phi,\;\; 1/y = \cos\phi,\;\; x/y = \cot\phi$$

All these relations are collectively called trigonometric functions.

Referring to Fig. 2.2, one can see that the point P' can be represented also by polar coordinates r (the radial coordinate) and ϕ (the angular coordinate). The relations between the polar coordinates and Cartesian coordinates are given by

$$\begin{aligned}x &= r\cos\phi\\ y &= r\sin\phi\end{aligned} \tag{2.10}$$

Further, from Eq. (2.8), we have $\cos 0 = \cos 2\pi = 1$ and $\cos\pi = -1$. Therefore, it can easily be visualized that as P' moves along the circle from 0 to 2π, its projection P moves along the x-axis from +1 to -1 and back to +1. The cycle is repeatable between 2π and 4π and so on. The function $f(\phi) = \cos\phi$ can be represented graphically as in Fig. 2.3 and it can be seen that

$$f(\phi + 2n\pi) = f(\phi) \tag{2.11}$$

where $n = 1, 2, 3, \ldots$ $f(\phi) = \cos\phi$ is, therefore, a periodic function, 2π being the period. Similarly, $\sin\phi$ is also a periodic function (Fig. 2.3). Fig. 2.2 also shows that

$$\sin^2\phi + \cos^2\phi = 1 \tag{2.12}$$

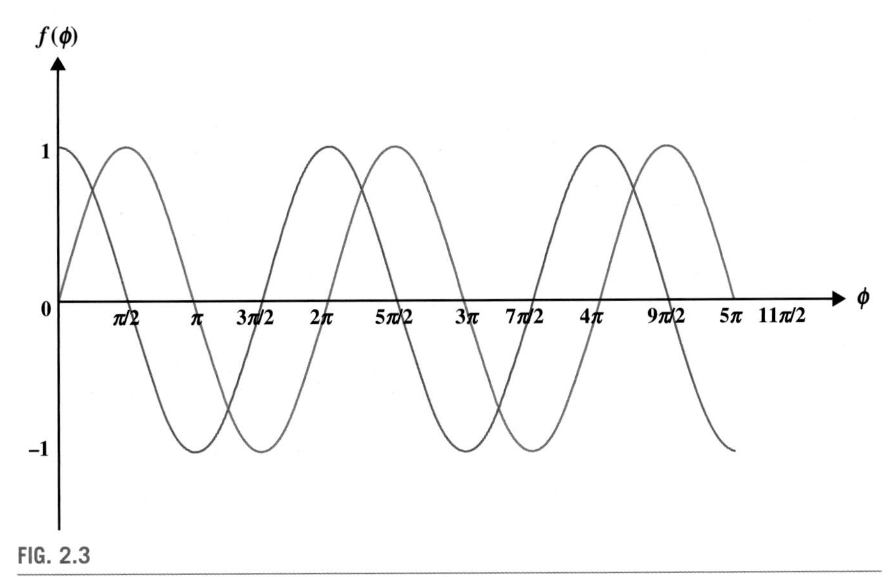

FIG. 2.3

Periodic functions: $\sin\phi$ *(green)* and $\cos\phi$ *(red)*.

2.4 Exponential and logarithm

We have seen that a polynomial contains terms like $a_n x^n$ where x is the variable and n is a fixed number (an integer). However, there can be occasions when a function will contain terms like n^x where n is a fixed number ($n > 0$ and $n \neq 1$) and the variable is in the exponent.

As for example, in a steadily growing and dividing bacterial culture, the number of cells after x generations is given by $N = N_0 2^x$, where N_0 is the initial number of cells. Such functions are called exponential functions. The exponentials most commonly used are 10^x, e^x (where $e = 2.7183\ldots$) and 2^x. All of them follow the same rules of manipulations

$$n^{x+y} = n^x \cdot n^y$$

and

$$n^{xy} = (n^x)^y = (n^y)^x \tag{2.13}$$

Logarithm is the inverse of exponential. If $y = a^x$, then by definition $x = \log_a y$. Log base e is known as natural logarithm and, therefore, written as ln. So we have

$$\ln(e^x) = x; \;\; e^{\ln x} = x \quad (x > 0)$$

and

$$\log(10^x) = x; \;\; 10^{\log x} = x \quad (x > 0) \tag{2.14}$$

The rule for conversion of log and ln is

$$\ln x = (\ln 10)(\log x) = (2.3026\ldots)\log x \tag{2.15}$$

2.5 Complex numbers

As we shall see in the next chapter, mathematical analysis of a periodic system has been well facilitated by the introduction of what are known as complex numbers and complex functions. To understand what a complex number is, let us consider a function

$$f(z) = z^2 - 3z + 2 \tag{2.16}$$

For $f(z) = 0$, Eq. (2.16) becomes

$$z^2 - 3z + 2 = 0 \tag{2.17}$$

which is a quadratic equation whose roots (solutions) are given, in accordance to the Eq. (2.7), as

$$z_1 = 1 \text{ and } z_2 = 2 \tag{2.18}$$

However, for another function

$$g(z) = z^2 - 2z + 2 = 0 \tag{2.19}$$

the solutions will be given as

$$z_1, z_2 = 1 \pm \sqrt{-1} \tag{2.20}$$

The problem can be understood by plotting the functions $f(z)$ and $g(z)$ against z (Fig. 2.4). It is seen that $f(z)$ intersects the z-axis (a real line) at 1 and 2 and the roots are, therefore, real. However, $g(z)$ does not intersect the z-axis.

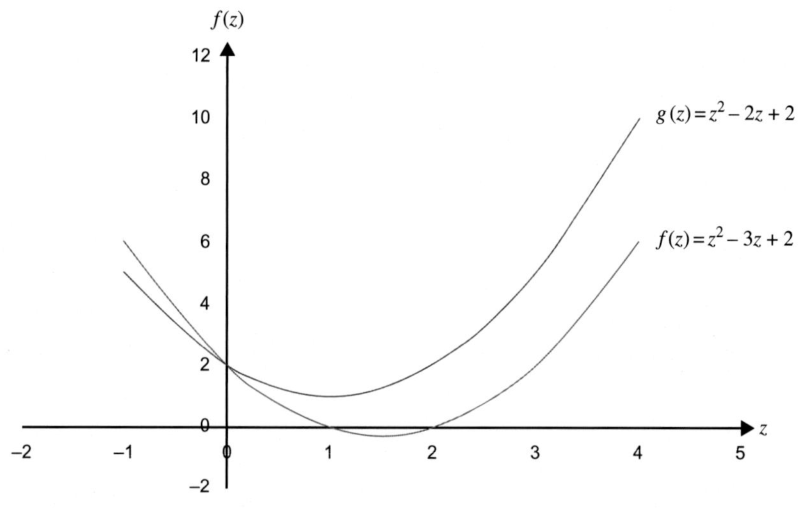

FIG. 2.4

Graphical solution of equations.

We may note here that that the real line or continuum can be considered to be composed of an infinite number of points, each of the points represented by a real number. The set R of real numbers includes both rational and irrational numbers. It is obvious that while the first term in Eq. (2.20) is real, the second term is not. The problem has been solved by introducing the concept of an imaginary number (or the basic complex number). An imaginary number has been defined as

$$i=\sqrt{-1} \tag{2.21}$$

and a generalized complex number can be written as

$$z=x+iy \tag{2.22}$$

where x and y are real numbers. x is the real part denoted by $Re(z)=x$ and y is the imaginary part denoted by $Im(z)=y$. Just as a real number is visualized as a point on an infinite straight line, a complex number can be considered as a point on an infinite (complex) plane (Fig. 2.5).

$z^*=x-iy$ is called the complex conjugate of $z=x+iy$. The magnitude or modulus $|z|$ of a complex number is given by

$$|z|^2=zz^*=x^2+y^2 \tag{2.23}$$

and the argument of the complex number is denoted by

$$\arg(z)=\tan^{-1}\left(\frac{y}{x}\right) \tag{2.24}$$

With reference to Fig. 2.5, Eq. (2.22) can be written as

$$z=r(\cos\phi+i\sin\phi) \tag{2.25}$$

Later in this chapter, we shall see that

$$e^{i\phi}=\cos\phi+i\sin\phi$$

so that a complex number can be represented in the polar form

$$z=re^{i\phi} \tag{2.26}$$

where r and ϕ can be identified with $|z|$ and $\arg(z)$, respectively. Polar representation of a complex number is easier to manipulate.

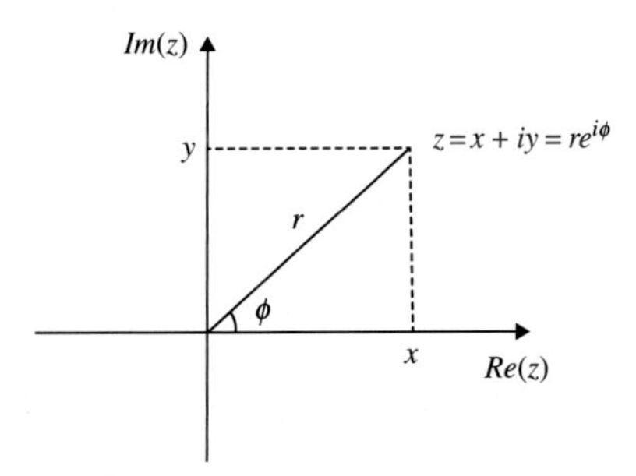

FIG. 2.5

Representation of a complex number z.

2.6 Vector

Certain entity can be completely specified by a single number, called its magnitude, together with the unit in which it is measured. Such entities, called scalar, abound in the physical world. Examples are mass, temperature, energy, and so on. However, several entities, called vector, require both magnitude and direction to be specified completely. We shall find a number of examples of a vector in the following chapter.

A vector can be denoted by **a**; its magnitude is denoted by $|\mathbf{a}|$. A unit vector, that is, a vector whose magnitude is unity, in the direction of **a** is denoted by

$$\hat{\mathbf{a}} = \frac{\mathbf{a}}{|\mathbf{a}|} \tag{2.27}$$

Multiplication of a vector by a scalar, say λ, gives a vector $\lambda\mathbf{a}$ in the same direction as the original, but with the magnitude changed λ-fold.

In the three-dimensional Cartesian coordinate system (x, y, z), $\hat{\mathbf{i}}$, $\hat{\mathbf{j}}$, and $\hat{\mathbf{k}}$ are considered to be orthonormal unit vectors in the directions of x, y, and z, respectively. A set of vectors S is said to be orthonormal if every vector in S has magnitude unity and the vectors are mutually orthogonal. The orthonormality of $\hat{\mathbf{i}}$, $\hat{\mathbf{j}}$, and $\hat{\mathbf{k}}$ can be expressed as

$$\begin{aligned} \hat{\mathbf{i}}\cdot\hat{\mathbf{i}}=1, \quad \hat{\mathbf{i}}\cdot\hat{\mathbf{j}}=0, \quad \hat{\mathbf{i}}\cdot\hat{\mathbf{k}}=0 \\ \hat{\mathbf{j}}\cdot\hat{\mathbf{i}}=0, \quad \hat{\mathbf{j}}\cdot\hat{\mathbf{j}}=1, \quad \hat{\mathbf{j}}\cdot\hat{\mathbf{k}}=0 \\ \hat{\mathbf{k}}\cdot\hat{\mathbf{i}}=0, \quad \hat{\mathbf{k}}\cdot\hat{\mathbf{j}}=0, \quad \hat{\mathbf{k}}\cdot\hat{\mathbf{k}}=1 \end{aligned} \tag{2.28}$$

Any vector in the three-dimensional space can be considered as a linear combination of the unit vectors

$$\mathbf{a} = \hat{\mathbf{i}}a_x + \hat{\mathbf{j}}a_y + \hat{\mathbf{k}}a_z \tag{2.29}$$

where a_x, a_y, and a_z are the magnitudes of the vector **a** in the x, y, and z directions, respectively.

The scalar product (also called the dot product, different from multiplication by a scalar) of two vectors **a** and **b** is defined as

$$\mathbf{a}\cdot\mathbf{b} = |\mathbf{a}||\mathbf{b}|\cos\phi \quad 0 \le \phi \le \pi \tag{2.30}$$

where ϕ is the angle between the two vectors. As the name implies, the scalar product has only magnitude and no direction.

In contrast, the vector product (also called cross product) of two vectors **a** and **b** is a third vector denoted as

$$\mathbf{c} = \mathbf{a}\times\mathbf{b} \tag{2.31}$$

whose magnitude is given by

$$|\mathbf{c}| = |\mathbf{a}||\mathbf{b}|\sin\phi \quad 0 \le \phi \le \pi \tag{2.32}$$

The vector **c** is orthogonal to both **a** and **b** and its direction can be conceptualized in analogy to the motion of a corkscrew. As the corkscrew turns **a** up to **b**, it advances in the direction of **c** (Fig. 2.6).

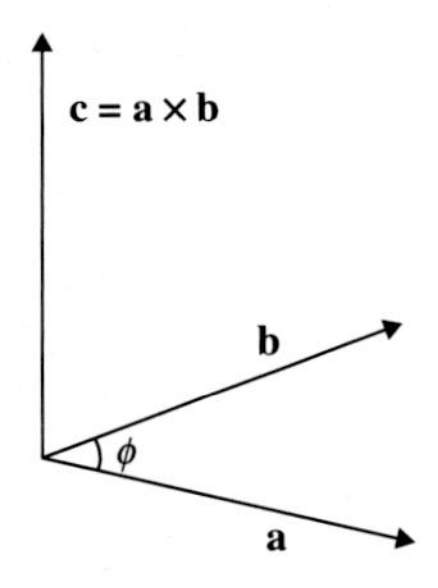

FIG. 2.6
The vector product.

The definition of a vector can be generalized to n-dimension. An n-dimensional vector is defined as an ordered array of n real numbers r_1, r_2, …, r_n and expressed as

$$\mathbf{r} = [r_1, r_2, \ldots, r_n] \tag{2.33}$$

r_j ($j = 1, 2, \ldots, n$) being the components of the vector $\mathbf{r}$.

In Eq. (2.33), the elements of vector $\mathbf{r}$ are arranged in a row, hence it is called a row vector. The number of columns, n, in the row vector gives the dimension of the vector.

On the other hand, in an n-dimensional column vector, the elements are arranged in a column with n rows

$$\mathbf{r} = \begin{bmatrix} r_1 \\ r_2 \\ \vdots \\ r_n \end{bmatrix}$$

The scalar product of two vectors $\mathbf{u}$ and $\mathbf{v}$ can be defined by taking a row vector

$$\mathbf{u} = [u_1, u_2, \ldots, u_n]$$

and a column vector

$$\mathbf{v} = \begin{bmatrix} v_1 \\ v_2 \\ \vdots \\ v_n \end{bmatrix}$$

both of the same dimension, as

$$\mathbf{u} \cdot \mathbf{v} = u_1 v_1 + u_2 v_2 + \cdots + u_n v_n \tag{2.34}$$

2.7 Matrix

Analysis of a wide range of physical problems involves the solution of linear equations. Simple linear equations are $ax + by = c$ and $ax + by + cz = d$.

The generalized form of a linear equation of n variables $x_1, x_2, \ldots, x_n$ can be written as

$$a_1x_1 + a_2x_2 + \cdots + a_nx_n = b \tag{2.35}$$

where $a_1, a_2, \ldots, a_n$ and b are all constants. It is to be noted that all the variables appear in a linear equation only to the first power and do not appear as arguments of trigonometric, logarithmic, or exponential functions. Further, a linear equation does not involve any products of the variables.

A system of linear equations, or briefly a linear system, is a finite set of linear equations involving the same variables. The generalized form of a linear system can be written as

$$\begin{aligned} a_{11}x_1 + a_{12}x_2 + \cdots + a_{1n}x_n &= b_1 \\ a_{21}x_1 + a_{22}x_2 + \cdots + a_{2n}x_n &= b_2 \\ &\vdots \\ a_{m1}x_1 + a_{m2}x_2 + \cdots + a_{mn}x_n &= b_m \end{aligned} \tag{2.36}$$

The solution to a system of linear equation is a sequence of n numbers $s_1, s_2, \ldots, s_n$ for the variable x_i ($i = 1, 2, \ldots, n$) such that each equation in the system is satisfied.

Eq. (2.36) can be compactly represented by

$$\begin{bmatrix} a_{11} & a_{12} & \cdots & a_{1n} \\ a_{21} & a_{22} & \cdots & a_{2n} \\ & \vdots & & \\ a_{m1} & a_{m2} & \cdots & a_{mn} \end{bmatrix} \begin{bmatrix} x_1 \\ x_2 \\ \vdots \\ x_n \end{bmatrix} = \begin{bmatrix} b_1 \\ b_2 \\ \vdots \\ b_m \end{bmatrix} \tag{2.37}$$

along with the rule of multiplication

$$\begin{bmatrix} a_{11} & a_{12} & \cdots & a_{1n} \\ a_{21} & a_{22} & \cdots & a_{2n} \\ & \vdots & & \\ a_{m1} & a_{m2} & \cdots & a_{mn} \end{bmatrix} \begin{bmatrix} x_1 \\ x_2 \\ \vdots \\ x_n \end{bmatrix} = \begin{bmatrix} a_{11}x_1 + a_{12}x_2 + \cdots + a_{1n}x_n \\ a_{21}x_1 + a_{22}x_2 + \cdots + a_{2n}x_n \\ \vdots \\ a_{m1}x_1 + a_{m2}x_2 + \cdots + a_{mn}x_n \end{bmatrix} \tag{2.38}$$

We already know that

$$\mathbf{x} = \begin{bmatrix} x_1 \\ x_2 \\ \vdots \\ x_n \end{bmatrix} \text{ and } \mathbf{b} = \begin{bmatrix} b_1 \\ b_2 \\ \vdots \\ b_m \end{bmatrix}$$

are column vectors.

$$\mathcal{A} = \begin{bmatrix} a_{11} & a_{12} & \cdots & a_{1n} \\ a_{21} & a_{22} & \cdots & a_{2n} \\ & & & \vdots \\ a_{m1} & a_{m2} & \cdots & a_{mn} \end{bmatrix}$$

is defined as an $m \times n$ matrix. $\mathcal{A}$ will be a square matrix if $m = n$. The numbers a_{ij} are called the elements of the matrix $\mathcal{A}$. Eq. (2.37) can also be written as

$$\mathcal{A}\mathbf{x} = \mathbf{b} \tag{2.39}$$

2.8 Calculus

Most often, it is not enough to just describe a physical system by a function of one or more variables. In order to understand the dynamics of the system, it is important to determine how the function changes with respect to the variable(s) on which it depends. The branch of mathematics that is concerned with changes in a function in a precise manner is calculus. It consists of two subbranches—differential calculus, which computes the change in the function (derivative) brought about by an infinitesimal (approaching zero) change in a variable, and integral calculus, which is concerned with the summation (integration) of the products of the derivatives and the infinitesimal changes in the variable within defined "limits."

2.8.1 Differentiation

Differentiation, as it has been indicated above, is a limiting process and hence restricted by the rule of limit. Therefore, before moving on to the definition of a derivative that is obtained in the process of differentiation, it will be reasonable to have a basic understanding of the concept of limit. The formal definition of the limit of a function $f(x)$ at $x=a$ as expressed by

$$\underset{x \to a}{Lim} f(x) = l$$

is as follows: given any real number $\varepsilon > 0$, there exists another real number $\delta > 0$ such that for $0 < |x-a| < \delta$, $|f(x) - l| < \varepsilon$.

Now, if x changes to $x+\Delta x$, where Δx is a very small amount, leading to a change Δf in the value of the function, that is, $\Delta f = f(x+\Delta x) - f(x)$, then the first derivative of $f(x)$ is defined as

$$f'(x) = \frac{df(x)}{dx} = \lim_{\Delta x \to 0} \frac{\Delta f}{\Delta x} = \lim_{\Delta x \to 0} \frac{f(x+\Delta x) - f(x)}{\Delta x} \quad (2.40)$$

provided that the limit exists and is independent of the direction from which ε approaches zero. If the limit exists at $x=a$, the function is said to be differentiable at a.

Since $f'(x)$ is also a function of x, its derivative with respect to x, that is, the second derivative of $f(x)$, can be defined as

$$f''(x) = \frac{d}{dx}\frac{df(x)}{dx} = \frac{d^2 f(x)}{dx^2} = \lim_{\Delta x \to 0} \frac{f'(x+\Delta x) - f'(\Delta x)}{\Delta x} \quad (2.41)$$

The derivative of $f(x)$ can be visualized graphically as the gradient or slope of the function at x. It can be seen from Fig. 2.7 that as $\Delta x \to 0$, $\Delta f/\Delta x$ approaches the gradient given by $\tan\phi$.

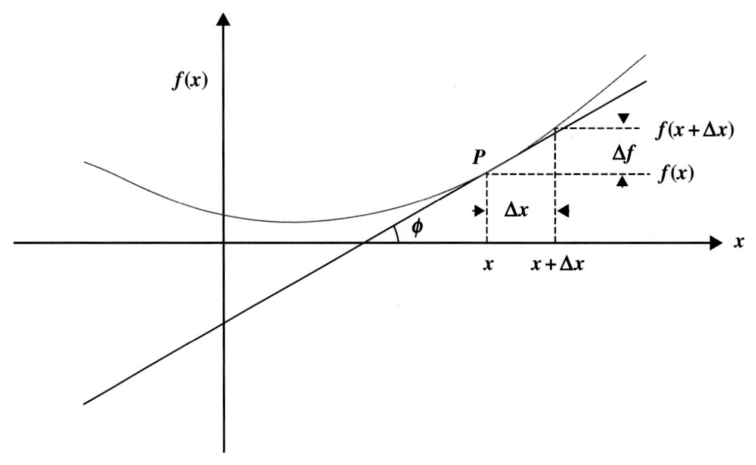

FIG. 2.7

The gradient or slope of a function $f(x)$ at x.

2.8.2 Integration

An integral has been conceived as the area under a curve in an $f(x)$ versus x plot. There are two kinds of integral—definite and indefinite. The definite integral of a function $f(x)$ sets the limits for integration and denoted as

$$\int_{\mathrm{a}}^{\mathrm{b}} f(x)dx$$

where $x=a$ is the lower limit and $x=b$ is the upper limit of integration. In order to arrive at a formal definition of definite integral, let us consider the graph of $f(x)$ as in Fig. 2.8. Let the finite interval $a \le x \le b$ be subdivided into a large number (say, n) of subintervals marked by intermediate points z_i such that $a=z_0<z_1<z_2<\cdots<z_i<\cdots<z_n=b$. $f(x_i)$ (z_i-z_{i-1}) can be visualized as the area of the ith rectangle. The sum of all $f(x_i)$ (z_i-z_{i-1}), where $i=1.2, \ldots, n$, can be expressed as

$$S=\sum_{i=1}^{n} f(x_i)(z_i-z_{i-1}) \tag{2.42}$$

As $n \to \infty$ and $(z_i - z_{i-1}) \to 0$, if S tends to a unique limit I_d, then

$$I_d=\int_a^b f(x)dx \tag{2.43}$$

is defined as the definite integral of $f(x)$ between the limits (a, b). $f(x)$ is called the integrand.

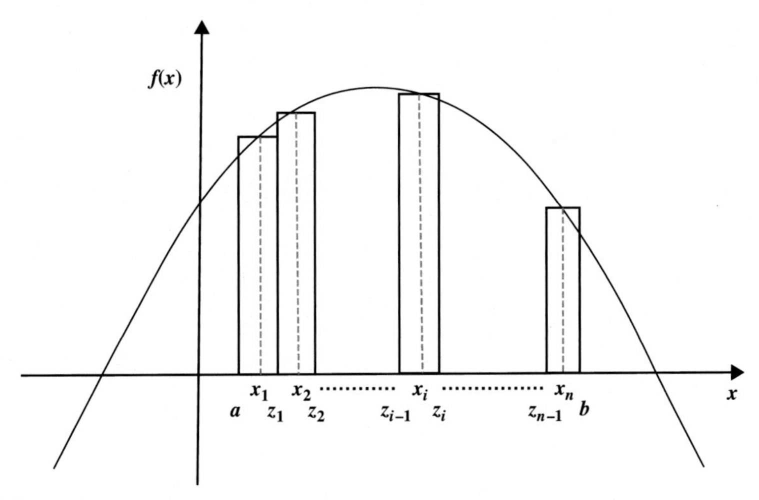

FIG. 2.8

Defining a definite integral by subdividing the interval $a \leq x \leq b$ into subintervals.

The indefinite integral can be considered as an inverse of differentiation. Thus, if $f(x)$ is the derivative of $F(x)$ with respect to x, that is

$$\frac{dF(x)}{dx} = f(x) \tag{2.44}$$

then $F(x)$ is the indefinite integral of $f(x)$ expressed as

$$F(x) = \int^{x} f(x)dx \tag{2.45}$$

where the lower limit is arbitrary. It should be noted that the differentiation does not have a unique inverse—any function $F(x)$ obeying Eq. (2.44) can be an indefinite integral of $f(x)$. However, any two such integrals will differ only by an arbitrary constant. Hence, an indefinite integral of $f(x)$ can be formally written as

$$\int f(x)dx = F(x) + c \tag{2.46}$$

where c is known as the constant of integration. c can be evaluated from what are known as boundary conditions.

2.8.3 Multivariate function

In Eq. (2.40), we have defined the derivative of a function one variable. This definition needs to be amended in the case of a function of two variables $f(x, y)$. Here, we introduce the concept of a "partial derivative" and the corresponding notations $\partial/\partial x$ and $\partial/\partial y$.

The partial derivative of $f(x, y)$ with respect to x is formally defined as

$$\left(\frac{\partial f}{\partial x}\right)_y = \lim_{\Delta x \to 0} \frac{f(x+\Delta x, y) - f(x, y)}{\Delta x} \tag{2.47}$$

provided that the limit exists. It can be seen that y is treated as a constant.

Similarly, the partial derivative of $f(x, y)$ with respect to y (x being held constant) can be expressed as

$$\left(\frac{\partial f}{\partial y}\right)_x = \lim_{\Delta y \to 0} \frac{f(x, y+\Delta y) - f(x, y)}{\Delta y} \tag{2.48}$$

also contingent on the existence of a limit.

Further extending the definition to the general n-variable function $f(x_1, x_2, \ldots, x_n)$

$$\frac{\partial f(x_1, x_2, \ldots, x_n)}{\partial x_i} = \lim_{\Delta x_i \to 0} \frac{[f(x_1, x_2, \ldots, x_i + \Delta x_i, \ldots, x_n) - f(x_1, x_2, \ldots, x_i, \ldots, x_n)]}{\Delta x_i} \tag{2.49}$$

provided that the limit exists.

Just as we have seen the derivatives of a function with multiple variables, the integral of a function with more than one variable can also be defined. Let us consider a function of two variables. For the integral of a function of one variable, we had divided the interval $a \le x \le b$ into a number of subintervals along a straight line. Here, we consider the limits of integration to be represented by a closed curve in a (two-dimensional) region R in the xy-plane and subdivide the region into N subregions ΔR_j of area ΔA_j, $j=1, 2, \ldots, N$. (x_j, y_j) is any point in ΔR_j. Let us consider the sum

$$S = \sum_{j=1}^{N} f(x_j, y_j) \Delta A_j \tag{2.50}$$

As $\Delta A_j \to 0$ and consequently $N \to \infty$, if S tends to a unique limit I, then

$$I = \int_R f(x, y) dA \tag{2.51}$$

is the surface integral of $f(x, y)$ over the (two-dimensional) region R. Considering $\Delta A_j = \Delta x_j \Delta y_j$, if both Δx_j and $\Delta y_j \to 0$, Eq. (2.51) can be written as a double integral of $f(x, y)$

$$I = \int_{x_1}^{x_2} \int_{y_1}^{y_2} f(x, y) dx\, dy \tag{2.52}$$

where (x_1, x_2) and (y_1, y_2) are the limits of x and y, respectively.

Similarly, in the case of a function of three variables $f(x, y, z)$, the triple integral is defined as

$$I = \int_R f(x, y, z) dV \tag{2.53}$$

or,

$$I = \int_{x_1}^{x_2} \int_{y_1}^{y_2} \int_{z_1}^{z_2} f(x,y,z) dx\, dy\, dz \tag{2.54}$$

contingent on the existence of a limit. R in Eq. (2.53) is three-dimensional and the V indicates integration over a volume.

2.8.4 Applications

1. Using derivatives, it is possible to determine the maxima and/or minima of a function. Let us consider the function $f(x) = 2x^3 + 3x^2 - 12x + 4$ plotted in Fig. 2.9 and, at first, try to understand the problem graphically. We can see that there is a maximum at point P and a minimum at point Q. (It may be noted that P and Q are not the overall maximum and minimum but rather local maximum and minimum, respectively; they are also called stationary points.) Nevertheless, at either of these points, the gradient of the curve representing the function becomes zero. Now, the gradient of the curve at any point x is given by the derivative

$$f'(x) = 6x^2 + 6x - 12 \tag{2.55}$$

which itself, as expected, is another function of x. The roots of the equation $f'(x) = 0$ are -2 and 1 which are also the points of intersection of $f'(x)$ with the x-axis. We can clearly see from Fig. 2.9 that $f(x)$ is maximum at -2 and minimum at 1.

Now, the second derivative of $f(x)$ is given by

$$f''(x) = 12x + 6 \tag{2.56}$$

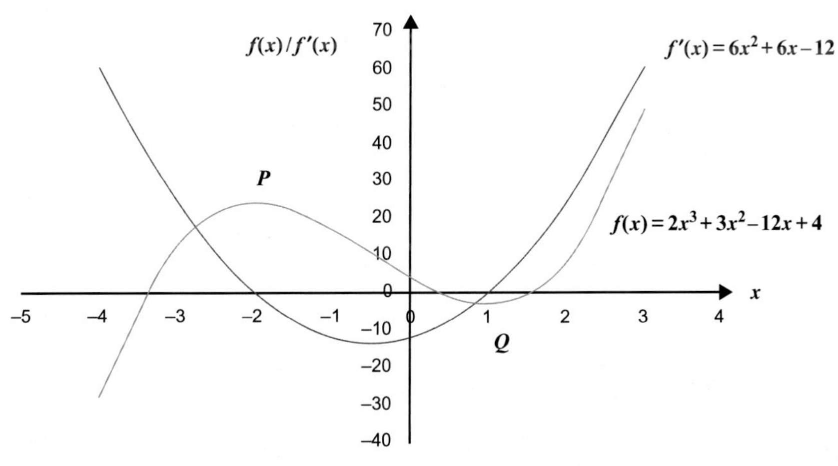

FIG. 2.9

Graphs of a function $f(x)$ and its derivative $f'(x)$. Points P and Q are, respectively, the maximum and minimum of $f(x)$.

which also represents the gradient of $f'(x)$. From Eq. (2.56), we find that $f''(-2)<0$ and $f''(1)>0$, which are also consistent with the gradients of $f'(x)$ at -2 and 1, respectively, as seen in Fig. 2.9. Hence, in summary

(a) for a maximum, $f'(x)=0$ and $f''(x)<0$ (negative) and

(b) for a minimum, $f'(x)=0$ and $f''(x)>0$ (positive)

2. Using definite integrals, we can find the mean value of a function $f(x)$ between two limits, say, a and b. Considering an infinitesimally small element dx, the "accumulation" of $f(x)$ over dx will be $f(x)dx$. The total "accumulation" of $f(x)$ between the limits a and b will then be given by

$$\int_a^b f(x)dx$$

and the mean value of $f(x)$ by

$$f_m = \frac{\int_a^b f(x)dx}{\int_a^b dx} \tag{2.57}$$

Since $\int_a^b dx = x\Big|_a^b = (b-a)$, Eq. (2.57) becomes

$$f_m = \frac{1}{b-a}\int_a^b f(x)dx \tag{2.58}$$

2.8.5 Vector operators

In Section 2.6, the concepts of scalar and vector quantities have been introduced. Sometimes, a particular scalar or vector quantity is defined not just at a point in space, but as a continuous "field" throughout a region in space (or rather the entire space). For an n-dimensional space, a scalar field may be denoted by $\varphi(x_1, x_2, \ldots, x_n)$, while a vector field by $\mathbf{a}\ (x_1, x_2, \ldots, x_n)$. Using the definition of partial derivative (Eq. 2.49), here we introduce certain differential operators, also known as vector operators, which are very useful in the analysis of atomic and molecular systems. We shall, however, restrict ourselves to the familiar three-dimensional space (x, y, z).

1. The vector operator ∇, called *del*, in the three-dimensional space (Cartesian coordinates) is denoted by

$$\nabla = \hat{\mathbf{i}}\frac{\partial}{\partial x} + \hat{\mathbf{j}}\frac{\partial}{\partial y} + \hat{\mathbf{k}}\frac{\partial}{\partial z} \tag{2.59}$$

The gradient of a scalar field $\varphi(x, y, z)$ will then be defined by

$$\text{grad}\,\varphi = \nabla\varphi = \hat{\mathbf{i}}\frac{\partial\varphi}{\partial x} + \hat{\mathbf{j}}\frac{\partial\varphi}{\partial y} + \hat{\mathbf{k}}\frac{\partial\varphi}{\partial z} \tag{2.60}$$

It can be seen that although $\varphi(x, y, z)$ is a scalar field, $\nabla\varphi$ is a vector field.

2. The divergence of a vector field $\mathbf{a}$ (x, y, z) is defined as the dot product of ∇ and $\mathbf{a}$ (x, y, z). That is

$$\text{div}\,\mathbf{a} = \nabla\cdot\mathbf{a} = \frac{\partial a_x}{\partial x} + \frac{da_y}{\partial y} + \frac{\partial a_z}{\partial z} \tag{2.61}$$

Clearly, $\nabla\cdot\mathbf{a}$ is a scalar field.

3. The curl of a vector field $\mathbf{a}$ (x, y, z) is defined by

$$\text{curl}\,\mathbf{a} = \nabla\times\mathbf{a} = \hat{\mathbf{i}}\left(\frac{\partial a_z}{\partial y} - \frac{\partial a_y}{\partial z}\right) + \hat{\mathbf{j}}\left(\frac{\partial a_x}{\partial z} - \frac{\partial a_z}{\partial x}\right) + \hat{\mathbf{k}}\left(\frac{\partial a_y}{\partial x} - \frac{\partial a_x}{\partial y}\right) \tag{2.62}$$

which is a vector field.

2.9 Series and transform

There are several examples in physical sciences where the function describing a system is expressed as a series. This facilitates the analysis of a relatively complicated function by means of approximation.

A series may contain a finite or infinite number of terms. If a series contains a sequence of terms u_1, u_2, u_3, …, the ith partial sum is defined as

$$s_i = \sum_{n=1}^{i} u_n \tag{2.63}$$

If the partial sum s_i converges to a limit as $i \to \infty$, that is

$$\lim_{i\to\infty} s_i = S \tag{2.64}$$

the infinite series

$$\sum_{n=1}^{\infty} u_n$$

is said to be convergent and the sum is given by the limit S.

In general, the terms of a series can be complex numbers. In such case, the partial sum S_i will also be complex and can be expressed as

$$S_i = X_i + iY_i \tag{2.65}$$

where X_i and Y_i are the partial sums of the real and imaginary terms separately. X_i and Y_i are real.

In a certain type of series, each term depends on a variable, say x, and the series assumes the general form

$$P(x) = a_0 + a_1 x + a_2 x^2 + a_3 x^3 + \cdots \tag{2.66}$$

where a_0, a_1, a_2, etc. are constants. This type of series, occurring frequently in physical problems, is called a power series. Using Taylor's theorem, a function can be expressed as a power series in x. Here, we are not going into the proof of the theorem. Suffice it to say that choosing a reference point such that $x - a = \Delta x$, which is not infinitesimal but sufficiently small, Taylor's series can be written as

$$f(x) = f(a) + (x-a)f'(a) + \frac{(x-a)^2}{2!}f''(a) + \cdots = \sum_{n=0}^{\infty} \frac{(x-a)^n}{n!} f^n(a) \tag{2.67}$$

If the reference point is set to be zero, the expansion Eq. (2.67) becomes a Maclaurin series

$$f(x) = f(0) + xf'(0) + \frac{x^2}{2!}f''(0) + \cdots = \sum_{n=0}^{\infty} \frac{x^n}{n!} f^n(0) \tag{2.68}$$

Eq. (2.68) can be used to expand various functions such as exponential, logarithmic, and trigonometric into infinite (power) series

For an exponential function we have

$$e^x = 1 + x + \frac{x^2}{2!} + \frac{x^3}{3!} + \cdots = \sum_{n=0}^{\infty} \frac{x^n}{n!} \tag{2.69}$$

The Maclaurin series for the trigonometric functions $\sin x$ and $\cos x$ are

$$\sin x = x - \frac{x^3}{3!} + \frac{x^5}{5!} \cdots + (-1)^n \frac{x^{2n+1}}{(2n+1)!} + \cdots \tag{2.70}$$

$$\cos x = 1 - \frac{x^2}{2!} + \frac{x^4}{4!} \cdots + (-1)^n \frac{x^{2n}}{2n!} + \cdots \tag{2.71}$$

If $x = i\varphi$, where φ is real, then

$$e^{i\varphi} = 1 + i\varphi - \frac{\varphi^2}{2!} - \frac{i\varphi^3}{3!} + \frac{\varphi^4}{4!} + \frac{i\varphi^5}{5!} - \cdots = \left(1 - \frac{\varphi^2}{2!} + \frac{\varphi^4}{4!} - \cdots\right) + i\left(\varphi - \frac{\varphi^3}{3!} + \frac{\varphi^5}{5!} - \cdots\right) \tag{2.72}$$

That is,

$$e^{i\varphi} = \cos\varphi + i\sin\varphi \tag{2.73}$$

Besides power series, a function may also be represented by a sum of sine and cosine terms. Such a representation is known as a Fourier series. One of the conditions a function has to fulfill in order to be expanded as a Fourier series is that it must be periodic.

In general, for a function $f(x)$ of period L, the Fourier series expansion is

$$f(x)=\frac{a_0}{2}+\sum_{s=1}^{\infty}\left[a_s\cos\left(\frac{2\pi sx}{L}\right)+b_s\sin\left(\frac{2\pi sx}{L}\right)\right] \tag{2.74}$$

where a_0, a_s, and b_s are the Fourier coefficients given by the integrals

$$a_s=\frac{2}{L}\int_{x_0}^{x_0+L}f(x)\cos\left(\frac{2\pi sx}{L}\right)dx \quad s=0,1,2,\ldots \tag{2.75}$$

$$b_s=\frac{2}{L}\int_{x_0}^{x_0+L}f(x)\sin\left(\frac{2\pi sx}{L}\right)dx \quad s=1,2,\ldots \tag{2.76}$$

x_0 is arbitrary and can be set to be zero.

The Fourier series in Eq. (2.74) can also be written in a complex form as

$$f(x)=\sum_{s=-\infty}^{\infty}c_s e^{i\left(\frac{2\pi s}{L}\right)x} \tag{2.77}$$

where Fourier coefficients are given by

$$c_s=\frac{1}{L}\int_{x_0}^{x_0+L}f(x)e^{-\frac{2\pi isx}{L}}dx \tag{2.78}$$

Sometimes, a pair of functions can be related by an expression

$$g(y)=\int_a^b f(x)K(x,y)dx \tag{2.79}$$

where $K(x, y)$ is called the kernel.

$g(y)$ can be considered as an integral transform of $f(x)$ by an "operator" $\mathcal{F}$ and can be written as

$$g(y)=\mathcal{F}f(x) \tag{2.80}$$

One of the ways such a relation (Eq. 2.80) can be attained is by a generalization of the Fourier series including in it functions defined over an infinite interval and with no apparent periodicity. This generalization is known as Fourier transform which is defined as

$$g(\omega)=\frac{1}{\sqrt{(2\pi)}}\int_{-\infty}^{\infty}f(t)e^{-i\omega t}dt \tag{2.81}$$

where we find $e^{-i\omega t}$ to be the kernel of the transform.

The inverse Fourier transform is then given by

$$f(t)=\frac{1}{\sqrt{(2\pi)}}\int_{-\infty}^{\infty}g(\omega)e^{i\omega t}d\omega \tag{2.82}$$

Both $f(t)$ and $g(\omega)$ in Eqs. (2.81), (2.82) are functions of continuous variables. Often, physical problems are concerned with values given at a discrete set of points. In such cases, we have to consider Fourier transform as an operation with N number of inputs and N number of outputs. Therefore, changing integration into finite summation, the discrete Fourier transform (DFT) g_m of a function f_n of discrete variables can be defined as

$$g_m = N^{-1/2} \sum_{n=0}^{N-1} e^{-2\pi imn/N} f_n \quad m = 0, 1, \ldots, N-1 \tag{2.83}$$

and the inverse transform as

$$f_n = N^{-1/2} \sum_{m=0}^{N-1} e^{2\pi imn/N} g_m \quad n = 0, 1, \ldots, N-1 \tag{2.84}$$

We shall come across expressions similar to (2.83), (2.84) in Chapter 6. It will be seen that if f_n is defined in the "reciprocal" space, then g_m will be given in real space by the summation over the entire volume of the reciprocal space.

Sample questions

1. Consider a function $f(z) = 2 + az + 2z^2$. For what values of a the roots of the equation $f(z) = 0$ will be complex?

2. Using calculus, show that the maxima of the function $f(\phi) = a \sin\phi$ are at $\phi = (2n + ½)\,\pi$, where $n = 0, 1, 2, \ldots$

3. Show that the vectors

$$\begin{bmatrix} 1 \\ -1 \\ 0 \end{bmatrix}, \begin{bmatrix} 1 \\ 1 \\ \sqrt{2} \end{bmatrix}, \begin{bmatrix} 1 \\ 1 \\ -\sqrt{2} \end{bmatrix}$$

are mutually orthogonal.

4. Using the chain rule of differentiation

$$\frac{df(y)}{dx} = \frac{df(y)}{dy}\frac{dy}{dx}$$

and considering

$$\frac{d(\ln y)}{dy} = \frac{1}{y}$$

find the derivative with respect to x of the function $f(x) = e^x$.

5. Find the mean value f_m of the function $f(x) = x^2 + 2x + 2$ between the limits $x = 2$ and $x = 4$.

6. Determine by integration the area enclosed between two concentric circles of radii a and b, respectively.

7. Write down the cross products of the units vectors in the Cartesian system, and using these, determine the area of a parallelogram with sides $\mathbf{a} = \hat{\mathbf{i}} + 2\hat{\mathbf{j}} + 3\hat{\mathbf{k}}$ and $\mathbf{b} = 3\hat{\mathbf{i}} + 2\hat{\mathbf{j}} + \hat{\mathbf{k}}$.

8. Let us consider a matrix A. The transpose A^T of the given matrix is formed by interchanging its rows and columns. The complex conjugate A^* is obtained by taking the complex conjugate of each of the elements of A, while the Hermitian conjugate, or adjoint, $A^\dagger$ is the transpose of A^* or, equivalently, the complex conjugate of A^T.
If

$$A = \begin{pmatrix} 2 & 1+i & i \\ 1+i & 3 & 2i \end{pmatrix}$$

write down its transpose, complex conjugate, and adjoint.

9. Using Taylor expansion show that

$$e^x \approx 1 + x \text{ for } x \ll 1$$

10. Find the Fourier transform of the following functions:

$$\text{(a) } f(t) = 0 \text{ for } t < 0$$
$$= Ae^{-kt} \text{ for } t \geq 0 \ (k > 0)$$

$$\text{(b) } f(t) = \begin{cases} 0 & \text{for } t < 0 \\ e^{-t/\tau}\sin\omega_0 t & \text{for } t \geq 0 \end{cases}$$

Hint:
$\sin\,\omega_0 t = (e^{i\omega_0 t} - e^{-i\omega_0 t})/2i$

References and further reading

Warren, W.S., 2001. The Physical Basis of Chemistry, second ed. Harcourt Academic Press.
Riley, K.F., Hobson, M.P., Bence, S.J., 2006. Mathematical Methods for Physics and Engineering, third ed. Cambridge University Press, Cambridge, UK.
Arfken, G.B., Weber, H.J., Harris, F.E., 2013. Mathematical Methods for Physicists: A Comprehensive Guide, seventh ed. Elsevier Academic Press.
Strang, G., 2017. Calculus, third ed. Wellesley-Cambridge Press, Wellesley, MA.

CHAPTER

Physical basis of chemistry

3

Richard Feynman, Nobel Laureate in Physics in 1965, had said, "there is nothing that living things do that cannot be understood from the point of view that they are made of atoms acting according to the laws of physics." On the one hand, physics has provided very powerful techniques such as X-ray crystallography, a wide variety of spectroscopy, and electron microscopy to investigate the biological system at the atomic level. At the same time, physics has also placed chemistry on a sound theoretical foundation from where it has been able to characterize complex molecules and unravel intricate molecular mechanisms of life.

3.1 Classical mechanics

3.1.1 Matter and motion

Let us begin by considering an object in motion. The simplest object would be a dimensionless particle. At any particular time t, the location of the particle can be denoted by the *position* vector $\mathbf{r}$:

$$\mathbf{r} = x\hat{\mathbf{i}} + y\hat{\mathbf{j}} + z\hat{\mathbf{k}} \tag{3.1}$$

where $\hat{\mathbf{i}}$, $\hat{\mathbf{j}}$, and $\hat{\mathbf{k}}$ are Cartesian unit vectors. (Vectors are represented by bold letters.)

It should be noted here that vectors are physical quantities which have both magnitude and direction as opposed to scalars which have only magnitude. We shall come across both kinds of quantities as we proceed.

It may happen that the position $\mathbf{r}$ of the particle changes with time t—this change in the position of the particle with time is called its *motion*. The rate of change of position of the particle at any instant of time t is denoted by the *velocity* vector

$$\mathbf{v} = \frac{d\mathbf{r}}{dt} \tag{3.2}$$

Eq. (3.2) also implies that when $\mathbf{v} = 0$, the particle is motionless or at rest with respect to a particular reference point or frame. It also means that $\mathbf{v}$ is nonzero if the magnitude or direction or both of $\mathbf{r}$ change with time.

Now the question is—can and, if so, how the velocity changes with time? The question was first answered by Galileo in his "principle of inertia"—left alone, and not disturbed, a particle continues to either be at rest or move with a constant velocity.

Fundamentals of Molecular Structural Biology. https://doi.org/10.1016/B978-0-12-814855-6.00003-1

However, in reality, an object is never "left alone." It interacts with other objects in its environment. The interaction of an object with another object is described in terms of a force **F** which, like position and velocity, is a vector. What are these forces?

3.1.2 Fundamental forces of nature

There are four fundamental forces of nature—gravitational, electromagnetic, strong, and weak. Of these, the strong and weak interactions are extremely short range (10^{-15} to 10^{-13} cm). The strong interaction holds together the constituents of the nucleus, whereas the weak interaction is involved in the radioactive decay of certain elementary particles to other elementary particles. We need not consider these two interactions any further in our present study.

The electromagnetic or electrical force between two charged particles 1 and 2 with charges q_1 and q_2, respectively, separated by a distance $\mathbf{r}$ is given by

$$\mathbf{F} = \frac{q_1 q_2 \mathbf{r}}{4\pi\varepsilon_0 r^3} \tag{3.3}$$

where $\varepsilon_0 = 8.854 \times 10^{-12}\,\mathrm{C\,N^{-1}\,m^2}$ is the permittivity of free space (C, Coulomb; N, Newton). Eq. (3.3) can also be written as

$$\mathbf{F} = q_2 \mathbf{E} \tag{3.4}$$

where

$$\mathbf{E} = \frac{q_1 \mathbf{r}}{4\pi\varepsilon_0 r^3}$$

is said to be the electric field due to charge q_1 experienced by charge q_2. **F** is also called the Coulombic force, while Eq. (3.3) is the mathematical expression of Coulomb's law.

The gravitational force has a mathematical form similar to the electrical force. Between two particles 1 and 2 of masses m_1 and m_2, respectively, the gravitational force is expressed as

$$\mathbf{F} = -\frac{G m_1 m_2 \mathbf{r}}{r^3} \tag{3.5}$$

where $G = 6.672 \times 10^{-11}\,\mathrm{N\,m\,kg^{-2}}$ is called the gravitational constant. The negative sign indicated that the gravitational force is always attractive in contrast to the electric force which can be attractive as well as repulsive. Also, we can consider the gravitational field produced by m_1 as

$$\mathbf{C} = -\frac{G m_1 \mathbf{r}}{r^3}$$

so that the force on m_2 can be written as

$$\mathbf{F} = m_2 \mathbf{C} \tag{3.6}$$

3.1.3 Newton's laws

In the 17th century, Sir Isaac Newton used the concept of force to describe the motion of a particle by three laws. Newton's laws are valid when the velocity of the particle does not approach the speed of light and the size of the particle does not go down to the atomic level. These are the domains dealt with by relativistic mechanics and quantum mechanics, respectively. Newton's laws lie in the domain of classical mechanics.

The *first law* is essentially a restatement of Galileo's principle of inertia—if there is no net force acting on a particle, it will continue to be either at rest or move with a constant velocity.

It should be mentioned here that Galileo's principle of inertia or Newton's laws are valid in any inertial frame of reference, that is, a frame which itself is either "at rest" or moving with a uniform velocity.

Newton's first law provides a qualitative description of a force—it is something that can change the motion (velocity) of a particle. The rate of change of velocity at an instant of time t is known as *acceleration* and is given by

$$\mathbf{a} = \frac{d\mathbf{v}}{dt} \tag{3.7}$$

Also, we can introduce another quantity here related to velocity, that is, *linear momentum* which is given by the product of the mass of the particle and its velocity

$$\mathbf{p} = m\mathbf{v} \tag{3.8}$$

According to Newton's *second law*, the rate of change of linear momentum is equal to the force causing the change, that is,

$$\mathbf{F} = \frac{d\mathbf{p}}{dt} = \frac{d(m\mathbf{v})}{dt} \tag{3.9}$$

If mass m is constant, the second law can also be expressed as

$$\mathbf{F} = m\mathbf{a} = m\frac{d\mathbf{v}}{dt} = m\frac{d^2\mathbf{r}}{dt^2} \tag{3.10}$$

As mentioned earlier, the force that acts on a particle to change its momentum is due to another particle in its environment. Considering two particles 1 and 2, if $\mathbf{F}_{12}$ be the force on particle 1 due to particle 2 causing a change in the momentum $\mathbf{p}_1$ of particle 1 and $\mathbf{F}_{21}$ be the force on particle 2 due to particle 1 causing a change in the momentum $\mathbf{p}_2$ of particle 2, then according to Newton's *third law*

$$\mathbf{F}_{12} = -\mathbf{F}_{21} \tag{3.11}$$

In other words, if a particle exerts a force on another, the second exerts on the first a force equal in magnitude and opposite in direction.

Considering a system containing only two particles and not acted upon by any external force, the total momentum is given by

$$\mathbf{P} = \mathbf{p}_1 + \mathbf{p}_2 \tag{3.12}$$

Combining Eq. (3.11) and Eq. (3.12), we have

$$\frac{d\mathbf{P}}{dt} = 0$$

or

$$\mathbf{P} = \text{constant} \tag{3.13}$$

This is the mathematical expression of the *law of conservation of total linear momentum* of an isolated system.

3.1.4 Circular motion and angular momentum

Now, let us consider that a particle of mass m at a position **r** relative to a point O (the origin) is constrained by a force (centripetal force) to move in a circular path with instantaneous velocity **v** (velocity at a particular instant of time t) (Fig. 3.1). If the magnitude of the velocity (speed) be constant, the particle is said to be in uniform circular motion around O. The movement of the particle can be described in terms of the angle θ measured counterclockwise between the radial line from the origin to the particle and a fixed reference line, say Oy, also through the origin (Fig. 3.1). The magnitude of the angular velocity is given by $\omega = \mathrm{d}\theta/\mathrm{d}t$. It can be very simply shown that $v = r\omega$.

We have seen above that the linear momentum **p** of the particle is given by Eq. (3.8). The *angular momentum* of the particle relative to the origin is defined as

$$\mathbf{L} = \mathbf{r} \times \mathbf{p} = \mathbf{r} \times m\mathbf{v} \tag{3.14}$$

In the present case (uniform circular motion), **r** is perpendicular to **v**. Hence, the magnitude of the angular momentum can be expressed as

$$|\mathbf{L}| = mvr = mr^2\omega \tag{3.15}$$

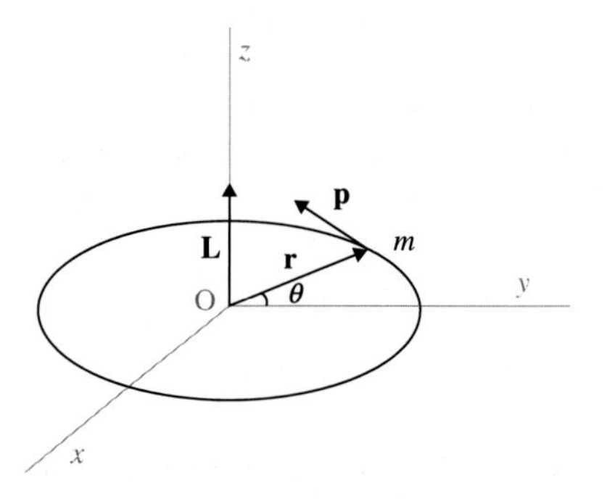

FIG. 3.1

Circular motion of a particle.

3.1.5 Work and energy

Simply speaking, energy of a system is a measure of its capacity to do work. There are different forms of energy—electrical, chemical, gravitational, mechanical, etc. There are two kinds of mechanical energy—kinetic and potential.

When a force **F** acting on a particle for a certain time interval causes its change of position or displacement **s**, it is defined that the force does a work given by

$$W = \mathbf{F} \cdot \mathbf{s} \tag{3.16}$$

Following Newton's second law, it can be shown that

$$W = \frac{1}{2}mv_f^2 - \frac{1}{2}mv_i^2 \tag{3.17}$$

where v_i and v_f are the initial and final velocities, respectively.

Now,

$$K = \frac{1}{2}m\mathbf{v} \cdot \mathbf{v} = \frac{1}{2}mv^2$$

is defined as the *kinetic energy* of a particle moving with a velocity v. Then

$$W = K_f - K_i \tag{3.18}$$

where K_i and K_f are the initial and final kinetic energies, respectively. Eq. (3.18) is the mathematical expression of *work-energy theorem*—the work done by a force acting on a particle is equal to the change in its kinetic energy.

Potential energy, denoted by U, is related to the configuration of a system. In the case of a multicomponent system, configuration means how the components are arranged with respect to each other. For a particle, it simply means the position.

Let us consider a force $\mathbf{F}(\mathbf{r})$ that does a work W on a system to change its configuration. If the amount of work done is dependent only on the initial and final configurations and not the path of transition, the force is said to be *conservative*. For a conservative force, the change in potential energy of a particle is given by

$$\Delta U = U(r_f) - U(r_i) = -W = \int_{\mathbf{r}_i}^{\mathbf{r}_f} \mathbf{F}(\mathbf{r}) \cdot d\mathbf{r} \tag{3.19}$$

which can be also written as

$$\mathbf{F}(\mathbf{r}) = -\nabla U(\mathbf{r}) = -\left[\hat{\mathbf{i}}\left(\frac{\partial U}{\partial x}\right) + \hat{\mathbf{j}}\left(\frac{\partial U}{\partial y}\right) + \hat{\mathbf{k}}\left(\frac{\partial U}{\partial z}\right)\right] \tag{3.20}$$

That is to say, the force on a particle at a given position is given by the negative gradient of the potential energy at that position.

Combining the kinetic energy and the potential energy of a system, we have its total mechanical energy given by

$$E = K + U \tag{3.21}$$

Now if we consider an isolated system on which no external force acts, but the particles within the system exert conservative forces on one another

$$\Delta K = -\Delta U \tag{3.22}$$

and

$$\Delta E = 0$$

or

$$E = \text{constant} \tag{3.23}$$

This is the *law of conservation of mechanical energy.*

3.1.6 Oscillatory systems

In the study of different areas of physics, we come across various oscillatory systems—a mass on a spring, a swinging pendulum, a vibrating tuning fork generating sound waves, vibrating electrons in an atom producing electromagnetic waves, etc. The motion of all such oscillatory systems can be represented by a common differential equation.

Let us consider a simple oscillatory system (Fig. 3.2), that is, a particle subject to a force given by

$$\mathbf{F}(\mathbf{r}) = -k\mathbf{r} \tag{3.24}$$

where k is a constant and $\mathbf{r}$ is the displacement of the particle from equilibrium position (say, $\mathbf{r}=0$). It is apparent from Eq. (3.24) that if the particle be displaced to either side of the equilibrium position, the "restoring" force will act towards the equilibrium position.

Combining Eq. (3.24) with Eq. (3.10), we have the guiding differential equation for an oscillatory motion

$$\left(\frac{d^2\mathbf{r}}{dt^2}\right) + \left(\frac{k}{m}\right)\mathbf{r} = 0 \tag{3.25}$$

If we consider the oscillation in y—direction only, the above differential equation can be written as

$$\left(\frac{d^2y}{dt^2}\right) + \left(\frac{k}{m}\right)y = 0 \tag{3.26}$$

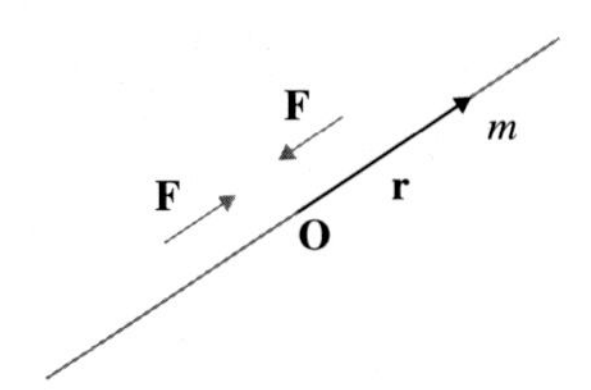

FIG. 3.2

A simple oscillatory system.

whose solution is given as

either $$y = y_m e^{i(\omega t + \varphi)} \quad (3.27)$$

or $$y = y_m \sin(\omega t + \varphi) \quad (3.28)$$

where y_m is the amplitude, ω the angular frequency, and φ the phase constant of the oscillation. It can be found from Eq. (3.27) or Eq. (3.28) that $\omega^2 = k/m$ and therefore the period of oscillation

$$T = \frac{2\pi}{\omega} = 2\pi\sqrt{\left(\frac{m}{k}\right)} \quad (3.29)$$

3.2 Wave motion

Mechanical waves, including sound waves, arise from the oscillation of particulate matter in an elastic medium. Electromagnetic waves, such as light waves, are generated by oscillatory electric and magnetic fields. Yet both can be represented by similar mathematical relations.

3.2.1 Mechanical waves

Interaction between the atoms in a medium is responsible for the propagation of mechanical waves. Each atom oscillates about an equilibrium position and, through the force of interaction, transmits this oscillatory motion to its neighboring atoms. There is no net displacement of the particles in the direction of the wave. The oscillatory motion of the particles is perpendicular to the direction of wave propagation for a transverse wave and along the direction of propagation for a longitudinal wave.

Let us consider a transverse wave traveling (say, in a string) with a speed v in the x-direction. At a time t, the wave is represented by

$$y = f(x - vt) \quad (3.30)$$

which is called a wave function. As it has been said that the transmission of oscillatory motion is due to the force of interaction between neighboring particles, Newton's second law can be applied to derive a differential equation that describes the behavior of the wave:

$$\frac{\partial^2 y}{\partial x^2} = \frac{1}{v^2}\frac{\partial^2 y}{\partial t^2} \quad (3.31)$$

This is called the wave equation. The general form of the wave equation is applicable not only to mechanical waves, but, as we shall see later, to electromagnetic waves as well. For a sinusoidal wave, the solution to Eq. (3.31) becomes

$$y = y_m \sin(kx - \omega t) \quad (3.32)$$

where $k = 2\pi/\lambda$ is the wave number and $\omega = 2\pi/T = 2\pi f$ is the angular frequency. λ, T, and f are, respectively, the wavelength, period, and frequency of the wave.

3.2.2 Electromagnetic wave

Just as mechanical waves are propagated by oscillating atoms in an elastic medium, electromagnetic waves are propagated by periodically varying electric and magnetic fields. The problem of electromagnetic wave propagation can, therefore, be addressed by the application of Maxwell's equations which are the basic equations of electromagnetism. Maxwell's equations describe (a) the relation between charge q and the electric field **E**, (b) the nature of the magnetic field **B**, (c) the effect on the electric field **E** of a changing magnetic field, and (d) the effect on the magnetic field **B** of a current or changing electric field.

Now the question is how an electromagnetic wave is generated? It is known that a static electric charge produces electric field lines, while a steady electric current, that is, a charge moving with a uniform speed, produces magnetic field lines. An energy density is associated with the electric and magnetic fields. However, in the steady state, there is no transport of energy or momentum and no propagation of electromagnetic waves.

For wave propagation, the charge needs to be accelerated. This is achieved by an oscillating charge in an electric dipole antenna. The oscillatory **E** and **B**, thus produced, are in phase and perpendicular to one another. From Maxwell's equations, one can derive the wave equations for **E** and **B** as

$$\frac{\partial^2 \mathbf{E}}{\partial x^2} = \frac{1}{c^2}\left(\frac{\partial^2 \mathbf{E}}{\partial t^2}\right) \tag{3.33a}$$

$$\frac{\partial^2 \mathbf{B}}{\partial x^2} = \frac{1}{c^2}\left(\frac{\partial^2 \mathbf{B}}{\partial t^2}\right) \tag{3.33b}$$

where $c = 1/\sqrt{(\mu_0 \varepsilon_0)}$ is the speed of the waves; ε_0 is the electric constant, and μ_0 is the magnetic constant. Eqs. (3.33a) and (3.33b) are similar in form to the wave equation we have seen for mechanical waves (Eq. 3.31). For a sinusoidal wave, the solutions to Eqs. (3.33a) and (3.33b) are respectively

$$E(x,t) = E_m \sin(kx - \omega t) \tag{3.34a}$$

and

$$B(x,t) = B_m \sin(kx - \omega t) \tag{3.34b}$$

where $k = 2\pi/\lambda$ and $c = \omega/k$. E_m and B_m are the electric and magnetic field amplitudes. The intensity of the electromagnetic wave is

$$I = \frac{1}{2\mu_0 c}E_m^2 \tag{3.35}$$

The wave equation (Eq. 3.31) is a linear differential equation since the function $y(x, t)$ is in the first power of x and t. For such a differential equation if $y_1(x, t)$ and $y_2(x, t)$ are two different solutions, then a linear combination of the two

$$y(x,t) = c_1 y_1(x,t) + c_2 y_2(x,t) \tag{3.36}$$

where c_1 and c_2 are arbitrary constants, is also a solution. This is the basis of the *principle of superposition* which states that if two or more waves pass through a point in space, they are algebraically added to produce the resultant wave. For n number of waves, Eq. (3.36) can be written as

$$y(x,t) = \sum_{j=1}^{n} c_j y_j(x,t) \tag{3.37}$$

3.2.3 Interference of light waves

When two or more waves superpose, under certain conditions they produce an intensity pattern that does not change with time. The phenomenon is known as interference. It is required that the phase relationship between the waves should remain unchanged with time. Such waves are said to be coherent. Coherence is a necessary condition for interference.

Let us consider the interference of two light waves. Firstly, we must note that, for reasons not discussed here, waves from two different sources are incoherent. However, coherence can be achieved by dividing a light beam from a single source into two components which can then be made to interfere.

A beam of light (preferably from a laser) is made to pass through two narrow parallel slits S_1 and S_2 (Fig. 3.3A). The light will be bent, or *diffracted*, as it passes through the slits. Each slit will, therefore, act as a line (or in two-dimension, a point) source of light. If waves emanating from the two slits meet at a point P, they can interfere (Fig. 3.3A).

Let the electric field components of the two waves at P be represented as

$$E_1 = E_0 \sin \omega t \tag{3.38a}$$

and

$$E_2 = E_0 \sin(\omega t + \varphi) \tag{3.38b}$$

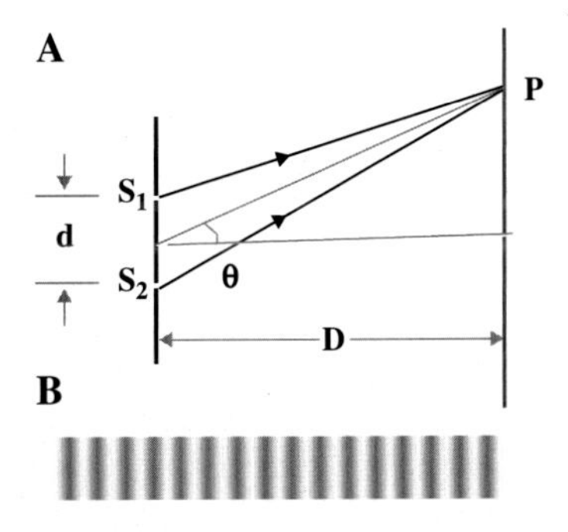

FIG. 3.3

(A) Interference of coherent light rays diffracted from two slits and (B) interference fringes.

where φ is the phase difference between the two waves arising out of path-length difference. Let P be any point of a screen at a distance D from the slits and d be the slit separation. If $D \gg d$, the two waves can be considered as parallel. In that case, it can be shown that the resultant wave function is

$$\begin{aligned} E = E_1 + E_2 &= E_0 \sin \omega t + E_0 \sin(\omega t + \varphi) \\ &= \left[2E_0 \cos \frac{1}{2}\varphi\right] \sin\left(\omega t + \frac{1}{2}\varphi\right) \end{aligned} \tag{3.39}$$

and the intensity of the resultant wave at P is

$$I = 4I_0 \cos^2 \frac{1}{2}\varphi \tag{3.40}$$

where I_0 is the intensity of the light reaching P from either slit separately.

If the position of P on the screen is indicated by the angle θ (Fig. 3.3A), then

$$\varphi = \frac{2\pi}{\lambda} d \sin\theta \tag{3.41}$$

It can be seen from Eqs. (3.40), (3.41) that for intensity maxima

$$d \sin \theta = m\lambda \qquad m = 0, \pm 1, \pm 2, \ldots \tag{3.42}$$

and for intensity minima

$$d \sin\theta = \left(m + \frac{1}{2}\right)\lambda \qquad m = 0, \pm 1, \pm 2, \ldots \tag{3.43}$$

Thus, a pattern of bright and dark bands, which are called *interference fringes*, is produced on the screen (Fig. 3.3B).

The analysis can be extended from the two-slit system to a diffraction grating where the number of slits, that is, diffraction sources, is much larger than two (may be as large as 10^4) (Fig. 3.4). In that case, the intensity maxima will be given by

$$d \sin\theta = m\lambda \qquad m = 0, 1, 2, \ldots \tag{3.44}$$

where d is the separation between adjacent slits.

3.3 Kinetic theory of gases

The macroscopic properties of a system are expressed in terms of its pressure, volume, and temperature. The simplest system to study is that of an ideal gas enclosed in a volume V at pressure p and temperature T. Like all kinds of matter, the system under consideration is also made up of molecules. The pressure, volume, and temperature of an ideal gas are related by the equation

$$pV = NkT \tag{3.45}$$

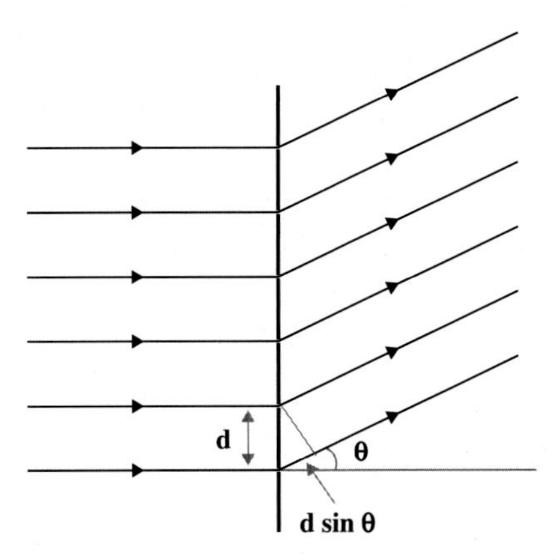

FIG. 3.4

Diffraction of coherent rays from a grating.

where N is the number of molecules contained in the volume and k is the Boltzmann constant. This is the ideal gas law. In terms of the number of moles n, Eq. (3.45) can also be written as

$$pV = nRT \tag{3.46}$$

where $R = kN_A$ is the molar gas constant, $N_A = 6.022140857 \times 10^{23}\,\text{mol}^{-1}$ being the Avogadro's number.

The kinetic theory of gases explains the macroscopic properties of a gas in terms of the molecular properties (microscopic behavior). It considers the molecules of an ideal gas as independent, randomly moving particles obeying Newton's laws of motion. The particles interact only with the walls of the container or one another by elastic collision.

Since the particles were in random motion, Maxwell assumed that the distribution of velocity vectors in the gas would be a Gaussian function, a function of the form e^{-X^2}. Considering only the magnitude (speed) v of the velocity vector, the distribution of molecular speed was calculated to be

$$N(v) = 4\pi N\left(\frac{m}{2\pi kT}\right)^{\frac{3}{2}} v^2 e^{-\frac{mv^2}{2kT}} \tag{3.47}$$

where m is the mass of each molecule. This is called the Maxwell-Boltzmann distribution of molecular speed. From Eq. (3.47), one can arrive at the ideal gas law given by Eq. (3.45) or (3.46).

3.4 Thermodynamics

Let us consider a system of particles and ignore the kinetic energy that may arise due to the motion of the entire system as also the potential energy due to an external field influencing the system as a whole. The system still possesses energy due to the random motion of the individual particles and the forces of interaction between these particles. This is said to be the *internal energy* of the system.

The internal energy can change in two ways—if work W is done on the system to cause redistribution of the individual particles and/or heat Q is transferred through the system boundary to change the random motion of the particles. If ΔE_{int} be the total change in internal energy, the law of conservation of energy can be expressed as

$$\Delta E_{int} = Q + W \tag{3.48}$$

This is the *first law of thermodynamics*. The first law introduces the concept of energy of a system (internal energy).

Now, it is a fact that all naturally occurring (irreversible) processes can proceed only in one direction. The first law of thermodynamics does not provide any clue to the direction of such a process. It is the second law of thermodynamics that deals with the direction of a naturally occurring process. The second law introduces yet another important concept—entropy—whose change ΔS determines the direction of an irreversible process.

The change of entropy is formally defined for a reversible process as

$$\Delta S = \int_i^f (dQ/T) \tag{3.49}$$

where dQ is the increment of heat energy at temperature T, while i and f represent, respectively, the initial and final states of the system. If the process takes place at a constant temperature (isothermal), Eq. (3.49) becomes

$$\Delta S = Q/T \tag{3.50}$$

Although the definition is for a reversible process, the entropy of an irreversible process can also be calculated by indirect means.

Based on the concept of entropy, the second *law of thermodynamics* states that, in a closed system, the entropy either increases for an irreversible process or remains unchanged for a reversible process. That is,

$$\Delta S \geq 0 \tag{3.51}$$

3.5 Quantum physics

Wave and particle are classical concepts; just as classical mechanics has been able to explain a number of properties of matter, wave theory has been consistent with a large number of experiments with light. However, with experiments to explore newer domains and observe newer phenomena, the limitations of the classical concepts became too evident.

3.5.1 Light as photons

Visible light is only a part of the entire electromagnetic spectrum. One of the three modes of energy transfer is electromagnetic radiation. Objects absorb, reflect, and emit electromagnetic radiation. On the higher wavelength side of the visible region of the electromagnetic spectrum is the infrared region, while on the smaller wavelength side is the ultraviolet region. Radiation emitted by an object due to its temperature is called thermal radiation, which is essentially infrared in nature.

The properties of thermal radiation were studied by using what is known as a *blackbody*. It is an idealized object radiating 100% of its energy and also absorbing 100% of any incident radiation. The emitted radiation does not depend on the material of the object.

A system close to a blackbody can be simulated by forming a cavity in a solid body and maintaining its walls at a uniform temperature. If a small hole is drilled in the wall, the radiation emerging from the cavity through the hole will not depend on the material or shape of the cavity, but only on its temperature. Therefore, the radiation is also called *cavity radiation*.

The nature of cavity radiation can be studied by measuring the spectral emittance. The power emitted per unit area of the blackbody for wavelengths between λ and $\lambda + d\lambda$ is $R(\lambda, T)d\lambda$, where $R(\lambda, T)$ is the *spectral emittance*. At a certain temperature T, the observed spectral emittance (plotted as a function of frequency of radiation instead of wavelength) is shown in Fig. 3.5.

Using the classical concept of light as a wave (electromagnetic), Rayleigh and Jeans arrived at an expression for the spectral emittance of blackbody radiation

$$R(\lambda, T) = \frac{2\pi c k_B T}{\lambda^4} \tag{3.52}$$

where c is the speed of light and k_B the Boltzmann constant

$$k_B = 1.38 \times 10^{-23}\ \mathrm{J\ K^{-1}} \tag{3.53}$$

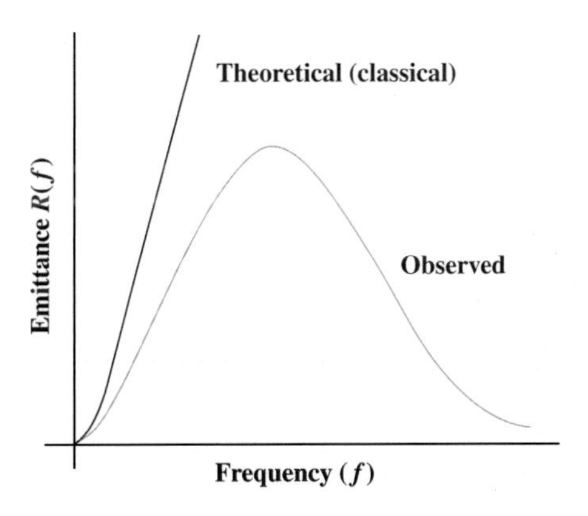

FIG. 3.5

Blackbody radiation.

It can be seen in Fig. 3.5 that the curve theoretically derived (based on classical concept) does not match the experimentally observed curve. It diverges at higher values of frequency (as $\lambda \rightarrow 0$)—the problem being known as the *ultraviolet catastrophe*.

Max Planck made a radical departure from the classical concept and introduced the concept of *quantized energy*. His formulation was based on the assumption that atoms in the cavity walls can exchange energy with the electromagnetic radiation only in discrete amounts that were integral multiples of hf where f is the frequency of the radiation these oscillators absorb and emit and h is the *Planck constant*. Planck's radiation law expresses spectral radiancy in terms of wavelength and frequency respectively as

$$R(\lambda, T) = \frac{2hc^2}{\lambda^5} \frac{1}{e^{\frac{hc}{\lambda k_B T}} - 1} \tag{3.54a}$$

and

$$R(f, T) = \frac{2hf^3}{c^2} \frac{1}{e^{\frac{hf}{k_B T}} - 1} \tag{3.54b}$$

The concept of quantization of energy in its exchange between matter and radiation was taken forward by Albert Einstein to the description of light as discrete bundles of energy or photons. The energy of a photon is given as

$$E = hf \tag{3.55}$$

Further, behaving as a particle the photon is assigned a linear momentum with magnitude

$$p = h/\lambda \tag{3.56}$$

3.5.2 Matter as waves

It is quite logical to think that if light can behave as both wave and particle, matter, such as an electron, should display wave nature and be assigned a wavelength and a frequency. Based on this symmetry argument, Louis de Broglie proposed that an electron with an energy E and linear momentum p can be conceptualized as a matter wave with wavelength and frequency given by

$$\lambda = h/p \tag{3.57}$$

and

$$f = E/h \tag{3.58}$$

Now, the simplest way to describe the matter wave related to a particle with a defined momentum p_0 (and, therefore, defined wavelength λ_0 or wave number $k_0 = 2\pi/\lambda_0$) would be a perfect sinewave moving along the x-axis which at a time $t = 0$ can be represented as

$$\psi(x) = \psi_0 \sin k_0 x \tag{3.59}$$

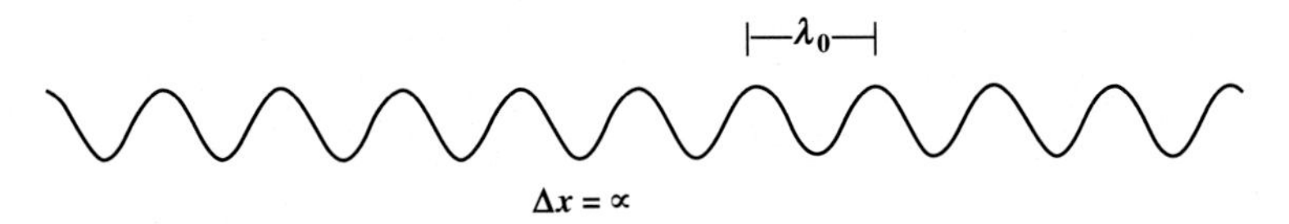

FIG. 3.6

A particle represented by an infinitely long sinewave.

and shown as an infinitely long wave (Fig. 3.6). In that case, the particle can be found anywhere between $x=-\infty$ to $x=+\infty$. In other words, the particle is completely delocalized or there is absolute uncertainty regarding the position of the particle.

If the particle be described by the superposition of a large number of sinusoidal waves with wave number spread Δk about k_0 (corresponding to a spread in the momentum Δp), a wave packet is formed and the particle thus described appears to be localized within Δx (Fig. 3.7). Evidently, a decrease in the uncertainty Δx can occur only at the cost of an increase in the uncertainty Δp. This is essentially *Heisenberg's uncertainty principle*, which states that it is impossible to determine the position and the momentum of a particle with unlimited precision. Mathematically, it can be expressed as

$$\Delta x \cdot \Delta p_x \geq \frac{h}{2\pi} \tag{3.60}$$

3.5.3 Schrödinger equation

A mathematical theory of the matter wave was developed by Erwin Schrödinger. This is known as *quantum theory*, *quantum mechanics*, or *wave mechanics*. Schrödinger introduced the *wave function* Ψ that would obey a fundamental equation of quantum mechanics known as the *Schrödinger equation*, which in one-dimension is written as

$$i\hbar\frac{\partial\Psi}{\partial t} = -\frac{\hbar^2}{2m}\frac{\partial^2\Psi}{\partial x^2} + V(x)\Psi \tag{3.61}$$

where $\hbar = h/2\pi$ and potential $V(x)$ is assumed to be time-independent. The equation is known as the *time-dependent Schrödinger equation*. For a particle moving in x-direction, it can be shown that

$$\Psi(x,t) = \Psi_0\, e^{i(kx-\omega t)} \tag{3.62}$$

is a solution to the Schrödinger equation, and hence, the wave function. It is not possible to know the precise location of a quantum particle. Therefore, Max Born introduced the concept of *probability density* expressed as

$$P(x) = \Psi\Psi^* \tag{3.63}$$

so that the probability of finding the particle between x and $x+dx$ is given by $\Psi\Psi^* dx$.

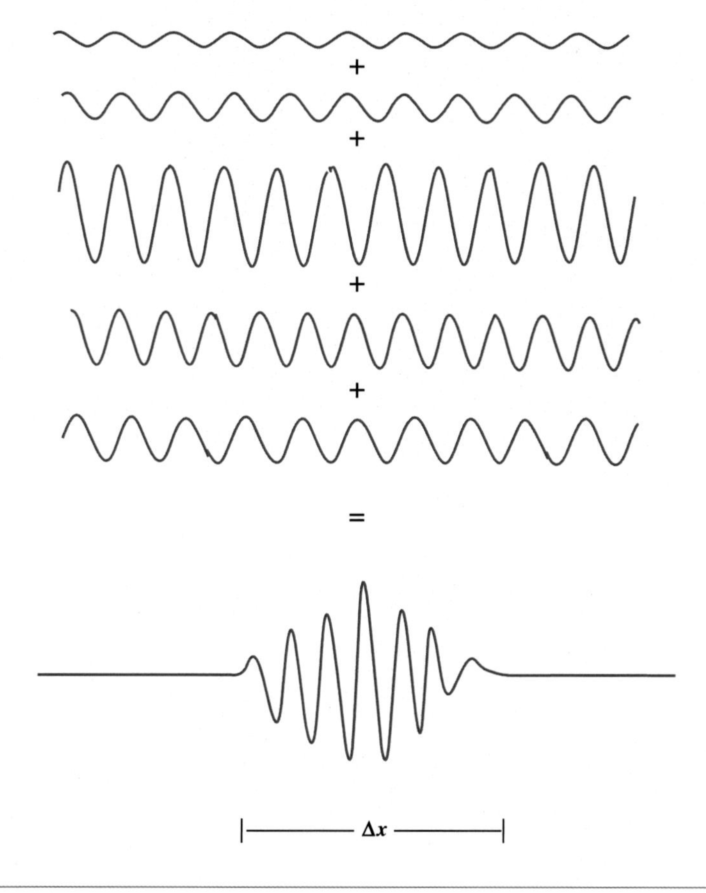

FIG. 3.7

A particle described by a wave packet.

Separating the time and space variations of the wave function, we can derive from Eq. (3.61) the *time-independent Schrödinger* equation

$$-\frac{\hbar^2}{2m}\frac{d^2\psi(x)}{dx^2}+V(x)\psi(x)=E\psi(x) \tag{3.64}$$

where $E=\hbar\omega=hf$, the energy of the particle. For a particle, not under the influence of any potential $[V(x)=0]$, the total energy is entirely kinetic, that is, $E=K=p^2/2m$. In such case, Eq. (3.64) becomes

$$\frac{\hbar^2}{p^2}\frac{d^2\psi(x)}{dx^2}+\psi(x)=0 \tag{3.65}$$

The solution to this equation is

$$\psi(x)=\psi_0 e^{ikx} \tag{3.66}$$

where $k=2\pi/\lambda$. This is the wave function for a free particle traveling in the x-direction.

3.5.4 Barrier tunneling by free electron

What happens when a particle moving in the x-direction with energy E hits a potential energy barrier of height V_0 such that $V_0 > E$ (Fig. 3.8A)? Classically speaking, the particle would rebound from the barrier. In quantum mechanics, the answer to the question is given by the wave nature of the particle and Schrödinger equation. Let a be the thickness of the barrier so that the potential energy

$$V(x) = \begin{cases} 0 & x<0 \\ V_0 & 0<x<a \\ 0 & x>0 \end{cases} \tag{3.67}$$

With these conditions, the solutions to the time-independent Schrödinger equation will be obtained as

$$\begin{aligned} \psi(x) &= \psi_1 e^{ikx} + \psi_2 e^{-ikx} && \text{to the left of the barrier } (x<0) \\ \psi(x) &= \psi_3 e^{-k'x} + \psi_4 e^{+k'x} && \text{within the barrier } (0<x<a) \\ \psi(x) &= \psi_5 e^{ikx} && \text{to the right of the barrier } (x>a) \end{aligned} \tag{3.68}$$

where $\psi_1, \ldots, \psi_5$ are amplitude constants, $k\hbar = (2mE)^{1/2}$ and $k'^2 = 2m(V_0 - E)/\hbar^2$. The corresponding probability densities $P(x)$ in the three regions are shown in Fig. 3.8B. These solutions suggest that the quantum mechanical particle (electron or any other particle with a small mass) moving in the x-direction as a wave with an amplitude ψ_1 will be partly rebounded from the barrier as a wave with an amplitude ψ_2, where $|\psi_2| < |\psi_1|$. However, there is also a finite probability with probability density

$$P(x) = |\psi_3|^2 e^{-2k'x} \tag{3.69}$$

that the particle will penetrate and "leak" through the barrier and appear on the other side as a wave with small but constant amplitude. This is the phenomenon of quantum tunneling which is important in many chemical reactions and chemical

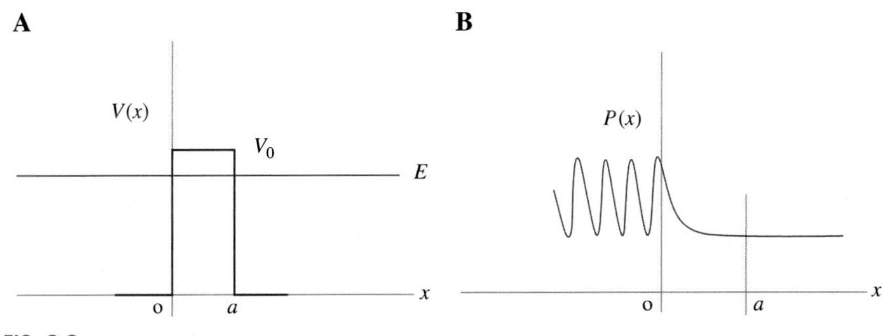

FIG. 3.8

Barrier tunneling by electron.

processes at relatively lower temperatures. The transmission coefficient is found to be essentially dominated by the factor $e^{-2k'a}$, indicating that it is sensitive to the thickness of the barrier and mass of the particle. For an infinitely tall barrier ($V_0 \rightarrow \propto$), the particle will be totally (100%) reflected.

3.5.5 Bound electron—Hydrogen atom

In contrast to a free electron on which no net force acts, a bound electron is confined to a limited region of space, say within an atom, by a force/forces acting on it. One can say that the electron is trapped in a potential energy well or, in short, a potential well.

The hydrogen atom has been the simplest system to study the movement of a bound electron. Bound by the Coulombic interaction with the nucleus (a single proton in this case), the electron moves in three-dimension. Its potential energy function is

$$U(r) = -\frac{1}{4\pi\epsilon_0}\frac{e^2}{r} \tag{3.70}$$

where r is the radial distance of the electron from the nucleus and e is the magnitude of (negative) charge of the electron.

Classical physics failed to explain the motion of the electron in stable orbits around the nucleus. Based on a combination of classical and quantum perspectives, Niels Bohr proposed a model of the hydrogen atom. Introducing the concept of a "*quantum state*," he postulated that electrons in an atom move in stable (quantum) states of fixed energy without radiating. Further, he made a second postulate according to which an electron moving from one quantum state (energy E_i) to another (energy E_f) emits or absorbs radiation at a fixed frequency f given by

$$hf = E_f - E_i \tag{3.71}$$

In order to calculate the radii of the orbits, Bohr quantized the magnitude of the angular momentum of the electron in a stable orbit as

$$mv_n r_n = n\frac{h}{2\pi} \quad n = 1,2,3,\ldots \tag{3.72}$$

The radii r_n was then calculated to be

$$r_n = \left(\frac{\epsilon_0 h^2}{m\pi e^2}\right)n^2 = a_0 n^2 \tag{3.73}$$

where m is the mass of the electron, ε_0 is the electric constant, and n is the quantum number. a_0 is called the *Bohr radius*. The energy of the nth state was calculated to be

$$E_n = -\frac{me^4}{8\varepsilon_0^2 h^2}\frac{1}{n^2} \tag{3.74}$$

Bohr model of the hydrogen atom, based on some ad hoc postulates, had its limitations. Quantum mechanics put the model on a stronger theoretical foundation.

As a prelude to the quantum mechanics of the hydrogen atom, it may be worthwhile to understand the quantum behavior of an electron localized in an infinite potential (Fig. 3.9) well-described by

$$V(x) = \begin{cases} \infty & x < 0 \\ 0 & \text{for } 0 \leq x \leq \text{a} \\ \infty & x > \text{a} \end{cases} \tag{3.75}$$

It can be shown that the only possible solutions of the Schrödinger equation are

$$\psi(x) = \begin{cases} 0 & x < 0 \\ A \sin\left(\frac{n\pi x}{a}\right) \quad n = 1,2,3,\ldots & \text{for } 0 \leq x \leq a \\ 0 & x > 0 \end{cases} \tag{3.76}$$

where A is the amplitude of the wave function and n is the principal quantum number. A can be determined from the *normalization* condition

$$\int_0^a P(x)dx = 1 \tag{3.77}$$

which implies that whatever the quantum state of the electron, it must be found within the infinite well. It can be shown that

$$A = \left(\frac{2}{a}\right)^{\frac{1}{2}} \tag{3.78}$$

so that the wave function within the infinite potential well becomes

$$\psi(x) = \left(\frac{2}{a}\right)^{\frac{1}{2}} \sin\left(\frac{n\pi x}{a}\right) \quad n = 1,2,3,\ldots \tag{3.79}$$

The electron can exist in discrete quantum states with the energy given by

$$E_n = -n^2\left(\frac{h^2}{8ma^2}\right) \quad n = 1,2,3,\ldots \tag{3.80}$$

In other words, the *energy* becomes *quantized*. Thus, localization of a bound electron leads to quantization.

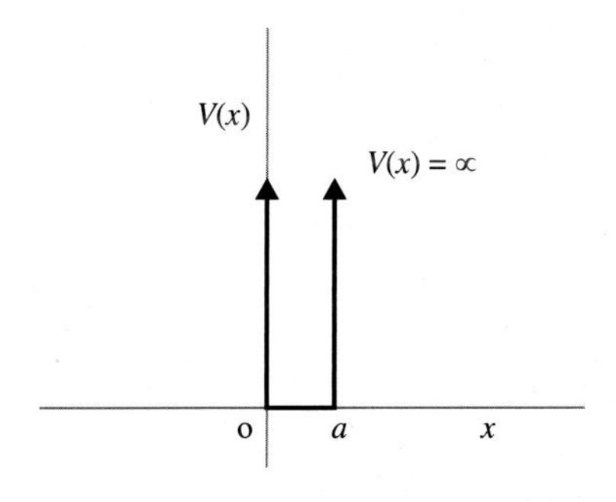

FIG. 3.9

Potential well.

In quantum mechanics, the electron in a hydrogen atom is not seen to be moving in a circular or elliptical orbit. Instead, the solution of the Schrödinger equation for hydrogen atom yields a set of wave functions which define three-dimensional regions of space of various sizes, shapes, and orientations where the electron resides. These regions of space, or sometimes the wave functions themselves, are referred to as *orbitals.*

Since the potential energy for the hydrogen atom depends only on the radial distance r of the electron from the nucleus (proton), the Schrödinger equation needs to be solved in spherical coordinates (r, θ, φ) instead of (x, y, z). Therefore, in contrast to a single quantum number that appears in the Bohr model, the quantum mechanical model introduces three quantum numbers n, l, and m where

$$n = 1,2,3,\ldots$$

$$l = 0,1,2,\ldots,n-1$$

$$m = -l, -l+1, -l+2,\ldots,0,\ldots,l-2,l-1,l \tag{3.81}$$

n is the *principal quantum number*. It determines the overall size of the orbital and the energy of the electron

$$E_n = -\frac{me^4}{8\epsilon_0^2 h^2}\frac{1}{n^2} \tag{3.82}$$

As we can see, the quantum mechanical calculations give the same energies as those obtained using the Bohr model.

The second quantum number l is called the *angular momentum quantum number*. It determines the shape of the orbital. The third quantum number, m_l, called the *magnetic quantum number*, determines the orientation of the orbital in space.

It is known that all moving charges produce magnetic fields. Although the quantum mechanical model of the atom does not envisage the electron to be moving in circular or elliptical orbits, yet the magnetic quantum number is indicative of the magnetic fields produced by the atomic electrons.

Experiments proved that electrons also produce tiny magnetic fields independent of those arising from their orbital motion. Therefore, besides its orbital angular momentum, the electron is said to possess an intrinsic angular momentum s. All electrons, whether free or bound, have $s = ½$. Accordingly, the *spin magnetic quantum numbers* of electrons are

$$m_s = -1/2, +1/2 \tag{3.83}$$

3.6 Some elements of statistics

3.6.1 Microscopic versus macroscopic

As we have understood to a large extent, the behavior of matter at the microscopic or atomic level can be analyzed by using the laws of quantum mechanics. However, it is also true that excepting the case of single-particle analysis, chemical reactions and interactions, whether in the inanimate world or in biological systems, are studied at the macroscopic level. The macroscopic or bulk regime is described in terms of measurable quantities such as temperature and pressure and it is not possible to keep track of the behavior of individual particles. Nonetheless, we can apply statistics to connect the macroscopic world to the microscopic.

The macroscopic properties of a system are divided into two groups—extensive and intensive. Examples of extensive properties are the number of molecules in the system (N), the volume occupied by the system (V), and the total energy of the system (E). On the other hand, the pressure (p) and the molecular density (ρ) are intensive parameters obtained by averaging the contributions from all the molecules.

The connection between the macroscopic world and the microscopic world can be illustrated by presenting a statistical view of entropy. In order to do so, we introduce the concept of microstate that relates to the specification of the state of every individual particle of the system. Let us consider a system consisting of N noninteracting particles. In the classical picture, the state of the system can be described by specifying $3N$ positions (x_i, y_i, z_i), $i=1, 2, 3, \ldots, N$ and $3N$ momenta (p_{xi}, p_{yi}, p_{zi}), $i=1, 2, 3, \ldots, N$ of the particles. For a quantum mechanical system, the microstate is defined by a complete set of quantum numbers for each particle. The macrostate for a system is described by specifying the number of particles N_i occupying a particular quantum state i.

In a system consisting of a very large number of particles, each macroscopic state j can be attained by a large number w_j of microstates, w_j being called the thermodynamic probability. The probability of the macrostate j is then given by

$$p_j = w_j/\Omega \tag{3.84}$$

where $\Omega = \sum_j w_j$ is the total number of microstates of the system. A closed system tends to maximize Ω. Hence, connecting the macroscopic and microscopic worlds, the Boltzmann entropy equation states

$$S = k_B \ln \Omega \tag{3.85}$$

3.6.2 Normal distribution

Many physicochemical processes are random in nature and can, therefore, be described using random variables. Most often, the data generated by observing these random variables and frequencies can be represented by a bell-shaped curve which is also called a Gaussian curve. In such cases, the data are said to be normally distributed.

In normal distribution, the frequency of variables or probability density $f(x)$ falls off proportionally to the difference between the score and the mean (μ) and to the frequency itself. Expressed mathematically

$$\frac{df(x)}{dx} = -k(x-\mu)f(x) \tag{3.86}$$

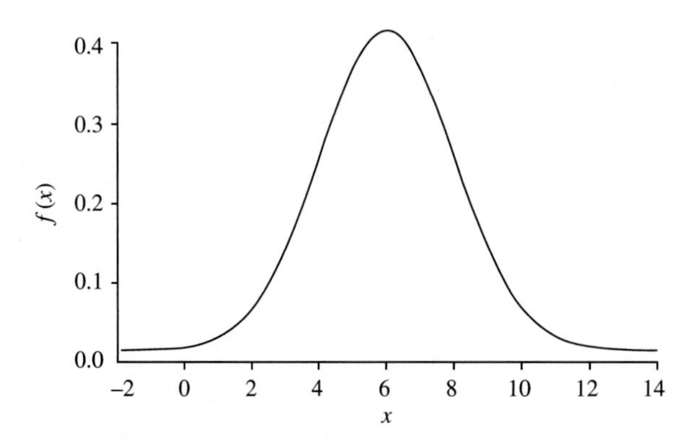

FIG. 3.10

Normal distribution for $\mu=6$ and $\sigma=2$.

Now, it may be remembered that normal distribution is a probability distribution and the total probability for the entire range of the variable is equal to one. Based on this fact and introducing the term variance σ such that

$$\sigma^2 = \int_{-\infty}^{\infty} (x-\mu)^2 f(x)dx \tag{3.87}$$

one can derive from Eq. (3.86) the probability distribution

$$f(x) = \frac{1}{\sigma\sqrt{(2\pi)}} e^{-\frac{1}{2}\left(\frac{x-\mu}{\sigma}\right)^2} \tag{3.88}$$

which is represented graphically in Fig. 3.10.

3.6.3 Boltzmann distribution

Now, one may ask what is the nature of probability distribution if a random process is subject to constraints. Let us consider a closed system consisting of a large number of particles contained in a volume V at temperature T. The particles will randomly collide against each other and reach an equilibrium with a certain distribution of energy. Let n_i be the number of particles in energy state E_i. The system is constrained by the total number of molecules $N=\Sigma n_i$ and the total energy $E=\Sigma n_i E_i$ that will remain unchanged. It is important to know how the total energy E is distributed among N particles of the system in equilibrium. The answer is given by the Boltzmann distribution function f_B as the probability of finding a particle in the energy state E_i

$$f_B = Ae^{-\beta E_i} \tag{3.89}$$

where A and β are constants. Rewriting β as

$$\beta = 1/k_B T \tag{3.90}$$

where T is the absolute temperature and $k_B = 1.38 \times 10^{-23}\,\mathrm{J K^{-1}}$ is the Boltzmann constant, Eq. (3.89) becomes

$$f_B = Ae^{-E_i/k_B T} \tag{3.91}$$

Sample questions

1. Show, by calculation, that the gravitational force between a proton and an electron is negligible compared with the electrostatic force between the two.
   Mass of proton, $m_H = 1.67 \times 10^{-27}$ kg; charge of proton $= +1.6 \times 10^{-19}$ C
   Mass of electron, $m_e = 9.11 \times 10^{-31}$ kg; charge of electron $= -1.6 \times 10^{-19}$ C
   Mean radius of hydrogen atom, $r = 5.3 \times 10^{-11}$ m
   $G = 6.67 \times 10^{-11}$ N-m^2/kg^2; $1/4\pi\varepsilon_0 = 8.99 \times 10^9$ N-m^2/C^2

2. 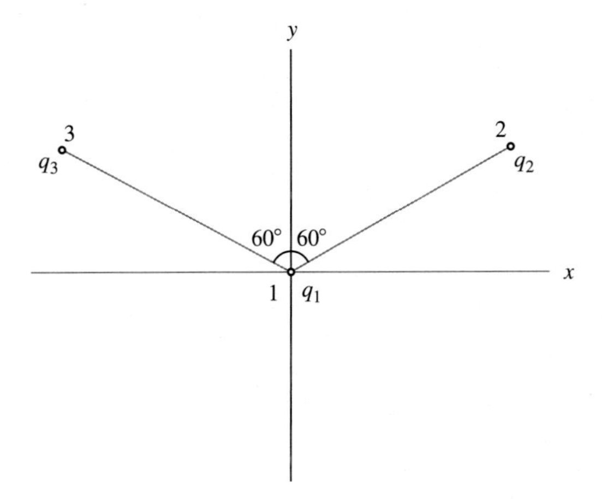

   Three point charges 1, 2, and 3 are located as shown in the figure. $q_1 = 1\,\mu$C, $q_2 = q_3 = 2\,\mu$C; $r_{12} = r_{13} = 1.5$ cm. Find the resultant force in the y-direction and show that there is no resultant force in the x-direction.
   $1/4\pi\varepsilon_0 = 8.99 \times 10^9$ N-m^2/C^2

3. In an X-ray tube, an electron leaves the filament with zero speed and reaches the metal target, which is 2 cm away, with a speed of $10^7\,\mathrm{m s^{-1}}$. If the acceleration is constant, what is the force on the electron? mass of electron, $m_e = 9.11 \times 10^{-31}$ kg

4. If a force **F** be applied on a particle of mass m at a position **r** with respect to a point O, the torque on the particle with respect to O is defined as $\boldsymbol{\tau} = \mathbf{r} \times \mathbf{F}$. Show that the torque is equal to the time rate of change of its angular momentum with respect to O.

5. From Newton's second law, prove the work-energy theorem for a nonconstant force F_x acting in the x-direction.

6. Consider a particle subject to a force $F_x = -kx$, where x is the displacement of the particle from its equilibrium position and k is a constant, undergoing a simple harmonic oscillation. Find the potential energy and the kinetic energy of the particle at point x and show that the total mechanical energy is conserved.

7.

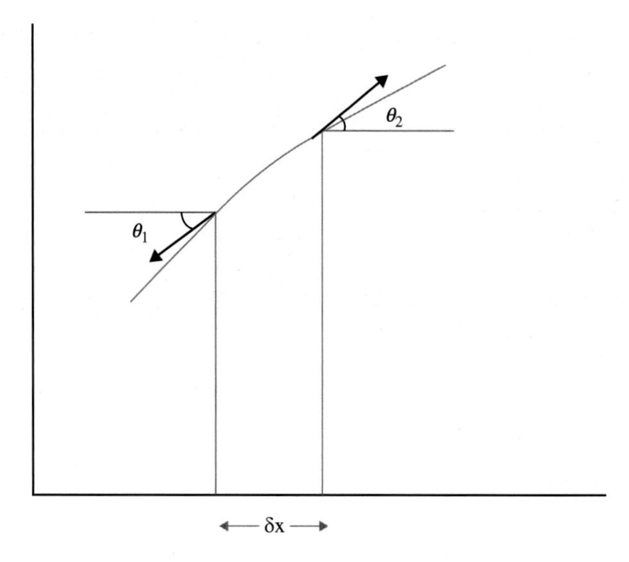

The figure represents a small element of length δx of a string through which travels a transverse wave at a certain instant of time t. The string is under tension F. On the basis of the figure and applying Newton' second law, derive the general form of wave equation.

8. The Maxwell speed distribution for an ideal gas containing N molecules at temperature T is

$$N(v) = 4\pi N\left(\frac{m}{2\pi kT}\right)^{3/2} v^2 e^{-mv^2/2kT}$$

so that the number of molecules with speed between v and $v+dv$ is given by $N(v)dv$. Let the interval be considered in terms of corresponding energies as E and $E+dE$ so that the number of molecules in the interval can also be expressed as $N(E)dE$. Consider a monoatomic gas so that the gas molecules only have translational kinetic energy ($E=(1/2)mv^2$). Show that the Maxwell-Boltzmann energy distribution is

$$N(E) = \frac{2N}{\pi^{1/2}} \frac{1}{(kT)^{3/2}} E^{1/2} e^{-E/kT}$$

9. Let us consider a system consisting of n moles of an ideal (say, monoatomic) gas. Heat is supplied to the system at a constant temperature (isothermal process). The internal energy of an ideal gas does not change during an isothermal process. How much heat the system needs to absorb to expand from an initial volume V_i to a final volume V_f? (the expression)

10. A monochromatic light of wavelength $\lambda = 600\,\text{nm}$ is incident on a diffraction grating containing 10^4 lines per centimeter. At what angle the spectral line will be seen in the first order ($m=1$)? If the grating contained 10^8 lines per centimeter, what is the wavelength of the electromagnetic radiation that would have produced the first spectral line at the same angle?

11. Einstein introduced the concept of photon to explain the photoelectric effect with the equation $hf = \varphi + k_{max}$ where f is the frequency of light falling on the metal surface with work function φ emitting electrons with maximum kinetic energy k_{max}. What is the maximum speed of an electron emitted from a sodium surface with work function $\varphi_{sodium} = 2.3\,eV$ when illuminated with light of wavelength 410nm?

12. Calculate the de Broglie wavelength of a dust particle of mass 2×10^{-9} kg moving with a speed of 10cm per second to show that the wavelength is much too small to be detected. In contrast, an electron whose kinetic energy is 100eV is easily detectable. Mass of electron, $m_e = 9.11 \times 10^{-31}$ kg; $1\,eV = 1.6 \times 10^{-19}$ J

13. An electron with 10eV energy approaches a potential barrier whose height is 20eV and thickness 1nm. Taking only the dominant factor $e^{-2k'a}$ into consideration, find out the order of the probability that the electron will tunnel through the barrier.

14. Solving the Schrödinger equation in spherical coordinates, the electron wave function for the ground state of the hydrogen atom is obtained as

$$\Psi(r) = \frac{1}{\sqrt{\pi a_0^3}} e^{-r/a_0}$$

where $a_0 = \frac{\varepsilon_0 h^2}{\pi m e^2}$ (m being the electron mass and ε_0 the electric constant) is called the Bohr radius. The radial probability density $P(r)dr$ gives the probability that the electron will be found between two spheres of radii r and $r + dr$. Show that the Bohr radius a_0 is the radius at which the electron is most likely to be found in the ground state of the hydrogen atom.

References and further reading

Bauer, W., Westfall, G.D., 2011. University Physics with Modern Physics, second ed. McGraw-Hill Education.

Resnick, R., Halliday, D., Krane, K.S., 2004. Physics. John Wiley & Sons, Inc, New York.

Tipler, P.A., Mosca, G., 2008. Physics for Scientists and Engineers with Modern Physics, sixth ed. W. H. Freeman and Company, New York.

CHAPTER

Chemical basis of biology

4

Molecular structural biology is all about understanding and explaining biological processes in the language of molecular structures and interactions. In the previous chapter, we have looked at the structure of the atom from classical and quantum mechanical perspectives. In this chapter, we shall deliberate on the physical forces and factors based on which atoms form molecular structures and the molecules interact among themselves.

4.1 From atoms to molecules

Molecules are formed when x number of atoms ($x = 2, 3, 4, \ldots$) are held together by what are known as *chemical bonds*. All x may not be the same type of atoms. The obvious question then would be why and how the bonds are formed.

4.1.1 Electron configuration of atoms

We know that atoms consist of positively charged protons in the nucleus (leaving aside the uncharged neutrons) and negatively charged electrons in the orbitals. When two atoms approach each other, a complex set of electrostatic interactions develop among the charged particles. The electrons of one atom are attracted by the nucleus, but repelled by the electrons of the other atom and vice versa. At the same time, the two nuclei also repel each other. If these interactions result in the reduction in energy of each atom and, thus, the total system, a chemical bond is formed. Therefore, the interactions should be understood in terms of the electron configurations of the atoms.

In Section 3.5.5, we have established four quantum numbers (n, l, m_l, m_s) for the hydrogen atom to identify the possible quantum states of its single electron. This quantum number principle can be extended to multielectron atoms as well. Further, in chemistry, it is customary to denote the angular momentum quantum number $l = 0, 1, 2, 3, 4, \ldots$ as s, p, d, f, g,..., respectively. As in the hydrogen atom, in a multielectron atom, too, increasing n is associated with increasing energy of the corresponding shell and, for a particular value of n, a subshell with a greater value of l is at a higher energy level. Thus,

$$E(4s) < E(4p) < E(4d) < E(4f)$$

Fundamentals of Molecular Structural Biology. https://doi.org/10.1016/B978-0-12-814855-6.00004-3

where E is the energy. However, when we compare two subshells with different principal quantum numbers, as for example 3d with 4s, we find that E (4s) < E (3d).

With an increase in the atomic number (denoted by Z), the number of electrons in the atom increases. It is expected that each electron would try to occupy the lowest energy state which, as stated above, is 1s (that is $n=1$, $l=0$). Does it mean that all would crowd into 1s? Not at all! Here, we have to take into account an extremely important quantum principle—*Pauli exclusion principle*. According to this principle, in a multielectron atom, no two electrons can have identical quantum numbers. Let us see how the exclusion principle works.

For a helium (He) atom, $Z=2$. The two He electrons, having the spin quantum numbers $m_s=+½$ and $-½$, can both occupy the 1s orbital in the ground (lowest energy) state. So, the electron configuration of He can be expressed as $1s^2$. However, in a lithium (Li) atom (Z=3), two out of its three electrons can occupy the 1s orbital, but the third electron has to find its place in the 2s orbital. Therefore, the electron configuration of Li is $1s^2 2s^1$. Following a similar reasoning, the electron configuration of other atoms can be obtained. In Table 4.1, the electron configurations of some of the atoms present in biological systems have been presented.

4.1.2 Atomic interactions—Chemical bonds

It is not difficult to appreciate that, for interactions between two atoms, the electrons in the outermost principal energy level (highest n value) of each atom are the most crucial. These are the *valence* electrons of an atom. (For the atoms of transition elements, the outermost d electrons are also considered as valence electrons, although they are not present in an outermost principal energy level.) All electrons in an atom other than the valence electrons are called *core* electrons. The valence electrons are held in an atom most loosely and, therefore, their movement is the least restricted.

Table 4.1 Electron configuration of biologically important atoms

Atom	Atomic number	Electron configuration
H	1	$1s^1$
C	6	$1s^2 2s^2 2p^2$
N	7	$1s^2 2s^2 2p^3$
O	8	$1s^2 2s^2 2p^4$
P	15	$1s^2 2s^2 2p^6 3s^2 3p^3$
S	16	$1s^2 2s^2 2p^6 3s^2 3p^4$
K	19	$1s^2 2s^2 2p^6 3s^2 3p^6 4s^1$
Ca	20	$1s^2 2s^2 2p^6 3s^2 3p^6 4s^2$

In the electron configuration of O represented as

$$1s^22s^22p^4$$

$2s^2$ and $2p^4$ are the valence electrons. In Lewis structure, the valence electrons are represented by dots surrounding the symbol of the element. Thus, for O, the Lewis structure is

The atoms with eight valence electrons, represented by eight dots or an "octet" ("duet" in case of helium), are the most stable. The *Lewis theory* considers a chemical bond as the sharing or transfer of valence electrons to attain stable electron configurations of the concerned atoms. Since stable configuration implies eight electrons in the outermost shell, this is also known as the *octet rule*. If two electrons are shared between the interacting atoms, a *covalent bond* is formed. On the other hand, if an electron is transferred from one atom to the other, the two atoms are held together by an *ionic bond*.

4.1.3 Electron sharing principles—Electronegativity

What determines if the electrons would be shared between the bonding atoms or transferred from one to the other? Primarily, it is the difference in "electronegativity" between the bonding atoms. *Electronegativity* is a measure of the ability of an atom to attract a bonding pair of electrons. The property was quantified by Linus Pauling based on comparative bond energies of homo- and heteronuclear diatomic molecules. Fluorine (F), the most electronegative element in the periodic table, was assigned an electronegativity of 4.0.

The force of attraction towards one of the two nuclei as experienced by the bonding pair of electrons depends on (a) the number of protons in the nucleus, (b) the distance from the nucleus, and (c) the extent of screening by the inner/core electrons. For example, carbon (electron configuration: $1s^22s^22p^2$) and fluorine (electron configuration: $1s^22s^22p^5$) both have their valence electron in the second shell ($n=2$) screened from their respective nuclei by $1s^2$ electrons. However, the fluorine nucleus has nine protons compared with six of carbon. Evidently, a bonding pair of electrons will experience a greater force of attraction from the nucleus of fluorine which, therefore, is more electronegative than carbon.

Again, considering hydrogen fluoride and hydrogen chloride, one can find that in each case the effective charge pulling the bonding pair of electrons towards the center of fluorine or chorine is +7. Nevertheless, fluorine has the bonding pair in the second shell ($n=2$), while chlorine has it in the third ($n=3$). So, fluorine, whose bonding pair is closer to the nucleus experiencing a greater force of attraction, is more electronegative than chlorine.

4.1.4 Polarity of bond—Molecular dipole

Let us consider a bond between two atoms, A and B, of equal electronegativity (e.g., two chlorine atoms) formed by sharing a pair of electrons. The bonding pair of electrons is equally attracted by the two chlorine atoms and, hence, placed exactly halfway between the two atoms. The bond is purely *covalent* or *nonpolar* (Fig. 4.1).

In accordance with the octet rule, two atoms can share even more than a single pair of electrons. Two atoms of oxygen share two pairs of electrons and form a double bond between them (Fig. 4.2). Both the pairs are positioned halfway between the oxygen atoms. Similarly, two atoms of nitrogen equally share three pairs of electrons to form a triple bond (Fig. 4.2). Of the three types of bond, the triple bond is the shortest and the strongest.

What happens when two atoms, A and B, with some difference in electronegativity form a bond by sharing a pair of electrons—for example, hydrogen (H) and chlorine (Cl) forming HCl? From the Lewis structure, the electron pair appears to be equally shared. Yet, due to the difference in electronegativity, the electron density is greater towards the Cl-end which, consequently, becomes slightly negative (δ^-). Simultaneously, the H-end becomes δ^+. The bond is *polar covalent*; the molecule is a *dipole* (Fig. 4.3). Evidently, the degree of polarity depends on the electronegativity difference between the bonding atoms.

Like HCl, there are several other molecules which have no net charge, but consist of equal positive and negative charges separated by a distance. Hence, they are electric dipoles. A molecular dipole is characterized by a physical quantity called the

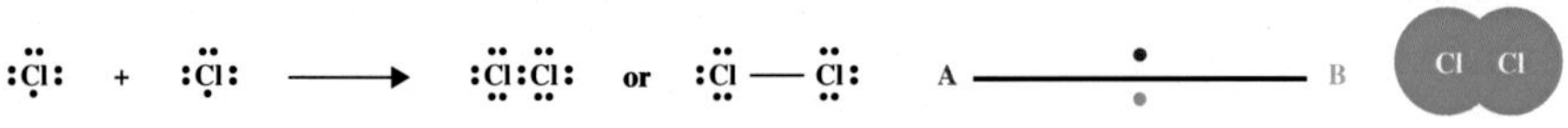

:Cl: + :Cl: ⟶ :Cl:Cl: or :Cl—Cl: A ——•—— B Cl Cl

FIG. 4.1

Pure covalent (nonpolar) bond between two chlorine atoms.

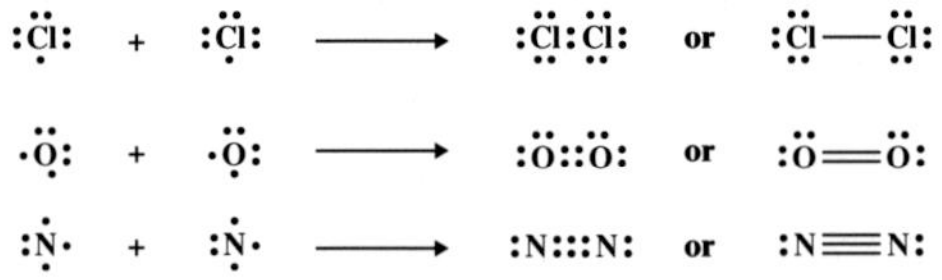

:Cl: + :Cl: ⟶ :Cl:Cl: or :Cl—Cl:

·O: + ·O: ⟶ :O::O: or :O=O:

:N· + :N· ⟶ :N:::N: or :N≡N:

FIG. 4.2

Single, double, and triple covalent bonds.

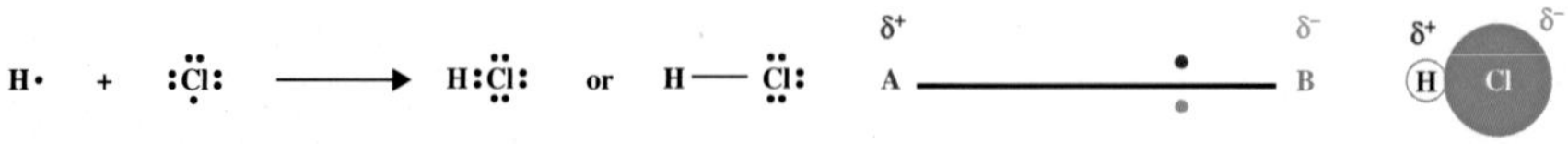

H· + :Cl: ⟶ H:Cl: or H—Cl: δ^+ A ——•—— B δ^- δ^+ H Cl δ^-

FIG. 4.3

Polar covalent bond between hydrogen and chlorine atoms.

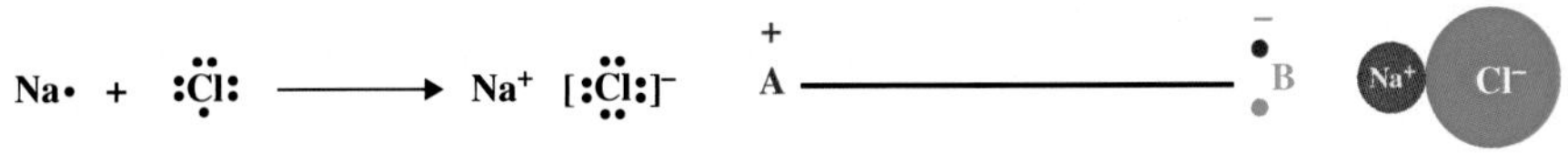

FIG. 4.4

Ionic bond between sodium and chlorine atoms.

electric dipole moment (or simply dipole moment). The *dipole moment* is defined (in magnitude) as

$$\mu = qd \tag{4.1}$$

where q is the magnitude of each of the separated charges and d the separating distance. (It may be noted that the dipole moment is a vector with its direction pointing from the negative charge to the positive charge along the line joining the charges.)

In an extreme case, where there is a large difference in electronegativity between two atoms, such as sodium (Na) and chlorine (Cl), an electron is almost completely transferred (in this case from Na to Cl). Ions are formed, and the bond becomes *ionic* (Fig. 4.4). Thus, covalent, polar covalent, and ionic bonds can be placed on a continuum of bond types based on electronegativity difference between the bonding atoms.

4.2 Bonding theories

Lewis structures do not provide geometrical or structural information on a molecule. Nevertheless, they remain a fundamental “shorthand” for arrangements of bonds. From a quantum mechanical perspective, representation of electrons as dots in Lewis theory is an oversimplification. Ideally, a complete quantum mechanical treatment of covalent interactions should give a more accurate depiction of molecular structure. However, even for the simplest molecule, the Schrödinger equation cannot be solved analytically, and approximations become necessary. Two such approximation methods are provided by (a) the valence bond (VB) theory and (b) the molecular orbital (MO) theory.

Both VB and MO theories treat electrons quantum mechanically and can, therefore, predict several properties of molecules such as bond length, bond strength, molecular geometry, and dipole moment. These theories are based on the premise that electrons reside inside quantum mechanical orbitals, and hence, represented by quantum mechanical wave functions.

4.2.1 Valence bond theory

In VB theory, the electrons reside in orbitals localized on individual atoms. These are essentially atomic orbitals—either the standard s, p, d, f orbitals or a combination (hybrid) of two or more standard orbitals. When two atoms approach each other, the electrons and nucleus of one interact with those of the other. The interaction

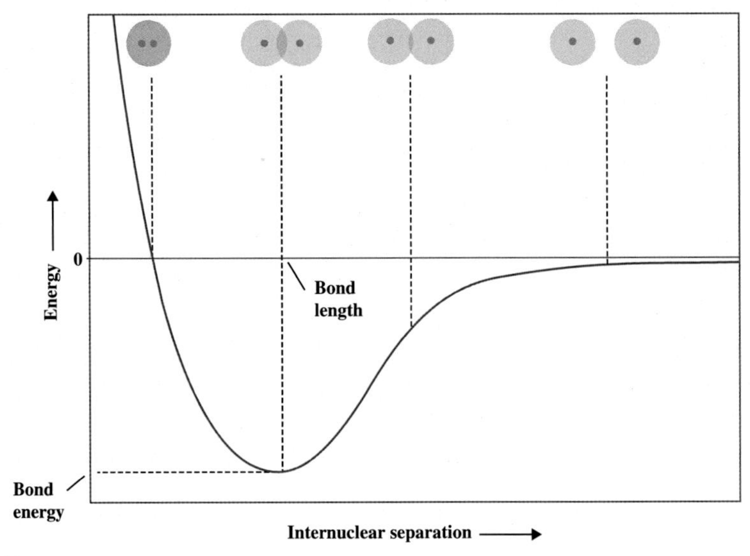

FIG. 4.5

Interaction energy for a hydrogen molecule. Variation of energy for different interatomic (1s-1s) separation.

Adapted from: Tro, N.J., Chemistry: A Molecular Approach, 2nd ed.

energy is determined as a function of the internuclear distance between the two atoms. Fig. 4.5 shows the calculated interaction energy as two hydrogen atoms approach each other. The lowest energy, called bond energy, gives the bond length which allows a significant amount of overlap between their 1s orbitals. This is the most stable point on the curve. In general, for any two atoms, the interaction energy is negative (i.e., stabilizing) when the interacting atomic orbitals contain two electrons that can align with opposite spins (spin-pair). The overlapping orbitals determine the shape of the molecule.

4.2.2 Molecular orbital theory

In Chapter 3, we have introduced atomic orbitals in regard to the hydrogen atom. In MO theory, the electrons belong to the entire molecule, that is, they are completely delocalized into molecular orbitals. The approximation method applied in the MO theory for the solution of wave equations is called the linear combination of atomic orbitals' molecular orbital (LCAO-MO) method. The quantum mechanical wave function is expressed in terms of the molecule. The molecular orbital represented by ψ is expressed as a linear superposition of atomic orbitals φ_is:

$$\psi = \sum_{i=1}^{n} c_i \varphi_i \tag{4.2}$$

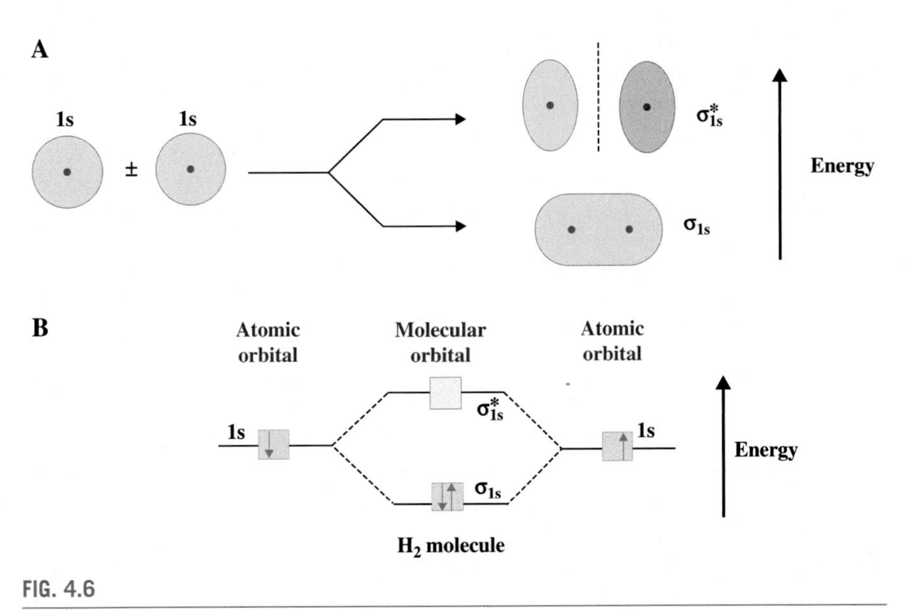

FIG. 4.6

Molecular orbital of hydrogen. (A) Shape of orbital and (B) energy of orbital.

Adapted from: Tro, N.J., Chemistry: A Molecular Approach, 2nd ed.

where c_i is a constant known as mixing coefficient.

Let us consider a simple example of the H_2 molecule. Each of the two H-atoms (A and B) has a single electron in the 1s atomic orbital. We know that electrons behave like waves. Therefore, two molecular orbitals may form by the constructive interference of 1s(A) and 1s(B) orbitals, both in the same phase, and the destructive interference of 1s(A) and 1s(B) in opposite phases (Fig. 4.6).

$$\psi_{\pm} = 1s(A) \pm 1s(B) \tag{4.3}$$

ψ_{+}, also designated as σ_{1s} molecular orbital (in accordance with its shape), is at a lower energy level compared with either of the two 1s atomic orbitals. Hence, it is called a *bonding orbital*. In contrast, ψ_{-} is designated as $\sigma^*{}_{1s}$ molecular orbital, which has a higher energy than the 1s atomic orbitals. $\sigma^*{}_{1s}$ is called the *antibonding orbital*. Since electrons can move from higher energy atomic orbitals to the lower energy σ_{1s} bonding molecular orbital, the two H-atoms can lower their overall energy by forming an H_2 molecule (Fig. 4.6).

4.3 Noncovalent interactions

There are innumerable inorganic molecules whose formation is a result of ionic bonding between their constituent atoms. In contrast, the atoms in organic molecules, particularly biomacromolecules, are held together primarily by covalent bonds.

Nevertheless, as we shall see in later chapters, even for the biomolecules, acquisition of three-dimensional structure and intermolecular binding are almost exclusively dependent on noncovalent interactions which are much weaker than the covalent bond.

4.3.1 Ion-ion interactions

The interaction between two ions (ion-ion interaction) is guided by Coulomb's law (Eq. 3.3 in Chapter 3). For two ions with charges q_1 and q_2, the interaction can be expressed in terms of the potential energy

$$V = \frac{q_1 q_2}{4\pi\epsilon_0\epsilon_r r} \quad (4.4)$$

where ε_r is the relative permittivity or dielectric constant of the medium. As the potential energy varies with $1/r$, it can be said to be a "long-range interaction" (range ~50 nm). When a positively charged cation and a negatively charged anion come close enough in space such that their electrostatic interaction potential is higher than the thermal energy (RT), they form a complex known as an *ion pair* and the interaction is also called ion-pair interaction.

4.3.2 Ion-dipole interactions

We have already seen that some molecules behave as electric dipoles. They carry no net charge (which may tend to suggest that they cannot interact with charged ions), but the nature of electron distribution within these molecules is asymmetric. As a result, partial charges appear at the two ends of the molecules and an ion is able to electrostatically interact with these charges. The interaction is aptly called ion-dipole interaction. Fig. 4.7 shows an ion with charge Ze interacting with a dipole represented by two charges $+q$ and $-q$, separated by a distance d. The interaction energy can be expressed as

$$U_{id} = \frac{Ze\mu\cos\theta}{4\pi\epsilon_0\epsilon_r r^2} \quad (4.5)$$

where $\mu = qd$ is the dipole moment and θ is the angle between the dipole moment vector and the vectoral distance (**r**) between the ion and the dipole.

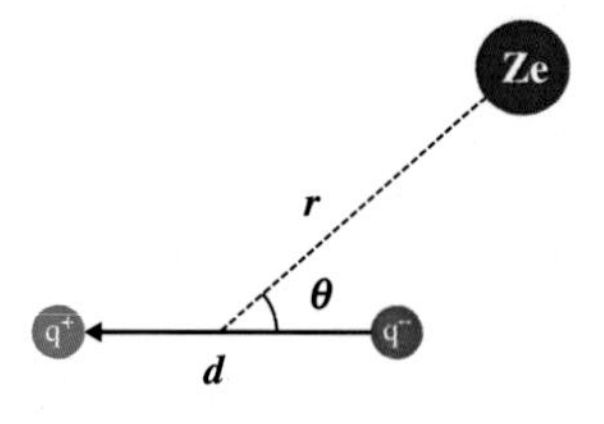

FIG. 4.7

Interaction of an ion (Ze) with a dipole (μ).

4.3.3 Ion-induced dipole interactions

At first, it may appear that neutral, nonpolar molecules do not interact with an ion since, besides the fact that they have no net charge, their electrons are symmetrically arranged about a positively charged center. However, some of these molecules (ethane, for example) are fairly polarizable, that is, their charges are displaced by an electric field (E) and a dipole moment μ_{ind} is induced such that

$$\mu_{id} = 4\pi\varepsilon_0\alpha E \tag{4.6}$$

where α is the polarizability. If, for example, a cation approaches a polarizable nonpolar molecule (Fig. 4.8), some electrons move towards the cation-proximal end, while the distal end becomes deficient in electrons. The ion interacts with the induced dipole. If an ion of charge Ze induces the dipole, the interaction potential is given by

$$U_{i-id} = -\frac{\alpha(\mathrm{Ze})^2}{8\pi\epsilon_0\epsilon_r r^4} \tag{4.7}$$

4.3.4 Van der Waals interactions—Leonard-Jones potential

Three types of interaction, all electrostatic in nature, are collectively called van der Waals interactions—permanent dipole-dipole, permanent dipole-induced dipole, and induced dipole-induced dipole. These attractive forces are effective between two molecules when they are sufficiently close to each other.

Dipole-dipole

Dipole-dipole interactions, as the term suggests, occur between two permanent molecular dipoles. For two dipoles, A and B, with dipole moments μ_A and μ_B, respectively, the interaction energy is given by

$$U_{d-d} = -\frac{2}{3kT}\left(\frac{\mu_A\mu_B}{4\pi\epsilon_0\epsilon_r}\right)^2\frac{1}{r^6} \tag{4.8}$$

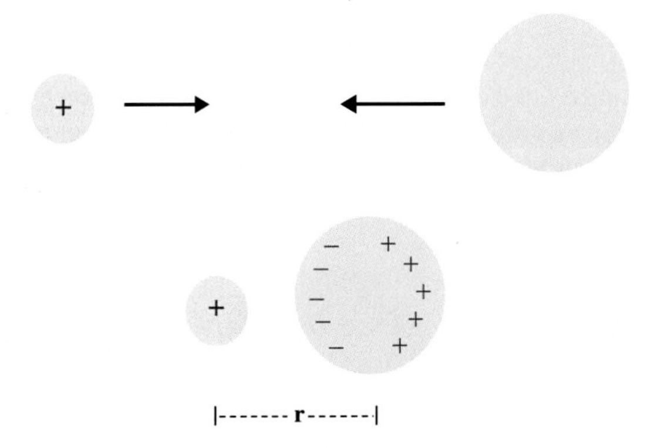

FIG. 4.8

Ion (Ze)-induced dipole (μ_{id}) interaction.

It is evident that the interactions are sensitive to temperature since electrostatic orienting forces between the residual charges at the dipole ends are countered by randomizing thermal forces.

Dipole-induced dipole

Like an ion, a permanent dipole can also induce a dipole moment in a neutral, nonpolar molecule. As a result, the permanent dipole (magnetic moment μ) is able to interact with the induced dipole. As expected, the dipole-induced dipole interaction is dependent on the polarizability (α) of the nonpolar molecule. However, the interaction is not sensitive to randomizing thermal forces. The interaction energy is given by

$$U_{d-id} = -\frac{2\mu^2\alpha}{(4\pi\epsilon_0\epsilon_r)^2 r^6} \tag{4.9}$$

Induced dipole-induced dipole—London forces

In view of the permanent dipole-induced dipole interaction, it may appear that interaction between two neutral, nonpolar molecules (where one cannot induce a dipole moment in the other) is unlikely. However, it needs to be appreciated that the electron distribution within a molecule is not static, but continuously fluctuating. The value of dipole moment assigned to a molecule is usually a time average. This implies that, at a particular instant of time, the electrons, even in a nonpolar molecule, may be asymmetrically distributed, thus producing an instantaneous or temporary dipole. The instantaneous dipole, in turn, induces an instantaneous dipole in its neighboring molecules generating induced dipole-induced dipole forces (Fig. 4.9). These are called London or dispersive forces. Based on quantum mechanics, the energy of an induced dipole-induced dipole interaction can be calculated as

$$U_{id-id} = -\frac{3}{2}\frac{I_A I_B}{I_A + I_B}\frac{\alpha_A\alpha_B}{(4\pi\epsilon_0\epsilon_r)^2}\frac{1}{r^6} \tag{4.10}$$

where I_A and I_B, and α_A and α_B are the respective ionization energy and polarizability of the two interacting molecules. (Ionization energy is the amount of energy required to remove an electron from an atom or molecule.)

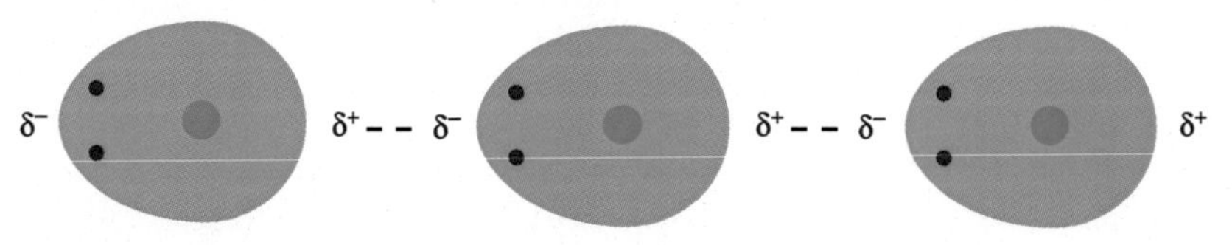

FIG. 4.9

Dispersion interaction between temporary dipoles in helium atoms.

The overall energy for van der Waals interactions between two molecules can be written as

$$U_{id-id} = -\left[\frac{2}{3kT}\left(\frac{\mu_A\mu_B}{4\pi\epsilon_0\epsilon_r}\right)^2 + \frac{2\mu^2\alpha}{(4\pi\epsilon_0\epsilon_r)^2} + \frac{3}{2}\frac{I_A I_B}{I_A + I_B}\frac{\alpha_A\alpha_B}{(4\pi\epsilon_0\epsilon_r)^2}\right]\frac{1}{r^6}$$

$$= -\frac{A}{r^6} \tag{4.11}$$

where A is a constant for a given molecule.

Leonard-Jones potential

Van der Waals interactions, as expressed in Eq. (4.11) and shown graphically in Fig. 4.10, are attractive in nature. The interactions drive molecules closer until they actually contact one another. Beyond this point, if the molecules try to move closer into each other and occupy the same space, a repulsive force comes into play between their electron clouds. The repulsion is not entirely electrostatic, but a manifestation of Pauli exclusion phenomenon. This energy of repulsion, when added to van der Waals attractive interactional energy, gives the Leonard-Jones potential:

$$U_{LJ} = -\frac{A}{r^6} + \frac{B}{r^{12}} \tag{4.12}$$

which is the total interaction energy of two closely interacting molecules (Fig. 4.10).

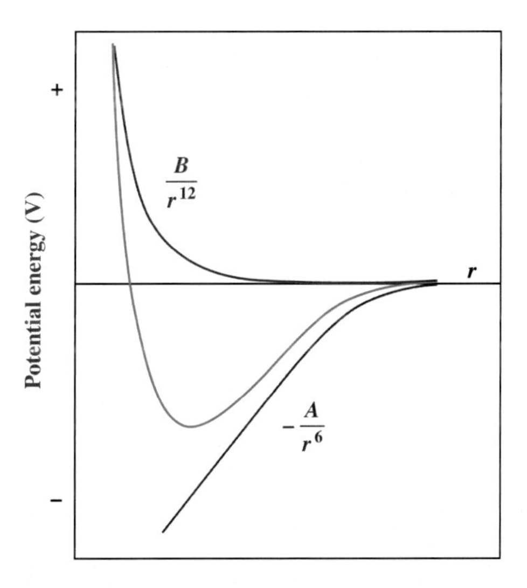

FIG. 4.10

Leonard-Jones potential *(green curve)* is a combination of attractive van der Waals ($-A/r^6$) and repulsive Pauli exclusion (B/r^{12}) interactions.

4.3.5 Hydrogen bond and water structure

The hydrogen bond (H-bond) is mixed in nature—it consists of both covalent and electrostatic (dipole-dipole interaction) components (Fig. 4.11). In a sense, it can be considered as a "super dipole-dipole" force. The covalent component of the hydrogen bond is the polar X—H bond, where X is an electronegative atom such as F, O, or N. X—H donates hydrogen to the bond and, therefore, is called the hydrogen bond donor. The bond formation requires an electronegative atom Y, which is also usually F, O, or N. A pair of nonbonding electrons available on Y acts as the hydrogen bond acceptor.

Basically, X being more electronegative than H, the X—H bond is polarized. As a result, a permanent dipole is formed with H as the positive end and X as the negative end. At the same time, a dipole is also created with the positive end at the nucleus and the negative end at the nonbonding electron pair of Y. The magnitude of the dipole moments can be influenced by the presence of other charged or polar molecules in the surroundings.

Due to a large difference in electronegativity between X and H, a fairly large partial positive charge (δ^+) resides on the latter. On the other hand, a fairly large negative charge (δ^-) is present on the nonbonding electron pair of Y, which is also quite small in size. The H atom can, therefore, approach the Y atom very closely and form an H-bond.

When X≡Y≡O, we have the water-water hydrogen bond. Each water molecule (H_2O) has two hydrogen atoms and two lone pairs of unbonded electrons at the oxygen. Therefore, it can potentially donate H-bonds to two neighbors and accept H-bonds from two donors, all four neighbors being arranged in a tetrahedral structure (Fig. 4.12A). In this way, water molecules form an H-bonded network (Fig. 4.12B).

In the crystalline ice form, each water molecule is engaged in four H-bonds. The liquid state of water, however, is not as ordered as the crystalline state and, on an average, a liquid water molecule is involved in ~3.5 intermolecular H-bonds.

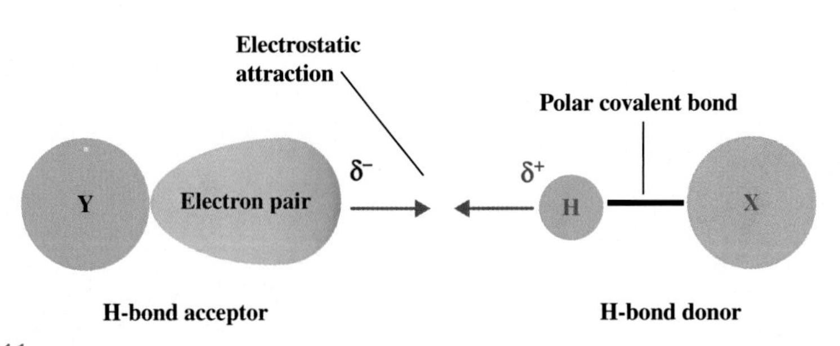

FIG. 4.11

Structure of a hydrogen bond.

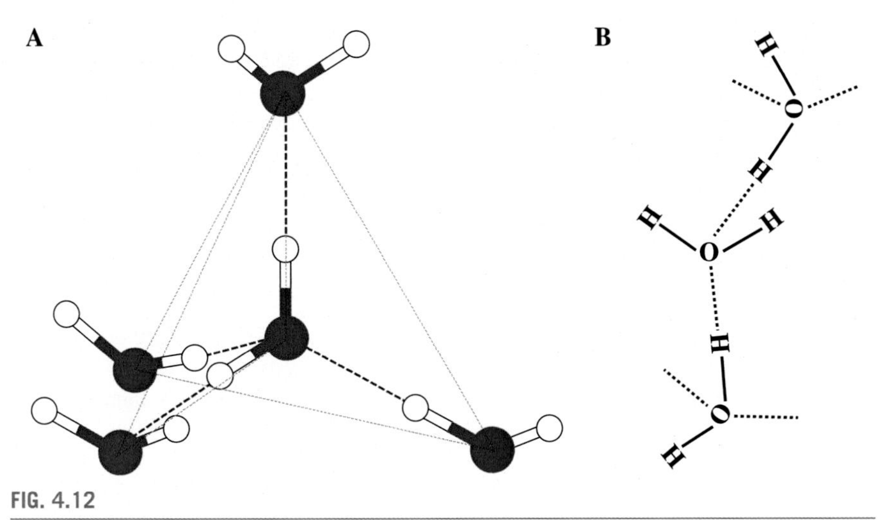

FIG. 4.12

Structure of liquid water. (A) Tetrahedral structure and (B) H-bonded network formed by water molecules.

Further, H-bonds in liquid water are relatively weaker than those in ice since the interaction between adjacent H-bonds is less likely in the liquid state.

4.4 Chemical thermodynamics—Free energy

Strictly speaking, a chemical reaction is a process that involves changes in the nature of chemical bonds. Let us consider the following processes: (a) formation of a hexagonal lattice of water molecules by hydrogen bonding, (b) solvation of NaCl in aqueous medium, and (c) conformational isomerization of a molecule involving rotation of component atom/atoms about a single bond. None of these processes qualify to be called a chemical reaction. Yet, all such processes, as well as typical chemical reactions, keep a biological system viable. Hence, in our discussion, we shall refer to them all as molecular processes.

All molecular processes, whether inside or outside the cell, obey the laws of thermodynamics. Further, since it is known that molecular structure determines the macroscopic property of a system, macroscopic parameters can be used to understand the thermodynamics of a molecular process. In Chapter 3, we have introduced the thermodynamic concept of energy of a system (internal energy) which can change by an amount of heat (Q) transferred to plus an amount of work (W) done on the system (the first law of thermodynamics). Further, we have seen that in a closed system, the entropy either increases for an irreversible process or remains unchanged for a reversible process (the second law of thermodynamics).

However, there are very few naturally existing systems which fulfill the criteria of a closed system. Hence, to determine the direction and equilibrium position of a system, we require a function that includes both energy and entropy components. Such a state function is the Gibbs free energy, G. We may note here that a state function or state variable describing a macroscopic system does not depend on the history of the system.

Before defining G, let us define the energy of a system in a different way by using the term enthalpy. The enthalpy (H) of a system is expressed as

$$H = E + pV \tag{4.13}$$

that is, as a sum of its internal energy and the product of its pressure and volume. Since E, p, and V are all state functions, H is also a state function. The change in enthalpy (ΔH) for any process occurring under constant pressure is then

$$\Delta H = \Delta E + p\Delta V \tag{4.14}$$

From Eq. (3.48), we know that $\Delta E = Q + W$. If Q_p be the heat transfer at constant pressure, the change in internal energy can be written as

$$\Delta E = Q_p + W \tag{4.15}$$

In addition, the work done on the system is given by

$$W = -p\Delta V \tag{4.16}$$

Combining Eqs. (4.14), (4.15), and (4.16), we have

$$\Delta H = \left(Q_p + W\right) - W = Q_p \tag{4.17}$$

Thus, ΔH is equal to the heat at constant pressure (Q_p).

Conceptually, ΔH and ΔE may appear to be similar; however, it needs to be remembered that ΔE is a measure of energy (heat as well as work) exchanged with the surroundings, whereas ΔH is the heat exchanged at constant pressure.

Having introduced enthalpy (H), we can now formally define Gibbs free energy (G) as

$$G = H - TS \tag{4.18}$$

where T is the temperature in kelvins and S is the entropy. At constant temperature

$$\Delta G = \Delta H - T\Delta S \tag{4.19}$$

Let us see how ΔG is related to the spontaneity of a process. Considering the universe to be constituted of the system, which may or may not be closed, and the surroundings, the second law of thermodynamics, which has been stated in Chapter 3, can be restated as: For any spontaneous process, the entropy of the universe increases, that is,

$$\Delta S_{\text{univ}} = \Delta S_{\text{sys}} + \Delta S_{\text{sur}} > 0 \tag{4.20}$$

where S_{sys}, S_{sur}, and S_{univ} are respectively the entropy of the system, surroundings, and universe. If Q_{sys} be the amount of heat exchanged between the system and surroundings during a physical or chemical process at temperature T, then

$$\Delta S_{sur} = -\frac{Q_{sys}}{T} \tag{4.21}$$

the negative sign indicates that the energy is "dissipated" into the surroundings. For a process occurring at constant pressure, we know that $\Delta H_{sys} = Q_{sys}$. Hence,

$$\Delta S_{sur} = -\frac{\Delta H_{sys}}{T} \tag{4.22}$$

Based on energy considerations, the second law of thermodynamics can also be conceptualized in the following way. For a spontaneous process (a) at least some energy is dispersed or dissipated and (b) the total organized energy of the universe decreases, while the total disorganized energy increases.

We can combine Eqs. (4.20) and (4.22), and write for process occurring at constant temperature and pressure

$$\Delta S_{univ} = \Delta S_{sys} - \frac{\Delta H_{sys}}{T} \tag{4.23}$$

or

$$-T\Delta S_{univ} = \Delta H_{sys} - T\Delta S_{sys} \tag{4.24}$$

Ignoring the subscript sys, the right-hand side of Eq. (4.24) appears to be same as that of Eq. (4.19), and therefore

$$\Delta G = -T\Delta S_{univ} \quad (T \text{ and } p, \text{ constant}) \tag{4.25}$$

So, the criterion for spontaneity, which has been expressed in Eq. (4.20), becomes

$$\Delta G < 0 \tag{4.26}$$

Now that we have a function whose change (ΔG) defines the criteria for spontaneity and which contains both energy and entropy components, it may be worthwhile to examine how changes in the components, that is, ΔH and ΔS, together with the temperature, determine the direction of a molecular process. It should be immediately obvious from Eq. (4.19) that if $\Delta H < 0$ (reaction is exothermic) and $\Delta S > 0$, ΔG will be negative and the process spontaneous at all temperatures. An example is the dissociation of nitrous oxide into nitrogen and oxygen. On the other hand, if $\Delta H > 0$ (reaction is endothermic) and $\Delta S < 0$, the reaction will be nonspontaneous at all temperatures.

In cases where both ΔH and ΔS are positive or both are negative, the change in free energy depends on temperature. Let us consider as an example the melting of ice [H_2O (s) $\rightarrow$ H_2O (l)]. One can calculate that for the process $\Delta H = 6.03 \times 10^3\,\text{J mol}^{-1}$ and $\Delta S = 22.1\,\text{J K}^{-1}\text{mol}^{-1}$. Using these values in Eq. (4.19), ΔG is found to be $-224\,\text{J mol}^{-1}$ at 10°C and $218\,\text{J mol}^{-1}$ at -10°C. Clearly, ice melts spontaneously at 10°C, but not at -10°C.

It is to be noted here that the nonspontaneity of a process does not imply impossibility. A process which is nonspontaneous in isolation can be made to occur by coupling it to another process which is spontaneous.

Entropy and hydrophobicity

Variation of entropy in respect to the structure of water gives rise to the phenomenon of hydrophobicity. Due to the relative weakness of the hydrogen bonds, a water molecule can maintain H-bonding interactions and, at the same time, preserve its translational and rotational degrees of freedom. Presence of polar molecules in aqueous solution does not affect the degrees of freedom of the water molecules since the polar molecules can also participate in H-bonding interactions.

What happens when a nonpolar solute (e.g., a hydrocarbon) and water are mixed together? The overall enthalpy of the interaction, which is primarily due to the hydrogen bonds, remains unchanged. In fact, the average number of H-bonds a single water molecule is engaged in goes up from 3.5 to about 4. However, water molecules in the bulk behave differently from those at the interface with the solute.

In the bulk, a water molecule can rotate and translate as before and continues to maintain H-bonding interactions. At the interface with the solute, the interactions are anisotropic or directional since nonpolar molecules do not form H-bonds. Water molecules become more structured, forming ice-like cages around the nonpolar molecule. This highly ordered structure, which curtails the degrees of rotational and translational freedom of water molecules, is entropically unfavorable.

Water tries to increase the entropy by driving nonpolar molecules out of the aqueous phase and thereby reducing the interface. This is the basis of hydrophobicity of nonpolar molecules. The aggregation of nonpolar substances in aqueous medium does not arise from any intrinsic attraction between the hydrophobic solute molecules. The hydrophobic effect is entirely a property of water.

4.5 Chemical kinetics

The thermodynamic criteria for spontaneity of a chemical reaction or a molecular process, as we have discussed, give the direction in which the reaction or process proceeds and the equilibrium position it attains. These criteria, however, do not automatically give an idea about the speed of the process.

4.5.1 Rate law—Arrhenius equation

The speed of a chemical reaction is usually expressed in terms of what is known as a rate law, indicating the dependence on the concentrations of reactants A, B, C,...

$$v = k[\mathrm{A}]^x[\mathrm{B}]^y \ldots \tag{4.27}$$

where the rate constant k contains the temperature dependence of the reaction rate. $x+y+\cdots$ is the reaction order. For a reaction involving a single reactant, Eq. (4.27) is reduced to

$$v=k[\mathrm{A}]^x \quad (4.28)$$

The dependence of the rate constant on temperature was formulated by Arrhenius based on experimental observations as

$$k=Ae^{-\frac{E_a}{RT}} \quad (4.29)$$

where R is the universal gas constant ($8.314\,\mathrm{J\,K^{-1}\,mol^{-1}}$), A is a constant called the frequency or preexponential factor, and E_a is the activation energy for the reaction. Fig. 4.13 represents a typical activation energy barrier separating the reactants and the products. The barrier (E_a) must be surpassed for the reactants to be transformed into products. The frequency factor is related to the number of times the reactants approach the energy barrier per unit time.

Arrhenius equation implies that, in a molecular system, a fraction of the molecules whose energy is at least E_a are able to get over the energy barrier. This fraction, according to the Boltzmann distribution function that has been discussed in Chapter 3 (Eq. 3.91), increases with temperature.

4.5.2 Collison theory

A semiempirical rate law for a bimolecular system can be derived based on a "simple collision theory" of chemical reaction. According to this theory, a reaction occurs as a consequence of a sufficiently energetic collision between two reactant molecules.

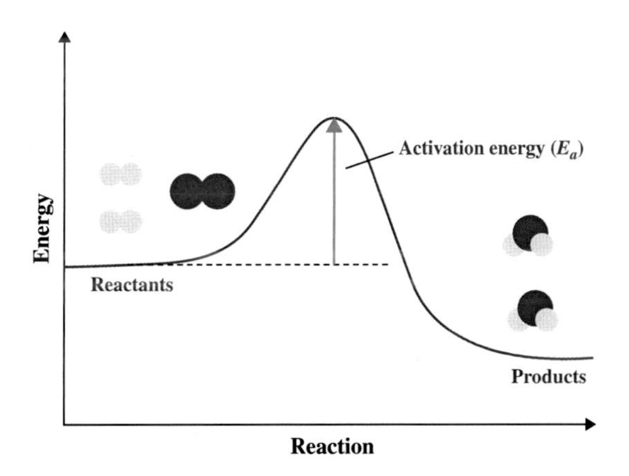

FIG. 4.13

A typical activation energy barrier.

Here, each approach to the activation barrier is seen as a collision between the two molecules. The derived rate constant is given by

$$k = pAe^{-\frac{E_a}{RT}} \tag{4.30}$$

where p and A are both constants. Eq. (4.30) is similar to Arrhenius equation excepting that the frequency factor has been split into two parts—A is the collision frequency and p is the orientation factor that depends on the relative orientations of the two colliding molecules.

4.5.3 Transition state theory

Arrhenius had suggested that a "transition state" exists between the reactants and the products. The transition state theory was theoretically formulated by Wagner and Eyring. In this theory, the reactants combine to form an activated complex at the peak of the reaction barrier and then decay into the products. For a bimolecular reaction, the rate constant as given by the transition state theory is expressed as

$$k = \left(\frac{k_B T}{h}\right) e^{-\frac{\Delta G}{RT}}$$

$$= \left(\frac{k_B T}{h}\right) e^{\frac{\Delta S}{R}} e^{-\frac{\Delta H}{RT}} \tag{4.31}$$

where k_B and h are, respectively, the Boltzmann constant and the Planck constant. Evidently, the activation energy has an enthalpic component (ΔH) as well as an entropic contribution (ΔS). The potential energy of the reactants, transition state, and products can be represented by a potential energy surface. Fig. 4.14 is a two-dimensional projection of such a surface with energy plotted against reaction coordinate.

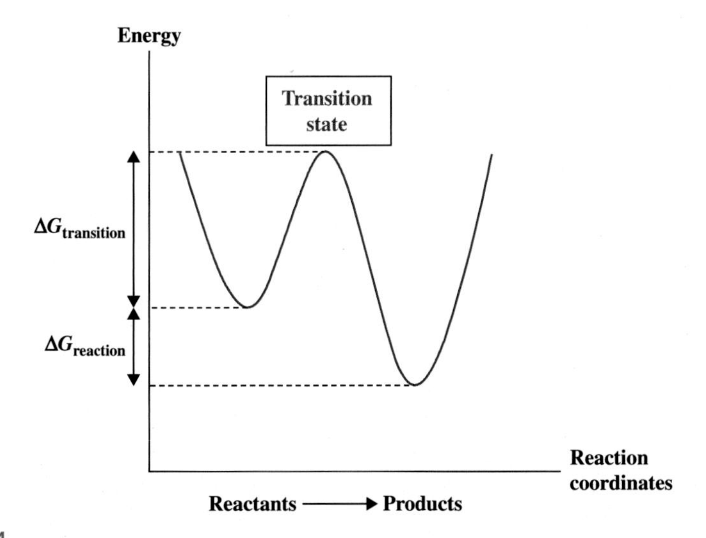

FIG. 4.14

Two-dimensional projection of the potential energy surface.

4.5.4 Quantum tunneling

The transition state theory is still in the realm of classical mechanics. Several quantum effects including "quantum tunneling" have been ignored. Quantum tunneling is associated with electron and proton transfer reactions. In these reactions, an electron or a proton moves from a molecule, atom, or functional group (donor) to another such chemical entity (acceptor). The phenomenon arises from the wave-particle dual nature of electrons and protons.

Let us consider a proton transfer reaction

$$R^1-O-H+{}^{-}O-R^2 \rightarrow R^1-O^-+H-O-R^2 \tag{4.32}$$

represented by a potential energy diagram (Fig. 4.15). Classically, the reactants are expected to gather enough energy (say, by O—H stretching vibration) to cross over the barrier. This is possible at higher temperatures. However, even at lower temperatures, the wave function describing O—H stretching vibration penetrates the barrier wall. Hence, there is a finite probability of finding the proton on the other side of the barrier, whether or not the system possesses sufficient thermal energy to overcome the barrier. So, for an electron or proton transfer reaction, there may be competing pathways—over the barrier (classical process) and through the barrier (quantum mechanical process).

4.5.5 Nucleophilic-electrophilic—S_N2 reaction

It has been already mentioned that a chemical reaction involves changes in the nature of chemical bonds. These changes are brought about by redistribution of electrons. In many chemical reactions, a molecule, atom, or ion that is electron-rich attacks sites

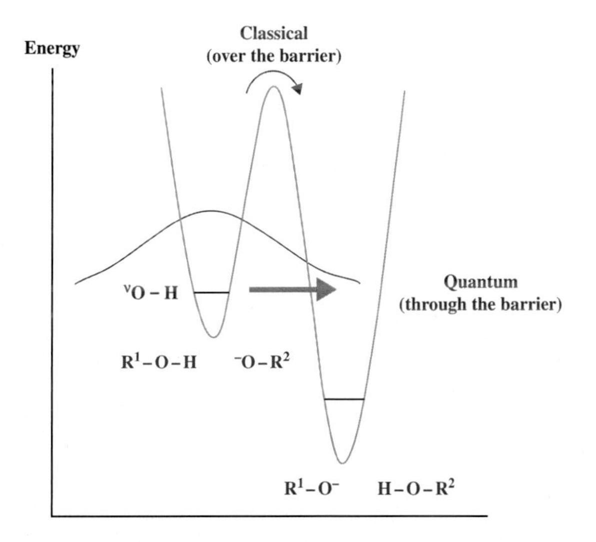

FIG. 4.15

Quantum tunneling. Transfer of proton through a potential energy barrier. ν, vibrational frequency.

with low electron density. Such a species is called a nucleophile. It carries the electrons that can be shared.

On the other hand, a molecule, atom, or ion that is electron-deficient or "electron poor" tends to accept electrons. It is called an electrophile. In general, a nucleophile donates a pair of electrons to an electrophile and forms a new covalent bond.

The nucleophile-electrophile interaction can be explained based on what is known as the frontier molecular orbital (FMO) theory. Instead of looking at the total electron density in the reactant species, the FMO theory focuses on the highest occupied molecular orbital (HOMO) and the lowest unoccupied molecular orbital (LUMO) as they are most likely to be involved in chemical reactivity. Electrons are most easily removed from the HOMO which, therefore, can donate electrons to form a bond (as a nucleophile). At the same time, it is energetically most facile to fill the LUMO with additional electrons. An electrophile is represented by its LUMO. The nucleophile-electrophile interaction can be seen essentially as a HOMO-LUMO interaction.

Let us consider a nucleophilic reaction

$$C_2H_5O^- + H_3C - I \rightarrow C_2H_5O - CH_3 + I^- \tag{4.33}$$

where the ethoxide ion (nucleophile) displaces (substitutes) the iodide ion (leaving group) from the methyl carbon. The general form of the reaction can be written as in Fig. 4.16. Such a mechanism in which the donation of an electron pair by a nucleophile to an atom displaces a leaving group from the same atom in one step (without any reactive intermediate) is called an S_N2 reaction. (S, N, and 2, respectively, symbolize substitution, nucleophilic, and bimolecular).

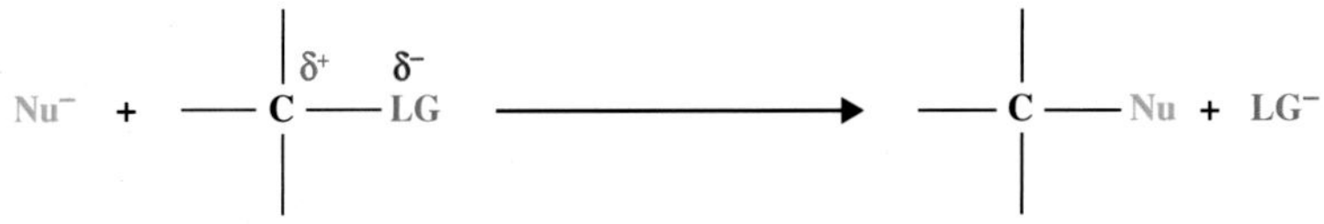

$$Nu^- + -\overset{|}{\underset{|}{C}}^{\delta+}-LG^{\delta-} \longrightarrow -\overset{|}{\underset{|}{C}}-Nu + LG^-$$

FIG. 4.16

General scheme of S_N2 reaction. LG, leaving group.

****

While discussing the underlying principles of structures and processes in this chapter, we have kept the illustrations limited to very simple and small molecules. As we move ahead, we shall find in chapters to follow that the biological system consists of small molecules as well as large molecules or macromolecules with a varying degree of complexity in their structures and the processes they are involved in. Nevertheless, the fundamental physicochemical principles we have learnt so far can be applied to the biological system as well, albeit with more sophisticated approaches.

Sample questions

1. What are the interactions that come into play as two atoms approach each other to form a bond? What is the criterion for bond formation?
2. Why are the valence electrons most crucial for the interaction between two atoms?
3. How does the octet rule explain the formation of both covalent and electrovalent bonds?
4. What are the criteria on which the electronegativity of an atom depends? How does electronegativity determine the nature of a bond?
5. Why do some molecules behave as electric dipoles?
6. Explain how pure covalent, polar covalent, and ionic bonds can be placed on a continuum of bond types.
7. What is the basic premise of the valence bond and molecular orbital theories?
8. How is the bond length determined by the valence bond theory?
9. What is the basic difference between the valence bond and molecular orbital theories?
10. What are the two molecular orbitals of H_2? How are they formed from the atomic orbitals of hydrogen?
11. How do some nonpolar molecules interact with an ion?
12. What is the basis of interaction between two neutral, nonpolar molecules?
13. What is the basis of tetrahedral structure of water molecules?
14. Explain why ice melts spontaneously at 10°C but not at −10°C.
15. The hydrophobic effect is a property of water and not of the nonpolar solute—explain.
16. Some reactions can occur at relatively lower temperatures even if the reactant molecules do not have sufficient energy to cross over the activation barrier—explain.
17. Explain nucleophilicity based on the frontier molecular orbital theory.

References and further reading

Bergethon, P.R., 2010. The Physical Basis of Biochemistry: The Foundation of Molecular Biophysics, second ed. Springer.

Cooksy, A., 2014. Physical Chemistry: Thermodynamics, Statistical Mechanics & Kinetics. Pearson Education, Inc.

Ley, D., Gerbig, D., Schreiner, P.R., 2012. Tunnelling control of chemical reactions—the organic chemist's perspective. Org. Biomol. Chem. 10, 3781–3790.

McMahon, R.J., 2003. Chemical reactions involving quantum tunneling. Science 299, 833–834.

Pollak, E., 2005. Reaction rate theory: what it was, where is it today, and where is it going? Chaos. https://doi.org/10.1063/1.1858782.

Richardson, J.O., 2018. Understanding chemical reactions beyond transition-state theory. Chimia 72 (5), 309–312.

Stan Tsai, C., 2007. Biomacromolecules: An Introduction to Structure, Function and Informatics. John Wiley & Sons.

Tro, N.J., 2011. Chemistry: A Molecular Approach, second ed. Prentice Hall.

Van Holde, K.E., Johnson, C., Ho, P.S., 2006. Principles of Physical Biochemistry, second ed. Pearson Education, Inc.

CHAPTER

5 Biomacromolecules

It has been mentioned earlier (Chapter 1) that all living organisms, plant or animal, are comprised of cells. Conspicuous diversity prevailing in their size, shape, appearance, and many other characteristics notwithstanding, all cells comprise proteins, nucleic acids (DNA and RNA), glycans (carbohydrates), and lipid assemblies. Of these, proteins, nucleic acids, and glycans are considered to be macromolecules—polymers built through the formation of covalent bonds between smaller molecules (the building blocks). Here, in this chapter, we shall mainly focus on the structures of nucleic acids and proteins. Limited elaborations on the others will be made in concerned chapters.

5.1 Nucleic acids

5.1.1 Nucleotides

DNA is made of four different kinds of deoxyribonucleotides, RNA from ribonucleotides. A nucleotide consists of three components: a pentose sugar, a nitrogenous base, and a phosphate group (Fig. 5.1). In ribonucleotides, the sugar is D-ribose, whereas in deoxyribonucleotides the sugar is 2′-deoxy-D-ribose (since it contains an H instead of an OH at the 2′ position). The carbon atoms in the sugar ring are labeled C1′-C5′ (Fig. 5.1A).

The planar nitrogenous base is derived from either pyrimidine or purine (Fig. 5.1B). A pyrimidine is an aromatic heterocyclic organic compound containing nitrogen atoms at positions 1 and 3 in the ring. A purine is a pyrimidine ring fused to an imidazole ring. Besides two nitrogen atoms of the pyrimidine ring, it also contains nitrogen atoms at positions 7 and 9.

The sugar (ribose or deoxyribose) and the base (purine or pyrimidine) together form a nucleoside (ribonucleoside or deoxyribonucleoside) (Fig. 5.2). A water molecule between the hydroxyl on C1′ and the base (N1 of pyrimidine or N9 of purine) is removed and a glycosidic bond (also called N-glycosidic bond) is formed (Fig. 5.2). Similarly, the phosphate is linked to the sugar by the removal of a water molecule between the phosphate and the hydroxyl on C5′, resulting in the formation of a 5′ phosphomonoester. Addition of 1, 2, or 3 phosphate(s) to a nucleoside creates a nucleotide. In other words, a nucleotide is an outcome of a glycosidic bond between the base and the sugar and a phosphomonoester bond between the sugar and the phosphoric acid (Fig. 5.2).

Fundamentals of Molecular Structural Biology. https://doi.org/10.1016/B978-0-12-814855-6.00005-5

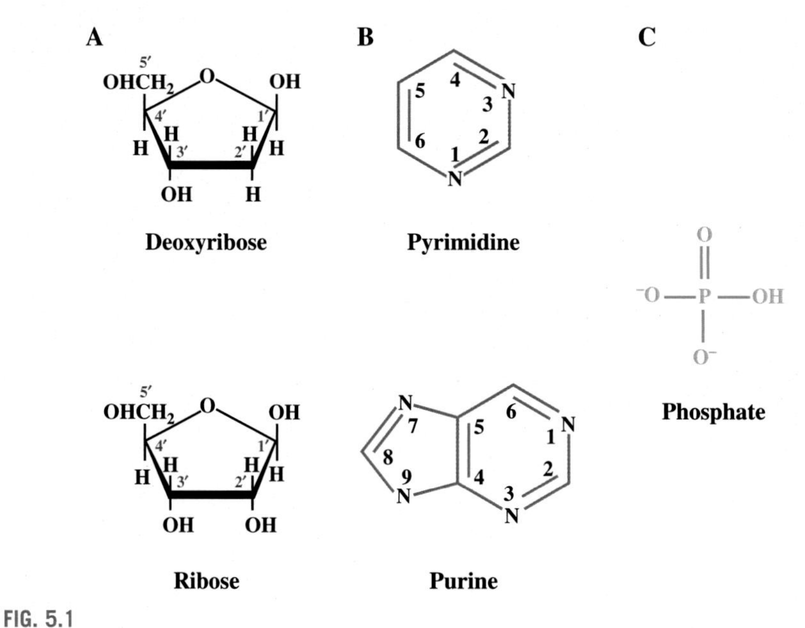

FIG. 5.1

Components of a nucleotide. (A) Pentose sugar, (B) nitrogenous base and (C) phosphate.

Nucleotides, in turn, can be linked to each other by the removal of a water molecule between the hydroxyl on C3′ (also called 3′ hydroxyl) of one nucleotide and the phosphate (5′ phosphate) attached to the 5′ hydroxyl of another nucleotide, thereby forming a phosphodiester linkage (Fig. 5.3). The phosphoryl group in between has the sugars on two sides esterified to it through a 3′ hydroxyl and a 5′ hydroxyl respectively.

Under certain conditions (discussed in Chapter 8), the nucleotides assemble one after another to synthesize a polynucleotide chain. The phosphodiester linkages create a repeating pattern of sugar-phosphate backbone of the chain. The asymmetry of the nucleotides and the manner in which they are linked result in an inherent polarity of the polynucleotide chain with a free 5′ phosphate or 5′ hydroxyl at one end and a free 3′ phosphate or 3′ hydroxyl at the other.

5.1.2 The DNA double helix

Polynucleotide chains constitute DNA and RNA; in DNA, the sugar is deoxyribose, whereas in RNA the sugar is ribose. In either case, the repeating sugar-phosphate backbone is a "regular" feature of the chain. However, the "irregularity" of the chain, and with it the enormous capacity of DNA or RNA to store genetic information, arises from the order of the bases. Every nucleotide in DNA or RNA has the same sugar, but each nucleotide has only one of four bases linked to it. In DNA, there are

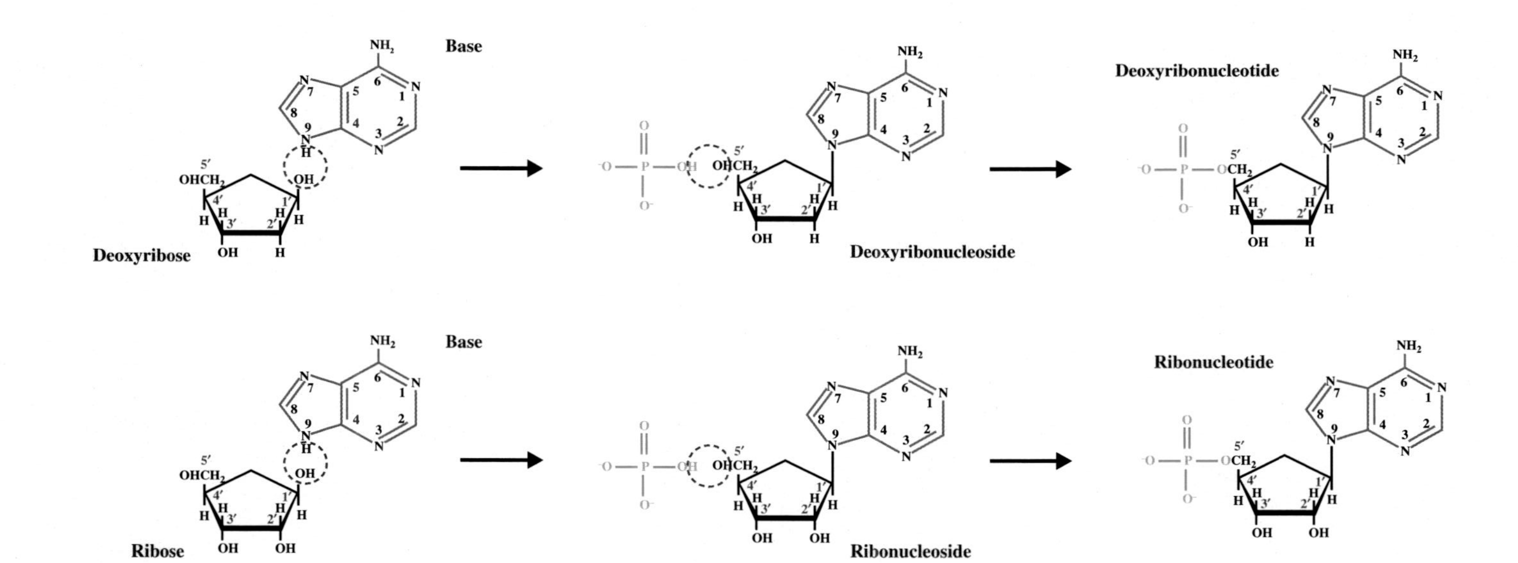

FIG. 5.2

Sugar, base, nucleoside, and nucleotide.

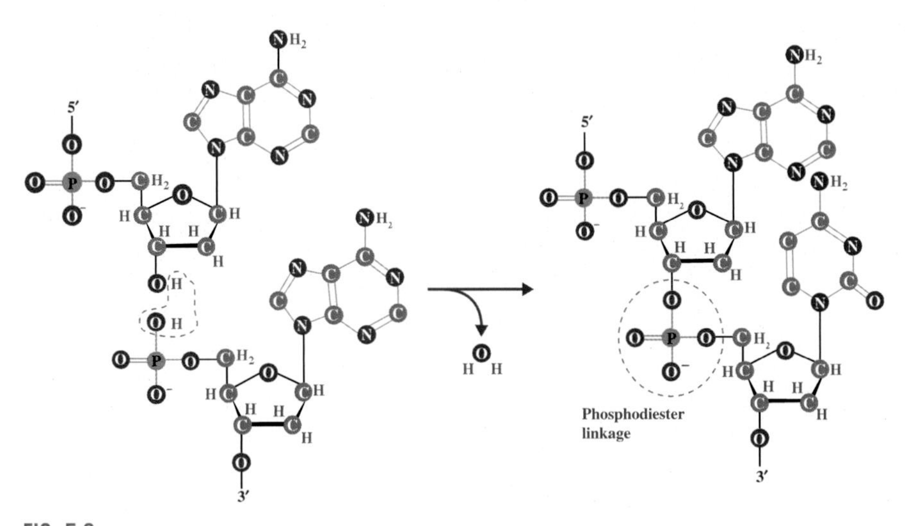

FIG. 5.3

Phosphodiester linkage and dinucleotide.

two purines, adenine (A) and guanine (G), and two pyrimidines, cytosine (C) and thymine (T). RNA contains uracil (U) instead of thymine (Fig. 5.4).

DNA and RNA molecules can be single-stranded (ss; each containing one polynucleotide chain) or double-stranded (ds; each containing two polynucleotide chains base-paired to each other). A majority of the RNA molecules in the living system are single-stranded; the genetic material of some viruses and small interfering RNA (siRNA) are double-stranded RNA molecules.

DNA is mostly double-stranded—two polynucleotide chains are intertwined around each other in the form of a double helix (Fig. 5.5). Each strand of the helix consists of alternating sugar and phosphate residues with the bases projecting inward. The two strands are held together, in an antiparallel fashion (that is, with opposite polarity), by weak, noncovalent interactions (discussed in Chapter 4) between pairs of bases. A can base pair with T, while C can base pair with G—this "Watson-Crick" (WC) base pairing is a result of the complementarity of both shape and hydrogen bonding (see Fig. 4.11) properties between A and T and between C and G (Fig. 5.6).

Complementary relationship is evident also between the sequences of bases on the two intertwined chains. As Fig. 5.7 shows, a sequence 5′-TCACG-3′ on one chain has a complementary sequence 3′-AGTGC-5′ on the opposite chain. Phosphodiester links consecutive sugars in each strand, while hydrogen bonds between complementary bases hold the two strands together. The separation between adjacent base pairs is usually 3.32 Å and they are rotated with respect of each other by ~36° (Fig. 5.7). It may also be noted that the two base pairs, A:T and G:C, have similar geometry; hence, the arrangement of the sugars is not differentially affected by the two base pairs (Fig. 5.6).

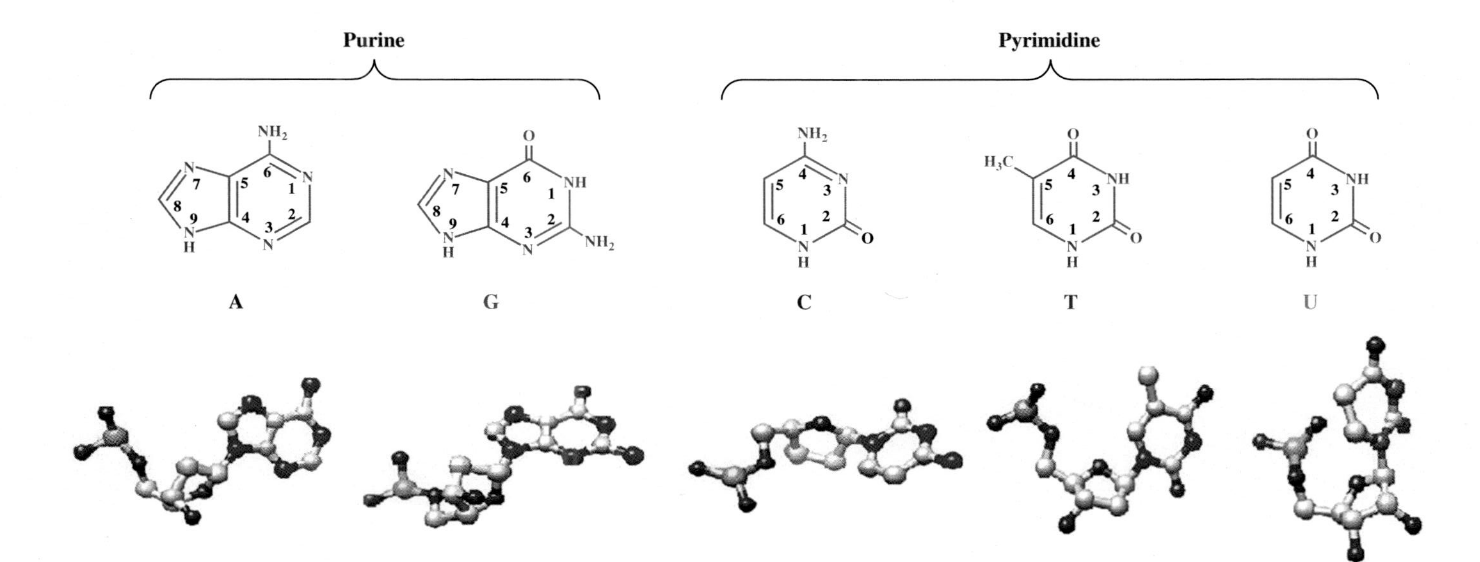

FIG. 5.4

Purines and pyrimidines: chemical (top panel) and three-dimensional (bottom panel) structures. Coloring scheme for atoms: carbon—*grey*; nitrogen—*blue*; oxygen—*red*; hydrogen—*white*; phosphorous—*orange*.

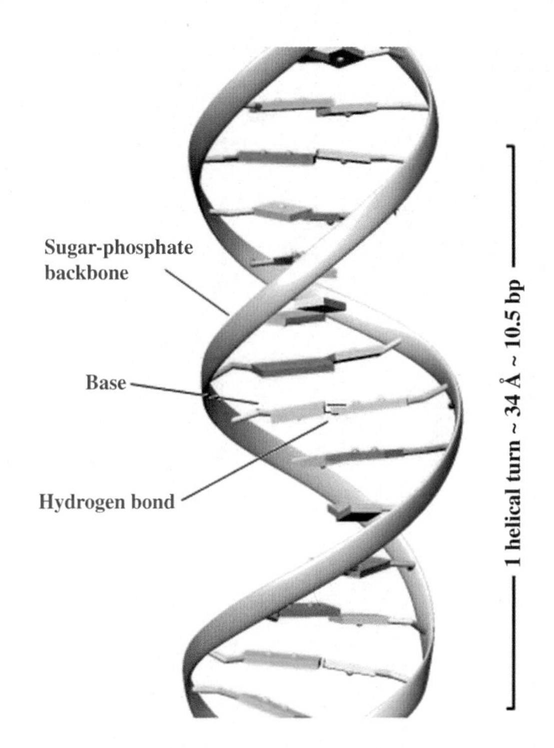

FIG. 5.5

Double helical structure of DNA. Coloring scheme for the bases: A—*coral*; C—*light blue*; G—*light green*; T—*light pink*.

Source: Ellenberger, T.E., et al., 1992. Cell 71, 1223.

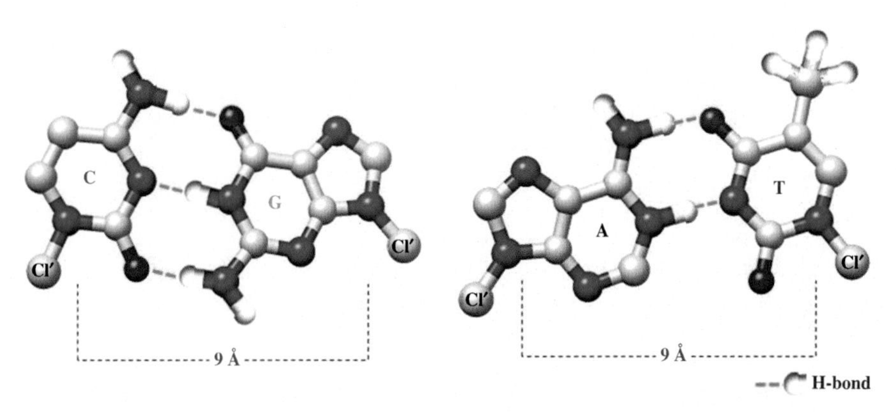

FIG. 5.6

A:T and G:C base pairs. Hydrogen bonds are shown by *green* dashes. Coloring scheme for atoms: carbon—*grey*; nitrogen—*blue*; oxygen—*red*; hydrogen—*white*.

Source: Larsen, T.A., Kopka, M.L., Dickerson, R.E., 1991. Biochemistry 30, 4443.

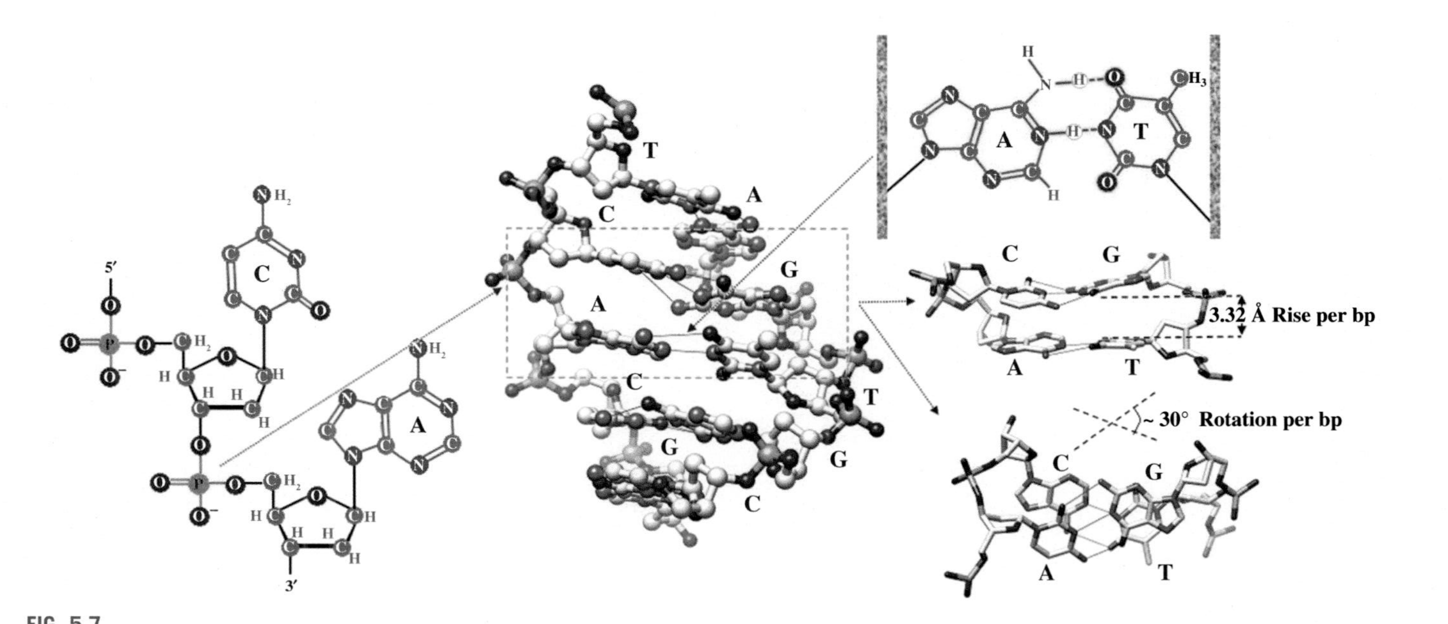

FIG. 5.7

Complementary sequences on opposite strands of a double helix. Some specifications of the double helix are shown. Coloring scheme for atoms: as in Fig. 5.6.

Source: Larsen, T.A., Kopka, M.L., Dickerson, R.E., 1991. Biochemistry 30, 4443.

Thermodynamics of double helix formation

The transition from a state with two separate strands to a double helix leads to a decrease in entropy. Yet, formation of double-stranded DNA is thermodynamically favorable over single-stranded DNA. It may be initially thought that the formation of double helix is primarily driven by hydrogen bonds between the paired bases. Nonetheless, prior to pairing, the edges of the unpaired bases are already involved in hydrogen bonding interactions with surrounding water molecules in the aqueous solution. Although the hydrogen bonds between unpaired bases in single-stranded DNA and water are enthalpically less favorable than those between paired bases in double-stranded DNA, annealing between two strands essentially replaces one set of hydrogen bonds with another; the overall contribution of hydrogen bonding to the stability of the double helix is only modest.

The major contribution to the stability of the double helix comes from base stacking. The bases are flat and relatively water-insoluble molecules. In the double-stranded structure, they stack over each other almost perpendicularly to the direction of the helical axis. Polar bonds are located on the edges of the bases, whereas their top and bottom surfaces are relatively nonpolar. Hence, when DNA is single-stranded, water forms ordered structures around the bases. Formation of duplex DNA releases the ordered water molecules, resulting in an increase in entropy. Base stacking also contributes enthalpically to the stability of double-stranded DNA by facilitating van der Waals interactions between the instantaneously formed dipoles on the hydrophobic surface of the bases.

Major and minor grooves

Although in the double helical structure of DNA the bases project inward, they are accessible through two grooves of unequal width—the major and the minor. To understand the basis of formation of the grooves, let us look at the geometry of the base pairs (Fig. 5.8). In each case, the glycosidic bonds that connect the bases to the sugars make an angle ($\sim$120° on the narrower side or 240° on the wider side) between themselves. As the base pairs stack on top of each other, being rotated by $\sim$36° at each step (Fig. 5.7), two unequal-sized spiral grooves are created—the large and narrow angles being responsible for the major and minor grooves, respectively.

Each groove exposes the edges of the stacked bases; geometrically, the major groove provides greater accessibility to proteins. However, an interacting molecule can distinguish between the base pairs only chemically. The chemical groups present on the edges of the base pairs are hydrogen bond donors (H_D), hydrogen bond acceptors (H_A), nonpolar hydrogens (Φ), and bulky hydrophobic surfaces (CH_3) (Fig. 5.8). It can be seen that each base pair exposes three such groups in the minor groove, while four in the major groove. Furthermore, the order of the groups in the minor groove for both A:T and T:A is H_A-Φ-H_A, and both G:C and C:G is H_A-H_D-H_A. Evidently, though it is possible to distinguish between an A:T and a G:C based on the pattern in the minor groove, distinction between A:T and T:A cannot be made.

On the other hand, the A:T, T:A, G:C, and C:A are characterized by the patterns (H_A-H_D-H_A-M), (M-H_A-H_D-H_A), (H_A-H_A-H_D-Φ), and (Φ-H_D-H_A-H_A), respectively,

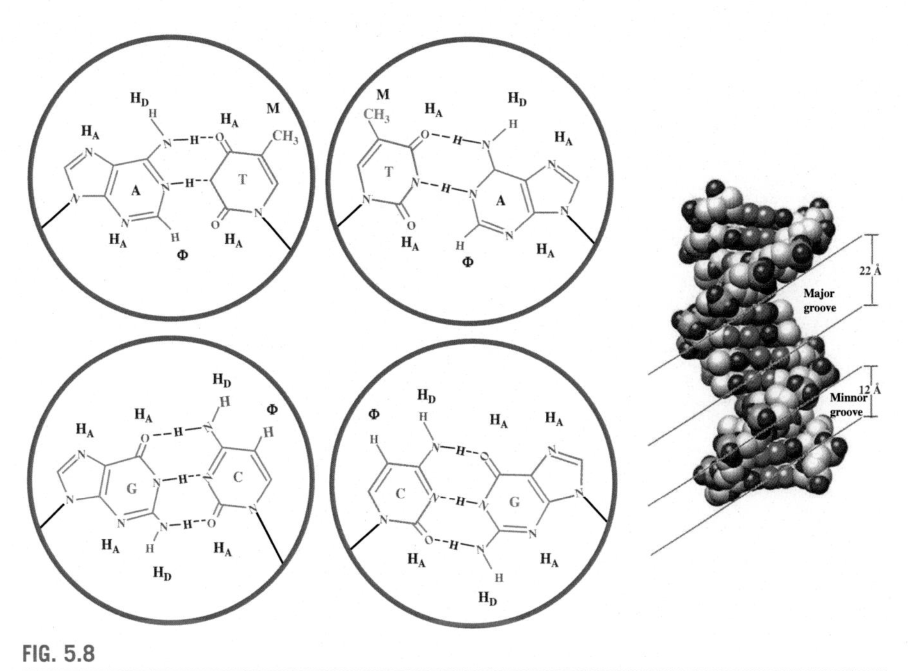

FIG. 5.8

The major and minor grooves of DNA. Exposed chemical groups: H_D—hydrogen bond donor; H_A—hydrogen bond acceptor; Φ—nonpolar hydrogen; M—methyl.

Source: Larsen, T.A., Kopka, M.L., Dickerson, R.E., 1991. Biochem 30, 4443.

in the major groove. Hence, based on the order of the chemical groups (which can be considered as the chemical code), each of the base pairs can be uniquely identified.

Multiple conformation of the double helix

The DNA double helix can adopt more than one conformation (Fig. 5.9). Early X-ray diffraction studies discovered two kinds of DNA structure in solution, designated as the B and A forms. The B form most closely resembles the average structure of DNA under physiological conditions. Experimentally, this form is observed at high humidity. There are about 10 base pairs per turn of the helix. As expected, the major groove is wide, and the minor groove is narrow. The A form, on the other hand, is observed at low humidity. It has about 11 base pairs per turn. Compared with the B form, its major groove is narrower but deeper, and its minor groove is wider and shallower.

DNA in the cell is not always as regular as the "idealized" B DNA. There are base pair to base pair variations in the structure. As for example, in some cases, two members of a base pair do not lie in the same plane, but "twisted" (called the "propeller twist") with respect to each other. Also, the rotation per base pair is not a constant along the length of the DNA. Yet, the B form is a good approximation to the structure of cellular DNA and helps explain various macromolecular interactions and functions. The A structure is found in certain protein-DNA complexes.

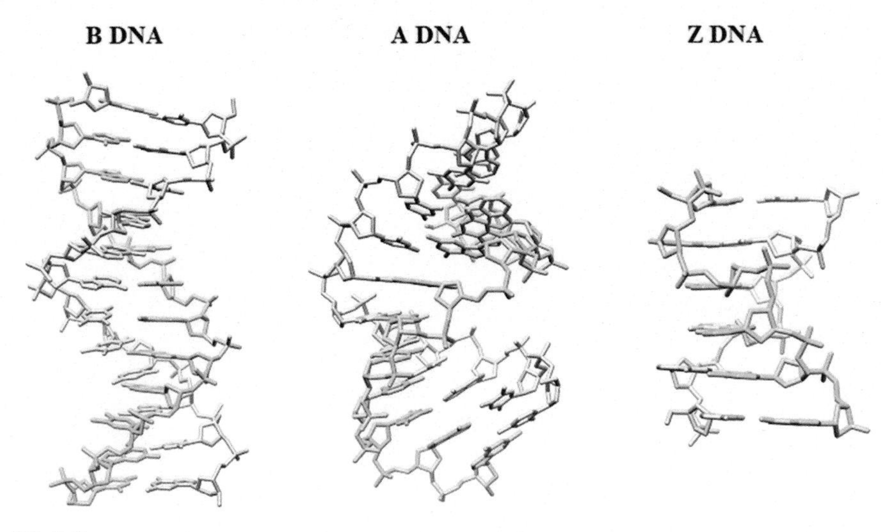

FIG. 5.9

Conformations of B, A, and Z forms of DNA.

Source: Larsen, T.A., Kopka, M.L., Dickerson, R.E., 1991. Biochemistry 30, 4443; Hardwick, J.S., et al., 2017. Nat. Struct. Mol. Biol. 24, 544; Tereshko, V., et al., 2001. Nucleic Acids Res. 29, 1208.

Helices of both B-form and A-form DNA are right-handed. Under certain conditions, DNA containing alternating purine and pyrimidine residues can adopt a left-handed structure. A prominent feature of the left-handed helix relates to the orientation of the purine base about the glycosidic bond. In nucleosides, rotation about the glycosidic bond allows the bases to occupy either of two principal positions—*syn* and *anti* (Fig. 5.10). In right-handed DNA, bases are always in the *anti* conformation. In left-handed DNA, pyrimidines do exist in the *anti* conformation, but the purine residues adopt the *syn* conformation. A purine-pyrimidine dinucleotide being the

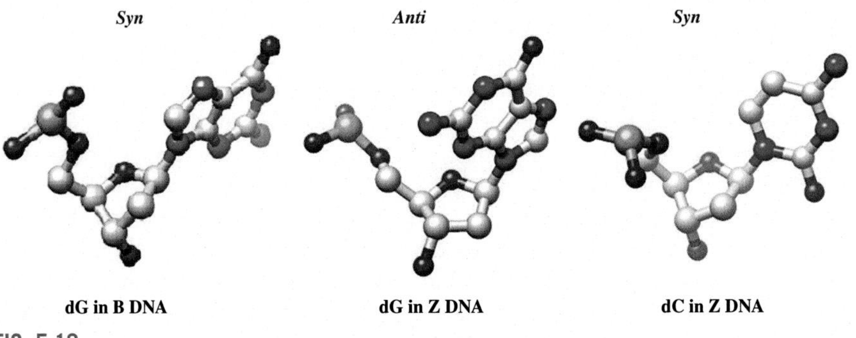

FIG. 5.10

Syn and *anti* orientations of guanine and cytosine.

Source: Larsen, T.A., Kopka, M.L., Dickerson, R.E., 1991. Biochemistry 30, 4443; Tereshko, V., et al., 2001. Nucleic Acids Res. 29, 1208.

fundamental repeating unit of left-handed DNA, alternating *syn-anti* conformations give its backbone a "zigzag" look; hence, it is called Z-DNA. Nevertheless, left-handed DNA helices are only infrequently found in the cell.

5.1.3 DNA superhelicity

Needless to reiterate, whatever be its form, B, A, or Z, DNA has a flexible structure. The precise molecular parameters of the double helix, such as the number of times the two chains are intertwined around each other, are dependent on ionic environment and the nature of the DNA-binding protein that forms a complex with it. A linear DNA molecule accommodates a change in this number since it can freely rotate.

For a covalently closed circular DNA (cccDNA) free rotation is not possible and, consequently, the number of times the two chains are intertwined cannot change. The topological problem of a cccDNA can be described by using the term "linking number" (*Lk*) which simply can be considered as the number of times one chain has to move through the other so that the two can be completely separated from each other.

Consider a "relaxed" cccDNA molecule created by simply joining the two ends together (that is, without any rotation of the double helix in either direction). The linking number of this molecule under physiological condition is denoted by Lk^0. Now if a nick is generated in one strand which is then rotated around the other before the nick is resealed (all this is enzymatically possible), the linking number of the molecule will be *Lk*. The difference between *Lk* and Lk^0, called the linking difference is expressed as

$$\Delta Lk = Lk - Lk^0$$

ΔLk is positive or negative depending on the direction of rotation, that is, whether the double helix is overwound or unwound. The double-helical DNA is now torsionally strained and it compensates the strain by winding around itself in the opposite direction. The DNA is said to be "supercoiled" or "superhelical." If $\Delta Lk > 0$, the superhelicity is positive; if $\Delta Lk < 0$, the DNA is said to be "negatively supercoiled." The measure of supercoiling, or the superhelical density, is defined by

$$\sigma = \Delta Lk / Lk^0$$

5.2 Proteins

5.2.1 Amino acids

Amino acids are the fundamental building blocks of proteins. They are organic compounds containing amino (—NH_2) and carboxyl (—COOH) functional groups, along with an amino acid-specific side chain (R group), covalently bonded to a central carbon atom (also called an α-carbon, or C_α) (Fig. 5.11). It is the side chain which makes one amino acid differ from another.

Twenty kinds of side chains are commonly found in proteins (Fig. 5.12). They differ in size, shape, charge, hydrogen bonding ability, hydrophobicity, and chemical reactivity. Hence, they interact differently with each other and with other molecules including water.

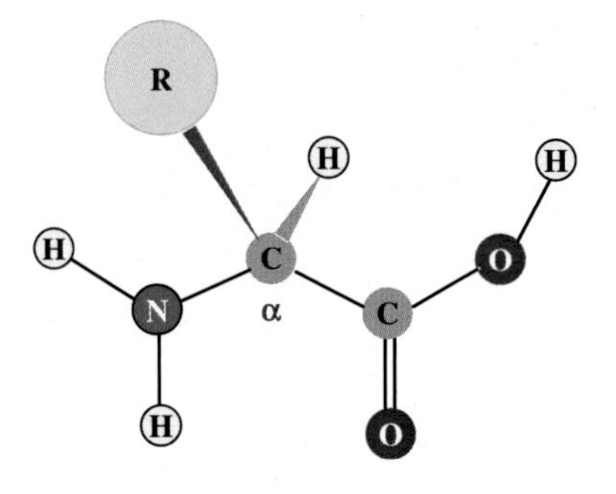

FIG. 5.11

The general chemical structure of an amino acid. α: α-carbon; R: side chain.

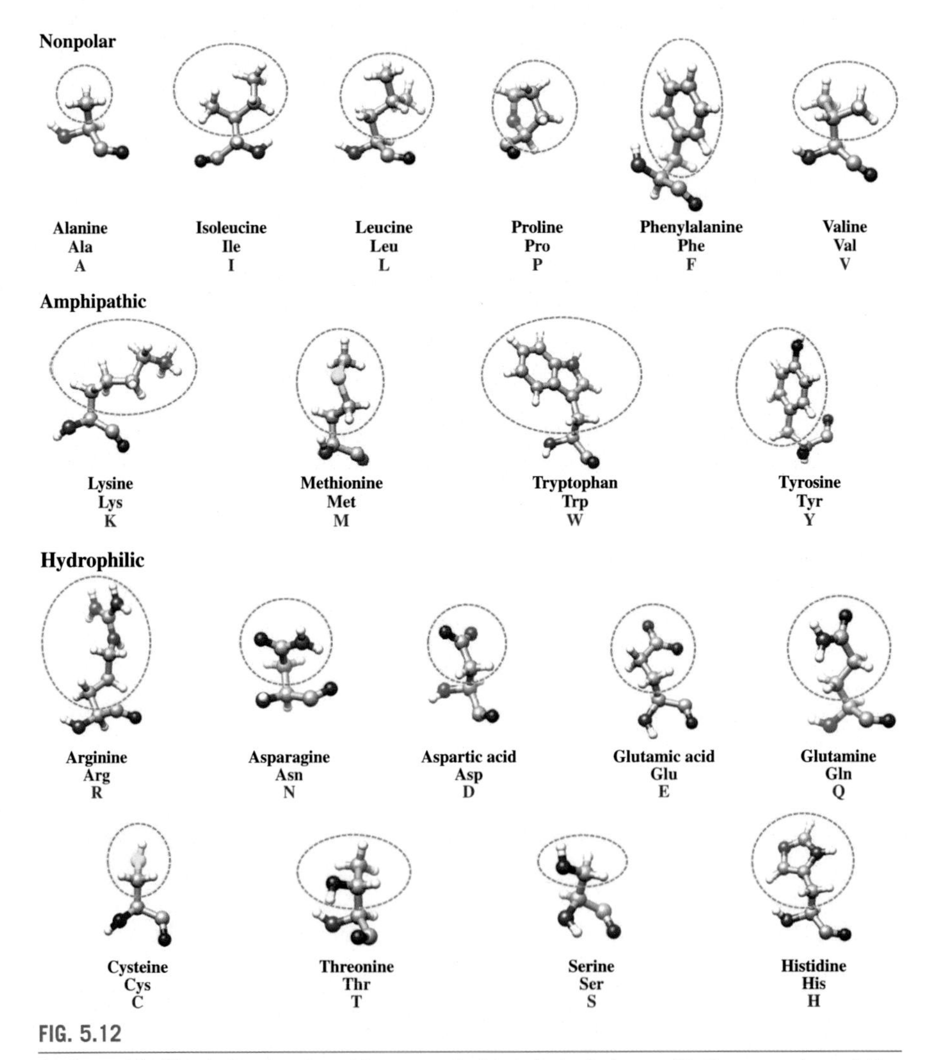

FIG. 5.12

Amino acid side chains (enclosed in green dashed circles) along with three- and one-letter symbols. Glycine (Gly, G) has not been included since it contains only a hydrogen in the side chain. The other 19 amino acids have been divided into three categories based on the property of the side chain—hydrophobic, amphipathic, and hydrophilic.

5.2.2 Peptide bond

Chemically, a covalent bond can be formed between the carboxyl group of one amino acid and the amino group of another by the loss of a water molecule. This is called an amide bond or a peptide bond (Fig. 5.13). The bond formation can be reversed by the addition of water (hydrolysis). The equilibrium of the reaction is more toward hydrolysis than synthesis. Hence, an input of free energy is required for its biosynthesis. In the cell, peptide bond formation is an enzymatically controlled process (discussed in Chapter 11).

The nature of the peptide bond is influenced by the phenomenon of resonance where electrons are delocalized over multiple atoms and a given molecule cannot be represented by a single valence structure. Two resonance structures are possible for the peptide bond (Fig. 5.14A). Due to resonance, the peptide bond possesses ~40% double-bond character which prevents rotation about this bond. As a consequence, the peptide group has a rigid, planar structure (Fig. 5.14B).

A peptide group usually adopts the trans conformation with successive C_α atoms being positioned on opposite sides of the peptide bond joining them. This trans peptide group forms the basic repeating unit of the polypeptide backbone. We can see that, unlike the peptide bond, the other two bonds in the unit, the N-C_α and C_α-C bonds, are pure single bonds. Two adjacent planar peptide groups are free to rotate about these bonds, provided there is no steric hindrance (Fig. 5.15). The two rotations are described by torsion angles (also called dihedral angles)—the rotation around N-C_α being designated as phi (ϕ), while that around C_α-C as psi (ψ). The backbone of a polypeptide can be considered as a linked sequence of planar peptide groups, with rotatable covalent bonds alternating with rigid ones (Fig. 5.16).

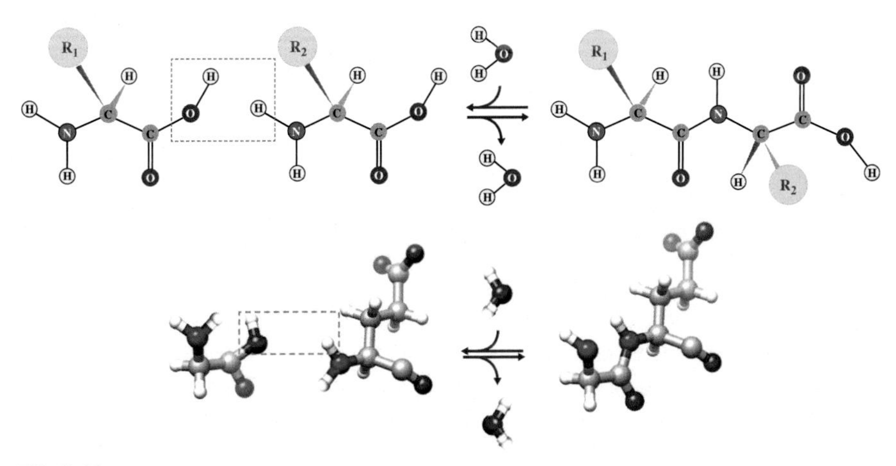

FIG. 5.13

Peptide bond formation. Upper panel: chemical structure; lower panel: three-dimensional structure. Coloring scheme for atoms as in Fig. 5.6.

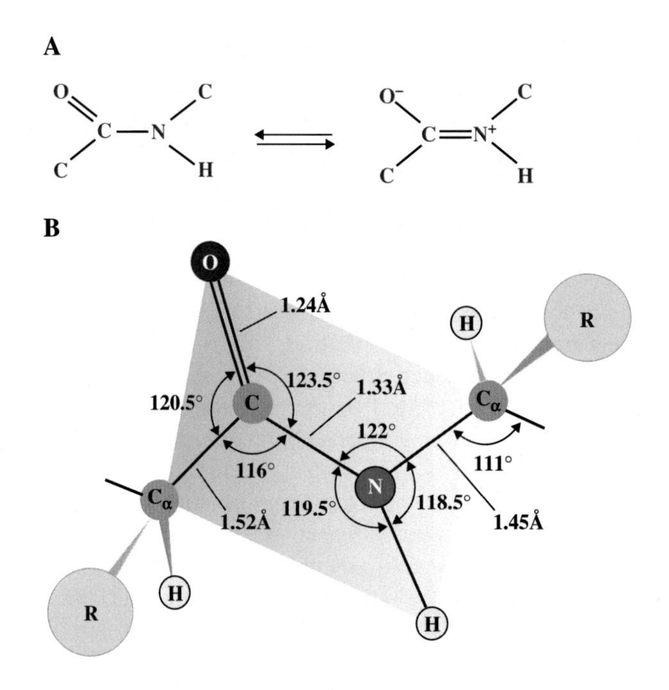

FIG. 5.14

The peptide bond. (A) Resonance structure; (B) peptide plane.

As already indicated, several (ϕ, ψ) combinations are restricted by the principle of steric exclusion. According to this principle, two atoms cannot be present in the same place at the same time. They cannot be closer to each other than the corresponding van der Waals distance. As a consequence, only certain values of ϕ and ψ are permitted. This is represented by a two-dimensional plot known as a Ramachandran diagram (named after its inventor, G.N. Ramachandran) (Fig. 5.17).

5.2.3 Secondary structures

Polypeptides are linear polymers; nevertheless, for most polypeptides, the structure is not like a random thread. Their structural organization can be described at different levels. Evidently, the sequence of amino acids in a polypeptide is considered to be its *primary structure*. Two periodic or repeating structures, initially proposed by Linus Pauling and Robert Corey, are found at the next hierarchical level—the α helix and the β sheet. These elements, composed of sequences of peptide groups with repeating ϕ and ψ values, are called *regular secondary structures*. The recurrence in most cases is only approximate; yet, it is so frequent that these elements are immediately recognizable in polypeptides with widely differing amino acid sequences.

The ϕ and ψ values of the α helix fall within the fully acceptable regions in the Ramachandran diagram (Fig. 5.17). It has a favorable hydrogen bonding pattern for a

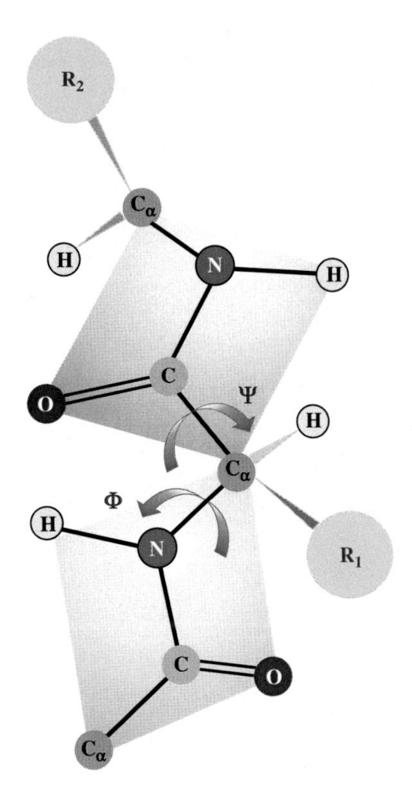

FIG. 5.15

Rotations of adjacent planar peptide groups.

regular secondary structure formation. In the α helix, the C═O group of the nth residue forms a hydrogen bond with the N—H group of the $(n+4)$th residue, resulting in a cylindrical structure (Fig. 5.18). Thus, except for the N—H group at the amino-terminal end and the C═O group at the carboxy-terminal end of the helix, all the main-chain N—H and C═O groups are hydrogen-bonded. The hydrogen bonds

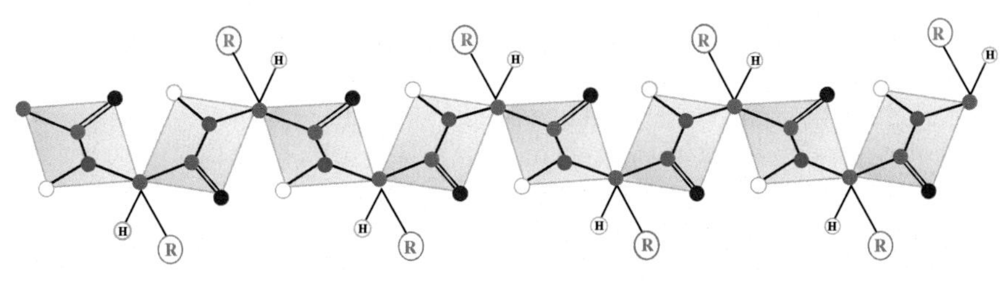

FIG. 5.16

Extended conformation of a polypeptide with the backbone shown as a linked sequence of planar peptide groups.

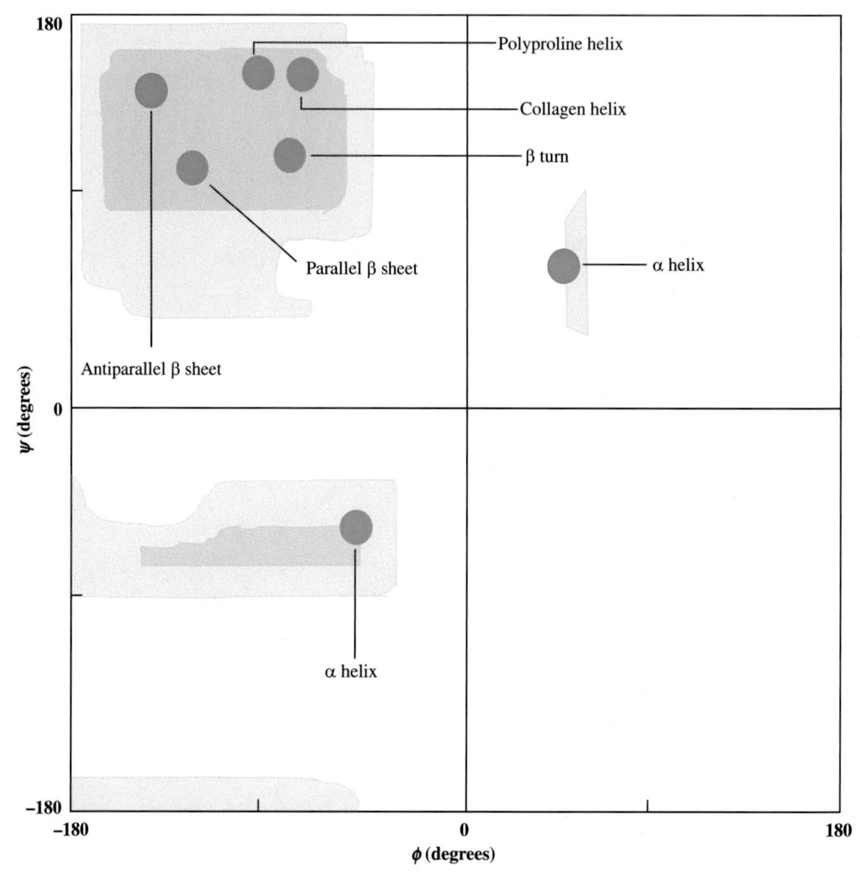

FIG. 5.17

A Ramachandran diagram. The shaded regions indicate the sterically permitted values of ϕ and ψ—the most favorable and borderline regions are shown in darker and lighter shades, respectively.

lie along the wall of the cylinder, while the side chains protrude outward, thereby avoiding any steric clash with the backbone as well as with each other. Each residue is separated from the preceding one along the helix axis by 1.5 Å and rotated by 100 degrees (Fig. 5.18A). There are 3.6 residues per turn and, hence, the distance the helix rises per turn along its axis (pitch) is 5.4 Å. The helix has a tightly packed core with its atoms in van der Waals contact (Fig. 5.19).

We may note here that each individual peptide group has an electrostatic dipole moment. In a randomly structured (i.e., unstructured) polypeptide chain, the dipole moments of the individual backbone amide groups are also randomly oriented. In contrast, all the amides of an α helix, and with them their dipole moments, are aligned in the same direction, approximately parallel to the helical axis.

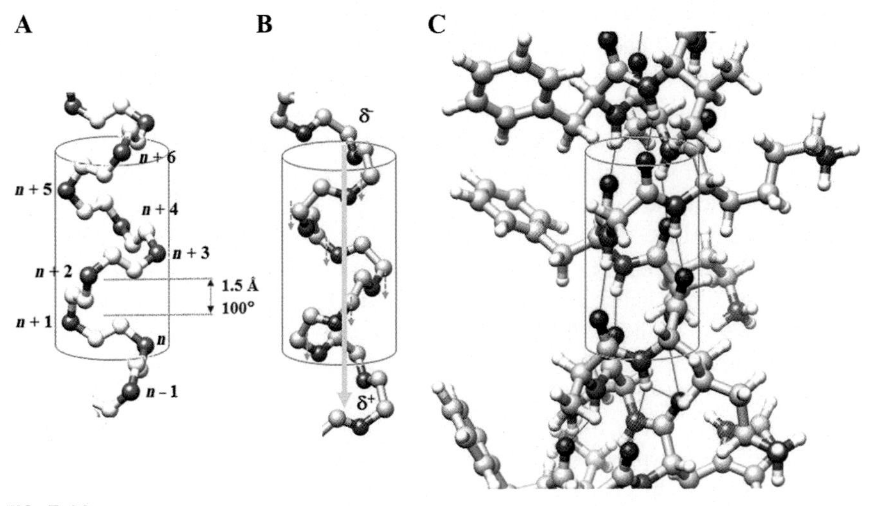

FIG. 5.18

The α helix. (A) The positions of C_α's indicating separation and rotation between adjacent residues. (B) The backbone with peptide dipoles. (C) The full hydrogen-bonded structure (hydrogen bonds in *red*).

Source: Gasell, J., Zasloff, M., Opella, S.J., 1997. J. Biomol. NMR 9, 127.

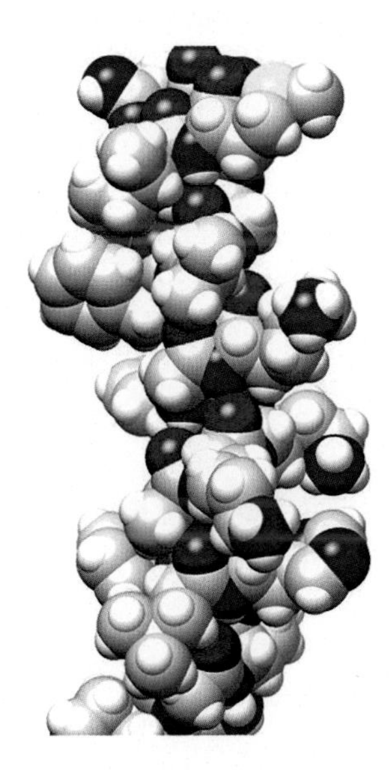

FIG. 5.19

Space-filling model of α helix showing van der Waals contacts between atoms. Coloring scheme for atoms: as in Fig. 5.6; additionally, sulfur is *yellow*.

Source: Gasell, J., Zasloff, M., Opella, S.J., 1997. J. Biomol. NMR 9, 127.

Hence, the individual dipoles together create a macrodipole with its positive end at the amino-terminal end of the helix (Fig. 5.18B).

Fig. 5.20 gives the view of a turn of an α helix from the top with the side chains projected outward. It can be seen that hydrophilic residues (histidine and lysine) are present on one side, while hydrophobic residues (phenylalanine and leucine) are present on the other side (Fig. 5.20). In many α helices, the pattern is repeated for the entire cylindrical surface to produce a hydrophilic face and a hydrophobic face on opposite sides of the helix. Such a helix is known as an amphipathic α helix.

The β sheet (sometime called the β pleated sheet) markedly differs in appearance from the rod like α helix. Rather than being tightly coiled, as in the α helix, a polypeptide chain (β strand) in the β sheet is almost fully extended (Fig. 5.21). The separation between adjacent amino acid residues along a β strand is ~3.5 Å in contrast to a distance of 1.5 Å along an α helix.

In a β sheet, two or more strands that may be widely separated in the primary structure of the polypeptide are arranged side by side with hydrogen bonds between them. Adjacent chains can run in the same direction (parallel β sheet) or in opposite direction (antiparallel β sheet). A β sheet can also be formed by a combination of parallel and antiparallel strands (Fig. 5.21). Hydrogen bonds are formed between N—H groups of one β strand with C=O groups of an adjacent strand and vice versa. Nevertheless, hydrogen bonds between parallel strands are not as linear as those between antiparallel strands. Also, edge strands make hydrogen bonds in a more complex manner. Structurally, β sheets are more diverse than α helices. They can be relatively flat; most β sheets, however, adopt a somewhat twisted shape (Fig. 5.21).

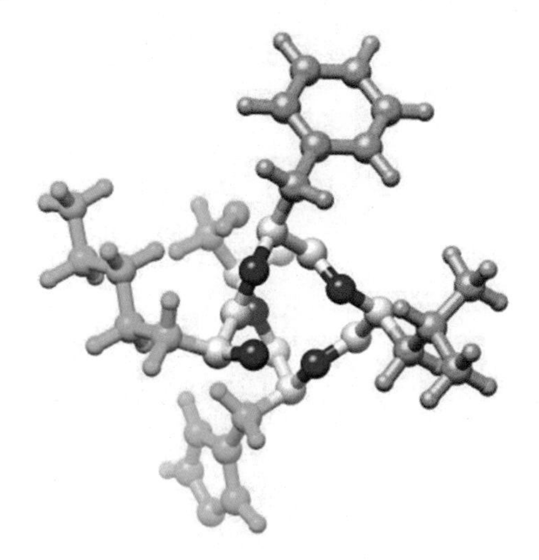

FIG. 5.20

The view of one turn of an amphipathic α helix from the top. Coloring: hydrophilic residues—*cyan*; hydrophobic residues—*coral*.

Source: Gasell, J., Zasloff, M. and Opella, S.J., 1997. J. Biomol. NMR 9, 127.

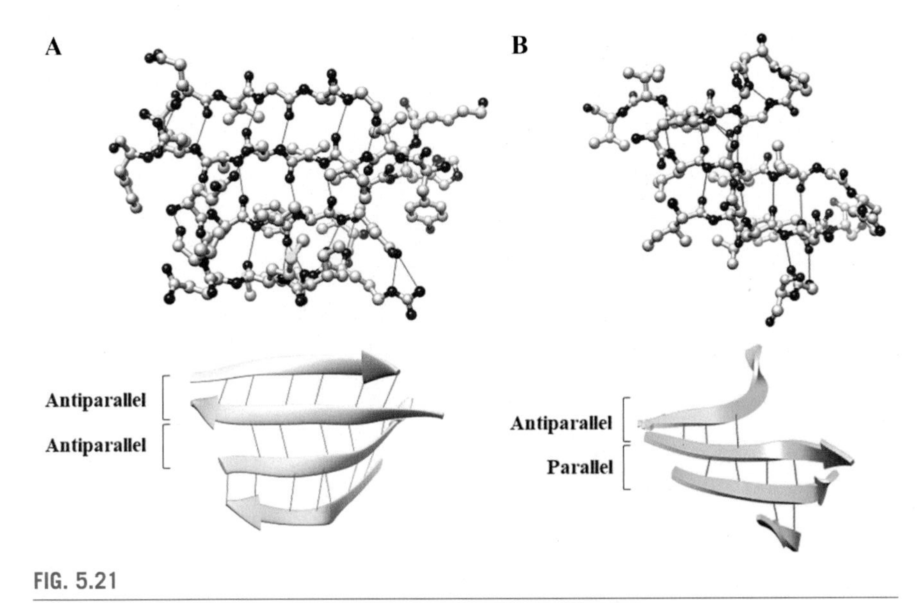

FIG. 5.21

The structure of β sheets: (A) antiparallel and (B) mixed (parallel + antiparallel).
Source: Brautigam, C.A., Steitz, T.A., 1998. J. Mol. Biol. 277, 363; Evans, R.J., et al., 2008. Proc. Natl. Acad. Sci. U. S. A. 105, 20695.

Two other secondary structural elements have been identified in polypeptide chains—the β turn and the Ω loop. Although nonperiodic, these common structures are well-defined and connect α helices and β sheets to form higher order structures. In a β turn (also called a reverse turn or hairpin turn), the C=O group of residue n of the polypeptide is hydrogen-bonded to the N—H group of residue $n+3$. This interaction abruptly reverses the direction of the chain (Fig. 5.22). A β hairpin specifically refers to a reverse turn connecting adjacent strands in an antiparallel β sheet. In many other cases, more elaborate structures are involved in chain reversal. These structures are known simply as loops or, to indicate their overall shape, as Ω loops (Fig. 5.23). Some of these loops help a polypeptide to interact with other molecules.

5.2.4 Supersecondary and tertiary structures

The tertiary structure of a polypeptide is a depiction of the folding of its secondary structural elements. It specifies the position of each atom of the polypeptide, whether located in the backbone or the side chain. The overall three-dimensional architecture of the molecule, synonymously called its conformation, refers to its secondary as well as tertiary structures. Intermediate between the two hierarchical levels are the motifs (supersecondary structures) which refer to the packing of adjacent secondary structural elements into distinct structural units to be correlated with function. Let us consider a few of the most commonly found motifs in polypeptides.

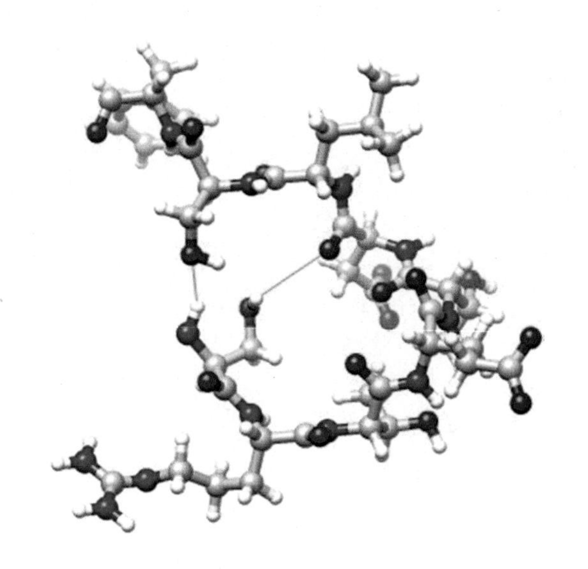

FIG. 5.22

β turn in a polypeptide chain. Hydrogen bonds are shown by *red* lines.

Source: Yu, H., Rosen, M.K., Schreiber, S.L., 1991. FEBS Lett. 324, 87.

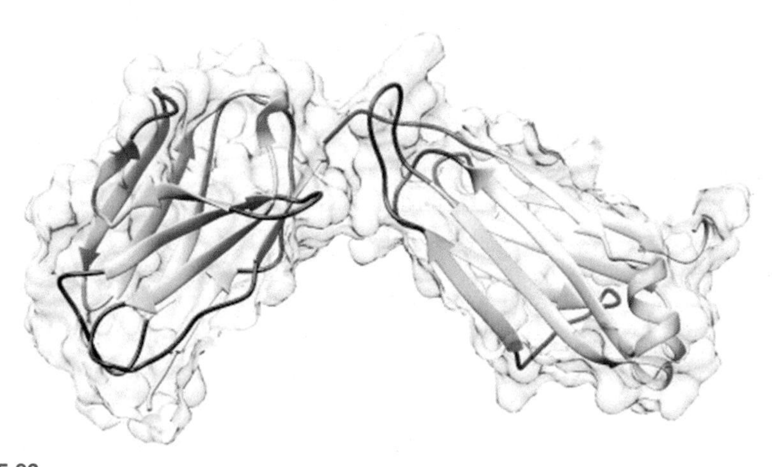

FIG. 5.23

Ω loops. Loops on the surface of a polypeptide shown in *firebrick color.*

Source: Harris, L.J., et al., 1997. Biochemistry 36, 1581.

Motifs

Antiparallel β strands, as we have seen, can be linked by a short β hairpin. In contrast, two parallel β strands must be connected by a longer region of the polypeptide chain which often contains an α helix (Fig. 5.24). This is a βαβ motif which is found in most polypeptides having a parallel β sheet.

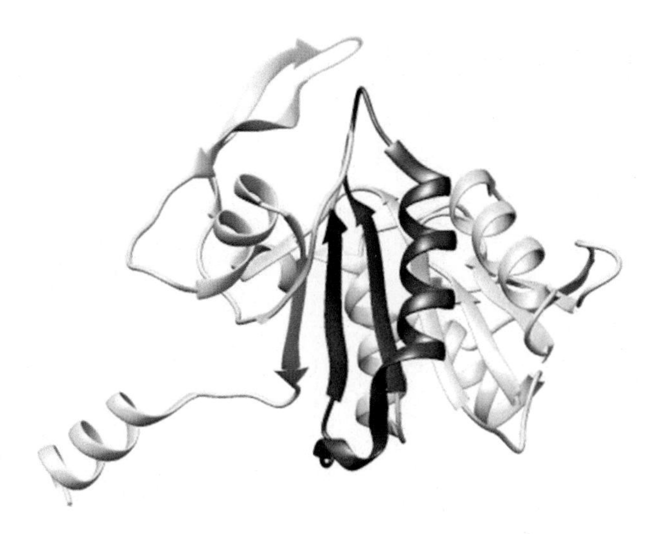

FIG. 5.24

βαβ motif shown in *firebrick color.*

Source: Lee, J.Y., et al., 2001. Nat. Struct. Mol. Biol. 8, 789.

The helix-turn-helix (HTH) motif (also called an αα motif), as we shall find in later chapters, is a common recognition element used by transcription regulators. As the name suggests, the motif is constituted of two almost perpendicular α helices connected by a 3–4 residue turn (Fig. 5.25). The latter may not always be a canonical β turn, nonetheless, the relative orientation of the helices is conserved.

One of the two helices of the HTH motif is the recognition helix which binds the major groove of DNA through hydrogen bonds and hydrophobic interactions with exposed bases. The other helix, which does not play an effective role in recognition, nonetheless, stabilizes the interactions.

The structure of a zinc-finger motif (also called zinc-coordinating or ββα motif) is characterized by an α helix and a short two-stranded antiparallel β-sheet (Fig. 5.26). Two pairs of cysteine and histidine residues tetrahedrally coordinate a Zn^{2+} ion. The α helix, called the recognition helix, makes sequence-specific contacts with bases in the major groove of DNA. The motif is often repeated in the polypeptide with two, three, or more 'fingers' binding regularly spaced major grooves. Widely present in eukaryotic transcription factors, the role of the zinc-coordinating motif is, however, not limited to DNA-binding. It has been found to be involved in several protein-protein interactions.

The leucine-zipper motif derives its name from the way it dimerizes involving the amino acid leucine. The motif, in fact, is a single α helix, about 60 amino acid long, containing a dimerization region and a DNA-binding region (Fig. 5.27A). The dimerization or zipper region consists of leucine or a similar hydrophobic amino acid at a regular interval. A localized amphipathic cylindrical surface is created (Fig. 5.27B). As a consequence, two copies of the helix make hydrophobic contacts at the interface

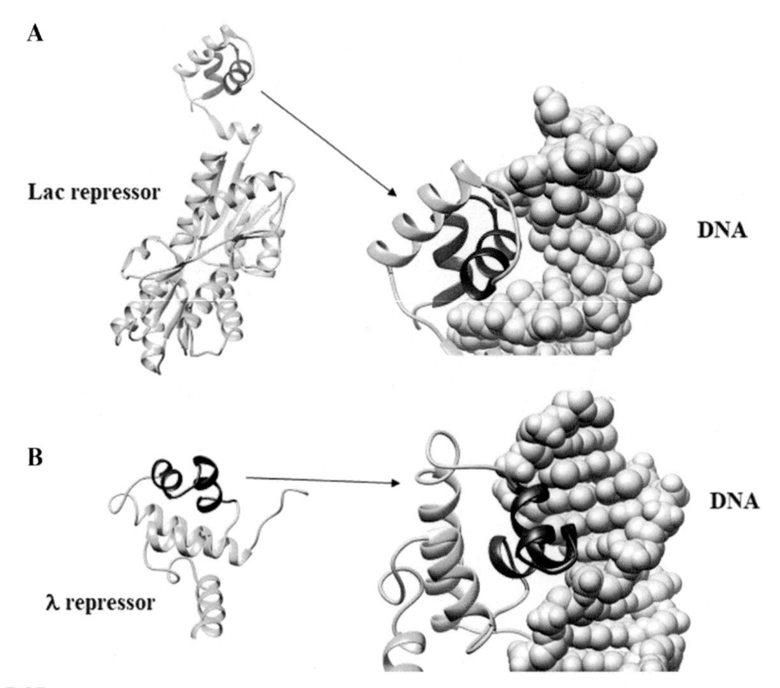

FIG. 5.25

Helix-turn-helix (HTH) motif. HTH motif of Lac repressor (A) and λ repressor (B) shown in *firebrick color* binds the DNA major groove.

Source: Bell, C.E., Lewis, M., 2000. Nat. Struct. Mol. Biol. 7, 209; Beamer, L.J., Pabo, C.O., 1992. J. Mol. Biol. 227, 177.

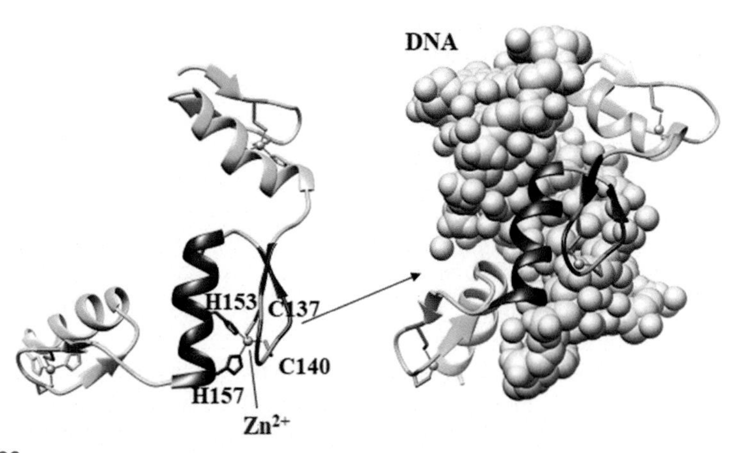

FIG. 5.26

A zinc-finger motif *(firebrick color)* binding DNA major groove.

Source: Elrod-Erickson, M., et al., 1996. Structure 4, 1171.

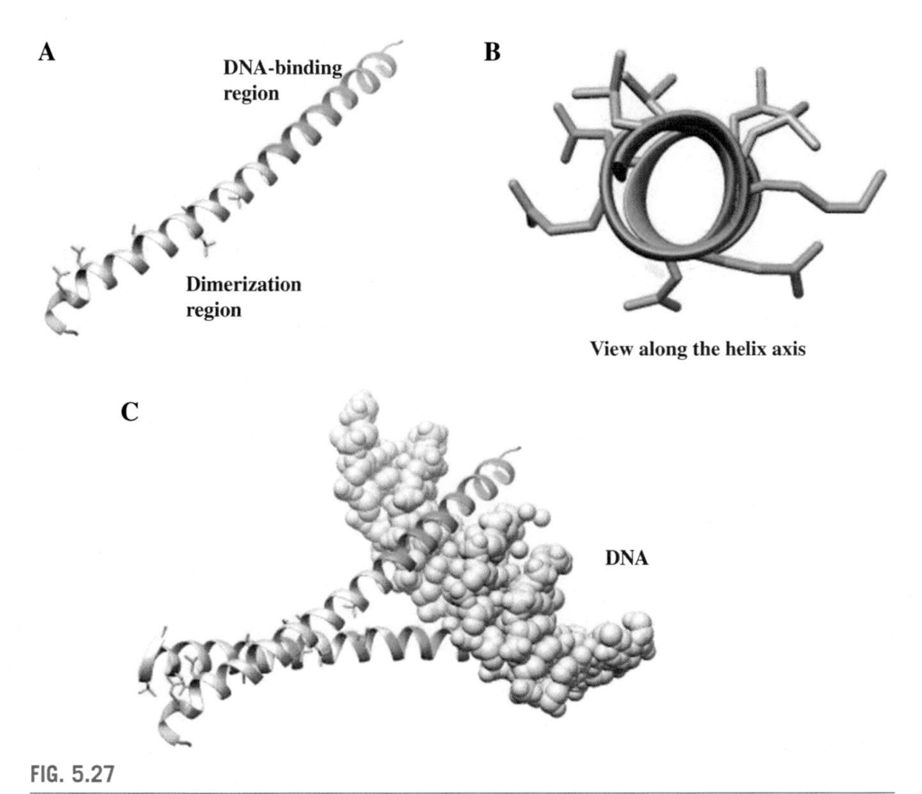

FIG. 5.27

A leucine-zipper motif. (A) Leucine residues in dimerization region shown in golden. (B) View from the top of an amphipathic cylindrical surface. Hydrophobic residues shown in *golden*; hydrophilic residues shown in *dodger blue*. (C) Basic region of motif *(dodger blue)* binding DNA major groove.

Source: Ellenberger, T. E. et al. 1992. Cell 71:1223.

through side-by-side packing. The amino-terminal DNA-binding region, on the other hand, is essentially a basic region. It binds the major groove of DNA (Fig. 5.27C).

The "catalytic triad," involved in many enzymatic actions, is often designated as a DD(D/E/H) motif (i.e., a sequence motif). However, three residues of the motif are not contiguous; the spacing between them and the order in which they are positioned may differ from polypeptide to polypeptide. Nevertheless, they structurally come close together in the molecule and form a catalytic unit with an approximately conserved geometry for similar function (Fig. 5.28).

Folds

Extended secondary structures and combination of structural motifs sometimes lead to the formation of novel folds. A large antiparallel β sheet can curve all the way around so that the last strand is hydrogen-bonded to the first and the

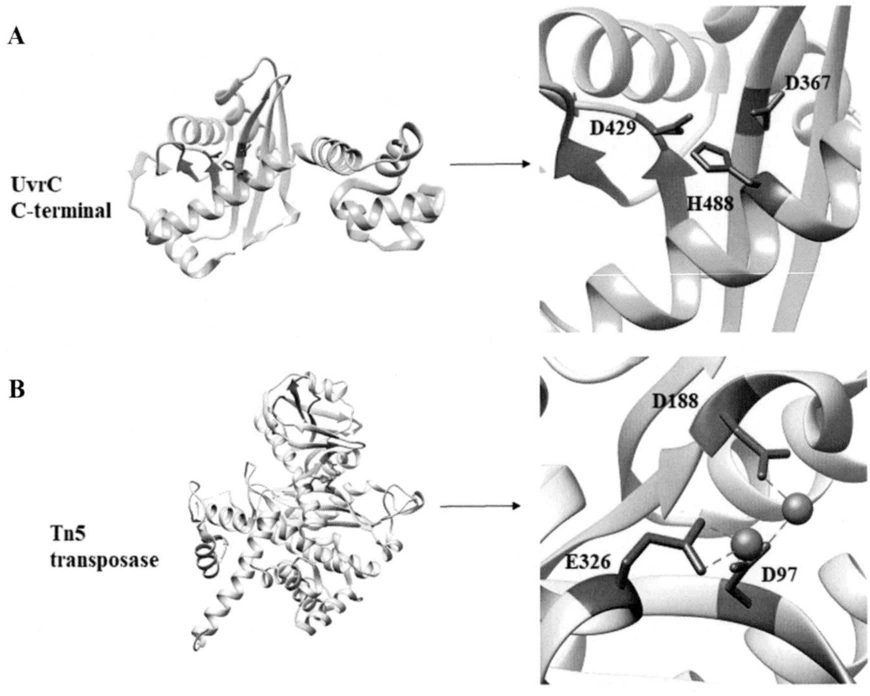

FIG. 5.28

Catalytic triad of (A) UvrC C-terminal (DDH in *orange*) and (B) Tn5 transposase (DDE in *orange*).

Source: Karakas, E., et al., 2007. EMBO J. 26, 613; Steiniger-White, M., Rayment, I., Reznikoff, W.S., 2004. Curr. Opin. Struct. Biol. 14, 50.

cylindrical surface is closed. This is a β barrel structure of the polypeptide (Fig. 5.29).

Under favorable conditions, two antiparallel α helices can pack against each other by interdigitation of their protruding side chains and with their axes slightly inclined. Such packing stabilizes coiled coil conformation of the helices. In some cases (as e.g., Ferricytochrome b_{562}), adjacent antiparallel α helices, located in the same polypeptide chain, form a four-helix bundle (Fig. 5.30). Electrostatic interactions make a limited contribution to the antiparallel packing of the helices.

An extended βαβ motif can form a βαβαβ unit, which is a parallel sheet of the β strands connecting α helices. Two such units combine to form what is known as a Rossmann fold (Fig. 5.31). βαβαβ often acts as a nucleotide-binding site; hence, it is also called a dinucleotide-binding fold.

The winged helix (WH) fold can be considered as an extension of the HTH motif. Of the three α helices (H1–H3) it contains, H2 and H3, together with a connecting turn (T), form the HTH motif. In addition, it also contains three β strands (S1–S3) and two characteristic loops (W1–2) having the appearance of "wings" (Fig. 5.32).

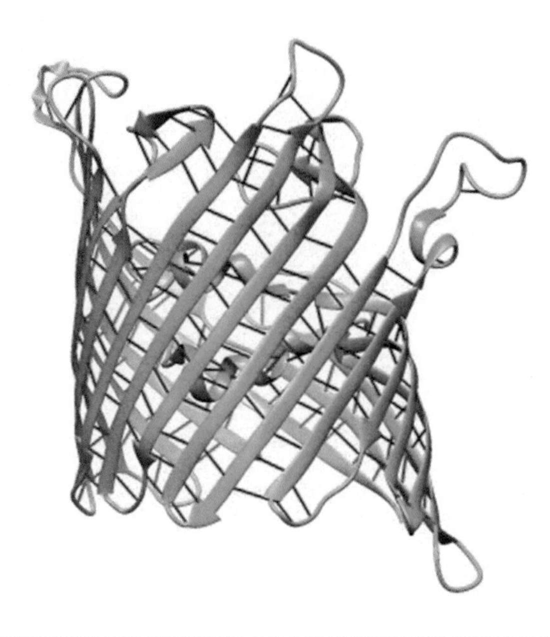

FIG. 5.29

A β barrel. Hydrogen bonds are shown by *red* lines.

Source: Tanabe, M., Nimigean, C.M., Iverson, T.M., 2010. Proc. Natl. Acad. Sci. U. S. A. 107, 6811.

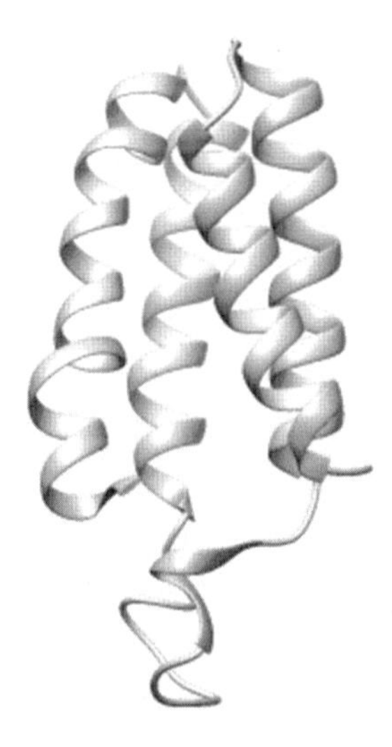

FIG. 5.30

A four-helix bundle motif.

Source: Arnesano, F., et al., 1999. Biochemistry 38, 8657.

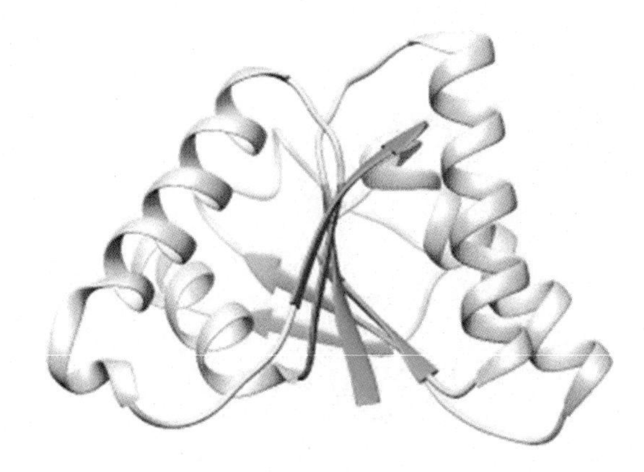

FIG. 5.31

A Rossmann fold.

Source: Patskovsky, Y., et al., PDB code: 3BQ9. https://doi.org/10.2210/pdb3BQ9/pdb.

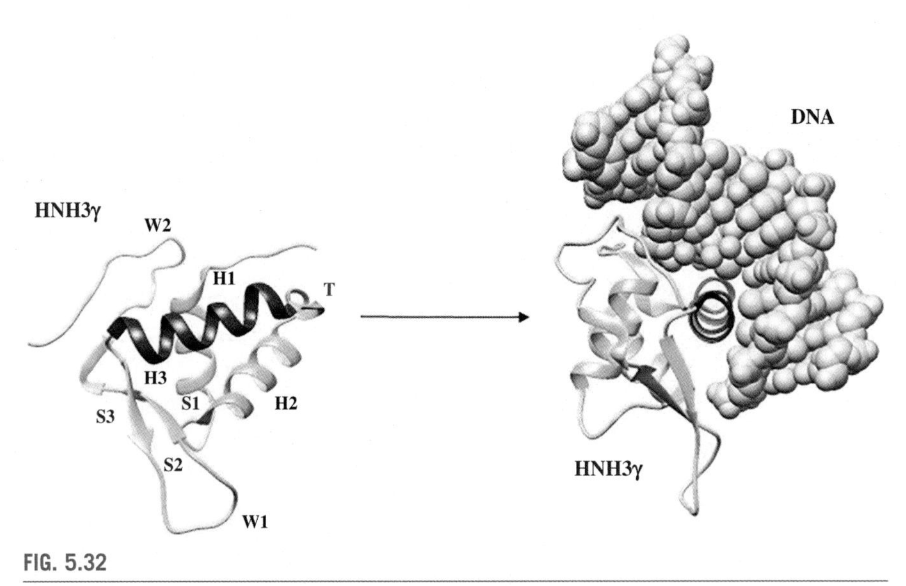

FIG. 5.32

A winged helix fold. The recognition helix *(colored firebrick)* binds DNA major groove.

Source: Clark, K.L., et al., 1993. Nature 364, 412.

The overall order of the secondary structure elements is then H1-S1-H2-T-H3-S2-W1-S3-W2. The recognition helix (H3) binds the major groove of DNA. The other elements provide additional support for DNA-binding. The WH fold plays a role also in the strand-separating action of helicases and protein-protein interactions.

Distant parts of a polypeptide may come together at intramolecular interfaces and interact to form the tertiary structure. Like the secondary structure, the tertiary

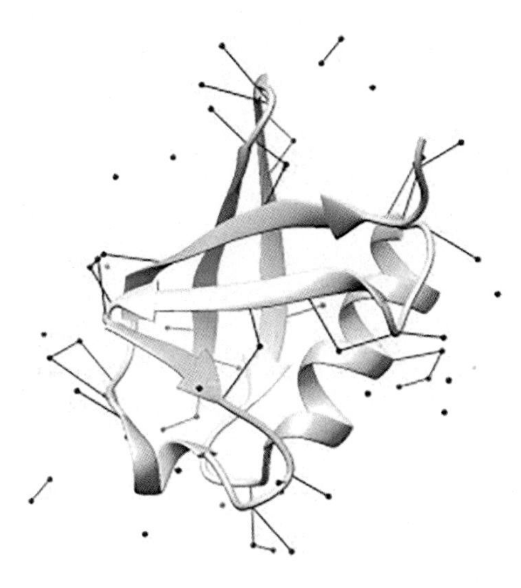

FIG. 5.33

Tertiary structure of ubiquitin stabilized by water. *Blue* dots are water molecules, *red* lines are hydrogen bonds.

Source: Vijay-Kumar, S., Bugg, C.E., Cook, W.J., 1987. J. Mol. Biol. 194, 531.

structure is also dependent on hydrogen bonding, hydrophobic effect and, to some extent, electrostatic interactions among its constituent atoms and with surrounding water molecules. In general, water-soluble polypeptides form a globular structure. Often, water molecules are bound to the surface of a folded polypeptide. Although in the course of secondary and tertiary structure formation some water molecules are released to rejoin the structure of liquid water, quite a few polar groups, particularly those on the surface of the molecule, remain in contact with water. The bound water molecules become part of the tertiary structure of the polypeptide (Fig. 5.33)

Domains

Most polypeptides are globular; tertiary interactions fold these polypeptide chains to form compact globular structure. Usually, the folded structure of relatively small polypeptides (each containing < 200 amino acid residues) is approximately spherical (a simple globular shape). The larger polypeptides fold into two or more globular clusters known as domains.

A domain is defined as a region of polypeptide structure that can fold independently of its neighboring sequences. Biochemically, domains have specific functions. In general, domains are categorized into five different groups based on their predominant secondary structure elements—(i) α, comprising entirely of α helices; (ii) β, containing β sheet; (iii) α/β, with β strands connected by α helices; (iv) $\alpha+\beta$, with separate β sheet and helical regions; and (v) cross-linked, where inconspicuous secondary structures are stabilized by disulfide bridges or metal ions.

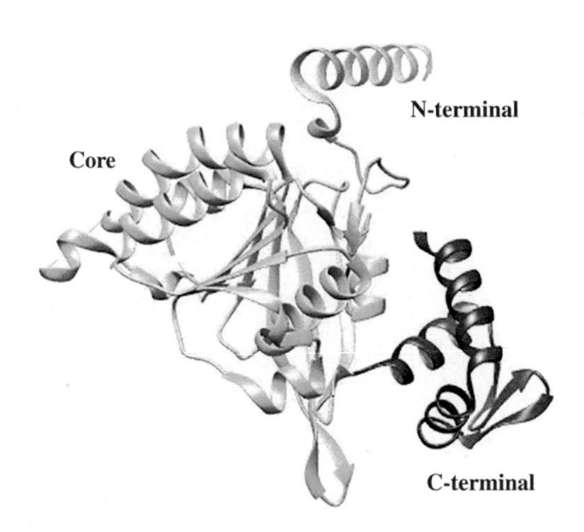

FIG. 5.34

Domain structure of RecA.

Source: Story, R.M., Weber, I.T., Steitz, T.A., 1992. Nature 355, 318.

Let us illustrate the domain structure of a polypeptide with the help of a RecA monomer. The polypeptide contains three structural domains—an N-terminal domain, a central core domain, and a C-terminal domain (Fig. 5.34). All three domains, though differing in size, are ($\alpha+\beta$) type. The central domain is involved in ATP-binding and ATP hydrolysis. It essentially consists of a twisted β sheet with eight β strands, flanked by eight α helices. The N-terminal domain is responsible for monomer-monomer interaction. It contains a long α helix and short β strand. The carboxy-terminal domain, which facilitates interfilament association, comprises three α helices and a three-stranded β sheet. The structural-functional properties of RecA will be further discussed in later chapters.

5.2.5 Quaternary structure

There are several proteins, each of which is an appropriately folded single polypeptide chain. In such cases, the terms polypeptide and protein become synonymous. Yet, most proteins, particularly those with molecular masses >100 kD, consist of more than one chain. Such proteins are called oligomers, while the individual chains are known as monomers or subunits. The polypeptide subunits associate to form a specific spatial arrangement known as the quaternary structure of a protein.

Association of two polypeptides is based on the principle of complementarity—of shape (geometric) and in binding interactions. An entire set of noncovalent interactions holds the subunits together. The complementarity of binding interactions requires that, at the interface, hydrogen-bond donors, nonpolar groups, and positive charges are opposite hydrogen-bond acceptors, nonpolar groups, and negative charges, respectively. Stable binding is achieved if the number of weak interactions is maximized by geometric complementarity of the interacting surfaces.

The geometric complementarity at a dimer interface is illustrated in Fig. 5.35. The "irregular" surfaces of two subunits of the *Escherichia coli* enzyme undecaprenyl pyrophosphate synthetase fit into each other quite elegantly. Hydrophobic effect plays a major role in stabilizing the dimeric structure.

Complementarity between interacting surfaces is also very conspicuous in coiled coil dimer of α helices (Fig. 5.36). In this helix-helix packing, hydrophobic side chains (mostly those of leucines) line up along one side of each helix forming "ridges" and intrude into the "grooves" on the interacting helix.

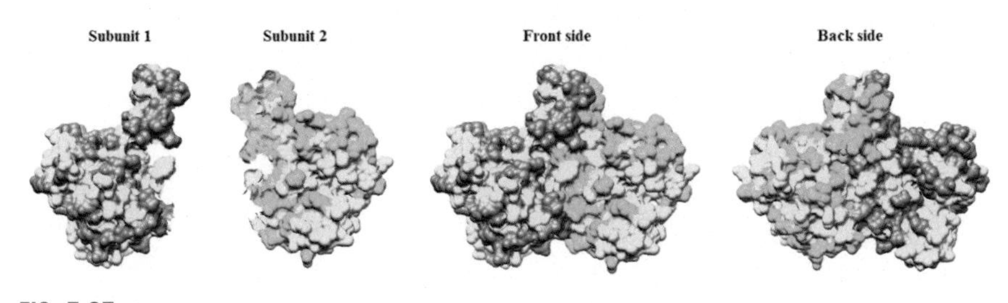

FIG. 5.35

Geometric complementarity at dimer interface. Two subunits of *E. coli* undecaprenyl pyrophosphate synthetase separately and two sides of the dimer have been shown. Hydrophobic residues have been colored *yellow*.

Source: Ko, T.P., et al., 2001. J. Biol. Chem. 276, 47474.

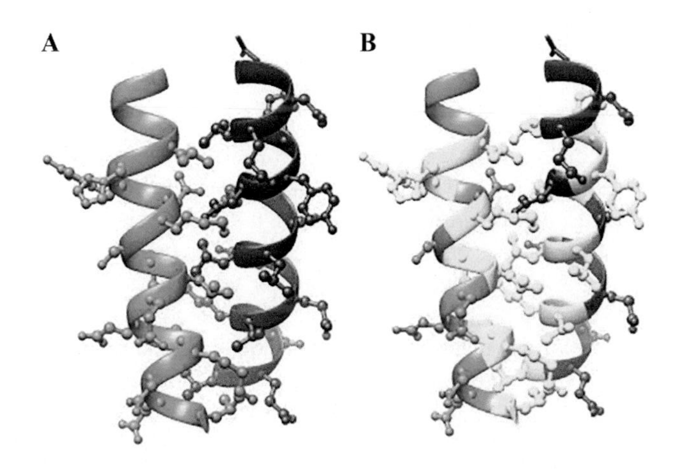

FIG. 5.36

Complementarity at coiled coil dimer interface. (A) Two helices (colored *dodger blue* and *firebrick*) intrude into each other by forming "ridges" and "grooves." (B) Hydrophobic residues of both helices colored *yellow*.

Source: Ellenberger, T.E., et al., 1992. Cell 71, 1223.

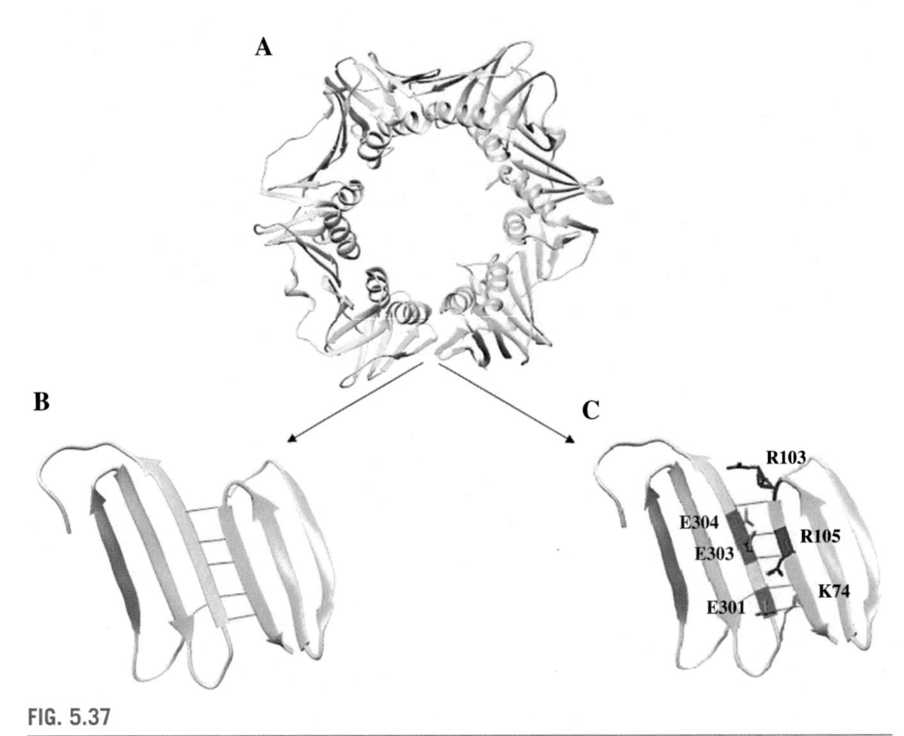

FIG. 5.37

β sheet-mediated dimerization. (A) *E. coli* β-sliding clamp with two monomers colored differently. (B) Hydrogen bonds between edge strands of β sheets shown by *red* lines. (C) Negative and positive residues at the interface colored *orange* and *blue*, respectively.

Source: Oakley, A.J., et al., 2003. Acta Crystallogr. Sect. D 59, 1192.

The subunits of a protein make maximal utilization of hydrogen-bonding potential at the interface. Sometimes, an extended β sheet can be formed between two subunits by hydrogen bonding and electrostatic interactions. As shown in Fig. 5.37, two monomers of the *E. coli* β-sliding clamp form a β-dimer interface where negatively charged amino acid residues on one side are aligned with positively charged residues on the other side. Additionally, hydrogen bonds are formed between the exposed edge strands of the two β sheets (Fig. 5.37).

The atomic-packing at the interface of two subunits is usually as tight as the interior of an individual monomer. Some water molecules are simply trapped when the subunits associate, yet, several other water molecules play a stabilizing role forming hydrogen bonds with polar backbone and side chain groups, as with one another (Fig. 5.38).

The three-dimensional structure of a protein, as determined by X-ray crystallography (next chapter), is only an average structure. We have seen that the secondary and tertiary structures are maintained by weak interactions. Under given conditions, some of these interactions may be broken and new interactions of comparable energy

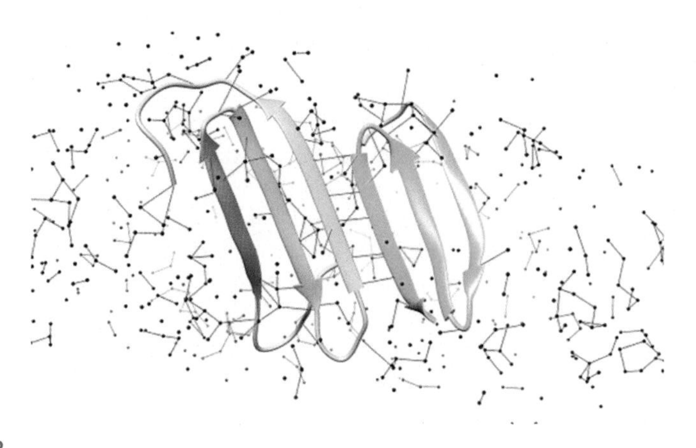

FIG. 5.38

Stabilization of β-sliding clamp dimer interface by water molecules. *Blue* dots are water molecules; *red* lines are hydrogen bonds.

Source: Oakley, A.J., et al., 2003. Acta Crystallogr. Sect. D 59, 1192.

formed. This gives the protein certain degree of flexibility. Furthermore, conformational fluctuations are inherent in a protein molecule due to the continuous motion of its constituent atoms.

5.2.6 Protein flexibility and dynamics

A protein must interact with other molecules (small molecules, nucleic acids, or other proteins) for its function. Molecular recognition plays a vital role in cellular processes. Originally, the concept of recognition of a ligand by a protein was based on a "lock and key" model in which all interacting molecules were considered to be "frozen and motionless." Subsequent experiments revealed the dynamic nature of protein-ligand interaction.

The flexibility and dynamic nature of protein molecules have led to the development of two alternative models for protein-ligand binding—induced fit and conformational selection. Both models seek to explain the mechanism of transition of a protein molecule from an unbound to a bound conformation with a ligand. In the induced fit model, the ligand first binds to the protein in its unbound conformation and "induces" the latter to shift to the bound state.

In the conformational selection model, the protein is in continuous transition between its stable unbound conformation and a less stable bound conformation even in the absence of a ligand. Statistically, the unbound conformation is more populated than the bound conformation. The ligand binds directly to the bound conformation. As a result, stability of the bound conformation is increased and there is a redistribution of population in its favor.

In certain cases of protein-ligand binding, the conformational model has been experimentally established. One example relates to the catalytic activity of the enzyme adenylate kinase (Adk). The active state of the enzyme, as determined by X-ray crystallography of its substrate-bound structure, is a high-energy closed conformation. More sophisticated experiments "captured" this scarcely populated high-energy state even in the absence of the ligand (Fig. 5.39). It was thereby proposed that the ligand-free "apo-enzyme" samples the scarcely populated functionally active closed conformation; substrate binding reorganizes the statistical weights in favor of the latter.

Another example suggests that conformational selection and induced fit are not mutually exclusive. The lysine-, arginine-, ornithine-binding (LAO) protein undergoes large-scale rearrangements from an inactive open to an active closed state upon binding its ligand. Simulation studies have identified a minor population of unliganded partially closed form of the protein in equilibrium with the open state. The partially closed state, which can weakly bind the substrate, is called the encounter complex state (Fig. 5.40). It has been proposed that the ligand can either induce the open state

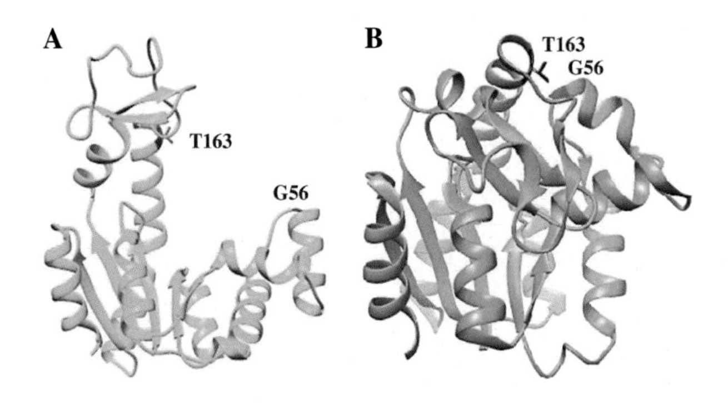

FIG. 5.39

Open (A) and closed (B) conformations of adenylate kinase.
Source: Muller, C.W., Schulz, G.E., 1992. J. Mol. Biol. 224, 159; Muller, C.W., et al., 1996. Structure 4, 147.

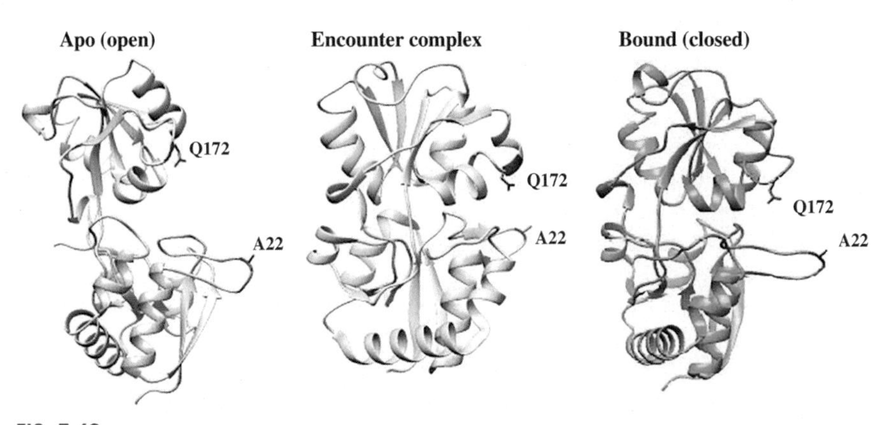

FIG. 5.40

Apo (open), encounter complex, and bound (closed) structures of LAO protein.
Source: Oh, B.H., et al., 1993. J. Biol. Chem. 268, 11348; Oh, B.H., Ames, G.F., Kim. S.H., 1994. J. Biol. Chem. 269, 26323; Vigourooux, A., Morera, S., PDB code: 6FT2. https://doi.org/10.2210/pdb6FT2/pdb.

rearrange itself to the encounter complex or bind directly to the encounter complex state. In either event, the induced fit mechanism is responsible for transition from the encounter complex to the closed state.

5.3 Carbohydrates

Simple sugars, known as monosaccharides, are the building blocks of carbohydrate polymers. Unlike DNA, RNA, and proteins which are linear polymers, carbohydrates are formed by both linear and branched combinations of a wide variety of naturally occurring monosaccharides. Further, each monosaccharide unit is capable of linking to another monosaccharide in a variety of different ways. As a result, carbohydrate polymers usually do not have an elegant hierarchical structure as found in proteins. Nevertheless, carbohydrate homopolymers, built with one or two kinds of monosaccharide units, do possess recognizable three-dimensional structures.

Monosaccharides, which are commonly known as sugars, are either aldehyde-aldoses or ketoses, depending on whether they possess an aldehyde or ketone functional group, with a general molecular formula $[C(H_2O)]_n$, where $n=3–7$. Based on the number of carbon atoms it contains, a monosaccharide can be called a triose ($C_3H_6O_3$), a tetrose ($C_4H_8O_4$), and so on. Glucose is a hexose, with six carbon atoms (Fig. 5.41), and an aldehyde-aldose monosaccharide. Fructose is also a hexose, but a ketose.

In solution, acyclic D-glucose is readily converted into cyclic five-member (example: β-D-glucofuranose) and/or six-member (example: α-D-glucopyranose) rings. One of the cyclic products is shown in Fig. 5.41.

Two monosaccharides can join together by means of an O-glycosydic bond to form a disaccharide (Fig. 5.42). Rotation is possible around both C—O and O—C of the glycosidic linkage. The spatial relation between the two residues is defined in terms of the dihedral angles ϕ and ψ (Fig. 5.42). When two to 20 monosaccharide residues are linked together, in a linear or branched mode, the resulting polymer is called an oligosaccharide. A carbohydrate polymer containing 20–100 residues is described as a polysaccharide (Fig. 5.42).

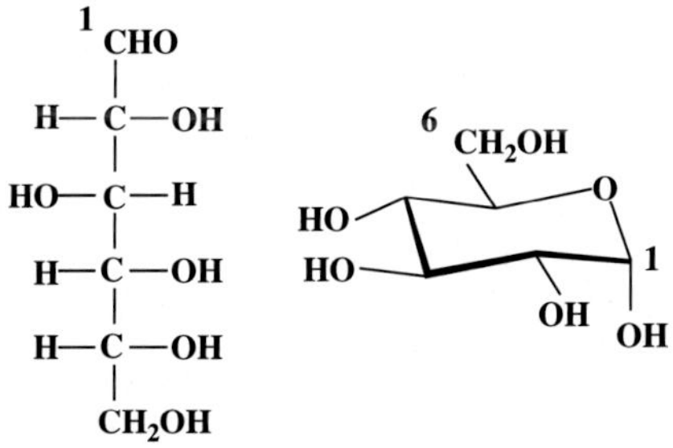

FIG. 5.41

D-Glucose and one of its cyclic products in solution.

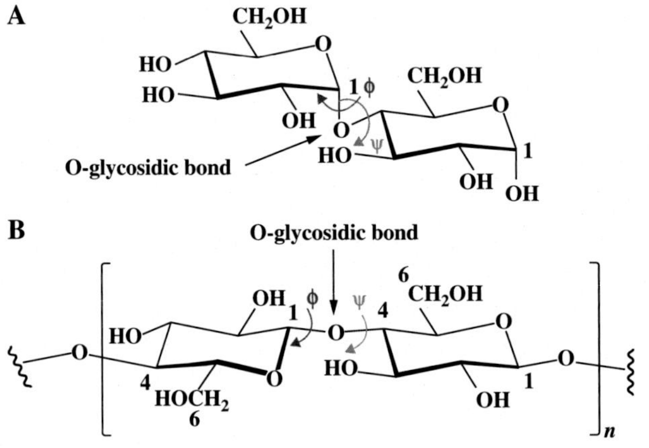

FIG. 5.42

A disaccharide (maltose) (A) and a polysaccharide (cellulose, containing n disaccharide repeats) formed by O-glycosidic bonds (B). Dihedral angles ϕ and ψ are shown.

Sample questions

1. How does base stacking contribute to the stability of the DNA double helix?
2. How are the major and minor grooves formed in DNA? How does the major groove ensure the specificity of protein-DNA interactions?
3. What is the basis of planarity of the peptide group?
4. What are the factors behind the formation of a β sheet?
5. What are the DNA-binding motifs found in a number of proteins? How do they bind DNA?
6. Why are certain α helices amphipathic?
7. How do polypeptides associate with each other in a multisubunit protein?
8. How are β sheets utilized in the formation of a dimer interface?
9. What are the two currently accepted models of protein-ligand binding? How do the models differ from each other?

References and further reading

Berg, J.M., Tymoczko, J.L., Gatto Jr., G.J., Stryer, L., 2015. Biochemistry, 8th ed. W. H. Freeman and Company.

Branden, C., Tooze, J., 1999. Introduction to Protein Structure, 2nd ed. Garland Science.

Fairman, J.W., Noinaj, N., Buchanan, S.K., 2011. The structural biology of β-barrel membrane proteins: a summary of recent reports. Curr. Opin. Struct. Biol. 21 (4), 523–531.

Harami, G.M., Gyimesi, M., Kovács, M., 2013. From keys to bulldozers: expanding roles for winged helix domains in nucleic-acid-binding proteins. Trends Biochem. Sci. 38 (7), 364–371.

Kovermann, M., et al., 2017. Structural basis for ligand binding to an enzyme by a conformational selection pathway. Proc. Natl. Sci. Acad. U. S. A. 114 (24), 6298–6303.

Levy, Y., Onuchic, J.N., 2006. Water mediation in protein folding and molecular recognition. Annu. Rev. Biophys. Biomol. Struct. 35, 389–415.

Miller, A., Tanner, J., 2008. Essentials of Chemical Biology: Structure and Dynamics of Biological Macromolecules. John Wiley & Sons.

Petsko, G.A., Ringe, D., 2004. Protein Structure and Function. New Science Press Ltd..

Purohit, A., et al., 2017. Electrostatic interactions at the dimer interface stabilize the *E. coli* β sliding clamp. Biophys. J. 113, 794–804.

Robinson, C.R., Sligar, S.G., 1993. Electrostatic stabilization in four-helix bundle proteins. Protein Sci. 2, 826–837.

Rohs, R., et al., 2010. Origins of specificity in protein-DNA recognition. Annu. Rev. Biochem. 79, 233–269.

CHAPTER

Structure analysis and visualization

6

In the previous chapter, we have described the structures of biological macromolecules. Here, in this chapter, the physical methods that are used to determine these structures will be considered. Three main experimental techniques that can be used to derive atomic-scale structural information of the macromolecules are X-ray crystallography, nuclear magnetic resonance (NMR) spectroscopy, and cryo-electron microscopy (cryoEM). No attempt will be made in this book to present a comprehensive account of any of the three techniques—only the basic principles will be discussed so that the reader can utilize this knowledge to venture into more advanced literature dedicated to each of these techniques.

Earlier, with the development and application of light microscopy, some of the biological structures such as bacteria (~1 μm), plant and animal cells (10–100 μm), cell nuclei (~5 μm), and mitochondria (0.5–10 μm) became directly visible. However, visible light microscopy had its limitation. As the nineteenth century scientist Ernst Abbe established, the resolving power of a microscope is proportional to half the wavelength of light used for visualization. In other words, as wavelength of visible light is in the range of 4000–7000 Å, the atomic and molecular fabric would remain "invisible" with visible light.

Atoms are around 0.1 nm or 1 Å in diameter and they are held in molecules by bonds whose length is typically 0.1–0.2 nm (1–2 Å). So, in order to "visualize" the atomic structure of a molecule, one has to go way down the wavelength spectrum. The wavelength of an electron accelerating through a potential difference of 100 V is close to 0.1 nm. We shall see in Section 6.3 how an electron microscope can be used to provide atomic level structural information.

6.1 X-ray crystallography

An alternative to the electron wave that has been more widely used so far to achieve atomic level resolution are the X-rays. These are electromagnetic radiation just like visible light; however, their wavelength is closer to that of electrons. X-ray microscopy is a formidable task and at a very nascent stage. Nonetheless, X-rays can be used in a different way to image atomic level structures based on a certain property of light diffraction. Diffraction and scattering are interrelated by wave-particle dualism. Diffraction is a wave phenomenon—when a wave encounters an obstacle there is a deviation in the direction of its propagation, accompanied by a change in its

Fundamentals of Molecular Structural Biology. https://doi.org/10.1016/B978-0-12-814855-6.00006-7

amplitude or phase, at the edge of the obstacle. This can be said as well for an electromagnetic wave interacting with an electron. The same phenomenon can also be seen as photon particles being deflected or scattered as a result of collision with an electron.

6.1.1 Scattering of X-rays by atoms

Here we shall consider elastic scattering of X-rays from electrons in which the incident radiation is absorbed by the scattering center and subsequently reemitted without any loss of energy. Let the point O in Fig. 6.1 represent a scattering center. When a monochromatic wave represented by the equation

$$y = A\cos(2\pi \nu t) \tag{6.1}$$

where A is the amplitude and ν the frequency, is incident on O in the direction as shown in the figure, it is scattered in all directions. The amplitude of the scattered wave at P (OP $= D$) is then given by

$$\eta(2\theta, D) = f_{2\theta}\frac{A}{D} \tag{6.2}$$

$f_{2\theta}$ is a constant of proportionality called the scattering length. The intensity of the scattered beam at P is

$$I_{2\theta} = f_{2\theta}^2 I_0 \tag{6.3}$$

where I_0 is the intensity of the incident beam on the scatterer.

Let us consider the scattering from a pair of identical scatterers located at points O_1 and O_2 and let P be a point at a distance very large compared to O_1O_2 (Fig. 6.2). The phase difference between the wave arriving at P from O_2 and that from O_1 is then

$$a_s = 2\pi \mathbf{r}\cdot\frac{(\mathbf{S}-\mathbf{S}_0)}{\lambda} \tag{6.4}$$

where $\mathbf{S}_0$ and $\mathbf{S}$ are two unit vectors along the directions of the incident and scattered waves, respectively, and $\mathbf{r}$ is the vector joining O_1 to O_2. Taking

$$\mathbf{s} = \frac{\mathbf{S}-\mathbf{S}_0}{\lambda}$$

we have

$$a_s = 2\pi \mathbf{r}\cdot\mathbf{s}$$

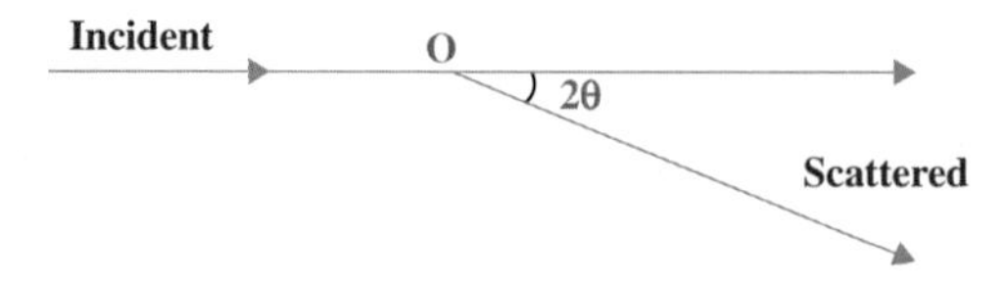

FIG. 6.1

Scattering of X-ray by a point scatterer.

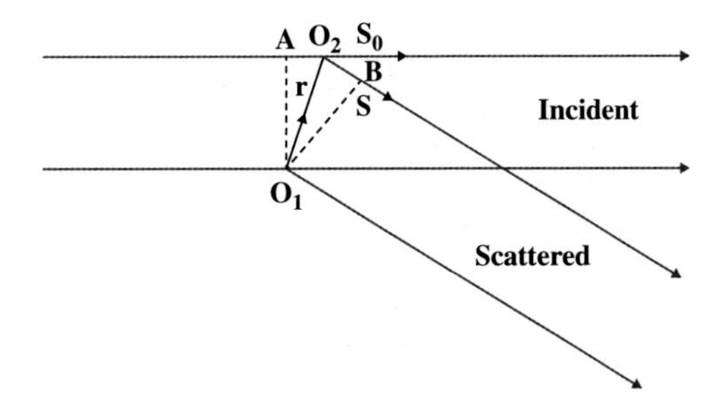

FIG. 6.2

Scattering by a pair of point scatterers.

If the positions of two identical scattering points O_1 and O_2 are given by vectors $\mathbf{r}_1$ and $\mathbf{r}_2$ with respect to the reference point O, the scattering amplitude can be expressed as

$$\eta_2 = \eta(2\theta, D)\{\exp(2\pi i\mathbf{r}_1 \cdot \mathbf{s}) + \exp(2\pi i\mathbf{r}_2 \cdot \mathbf{s})\} \tag{6.5}$$

where $\eta(2\theta, D)$ is the amplitude of scattering from a single unit.

This can be extended to the case of scattering from a general distribution of n scatterers where $\hat{\mathbf{r}}_\mathbf{j}$ denotes the position of jth scatterer. The amplitude of scattering corresponding to $\mathbf{s}$ at a large distance from the system of scatterers is then

$$\eta_n = \eta(2\theta, D)\sum_{j=1}^{n} \exp(2\pi i\mathbf{r}_\mathbf{j} \cdot \mathbf{s}) \tag{6.6}$$

for identical scatterers. If the scatterers are nonequivalent, the above equation is modified as

$$\eta_n(2\theta, D) = \sum_{j=1}^{n} [\eta(2\theta, D)]_j \exp(2\pi i\mathbf{r}_\mathbf{j} \cdot \mathbf{s})$$

$$= \frac{A}{D}\sum_{j=1}^{n} (f_{2\theta})_j \exp(2\pi i\mathbf{r}_\mathbf{j} \cdot \mathbf{s}) \tag{6.7}$$

Till now, we have considered free electrons as scatterers. However, in an atom the electrons are bound in discrete energy states and X-rays can be scattered in two ways. In one case, Thomson scattering, there is no change in the energy of the electrons and, therefore, the incident radiation. The scattered wave is coherent with the incident wave. In the other case, Compton scattering, the electrons can gain or lose energy by moving from one energy state to another, leading to incoherent scattering. Further, an atomic electron can be represented by a wave function ψ and electronic charge is not located at a point but distributed in accordance with

$$\rho(r) = |\psi|^2 \tag{6.8}$$

If the atom contains Z electrons, the total electron density will be given by

$$\rho_a = \sum_{j=1}^{Z} \rho_j(r) \tag{6.9}$$

The scattered amplitude from $\rho(r)$ is expressed as a fraction p_s of the amplitude that would be obtained due to a point electron at the origin. The amplitude of coherent scattering (Thomson scattering) due to the total electron density is expressed as the atomic scattering factor

$$f_a = \sum_{j=1}^{Z} (p_s)_j \tag{6.10}$$

It can be shown mathematically that under the conditions of diffraction of X-rays from crystals (that we are going to discuss next), incoherent scattering (Compton scattering) may generally be ignored.

6.1.2 Diffraction from a crystal

Having appreciated that X-rays can be scattered from atomic electrons, one can expect that by measuring the scattering from all the atoms in a molecule it should be possible to determine the structure of the molecule. However, it also needs to be realized that X-ray scattering from a single molecule will be extremely weak and can hardly be detected above the noise level. In order to be detectable, the signal requires to be amplified by periodic arrays of molecules (and thereby atoms) such as a crystal or a fiber.

A crystal is a periodic layout of a motif in a lattice—the motif can be a single atom, a small molecule, a macromolecule, or a combination of two or more of these. A typical protein crystal is known to contain more than 10^{14} aligned molecules. A crystal can also be seen as consisting of billions of identical units—called the unit cells—closely packed in three-dimensions (Fig. 6.3). A unit cell may contain one or more molecules.

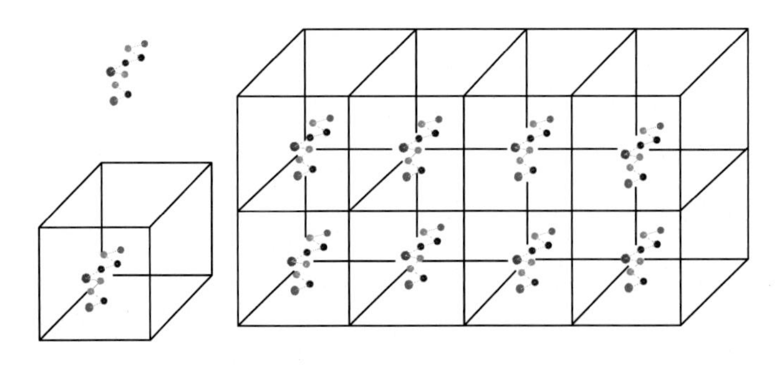

FIG. 6.3

Molecule, unit cell and crystal.

The scattered radiation from all the atoms will be coherent with respect to each other. Since a crystal organizes an enormous number of atoms in the same orientation, the scattered wave adds up in phase in certain directions while canceling out in others. As a result, the diffraction pattern is observed as an array of spots.

In order to analyze the diffraction of X-rays by a crystal, it will be useful to first look at an one-dimensional array of n atoms, each atom being separated from its neighbor by a vector distance $\mathbf{a}$. If we assume n to be odd and the atom at the center of symmetry as the coordinate origin, the amplitude of the scattered radiation corresponding to a scattering vector $\mathbf{s}$ at a distance large compared with the lattice dimension is given by

$$A_n = f_a \sum_{j=-\frac{1}{2}(n-1)}^{\frac{1}{2}(n-1)} \cos(2\pi j \mathbf{a}\cdot\mathbf{s}) \tag{6.11}$$

where f_a is the atomic scattering factor. From here, it can be shown that there will be significant diffracted intensity only when

$$\mathbf{a}\cdot\mathbf{s} = h \tag{6.12}$$

where h is an integer. Extending the concept to diffraction from a three-dimensional array of atoms which are separated by three vector distances $\mathbf{a}$, $\mathbf{b}$, and $\mathbf{c}$, we can arrive at the conditions of diffraction as

$$\begin{aligned} \mathbf{a}\cdot\mathbf{s} &= h \\ \mathbf{b}\cdot\mathbf{s} &= k \\ \mathbf{c}\cdot\mathbf{s} &= l \end{aligned} \tag{6.13}$$

which are known as the Laue equations. To find the scattering vector $\mathbf{s}$ that will satisfy the Laue equations for a given set of integers (hkl) let us consider the unit cell defined by the vectors $\mathbf{a}$, $\mathbf{b}$, and $\mathbf{c}$. We define a set of new vectors $\mathbf{a}^*$, $\mathbf{b}^*$, and $\mathbf{c}^*$ which satisfy the following set of relationship:

$$\begin{array}{lll} \mathbf{a}\cdot\mathbf{a}^* = 1 & \mathbf{a}\cdot\mathbf{b}^* = 0 & \mathbf{a}\cdot\mathbf{c}^* = 0 \\ \mathbf{b}\cdot\mathbf{a}^* = 0 & \mathbf{b}\cdot\mathbf{b}^* = 1 & \mathbf{b}\cdot\mathbf{c}^* = 0 \\ \mathbf{c}\cdot\mathbf{a}^* = 0 & \mathbf{c}\cdot\mathbf{b}^* = 0 & \mathbf{c}\cdot\mathbf{c}^* = 1 \end{array} \tag{6.14}$$

The vectors $\mathbf{a}$, $\mathbf{b}$, and $\mathbf{c}$ define a real space lattice—the corresponding reciprocal lattice is defined by $\mathbf{a}^*$, $\mathbf{b}^*$, and $\mathbf{c}^*$. The scattering vector $\mathbf{s}$ that will satisfy Laue equation will then be given by

$$\mathbf{s} = h\mathbf{a}^* + k\mathbf{b}^* + l\mathbf{c}^* \tag{6.15}$$

For a crystal with N atoms in the unit cell, the total scattered amplitude is given by

$$F_{hkl} = \sum_{j=1}^{N} f_j \exp(2\pi i \mathbf{r_j}\cdot\mathbf{s}) \tag{6.16}$$

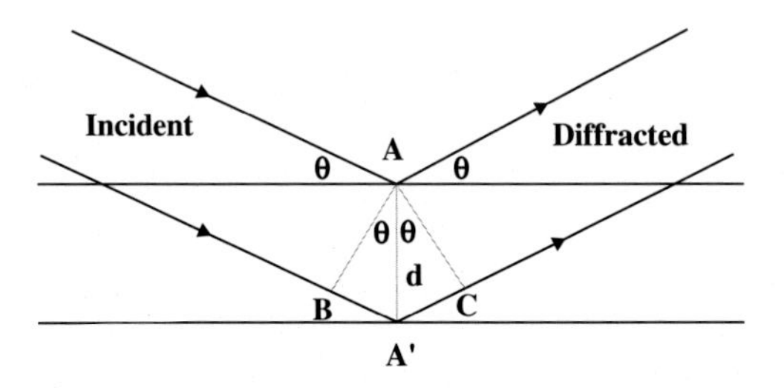

FIG. 6.4

Parallel rays reflected from points on neighboring planes.

where f_j is the scattering factor of the jth atom. F_{hkl} is a function of the content of a unit cell and is called the structure factor.

The actual positions of the X-ray diffraction spots are given by Bragg's law where diffraction is considered as partial reflection of the incident radiation by sets of parallel planes. The phenomenon can be simplified as in Fig. 6.4 where scattering from two atoms A and A′ located on two parallel planes separated by a distance d produces a phase difference of

$$(\mathrm{BA}' + \mathrm{A}'\mathrm{C})/\lambda - 2d \sin\theta/\lambda \tag{6.17}$$

Therefore, the scattered waves from the two atoms will reinforce each other if they are in the same phase, that is, if

$$2d \sin\theta = n\lambda \tag{6.18}$$

where n is an integer. This is the mathematical form of Bragg's law.

The above representation of Bragg's law can be extended to the case where there are N atoms in the unit cell and there will be multiple sets of reflecting parallel planes of spacing d_{hkl}. Now, if cell edges are denoted by the vectors **a**, **b**, and **c** and a scattering atom by coordinates (x, y, z) where x, y, and z being three fractional numbers in the range 0–1, then the position of the atom is

$$\mathbf{r} = x\mathbf{a} + y\mathbf{b} + z\mathbf{c} \tag{6.19}$$

and from Eq. (6.16) we have

$$\begin{aligned} F_{hkl} &= \sum_{j=1}^{N} f_j \exp\left[2\pi i\left(x_j\mathbf{a} + y_j\mathbf{b} + z_j\mathbf{c}\right)\cdot\left(h\mathbf{a}^* + k\mathbf{b}^* + l\mathbf{c}^*\right)\right] \\ &= \sum_{j=1}^{N} f_j \exp\left[2\pi i\left(hx_j + ky_j + lz_j\right)\right] \end{aligned} \tag{6.20}$$

Each diffraction spot can therefore be considered as created by a wave scattered from a plane (h, k, l) with an amplitude $|F_{hkl}|$ and relative phase α_{hkl}. Fourier transform will then give the electron density distribution as

$$\rho(\mathbf{r}) = \sum_{hkl} F_{hkl} \exp(-2\pi i \mathbf{r}\cdot\mathbf{s})$$
$$= \frac{1}{V}\sum_h \sum_k \sum_l |F_{hkl}| \exp(i\alpha_{hkl}) \exp[-2\pi i(hx+ky+lz)] \tag{6.21}$$

where V is the unit cell volume. h, k, and l are Miller indices which can be obtained from the position of the diffraction spot. The amplitude $|F_{hkl}|$ can be calculated from the measured intensity of the diffracted wave at the corresponding spot. However, the measurements are unable to provide any information on the relative phase α_{hkl}. This is precisely known as the "phase problem."

6.1.3 The phase problem

The phase problem was initially addressed by the Patterson function which is given as

$$P(\mathbf{r}) = \frac{1}{V}\sum_{hkl} |F_{hkl}|^2 \exp(-2\pi i \mathbf{r}\cdot\mathbf{s}) \tag{6.22}$$

Taking the real part, the above expression can also be written as

$$P(x,y,z) = \frac{1}{V}\sum_h \sum_k \sum_l |F_{hkl}|^2 \cos 2\pi(hx+ky+lz) \tag{6.23}$$

As indicated by the above equation, a map calculated using the Patterson function does not require the phase information. However, the map does not have peaks at the absolute atomic positions, but at the interatomic vectors. For molecules with relatively small number of atoms, it is possible to determine the original positions of the atoms from the observed Patterson peaks. Now, if there are N atoms in a unit cell, there will be N electron density peaks, but $N(N-1)$ interatomic vector peaks. So, for macromolecules (with $N > 1000$), the Patterson peaks will not be resolved from each other. Nonetheless, as we shall see next, the Patterson function will still be useful as a part of other methods to solve macromolecular structures.

Direct methods

The simplest way to solve the phase problem is by the application of direct methods which are based on spatial constrains on the electron density of the atoms in the crystal. In such case, it can be statistically shown that phase relationships exist between normalized structure factors and phases of three reflections (h, k, l), (h', k', l'), and (h–h', k–k', l–l') follow a triplet relation

$$\alpha_{-h,-k,-l} + \alpha_{h',k',l'} + \alpha_{h-h',k-k'\cdot l-l'} \cong 0 \tag{6.24}$$

Thus, if the phases of some reflections are either known, or given some initial trial values, then the phases of other reflections can be deduced. However, the application of direct methods in protein crystallography requires high resolution data (<1.2 Å) and they are more useful for small proteins (<1000 atoms).

Molecular replacement (MR)

Molecular replacement is useful when a structurally similar model is available, that is to say, a sequence identity of >25% and root-mean-square-deviation (rmsd) of <2.0 Å exist between the model and the unknown structure. From the atomic coordinates of the structurally similar model, amplitudes can be calculated and a Patterson map obtained. This Patterson map is matched with the Patterson map calculated from the structure-factor amplitudes of the new crystal to determine the orientation of the model in the new unit cell.

Isomorphous replacement

In the isomorphous replacement approach, heavy atoms are incorporated into protein crystals such that the shape and size of the unit cell and orientation of the protein in the cell remain unchanged. The heavy atom will enhance the scattered intensity significantly. The difference in the scattered intensities between that produced by the native crystal and the derivative crystal can be used to compute a Patterson map. Since there will be only a few heavy atoms, the Patterson map will be relatively easy to deconvolute. The contribution of the heavy atom to the structure factor can then be calculated. The structure factor of the derivative crystal $\mathbf{F}_{PH}$ is related to that for the native crystal $\mathbf{F}_P$ by the equation

$$\mathbf{F}_{PH} = \mathbf{F}_P + \mathbf{F}_H \tag{6.25}$$

where $\mathbf{F}_H$ is the heavy atom structure factor. The relation will, however, give two possible values for $\mathbf{F}_P$. This twofold phase ambiguity can be resolved by more than one heavy atom derivative in multiple isomorphous replacement (MIR).

Multiwavelength anomalous dispersion

Since the introduction of tunable (variable wavelength) beamlines of synchrotrons, multiwavelength anomalous dispersion (MAD) has replaced the MIR method to a large extent. We know that the electric component of an electromagnetic field induces an oscillation of electrons. In general, the electrons will oscillate with the same phase and scattering from an atom is largely independent of the wavelength. However, around the absorption edge (sharp discontinuity in the absorption spectrum) of an atom, where X-ray photon energy is close to the transition energy of some inner shell electrons, the scattering varies rapidly, both in amplitude and phase, with the wavelength. This is the phenomenon of anomalous scattering.

In the MAD technique, often selenomethionine (where selenium atoms replace the sulfur atoms) is used in place of methionine residues in a protein. The wavelength is varied around the absorption edge of selenium. The selenium atoms have a strong anomalous signal at the wavelengths of synchrotron X-rays based on which these atoms can be located. Subsequently, the phase information can be derived in the same way as for MR.

6.1.4 Temperature factor

The atoms in a crystal vibrate about their mean positions with amplitude that increases with temperature. However, the frequencies of vibrations are very low in the time-scale of measurements. Hence, the diffraction pattern is produced from a "frozen" crystal where the atoms are stationary but displaced from their mean positions. Consequently, the time-average of the pattern effectively smears an atom from a point to a sphere or ellipsoid of electron density. This is accounted for by including in Eq. (6.20) the temperature factor which for the jth atom is given by

$$B_j = 8\pi^2 \mu_j^2 \tag{6.26}$$

where μ is the mean square displacement in the three principal lattice directions. The structure factor can then be calculated as

$$F_{hkl,\text{calc}} = \sum_{j=1}^{N} f_j \exp\left[2\pi i\left(hx_j + ky_j + lz_j\right)\right] \exp\left(-\frac{B_j \sin^2\theta}{\lambda^2}\right) \tag{6.27}$$

Atoms in a small organic molecule have a mean displacement of about 0.1 Å, whereas protein atoms have mean displacement between 0.15 and 0.5 Å. Hence, protein atoms have higher B values.

6.1.5 Refinement

Once an electron density map is obtained from a crystal, an atomic model is attempted to be built by trial and error using dedicated computer graphics program. The backbone of the protein is "weaved" through the electron density map and the density of the side chains matched. Although the initial atomic model thus built is not expected to be accurate, it is possible to improve the model by a process called refinement. Structure-factor amplitude $|F_{hkl,\text{calc}}|$ is computed by using Eq. (6.27) and compared with the observed data$|F_{hkl,\text{obs}}|$ to obtain an R factor:

$$R = \frac{\sum_{hkl} \left[|F_{hkl,\text{obs}}| - |F_{hkl,\text{calc}}|\right]}{\sum_{hkl} |F_{hkl,\text{obs}}|} \tag{6.28}$$

The difference between the observed and calculated is minimized by least-squares adjustment of the model changing the parameters x, y, z, and B for each component atom. Completion of refinement requires the R factor to approach 0.2.

6.2 Nuclear magnetic resonance (NMR)

6.2.1 Basic principles

Structure determination by nuclear magnetic resonance (NMR) spectroscopy is based on a quantum property of the nucleus called the spin (discussed in Chapter 3). Like an electron, the nucleus also behaves as if it is spinning. Hence, it has an angular momentum. Also, the nucleus has a positive charge. Therefore, the spinning nucleus behaves like a current loop and generates a magnetic field (Fig. 6.5). It possesses a magnetic moment μ proportional to the spin I given as

$$\mu = \gamma(Ih/2\pi) \tag{6.29}$$

where γ, which is a characteristic of the nucleus, is called the gyromagnetic or magnetogyric ratio and h is the Planck constant. The direction of the magnetic moment is along the spin axis.

Most of the nuclei of biological interests such as ^{1}H, ^{13}C, ^{15}N, and ^{31}P have spin $I=1/2$. When the nucleus is placed in an external magnetic field B_0 (say) applied in the z-direction, it can exist in one of the two spin states corresponding to $I=+1/2$ (parallel or lower energy state) and $I=-1/2$ (antiparallel or higher energy state). This splitting of spin is called the Zeeman effect. μ is not exactly parallel to the z-axis, but precesses around B_0 with an angular frequency ω_0 (called the Larmor frequency). If μ_B be the component of the magnetic moment in the direction of the field B_0, then in the lower energy state, $E=-\mu_B B_0$, while in the higher energy state, $E=+\mu_B B_0$, the energy separation between the two states being (Fig. 6.6)

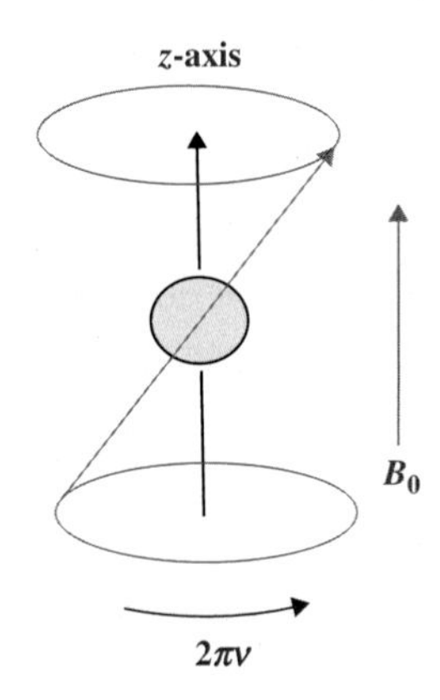

FIG. 6.5

A spinning nucleus in an external magnetic field.

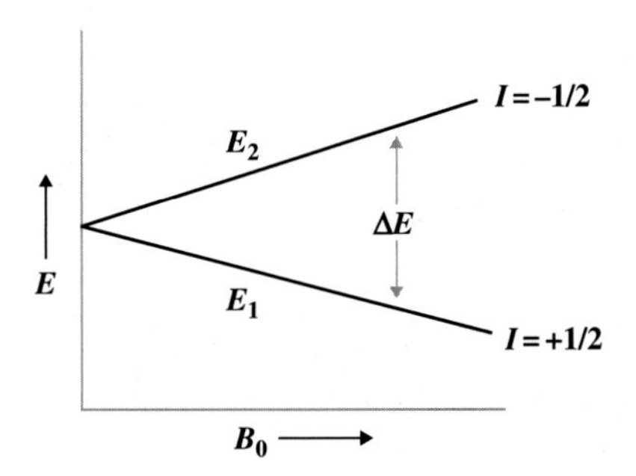

FIG. 6.6

Energy separation between the spin states.

$$\Delta E = 2\mu_B B_0 \tag{6.30}$$

Evidently, there will be a greater tendency of the magnetic moment to be aligned in the general direction of B_0 (or $+z$) over the opposite direction ($-z$). In the test sample where there are a huge number (in the order of Avogadro's number) of nuclei present, the distribution of population n between the two states is given by Boltzmann's law as

$$\frac{n\left(+\frac{1}{2}\right)}{n\left(-\frac{1}{2}\right)} = \exp\left(\frac{\Delta E}{kT}\right) \tag{6.31}$$

where k is Boltzmann's constant.

Now, if an oscillating electromagnetic field with an appropriate frequency be applied to the sample, there will be a tendency of the nuclei to absorb energy from the applied field and flip between the states. The resonant frequency will be given by

$$h\nu = \Delta E = 2\mu_B B_0 = (\gamma h B_0)/2\pi$$

or

$$\nu = (\gamma B_0)/2\pi \tag{6.32}$$

where h is the Planck constant.

6.2.2 Chemical shift

It is apparent from Eq. (6.31) that the resonance frequency of an atomic nucleus depends on the gyromagnetic ratio γ, which is a characteristic of the nucleus, and the magnetic field B_0. In a particular magnetic field, nuclei of different elements, possessing different gyromagnetic ratios, will produce signals at different frequencies. As for example, protons (^1H) will resonate at about ten times higher frequency than nitrogen nuclei (^{15}N). However, it may be noted that nuclei of the same type, when present in different electronic and structural environments, are also found to resonate at different frequencies in the same magnetic field. This is due to the variation in the local magnetic field.

It can be very easily conceived that the actual magnetic field B_{local} that a nucleus experiences is made up of the applied magnetic field B_0 and an additional contribution δB from the surrounding electrons and nuclei. To understand this phenomenon, we can consider the fact that the electron cloud that surrounds the nucleus has charge as well as motion and thereby a magnetic moment. Therefore, B_0 is modulated in the microenvironment of the nucleus. This electronic modification of the applied field is known as *shielding* which is quantitatively represented by σ. The actual field then becomes $B_0(1-\sigma)$ and the resonance frequency

$$\nu = \frac{\gamma B_0(1-\sigma)}{2\pi} \tag{6.33}$$

In other words, the resonance frequency undergoes a *chemical shift* as a result of shielding. The chemical shift is usually expressed in parts per million (ppm) by a dimensionless number

$$\delta = \frac{\nu_{\text{sample}} - \nu_{\text{ref}}}{\nu_{\text{ref}}} \times 10^6 \tag{6.34}$$

where ν_{ref} is a reference frequency.

Chemical shift measurements of nuclides $^1H_\alpha$, $^{13}C_\alpha$, $^{13}C_\beta$, and ^{15}N can be used to determine secondary structures of proteins. It has been shown that the chemical shifts of these nuclides are strongly correlated to the backbone dihedral angles. The procedure to identify α-helices, β-strands, and random coil regions in proteins based on chemical shift differences is known as chemical shift index (CSI).

6.2.3 *J*-Coupling

A more elegant method to determine dihedral angles is by the measurement of 3-bond scalar coupling constant (*J*). Scalar or *J*-coupling is an indirect dipole-dipole interaction between two nuclear spins mediated by bonding electrons. The nuclear spin polarization of one nucleus (say A) polarizes the surrounding electrons which, in turn, orient the coupling nucleus (say B) either parallel or antiparallel to A. Thus, the coupling leads to splitting of the spectral lines for both coupled spins by an amount J, which is called the coupling constant and denoted as $^nJ_{\text{AB}}$ where n (usually 1, 2, or 3) is the number of intervening bonds and A and B are the two coupled spins. Of particular interest is the 3-bond proton-proton (^{1}H-^{1}H) coupling in H-N-C_α-H where the two protons are separated by three chemical bonds (Fig. 6.7).

The *J*-coupling here is dependent on the dihedral angle φ (Fig. 6.7) and given by the Karplus relation

$$^3J = A\cos^2\varphi + B\cos\varphi + C \tag{6.35}$$

where A, B, and C are constants. Evidently, the coupling can provide conformational information on the peptide.

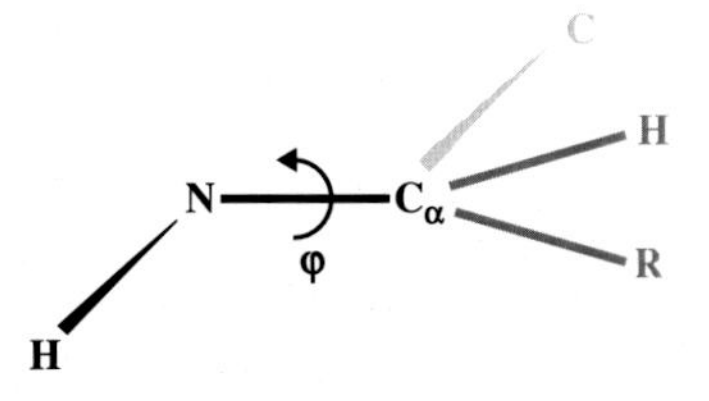

FIG. 6.7

Torsional angle (φ) between amide proton and alpha proton.

6.2.4 Nuclear overhauser effect (NOE)

The most effective way to determine distance constrains by NMR is based on ^1H-^1H NOE, which arises from dipole-dipole cross-relaxation through space. When two spins (normally protons) are very close in space (separated by <5 Å), their magnetic dipoles interact with each other and each spin affects the relaxation rate of the other. The cross-relaxation rate is proportional to $1/r^6$ where r is the interatomic distance.

The NOE can be used to estimate the distance between two nuclei (protons) in a protein. In particular, it is effective in determining the secondary structures since inter-strand proton-proton distances in a β sheet can be <4 Å, while in an α-helix the closest proton-proton distances are ~5 Å. Determination of tertiary distances by the application of NOE is somewhat more complex, though possible.

6.2.5 Structure computation

Once the NMR spectroscopic data are available, it is imperative to calculate the 3D structure of the molecule. Needless to say, since one is dealing here with a huge amount of data for a macromolecule (say protein), suitable computer programs are employed for this task. The NMR-derived conformational restraints—inter-proton distances from NOE spectroscopy (NOESY) and dihedral restraints from coupling constants and chemical shifts—constitute the input to these programs. Additionally, the computation also takes into account the restraints based on the covalent structure of the macromolecule such as bond lengths and bond angles and atomic properties—radius, mass, partial charge, etc.

Even before embarking on complex computational endeavor, one needs to keep in mind that the NMR-derived constraints, with their inherent limitations, cannot reach an exact and unique 3D structure. However, repeated calculations sampling the complete conformational space, permitted by and not violating the input constraints, can generate a sufficiently useful ensemble of structures.

The two most common methods for the computation of 3D structures are based on (a) distance geometry (DG) and (b) restrained molecular dynamics (rMD). The DG algorithm uses the NMR distance constraints data and chemical data such as bond length, van der Waals' radii, and planarity of aromatic rings to determine the ensemble of 3D structures. It works with a random distance matrix in an n-dimensional space (where n is a very large number) to determine the configurations that satisfy a given set of constraints. Subsequently, the n-dimensional distance space is projected on to a 3D space to arrive at a conformation compatible with the experimental data.

The second method, rMD, is a later development in structure determination. It is based on classical mechanics and involves the solution of Newtonian equations of motion for a many-particle system where the total potential energy is the sum of physical potentials (arising from chemical bonds, van der Waals forces, etc.) and pseudo-potentials from NMR-derived constraints (such as those related to inter-nuclei distances and dihedral angles). The motion of the molecule is simulated for a long time by "heating" so that the system acquires sufficient energy to overcome false local minima and samples a large region of the conformational space. Subsequently, by energy minimization (gradually "cooling"), the stable 3D structure is generated.

6.3 Cryo-electron microscopy

Although X-ray crystallography is still the most widely used method to determine the three-dimensional structure of macromolecules, lately a very powerful and promising technology, which is if not replacing but supplementing in a major way the X-ray crystallographic approach, is three-dimensional electron microscopy or cryo-electron microscopy (cryo-EM). Cryo-EM has two distinct advantages over X-ray crystallography. One, it can deal with molecules that are difficult to crystallize. Secondly, in contrast to the crystallographic approach where the crystals fix proteins in a single static posture, in cryo-EM a protein can be flash-frozen in several formations, thus offering a better clue to its working mechanisms.

6.3.1 Physical principles

Just as the light microscope extends the resolving power of the human eye to visualize objects, the electron microscope extends the resolving power of the light microscope. Both techniques are, in fact, based on the principle of wave propagation. In light microscopy, visible electromagnetic waves or photons are made to pass through the object and then refracted through optical lenses fabricated with glass to form an image (Fig. 6.8).

In an electron microscope, electrons are emitted by a source in a high vacuum and accelerated down the microscope column towards the sample by electrical potential differences in the range of 100–300 keV. Now, the electrons possess a wavelike character with their wavelength given by Louis de Broglie as

$$\lambda = h/p = h/mv \tag{6.36}$$

where $h = 6.626 \times 10^{-34}$ J s is the Planck constant, while m, v, and p are, respectively, the mass, velocity, and momentum of an electron. They interact with atoms as waves and diffracted either from the regular array of atoms on the surface of a

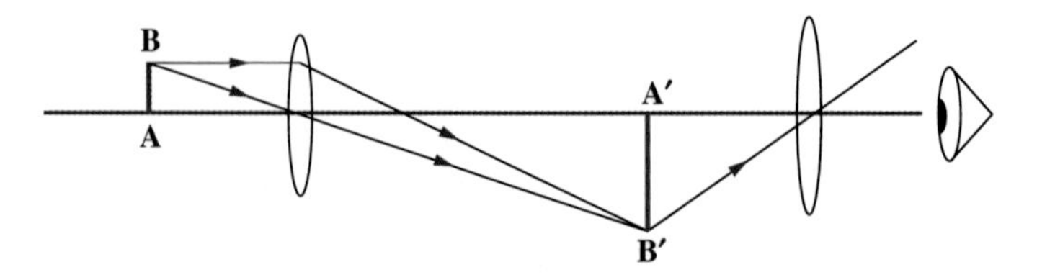

FIG. 6.8

Schematic diagram of a simple light microscope.

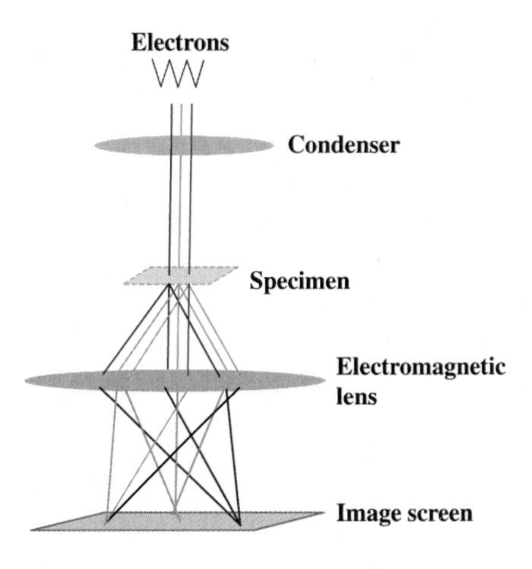

FIG. 6.9

Schematic diagram of a transmission electron microscope.

crystal or by atomic planes inside the sample. At the same time, the electrons also behave as particles, negatively charged, and can, therefore, be deflected by electric or magnetic fields. This property allows the electrons that pass through the sample to be focused by electromagnetic lenses to produce an image, as in a transmission electron microscope (Fig. 6.9).

We have seen earlier that the resolving power of an imaging technique depends on the wavelength of the imaging radiation or wave—shorter wavelengths giving higher resolutions. Visible light has wavelength in the range 4000–7000 Å, whereas an electron accelerated through a potential difference of 300 kV has a wavelength of ~0.02 Å. Clearly, the electron microscope can achieve several orders of higher resolution than the light microscope. Hence, it may be conceived that wavelength should not be a limiting factor for imaging a molecule in electron microscopy.

6.3.2 Problem of imaging biomolecules

Notwithstanding its resolving capability, there is a problem in using traditional electron microscopy to image biomolecules in their native state. This is essentially due to the nature of interaction of electrons with organic matter. As electrons pass through a specimen, individual electrons may or may not be scattered by the atoms of the specimen. The scattering can be elastic with no loss of energy of the electron or inelastic where energy lost by the electron is deposited in the specimen. The inelastic scattering leads to the damage of the specimen through breakage of chemical bonds and generation of free radicals.

Earlier, the problem of electron-induced damage was addressed by either negatively staining the sample surface with heavy atoms that are less radiation-sensitive or stabilizing proteins by embedding in sugars such as trehalose. However, the first method largely misses out the internal structural information of the sample. The second approach, though producing internal structural information of the unstained sample with low electron doses, at the same time leads to poor signal-to-noise ratio.

Two approaches, not mutually exclusive, have been adopted to improve signal-to-noise ratio without excessive specimen damage. In one, the specimen is frozen in either liquid nitrogen or liquid helium, which is the cryogen—hence the name cryo-electron microscopy. This was based on earlier observations that cryogenic temperatures could reduce the effects of radiation damage. A drop of the sample is put on an electron microscope grid and quickly plunged into the cryogen. If the sample is sufficiently thin and the plunging rapid, the sample is frozen-hydrated in a close-to-native state in vitreous ice. The process is also called vitrification.

The second approach to improve signal-to-noise ratio depends on averaging of thousands of low-dose images of identical units. This is analogous to the situation in X-ray crystallography where scattering by billions of molecules is averaged to produce useful structural information.

6.3.3 Three-dimensional electron microscopy

There are three variants of three-dimensional electron microscopy—cryo-electron tomography (CET), electron crystallography, and single particle analysis (SPA). Each of these variants has been used either singly or in combination with X-ray crystallography and NMR data.

Cryo-electron tomography

In general, tomography is an imaging strategy where a higher dimensional reconstruction of an object is computed from a set of lower dimensional projections. A special case in medical application is computerized axial tomography (CAT) scan where a three-dimensional model of a patient is derived from two-dimensional projections obtained with X-rays.

In traditional electron microscopy (EM), two-dimensional projections are produced from thin specimen slices. In CET, the plunge-frozen sample is loaded on the microscope and a series of projection images, known as "tilt-series," are recorded by incrementally varying the orientation of the sample relative to the incident beam (Fig. 6.10A).

From these two-dimensional projection images, a three-dimensional structure of the object is reconstructed by what is known as a "back projection" method. For each projection image, a "back projection" box is computationally created by smearing back the density along the direction of the electron beam (Fig. 6.10B). Back projection profiles of the entire tilt-series are then superimposed to produce the density distribution of the original object.

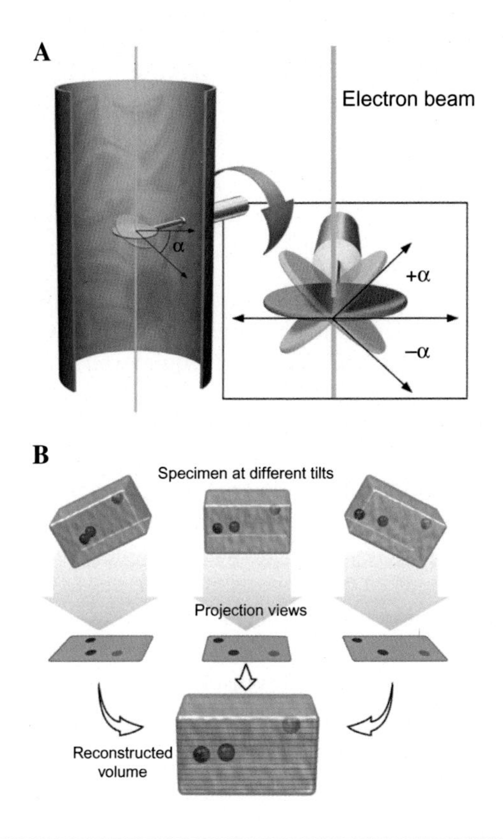

FIG. 6.10

Three-dimensional tomographic reconstruction. (A) Recording of "tilt-series" of image projections. (B) Three-dimensional reconstruction from "back projection" profiles.

Courtesy: Subramaniam, S., et al., 2003. National Cancer Institute, National Institutes of Health, Bethesda, Maryland, USA, ASM News 69, 240.

Electron crystallography

Electron crystallography is similar to X-ray crystallography in the way that in both cases a macromolecular crystal scatters a beam to produce a diffraction pattern from which the structure of the molecule can be determined. However, since electrons interact with the atoms in the crystal more strongly than X-rays, relatively smaller and thinner crystals can give useful diffraction data with an electron beam.

Electron crystallography has been successfully applied to determine the structure of proteins from two-dimensional (2D) crystals which are formed by single layers of molecules ordered in the x–y plane or a protein crystallized within a lipid bilayer. However, high energy electrons cause radiation damage to such 2D samples and each crystal can produce only a single diffraction pattern. Nonetheless, electron diffraction data from hundreds of individual crystals can be merged to determine the macromolecular structure.

Lately, technical improvements have made possible for electron crystallography to work with small three-dimensional crystals as well. This new technique, called microelectron diffraction (MicroED), can work with lower electron dose, thus reducing radiation damage to the sample. Therefore, it is possible to collect nearly 100 diffraction patterns from the same crystal.

Single-particle analysis

SPA is perhaps the most widely used variant of cryo-EM. It is a noncrystallographic method dealing with randomly oriented single molecules. Evidently, unlike electron crystallography, single-particle EM does not require well-ordered 2D crystals which are often not very easy to grow.

As in other cryo-EM techniques, single-particle imaging also requires vitrified samples. The specimen is plunge-frozen to form a thin layer or vitreous ice containing identical copies of the test molecule or molecular complex arranged in different orientations. Then, from a microscopic field containing a number of molecules/complexes, individual particles are selected for imaging. From the individual particle images, clusters are identified on the basis of the degree of relatedness. The related images are then averaged to obtain a characteristic projection. From all such characteristic projections, which correspond to a wide range of orientation relative to the electron beam, a 3D structure of the object is reconstructed. SPA has been further elaborated in Chapter 17.

6.4 Representation of structures

6.4.1 Structural data

Analysis of macromolecules, proteins in particular, using X-ray crystallography (XRC), nuclear magnetic resonance (NMR), or electron microscopy (EM) produces dataset containing atomic coordinates. The dataset is available in the Worldwide Protein Data Bank (wwPDB) as a PDB file. Fig. 6.11 shows a small part of the PDB file for *Escherichia coli* Klenow fragment containing the location of the atoms in Cartesian coordinates (x, y, z). The second column is the atom serial number, the third column the atom name (in up to four-letter code), and the fourth column the amino acid residue. The fifth and the sixth columns are, respectively, chain and residue number identifiers. The x, y, and z coordinates of the atoms are, respectively, presented in the next three columns. Occupancy and temperature factors, which will be discussed towards the end of the chapter, are indicated by the numbers in the tenth and eleventh columns, respectively.

The data provide significant insight into the three-dimensional structure of the molecule. However, in order to use the data efficiently to understand and address

```
ATOM     65  N   MET A 324     78.046  35.224  76.430  1.00 36.59           N
ATOM     66  CA  MET A 324     77.448  35.971  77.559  1.00 36.04           C
ATOM     67  C   MET A 324     76.128  36.560  77.121  1.00 36.94           C
ATOM     68  O   MET A 324     75.957  36.910  75.954  1.00 36.40           O
ATOM     69  CB  MET A 324     78.367  37.150  77.941  1.00 37.70           C
ATOM     70  CG  MET A 324     77.928  37.949  79.172  1.00 43.16           C
ATOM     71  SD  MET A 324     78.630  37.275  80.698  1.00 49.00           S
ATOM     72  CE  MET A 324     77.898  35.585  80.685  1.00 48.22           C
ATOM     73  N   ILE A 325     75.188  36.632  78.056  1.00 37.07           N
ATOM     74  CA  ILE A 325     73.924  37.282  77.737  1.00 33.37           C
ATOM     75  C   ILE A 325     74.148  38.757  78.133  1.00 33.80           C
ATOM     76  O   ILE A 325     74.785  39.061  79.157  1.00 32.70           O
ATOM     77  CB  ILE A 325     72.717  36.711  78.552  1.00 30.39           C
ATOM     78  CG1 ILE A 325     72.357  35.308  78.055  1.00 24.22           C
ATOM     79  CG2 ILE A 325     71.486  37.617  78.375  1.00 26.46           C
ATOM     80  CD1 ILE A 325     71.736  35.279  76.697  1.00 22.94           C
ATOM     81  N   SER A 326     73.763  39.658  77.238  1.00 32.52           N
ATOM     82  CA  SER A 326     73.907  41.072  77.507  1.00 30.04           C
ATOM     83  C   SER A 326     72.521  41.686  77.492  1.00 31.28           C
ATOM     84  O   SER A 326     71.511  40.982  77.424  1.00 29.50           O
ATOM     85  CB  SER A 326     74.795  41.751  76.456  1.00 30.61           C
```

FIG. 6.11

Part of a PDB file.

Source: Brautigam, T.A, Steitz, C.A., 1998. J. Mol. Biol. 277, 363–377.

biological questions, they need to be made visual with the help of appropriate 'metaphor'. Three of the most widely used 'metaphors' are lines, spheres, and ribbons.

6.4.2 Line representation

The line metaphor has been used for organic molecules for a long time. The bonds in the covalent structures are very well displayed by lines. Illustrated in Fig. 6.12 below for the organic molecule phosphomethylphosphonic acid adenylate ester (ACP) are its wireframe, stick, and ball-and-stick representations. The simplest representations are thin lines or wireframe in which the atoms are shown as dots and bonds as lines (Fig. 6.12A). The sticks are relatively thicker lines representing the bonds with atoms as "endcaps" (Fig. 6.12B). The ball-and-stick is a variation of the stick representation where the atoms are shown as small spheres (balls) (Fig. 6.12C). It is clear that in these representations the molecular electrons have not been considered.

The line representation has been later extended to macromolecules (proteins) as well. As in the case of small molecules, three-dimensional structures of biomolecules can also be represented by wireframe, stick, or ball-and-stick (Fig. 6.13). Each atom and each bond in the molecule are clearly visualized. However, with a large number of amino acids in a protein molecule and, correspondingly, even several-fold larger number of atoms, the line representation of a biomolecule is information-overloaded. As for example, *E. coli* DNA polymerase I Klenow

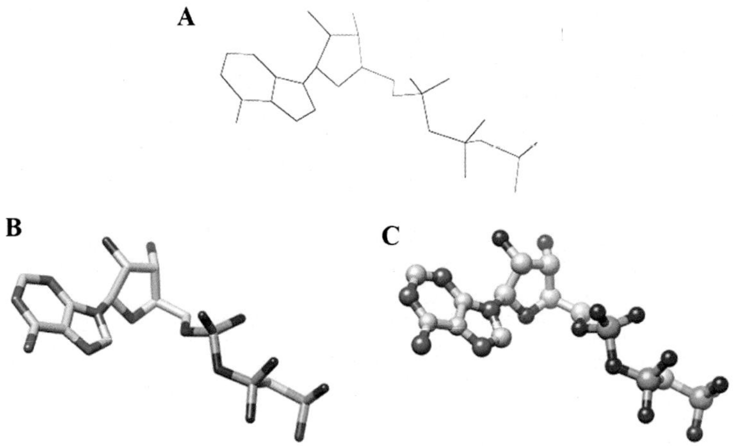

FIG. 6.12

(A) Wireframe, (B) stick, and (C) ball-and-stick representations of ACP.

fragment contains nearly 5000 atoms. As a consequence, it is difficult to interpret the three-dimensional structure particularly in relation to the function. In the ball-and-stick representation, the balls are not scaled according to the size of the atoms and, therefore, the dimension and geometry of interaction between two molecules will not be clear from this model.

6.4.3 Spheres—CPK representation

The distribution of electrons in the molecular atoms is accounted for in the space-filling representation. Here the radii of the spheres are proportional to the van der Waal (VDW) radii (discussed in Chapter 4) of the atoms. Space-filling models are also called Corey-Pauling-Koltun (CPK) model, which follows the CPK color scheme: black or gray—carbon, white—hydrogen, red—oxygen, blue—nitrogen, and orange—phosphorous. The space-filling model shows the overall shape and size of the molecule. However, it obscures the chemical bonds between the atoms and the internal structure of the molecule. Fig. 6.14 shows the space-filling model for ACP and the Klenow fragment.

6.4.4 Ribbon representation

Since both nucleic acids and proteins are primarily composed of linear chains that fold into three-dimensional structures, the topology of proteins is best depicted by the ribbon representation and nucleic acids by a little modified ribbon representation.

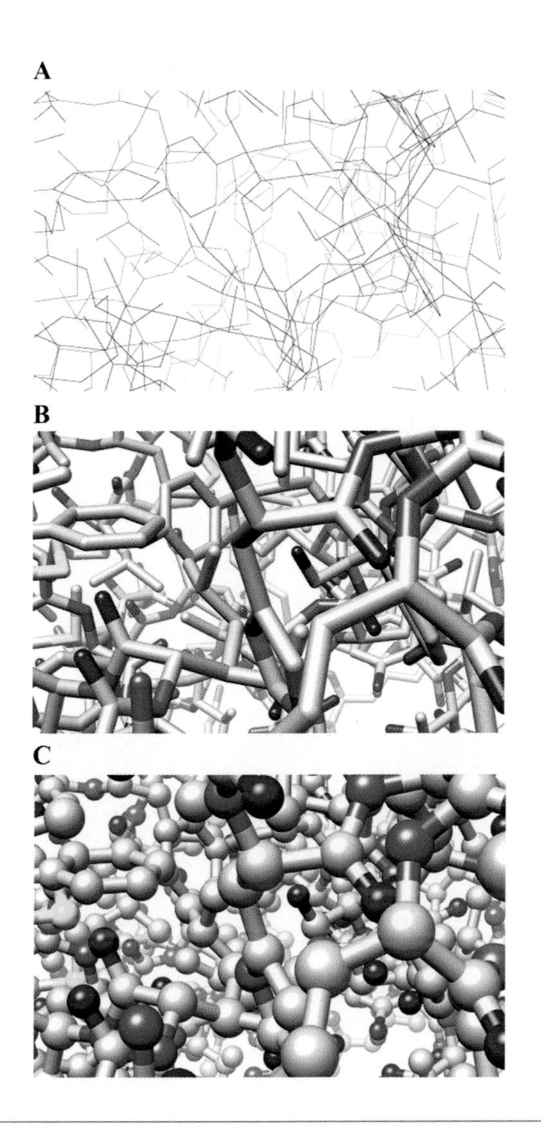

FIG. 6.13

(A) Wireframe, (B) stick, and (C) ball-and-stick representations of a part of *E. coli* Klenow fragment.

Source: Brautigam, T.A., Steitz, C.A., 1998. J. Mol. Biol. 277, 363–377.

For a protein, the basic ribbon hides the side chain atoms and interpolates itself as a smooth curve through the polypeptide backbone (Fig. 6.15A). It consists of coils (or thick tubes), arrows (with amino- to carboxy-polarity), and lines (or narrow tubes) representing α-helices, β-strands, and nonrepetitive coils (or loops), respectively. Further details of the atomic structure can be globally or selectively added to the basic framework. As for example, side chains can be added for selected amino acid residues (Fig. 6.15B).

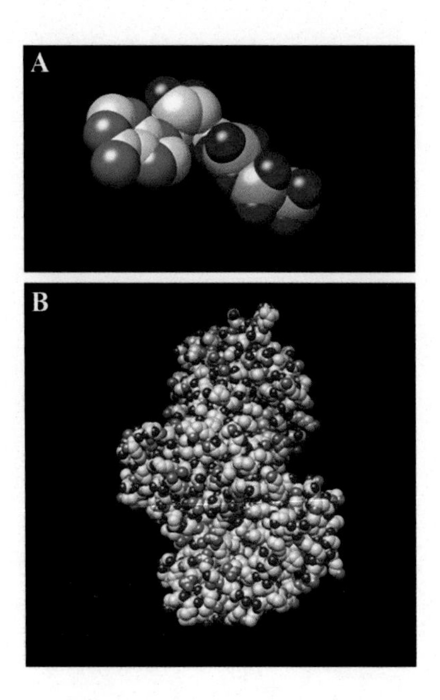

FIG. 6.14

Space-filling models of (A) ACP and (B) *E. coli* Klenow fragment.

Source: Brautigam, T.A., Steitz, C.A., 1998. J. Mol. Biol. 277, 363–377.

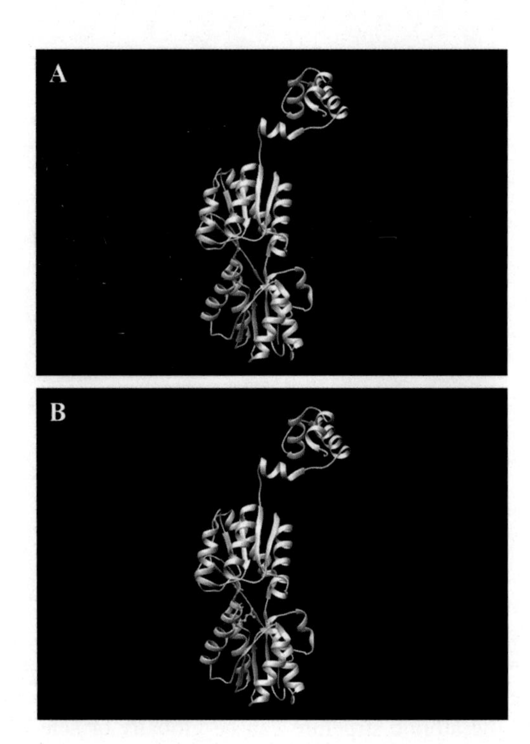

FIG. 6.15

Ribbon representation of Lac repressor structure: (A) basic ribbon and (B) side chains shown for selected residues *(cyan)*.

Source: Bell, C.E., Lewis, M., 2000. Nat. Struct. Biol. 7, 209.

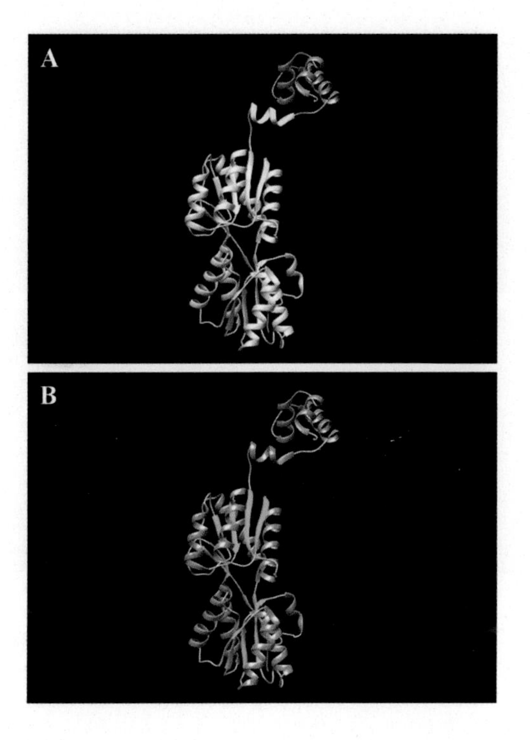

FIG. 6.16

Structural representation of Lac repressor colored by (A) different domains: head *(coral)*, hinge *(yellow)*, catalytic *(light blue)*; and (B) secondary structures: α-helices *(coral)*, β-strands *(green)*.

Source: Bell, C.E., Lewis, M., 2000. Nat. Struct. Biol. 7, 209.

Functional insight can be facilitated by using different colors for different secondary structure or different domains. Fig. 6.16A shows a basic ribbon for *E. coli* lac repressor protein monomer containing the head, hinge, and catalytic domains (devoid of the tetrameric domain). In Fig. 6.16B, the secondary structures have been distinguished by different colors.

6.4.5 DNA

The original DNA ladder structure of Watson and Crick is similar to the stick representation for protein structures (Fig. 6.17A). Space-filling models have also been widely used to indicate the shape of double helical DNA, particularly to compare *A*-, *B*-, and *Z*-forms of DNA (Fig. 6.17B). Sometimes, the sugar-phosphate backbone, which does not show much variability, is drawn as a smooth ribbon, whereas the bases, whose arrangements are more informative, are represented by sticks

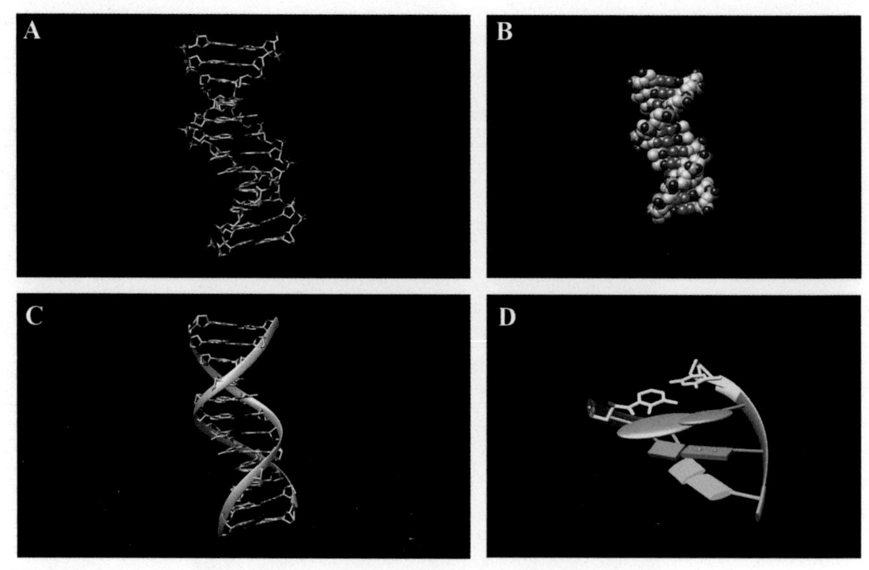

FIG. 6.17

Representation of DNA. (A) atomic model- stick representation, (B) space-filling model, (C) Backbone ribbon and stick representation of bases, and (D) different base representations—stick, filled ellipsoid, filled box, and filled elliptical tube.

Adapted from: Couch, G.S., Hendrix, D.K., Ferrin, T.E., 2006. Nucleic Acid Res. 34(4), e29.

(Fig. 6.17C). Further, since base stacking is a key structural feature in stabilization of nucleic acids, other options to represent the bases are also used to emphasize this feature (Fig. 6.17D).

6.4.6 Surface representation

The backbone, represented by either lines or ribbons, is useful for comparing protein structures. But neither of these can provide significant insight into macromolecular interactions—as for example, a protein with another protein or a small molecule.

Proteins interact via their surface and, therefore, to understand the detail of the interaction, specialized surfaces had to be defined with the space-filling model as the starting point. The surfaces were then refined by the incorporation of features that are responsible for the interactions.

Two best-defined surfaces are the solvent accessible surface developed by Lee and Richards and the molecular surface (also called the solvent excluded surface)

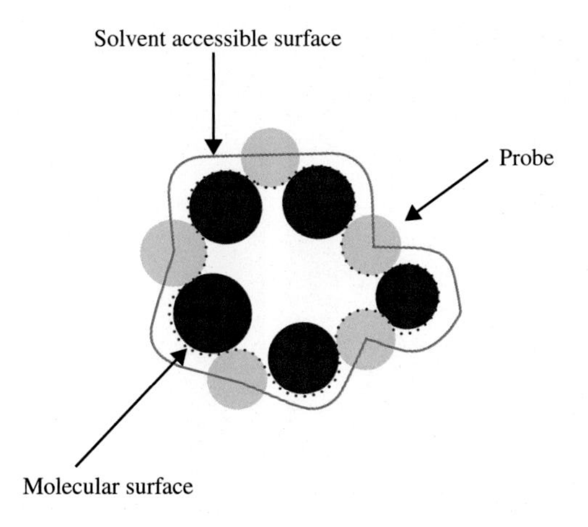

FIG. 6.18

Protein surface definitions. Spherical atoms shown in *brown* and spherical probes shown in *light green*.

introduced by Connolly (Fig. 6.18). For either of them, a space-filling representation of the macromolecule is first created. Then a spherical probe (usually representing a water molecule) is rolled over the surface (Fig. 6.18). The set of all points that the center of the probe can occupy forms the solvent accessible surface. On the other hand, to construct the molecular surface, the set of all points in the space-filling representation which the probe touches is added to the set of points on the probe surface in the "cracks and dips" between the contact regions.

6.4.7 Surface hydrophobicity and charge

It is known that the side chains of certain amino acids such as alanine, leucine, and methionine are nonpolar in nature. These amino acids, when present in close vicinity of each other, create hydrophobic patches on the surface of a protein. The hydrophobicity of a protein surface plays an important role in its stability due to favorable solvent-mediated free energy of aggregation as also in its interaction with the hydrophobic regions of other molecules. Hence, the "hydrophobicity representation" of the surface of a protein is informative in understanding its behavior. In Fig. 6.19A, the hydrophobicity surface of the Klenow fragment has been shown.

Similarly, we also know that amino acids such as arginine and lysine are basic or positively charged, whereas aspartic acid and glutamic acid are acidic or negatively charged in character. The electrostatic potential of a protein surface evidently depends on the distribution of such amino acids on the surface. Depiction of

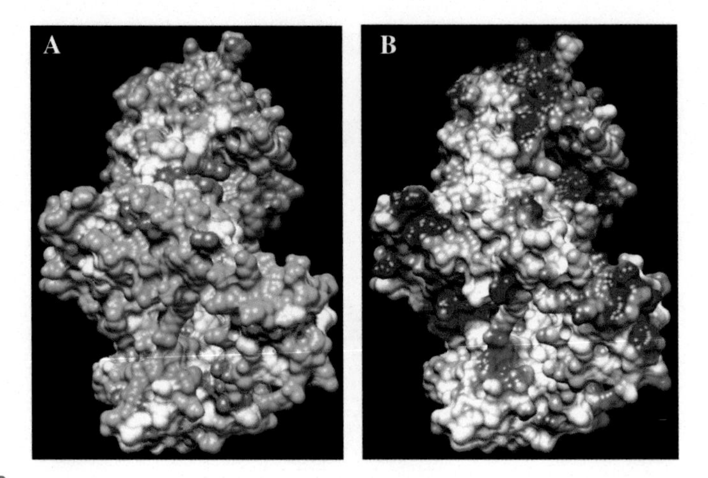

FIG. 6.19

Surface representations of the Klenow fragment. (A) Hydrophobic surface: *blue*—most hydrophilic and *orange*—most hydrophobic. (B) Electrostatic surface: *blue*—positive and *red*—negative.

Source: Brautigam, T.A., Steitz, C.A., 1998. J. Mol. Biol. 277, 363–377.

electrostatic or Coulombic potential by surface coloring (Fig. 6.19B) can give a clue to the mechanisms of interaction of the protein with charged groups of other molecules. In addition, increased solubility of a protein has been correlated with increased negative surface charge.

6.4.8 Volume representation

In the volume-based representation, a probe scans the entire three-dimensional space in and around the molecule under consideration. The method displays volumetric function $f(\mathbf{r})$ which is continuous in space. Among such functions, the electron density distribution $\rho(\mathbf{r})$, which is a scalar function obtained from X-ray crystallographic analysis, is relatively simpler to represent. Representation of vector quantities is far more challenging.

The electron density map spanning a small segment of *Thermus aquaticus* (*Taq*) DNA polymerase including the amino acid residue Tyr671 is shown in Fig. 6.20. This residue, which we shall see later in Chapter 8, has an important role in the polymerase action. Electron density is usually represented by isovalue meshes (surfaces) which appear like "chicken-wire." The region enclosed by the mesh surface possesses density higher than a given threshold. Use of the mesh instead of a solid surface allows the visualization of fit of the underlying atoms in the electron density map.

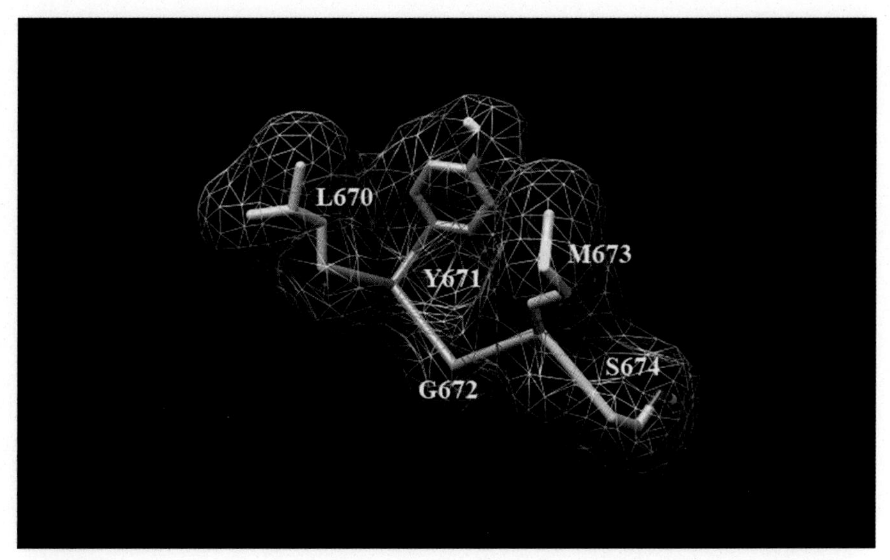

FIG. 6.20

Electron density map of a segment of *Taq* DNA polymerase.

Source: Korolev, S., et al., 1995. Proc. Natl. Acad. Sci. U. S. A. 92, 9264.

6.4.9 Occupancy

Macromolecular crystals are composed of a number of individual molecules symmetrically packed. Sometimes, the side chain of an amino acid or a main chain atom may have more than one conformation among individual molecules of the crystal. The proportion of each conformation in the crystal is denoted by the parameter "occupancy". As for example, 1.0 corresponds to 100% occupancy and 0.5/0.5 (50/50) indicates that two conformations are found at an equal frequency. In the high resolution crystal structure of myoglobin complexed with its ligand, two conformations are observed for each of the residues E8 and Y151 (Fig. 6.21).

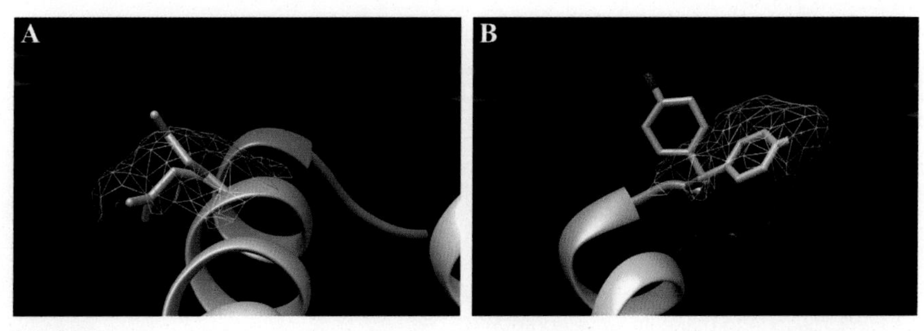

FIG. 6.21

Occupancy of (A) E8 and (B) Y151 represented in the high resolution structure of myoglobin.

Source: Vojtechovsky, J., et al., 1999. Biophys. J. 77, 2153.

6.4.10 Temperature factor

An ideal electron density distribution around an atom can be observed only if the atom is rigidly fixed in one position. However, due to vibration or flexibility of the atoms, the electron density distribution is usually wider than the ideal. This flexibility or displacement of atomic positions from an average value is incorporated into the atomic model by what is known as the "temperature factor" or the "*B*-factor". In particular, the atoms on the surface of a protein have high temperature factors since long side chains are relatively free to wag in the surrounding solvent.

****

What have been described in this chapter are the basic tools and techniques of macromolecular structure analysis and representation. Substantial progress has been made in this area aided by continued advances in computational capability. We shall see in the following chapters how such progress has transformed molecular biology from a descriptive study of the basic genetic phenomena to a more mechanistic investigation based on considerable structural insight. The boundary between biological and physical sciences has become ever more diffused than before.

Sample questions

1. Consider a unit cell with edges perpendicular to each other and denoted by the vectors **a**, **b**, and **c** where $|\mathbf{a}|=10\,\text{Å}$, $|\mathbf{b}|=8\,\text{Å}$, and $|\mathbf{c}|=12\,\text{Å}$.
   (i) Determine the parameters of the reciprocal cell.
   (ii) Determine the volume of the reciprocal cell.
   (iii) Express the coordinates of a single atom located at P (2.51, 3.75, 4.24 Å) as fractional numbers.
2. Determination of electron density distribution of a molecular crystal requires the quantitative information about the structure factor and the relative phase.
   (i) How is the structure factor obtained?
   (ii) What is the basis on which the Patterson function solves the phase problem? What are its limitations?
   (iii) How does the isomorphous replacement approach compute the Patterson function?
   (iv) What is the basis of inclusion of the temperature factor in the structure factor calculations?
3. The gyromagnetic ratio of ^{13}C is $6.73\times10^{7}\,\text{T}^{-1}\,\text{s}^{-1}$. What is the resonant frequency of ^{13}C in an external magnetic field of 21.2T?
4. What is the basis of determination of secondary structures of proteins by chemical shift measurements of $^{1}\text{H}_{\alpha}$, $^{13}\text{C}_{\alpha}$, or $^{13}\text{C}_{\beta}$?
5. What is the principle of nuclear overhauser effect (NOE)?
6. Through what potential difference an electron will have to be accelerated in an electron microscope to make its wavelength $\lambda=0.1\,\text{Å}$?
   Electron charge $=1.6\times10^{-19}\,\text{C}$
   Electron mass $=9.11\times10^{-31}\,\text{kg}$
   Planck constant $h=6.63\times10^{-34}\,\text{J s}$

7. What are the advantages of cryo-electron microscopy (cryo-EM) over X-ray crystallography?
8. Why is single-particle analysis (SPA) the most widely used variant of cryo-EM?
9. The enzyme DNA polymerase I Klenow fragment interacts electrostatically with DNA. Download the three-dimensional structure of the enzyme (1KLN) from the Protein Data Bank and depict the electrostatic surface to identify the possible site and nature of interaction.

References and further reading

Callaway, E., 2015. The revolution will not be crystallized. Nature 525, 172–174.

Cao, J., Pham, D.K., Tonge, L., Nicolau, D.V., 2002. Predicting surface properties of proteins on the Connolly molecular surface. Smart Mater. Struct. 11, 772–777.

Cheng, Y., Nikolaus, G., Penczek, P.A., Walz, T., 2015. A primer to single-particle cryo-electon microscopy. Cell 161 (3), 438–449.

Chou, J.J., Sounier, R., 2013. Solution nuclear magnetic resonance spectroscopy. In: Schmidt-Krey, I., Cheng, Y. (Eds.), Electron Crystallography of Soluble and Membrane Proteins, Methods and Protocols. In: Methods in Molecular Biology, vol. 955. Springer Science and Business Media, New York, pp. 495–517.

Egerton, R.F., 2005. Physical Principles of Electron Microscopy: An Introduction to TEM, SEM and AEM. Springer Science+Business Media, Inc., New York.

Goodsell, D.S., 2005. Visual methods from atoms to cells. Structure 13, 347–354.

Hendrickson, W.A., 1995. X-ray in molecular biophysics. Phys. Today 48 (11), 42–48.

Lambert, J.B., Mazzola, E.P., 2004. Nuclear Magnetic Resonance Spectroscopy: An Introduction to Principles, Applications, and Experimental Methods. Pearson Education Inc., New Jersey.

Milne, J.L.S., et al., 2013. Cryo-electron microscopy: a primer for the non-microscopist. FEBS J. 280 (1), 25–45.

Mura, C., McCrimmon, C.M., Vertrees, J., Sawaya, M.R., 2010. An introduction to biomolecular graphics. PLoS Comput. Biol. 6(8), e1000918. https://doi.org/10.1371/journal.pcbl.1000918.

O'Donoghue, S.I., et al., 2010. Visualization of macromolecular structure. Nat. Methods Suppl. 7 (3s), S42–S66.

Shi, D., Nannenga, B.L., Iadanza, M.G., Gonen, T., 2013. Three-dimensional electron crystallography of protein microcrystals. eLife 2, 1–17. https://doi.org/10.7554/eLife.01345 e01345.

Taylor, G., 2003. The phase problem. Acta Crystallogr. D 59, 1881–1890.

Wider, G., 2000. Structure determination of biological macromolecules in solution. BioTechniques 29, 1278–1294.

Woolfson, M.M., 1997. An Introduction to X-ray Crystallography, second ed. Cambridge University Press, Cambridge, UK.

CHAPTER 7

Protein folding

Proteins are basic components of all living cells. They are absolutely essential as structural ingredients and for various enzymatic and nonenzymatic activities of the cell. Chapter 5 has described the four levels of protein structure. In Chapter 11, we shall see how the ribosomal machinery synthesizes protein molecules by joining together amino acids into linear polypeptide chains.

7.1 The protein folding problem

We need to understand that, in order to carry out its assigned function, a protein molecule must adopt a specific three-dimensional form, called its "native" state as it emerges from the ribosome. Under "native" conditions of the cell, the majority of protein molecules exist in their native state. Protein folding is the process that steers a polypeptide chain from its linear amino acid sequence to a defined spatial structure characteristic of the native state of the protein.

Now the obvious question would be—what determines the native structure of a protein? The question was answered by Christian Anfinsen in the late 1950s in a series of experiments which demonstrated that the denaturation of a protein could be reversed (i.e., the denatured protein could be refolded to its native structure) without any auxiliary agent (Fig. 7.1).

Anfinsen's experimental enzyme was bovine pancreatic ribonuclease. The 124 amino acid (aa) residue-long protein has eight cysteines that form four disulfide bonds ($—CH_2—S—S—CH_2—$). A reducing agent was used to cleave the disulfide bonds in solution. Next, urea was added up to a concentration of 8 M to denature (or unfold) the protein. The denatured protein did not show any enzymatic activity. At this point, if urea was removed first, followed by the addition of an oxidizing agent to allow the disulfide bonds to reform, the product obtained was practically indistinguishable from the starting native protein and had regained full biological activity. In contrast, if the oxidizing agent was added first, followed by the withdrawal of urea, the product obtained was a mixture of many (or all) of the possible 105 isomeric disulfide bonded forms. The mixture regained hardly 1% of the activity of the native enzyme.

Evidently, Anfinsen's experiment established that the amino acid sequence of a protein determines its native structure and gave rise to the "Protein Folding Problem." In fact, three different but interrelated problems are associated with protein

Fundamentals of Molecular Structural Biology. https://doi.org/10.1016/B978-0-12-814855-6.00007-9

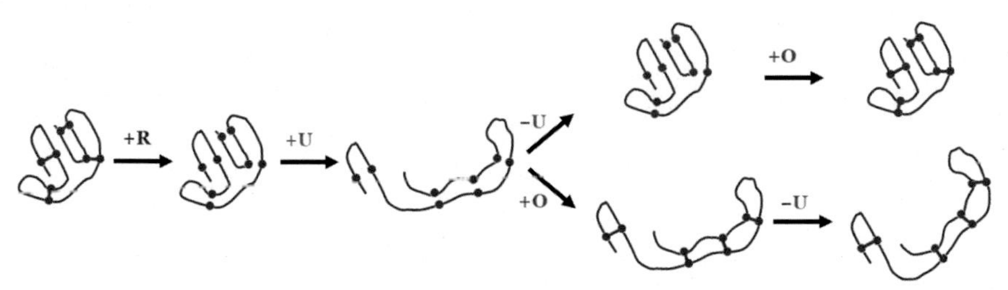

FIG. 7.1

Schematic representation of unfolding and refolding of bovine pancreatic ribonuclease. Symbols used: *R*, reducing agent; *U*, up to 8 M urea; *O*, oxidizing agent; *red dot*, cysteine; *red line*, disulfide bond; + indicates addition; – indicates withdrawal.

folding: (a) the "folding code" that would indicate the particular combination of interatomic forces dictating the native three-dimensional structure of the protein; (b) computational "structure prediction" of the protein from its amino acid sequence; and (c) kinetics and thermodynamics of the remarkably rapid "folding process." Here, in this chapter, we shall discuss the physical factors that determine the native structure of a protein, and the kinetics and thermodynamics of the folding process. The problem of computational structure prediction will be considered in Chapter 15.

Considering the fact that usually an unfolded polypeptide chain has a large number of degrees of freedom, the molecule can exist in a vast conformational space (with about 10^{143} possible conformations as estimated by Cyrus Levinthal). If it is assumed that a protein chain "walks" randomly in this conformational space to arrive at the correctly folded state, it would take a ridiculously long time. Yet, most proteins fold in milliseconds to seconds. This contradiction is known as Levinthal Paradox.

Nevertheless, to Levinthal, this was no paradox; according to his view, which proved to be correct, the assumption of random walk is not valid. Instead, he speculated that protein folding is speeded and guided by the rapid formation of local interactions, which subsequently determine further folding of the polypeptide. Levinthal did not specify what those interactions are and how do they speed and guide protein folding.

7.2 Physical forces in protein folding

We now know that protein folding is driven by forces exerted on the atoms of the amino acid chain. In particular, the side chains of the protein are more relevant here, since they make the difference between two proteins. Hence, the folding code must be written in the side chains, and not the backbone.

The forces on certain atoms or groups of atoms arise from interactions with other atoms of the protein itself as well as the solvent molecules (solvent-induced forces). The interactions can be in the form of H-bonds, ion pairs, van der Waals contacts, and water-mediated hydrophobic effect. There has been, and is still, a debate as to which one of these interactions is the dominant component in the folding code. However, the role of none of them has been fully discounted.

Electrostatic (or ion pair) interactions, as we have seen in earlier chapters, arise from the charged side chains of protein molecules. Folding is not likely to be dominated by these interactions since most proteins contain relatively few localized charged residues, and also protein stabilities are found to be mostly independent of pH (within a range) and salt concentration.

At one point of time, H-bonds were considered to be the dominant factors, since almost all possible H-bonding interactions are actually formed in native structures. H-bonds among backbone amides and carbonyls are certainly the key components of all secondary structures. Subsequently, it was found that the strength of an H-bond between a donor and an acceptor in a protein is effectively weakened by the formation of an H-bond with a solvent water molecule. Hence, H-bonds cannot be considered as a significant driving force so far as protein folding is concerned.

The energy of van der Waals interactions in tightly packed folded proteins was found to be comparable to that of hydrophobic interactions. Hence, the contribution of van der Waals interactions to protein folding has been recognized, but certainly not as a dominant factor.

On the other hand, considerable evidence emerged in favor of hydrophobicity as the major player in protein folding. Two compelling observations were: (a) involvement of large negative free energies with the transfer of nonpolar solutes from water to an organic solvent and (b) presence in most native proteins of hydrophobic core - nonpolar side chains tend to be buried in the core, sequestered from the polar environment of water molecules.

Yet, the issue of dominance is far from settled. Newer experimental and computational analyses found H-Bonds, both intraprotein as well as between the protein and solvent water molecules, to contribute at least as much as the hydrophobic effect. Further, when instead of individual atoms, hydrophobic and hydrophilic functional groups were considered as the sites of interactions, the forces on the hydrophilic groups were found to be significantly stronger than that on the hydrophobic groups. Leaving the question of dominance aside, it may be inferred that the hydrophilic and hydrophobic groups play complementary roles in accelerating the folding process.

7.3 Protein energy landscape

For a protein molecule, the completely folded native structure is the lowest energy state. The problem of folding is to attain this thermodynamically stable state. It is to be noted here that there is a clear distinction between protein folding and a simple chemical reaction. The latter proceeds from a reactant to a product through a pathway, which is essentially a succession of individual chemical structures. An unfolded protein, in contrast, is not a single microscopic structure. Hence, folding is a transition from disorder to order that requires the completely folded state to have minimum entropy as well.

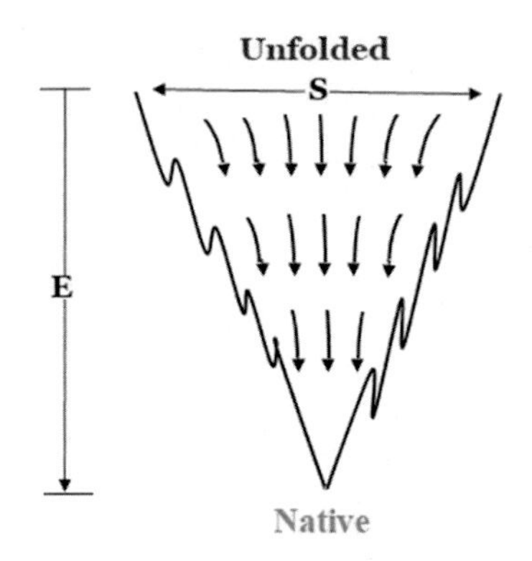

FIG. 7.2

An idealized funnel-shaped free energy landscape of proteins showing multiple pathways of the folding process. Symbols used: *E*, energy; *S*, entropy.

Therefore, the entropy-energy thermodynamics of folding can be explained by considering the nature of the conformational space of the protein. A mapping of the chain conformation to the intramolecular-plus-solvation free energy defines the energy landscape. Further, using statistical mechanical approaches, the density of states (DOS), that is, the number of conformational states at each energy level, can be determined. The logarithm of the DOS is the conformational entropy. However, since an enormous number of conformational states are involved, simplified models are required to understand a protein's DOS.

The folding process has been best described as an approach to, followed by a descent into, a funnel-shaped free energy landscape (Fig. 7.2). The unfolded state has the highest energy as well as entropy, while the completely folded state has the lowest energy and entropy. Folding proceeds through multiple pathways, each of which is a "trade-off" between energy and entropy. The energy landscape is accompanied by rules defining what configurations are available from a given configuration. Thus, the possibility of "random walk" is ruled out.

The loss of conformational entropy of the protein is compensated by an increase in solvent entropy due to the burial of nonpolar side chains and reorganization of the polar groups. In case of incomplete cancellation of energy and entropy, free energy barriers arise. As a result, proteins can be transiently trapped in local free energy minima as depicted by ruggedness of the landscape. This is known as kinetic trapping. Eventually, the kinetically trapped proteins reach the native state.

7.4 Folding in cellular environment

The cytosolic environment of the cell is fairly crowded with protein concentration of ~ 300–400 mg/mL. In this milieu, spontaneous protein folding is likely to be error-prone, inefficient, and time-consuming. Another problem is that due to the requirement of conformational flexibility for biological activity, proteins are generally only marginally stable and, therefore, susceptible to misfolding. Nonspecific interactions in the misfolded states often lead to toxic aggregate formation. This causes the loss of protein function and, further, accumulation of toxic protein species may lead to diseases such as Alzheimer's and Parkinson's. To overcome this challenge, the cell engages a network of molecular chaperones.

Chaperones assist in a number of cellular processes such as de novo protein folding, refolding of stress-denatured or aggregated proteins, and assembly of oligomeric proteins. Some of the heat-shock proteins (Hsps), whose expression is up-regulated in response to heat-induced folding stress, function as chaperones. Yet, there are chaperones which are abundant and functional under nonstress conditions.

The mechanisms of protein folding are intimately related to the question as to when and where in the cell does the process occur. It is thought that folding of a nascent polypeptide does not begin in the early stages of translation. Some secondary structures, such as α-helix, do form during the passage of a nascent chain through the ribosomal exit channel. However, it is when the polypeptide chain is exposed at the ribosomal surface and has enough time to accumulate sequence elements necessary for the formation of tertiary structures, that folding initiates outside the exit tunnel. This cotranslational folding is dependent upon the nature of the nascent chain, speed of translation, and interactions with relevant chaperones. Some chaperones can associate with the ribosome to assist the nascent polypeptide early in the folding process (cotranslationally), while there are others which do not associate with the ribosome but act at a later stage of translation or after chain release. Here, we elaborate on a few specific examples of chaperone to appreciate the diversity in the mechanisms of protein folding.

7.5 Some model chaperones

7.5.1 Trigger factor

Trigger Factor (TF) is known to be the only ribosome-associated chaperone in bacteria. It is also found in chloroplasts. Eukaryotic and archaeal ribosome-associated factors are structurally different from the bacterial TF. The 432 aa-long *Escherichia coli* TF consists of three domains (Fig. 7.3). The N-terminal domain (residues 1–112), also called the ribosome-binding domain (RBD), contains a GFRxGxxP motif (TF signature motif) in a helix-loop-helix element. This motif is necessary and sufficient for ribosome-binding.

The middle domain (residues 150–246), called the peptidyl-prolyl isomerase domain (PPD), is not essential for the general chaperone function, but possibly acts as an auxiliary substrate-binding site. The discontinuous C-terminal domain (residues 113–149 and 247–432) is also called the substrate-binding domain (SBD). It shapes the central body and two protruding helical arms of TF and, in the three-dimensional structure, is positioned between the RBD and PPD. In vitro, the SBD shows chaperone activity on its own.

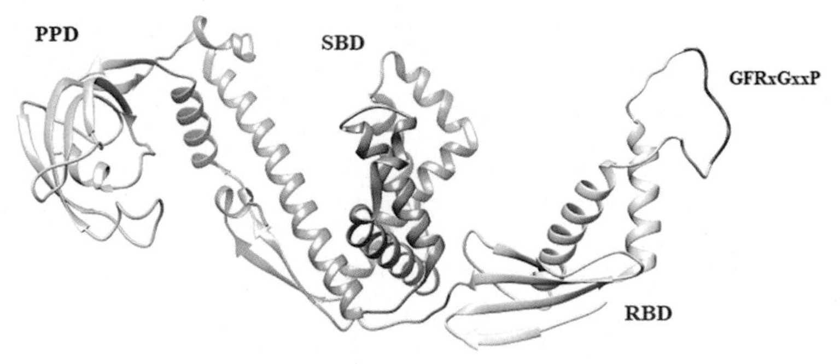

FIG. 7.3

E. coli TF domains and signature motif.

Source: Ferbitz, L., et al., 2004. Nature 431, 590.

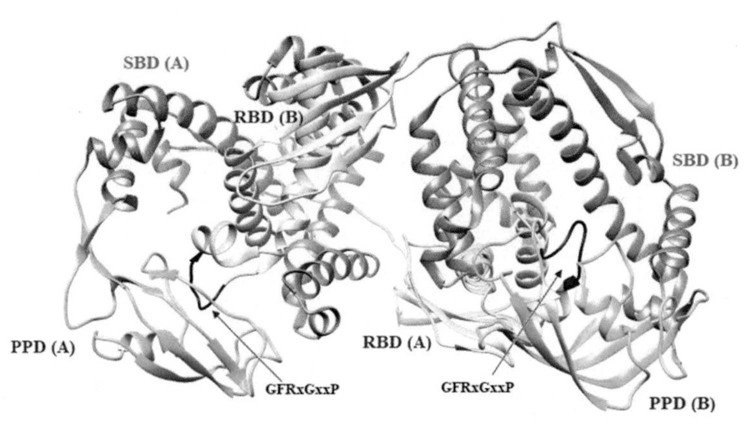

FIG. 7.4

TF dimer with domains of two subunits A and B labeled.

Source: Saio, T., et al., 2018. Oligomerization of a molecular chaperone modulates its activity. eLife 7, e35731. https://doi.org/10.7554/eLife.35731.

In solution, TF can dimerize (Fig. 7.4); however, in contrast to several other chaperones, such a GroEL (which we shall discuss shortly), that form stable oligomers, TF molecules exist in a dynamic equilibrium between the monomeric and dimeric forms. It binds to the ribosome as a monomer, but, in the cytosol, the dimer has a more potent antiaggregation activity.

Dimerization has a significant effect on the conformation of TF (Fig. 7.5). The ribosome-binding site as well as some of the substrate-binding sites are buried in the

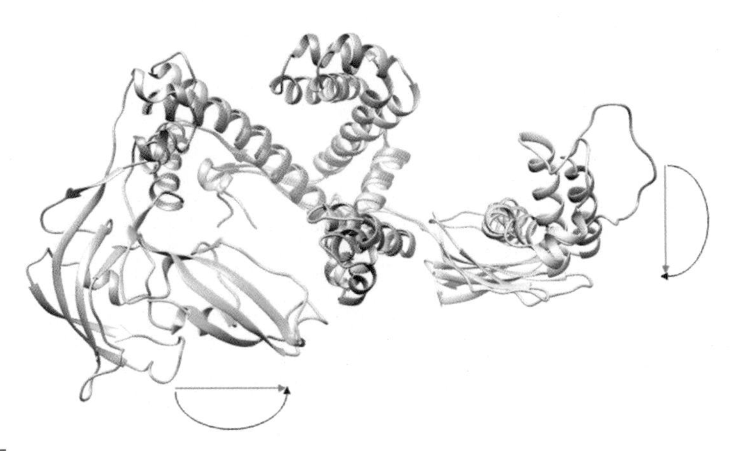

FIG. 7.5

Structural comparison of one subunit of the TF dimer (colored as in Fig. 7.4) and monomeric TF (colored *gray*). Directions of rotation and translation are indicated by curved *red* arrows and straight *green* arrows, respectively.

Source: Ferbitz, L., et al., 2004. Nature 431, 590; Saio, T., et al., 2018. Oligomerization of a molecular chaperone modulates its activity. eLife 7, e35731. https://doi.org/10.7554/eLife.35731.

dimer. The signature motif of one subunit is insulated by the PPD of the other subunit (Fig. 7.4). Hence, TF needs to monomerize for binding to the ribosome and initial binding to the nascent polypeptide chain.

In *E. coli*, TF is the most abundant chaperone having a two- to threefold molar excess relative to ribosomes. It transiently associates with ribosomes in a 1:1 stoichiometry using ribosomal protein L23 as the main docking site (Fig. 7.6). TF is positioned at the mouth of the ribosomal exit tunnel and is the first chaperone to interact with nascent polypeptide chains as they leave the ribosome and enter the cytosolic milieu. Ribosome-binding is crucial for TF function as point mutations in L23 that restrict TF's ribosome-binding also negatively affect its ability to associate with nascent polypeptide chains and assist in folding.

In the ribosome-bound monomeric state, TF exposes all its substrate-binding sites to the nascent protein. Binding to the nascent substrate delays its folding as well as prevents aggregation. As the chain grows, additional TF molecules associate with it. Although, compared with the monomer, the dimer has a lower affinity for the substrate, it has a more potent antiaggregation activity. This is due to the higher local concentration of TF subunits in the dimeric form. If a dimer first binds to a growing chain as it emerges from the ribosome exit tunnel, binding of the second molecule to the substrate by dimer dissociation is faster than that of a free monomer.

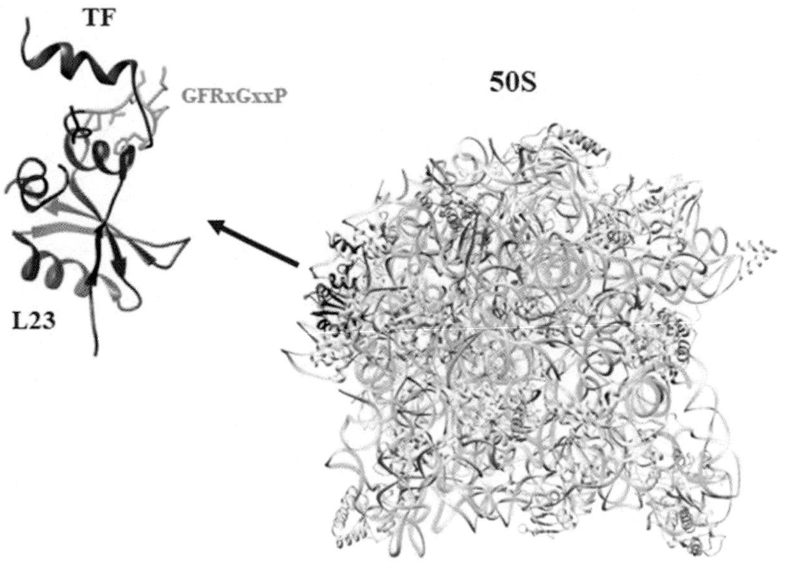

FIG. 7.6

TF binding to 50S ribosomal protein L23. 50S rRNAs have been colored *light green* and ribosomal proteins *light blue* (except L23 which has been shown in *red*). GFRxGxxP motif has been colored *cyan*; the rest of TF shown in *purple*.

Source: Ferbitz, L., et al., 2004. Nature 431, 590.

TF functions not only at the ribosome to prevent aggregation and premature folding of nascent polypeptides, but also in the cytosol. As a dimer, it associates with full-length proteins to assist in folding and complex formation.

There are multiple substrate-binding sites in TF. In fact, its entire inner cavity-forming surface, which has a mixed character, is utilized for substrate accommodation. The surface contains pockets of hydrophobic residues decorated by polar residues (Fig. 7.7) that can form H-bonds. TF uses various combinations of these hydrophobic and hydrophilic residues for promiscuous recognition of multiple substrates.

Binding of TF to one of its substrates, PhoA, has been demonstrated by NMR. *E. coli* alkaline phosphatase, PhoA, is a periplasmic enzyme containing 471 amino acids. Like many other periplasmic proteins, PhoA requires an oxidizing environment for folding. In the reducing cytosolic environment, it is found to be unfolded and interacting extensively with TF.

Four distinct sites in TF interact with PhoA. Three of these sites are located in the SBD, while one in the PPD. These binding sites in TF are almost exclusively populated by nonpolar residues. On the other hand, TF also prefers patches of residues in the substrate with elevated hydrophobicity. There are seven distinct TF-binding regions in the entire PhoA molecule. They are dominated by hydrophobic and,

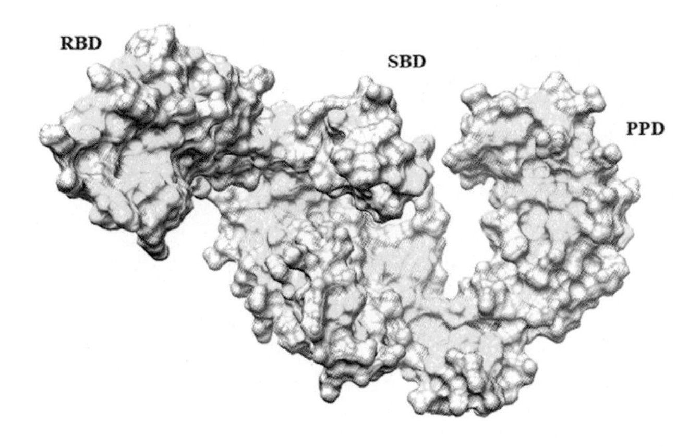

FIG. 7.7

Surface representation of TF. Hydrophobic and polar residues are shown in yellow and pink, respectively.

Source: Saio, T., et al., 2014. Structural basis for protein anti-aggregation activity of the Trigger Factor chaperone. Science 344(6184), 1250494.

particularly, aromatic residues. Three TF molecules are required to bind all the seven sites. Fig. 7.8 shows a TF molecule bound to a PhoA fragment containing two of these sites.

7.5.2 Spy

Spy is an ATP-independent periplasmic chaperone found in a range of enterobacteria and proteobacteria as well as some cyanobacteria. *E. coli* Spy is 161 aa residues in length and displays broad client specificity.

Spy forms a tightly bound α-helical homodimeric "cradle"—one surface of the dimer is concave the other being convex (Fig. 7.9). Each monomer consists of four α-helices (α1–α4). Helices α1, α2, and α3 fold into a hairpin. α1 and α2, which together form one arm of the hairpin, are connected by a flexible linker. The Spy dimer is formed through antiparallel coiled-coil interaction. The concave surface has an overall positive charge, but also contains four patches of hydrophobic residues (Fig. 7.10). Evidently, the binding surface of Spy is flexible and amphiphilic in nature. The flexibility and amphiphilicity confer Spy the dynamicity of binding to its client proteins. It is able to bind to several conformational states that appear along the folding trajectory of a client. Throughout the folding process, Spy remains continuously, yet loosely, bound to the client.

X-ray crystallographic and biophysical studies using a 10 kDa *E. coli* monomeric protein Im7 (Immunity protein 7) as a substrate (client) have elucidated the mechanism of chaperone activity of Spy. There are four key steps in the interactions

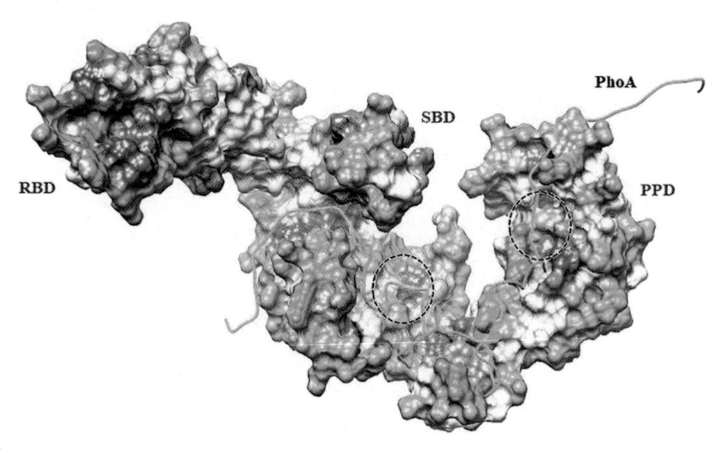

FIG. 7.8

PhoA fragment bound to TF hydrophobicity surface. Coloring scheme: *dodger blue*—most hydrophilic; *white*—no hydrophilicity/hydrophobicity; *orange*—most hydrophobic. Binding sites shown by dashed circles.

Source: Saio, T., et al., 2014. Structural basis for protein anti-aggregation activity of the Trigger Factor chaperone. Science 344(6184), 1250494.

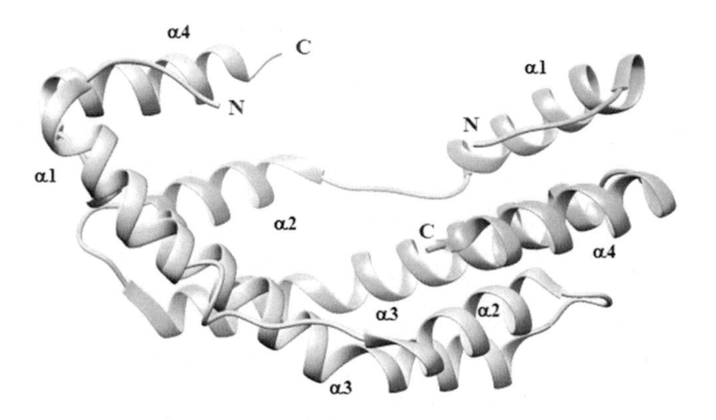

FIG. 7.9

Three-dimensional structure of Spy dimer. The two subunits are colored *light blue* and *light green.* N and C termini and the secondary structure elements of both subunits are labeled.

Quan, S., et al., 2011. Genetic selection designed to stabilize proteins uncovers a chaperone called Spy. Nat. Struct. Mol. Biol. 18, 262–269.

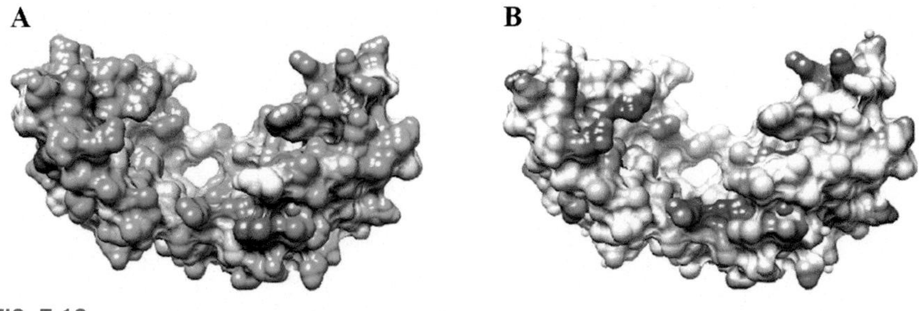

FIG. 7.10

Surface representation of Spy. (A) Hydrophobic (*dodger blue*—most hydrophilic; *white*—no hydrophilicity/hydrophobicity; *orange*—most hydrophobic. (B) Electrostatic (*blue*—positive; *white*—neutral; *red*—negative).

Quan, S., et al., 2011. Genetic selection designed to stabilize proteins uncovers a chaperone called Spy. Nat. Struct. Mol. Biol. 18, 262–269.

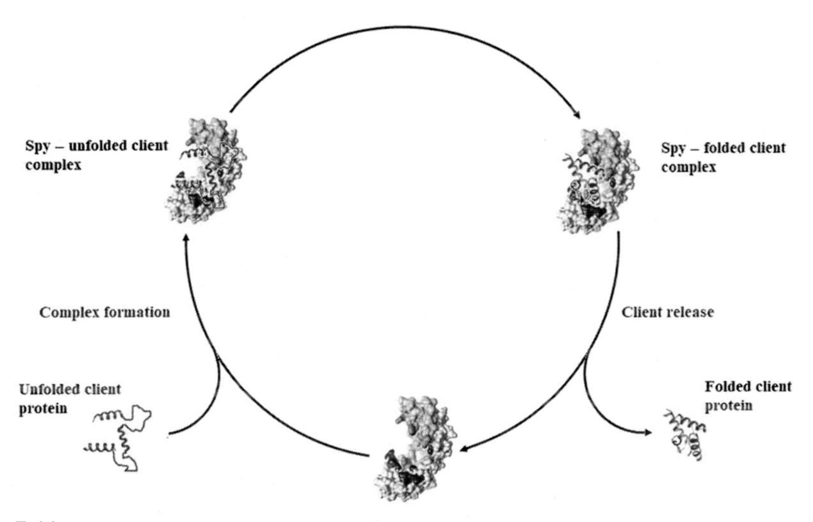

FIG. 7.11

Spy-client interaction cycle.

between Spy and Im7—initial binding, complex stabilization, folding, and release (Fig. 7.11). At first, Spy uses long-range electrostatic forces to bind unfolded Im7. This is immediately followed by the formation of hydrophobic contacts, which complement the electrostatic forces to stabilize the complex. Interestingly, it has been reported that Im7 interacts with only 38% of the hydrophobic residues, but with 61% of the hydrophilic residues in the cradle. The primary interacting sites are F29 at the N-terminus, R122 and T124 at the C-terminus, and R43, M46, K47, R61, and H96 on the concave surface (Fig. 7.12).

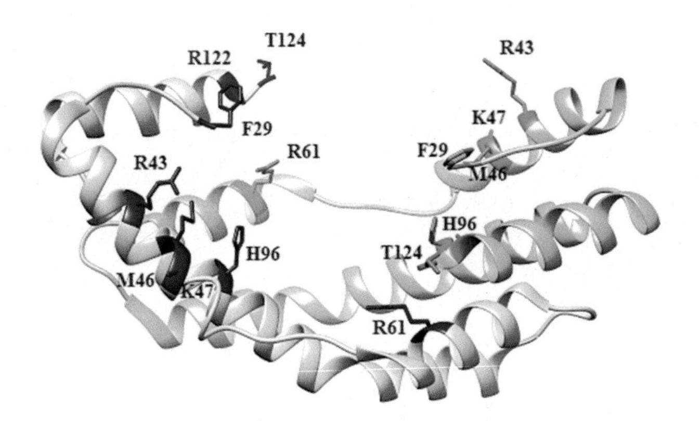

FIG. 7.12

Primary client interaction sites in Spy.

Quan, S., et al., 2011. Genetic selection designed to stabilize proteins uncovers a chaperone called Spy. Nat. Struct. Mol. Biol. 18, 262–269.

Various combinations of the hydrophobic and hydrophilic interactions enable the client to explore its folding landscape while bound to the chaperone. The client is bound as a heterogeneous ensemble with conformations representing different states of folding. As folding progresses from the unfolded to the native state, the interactions between Spy and the client evolve accordingly. The maximum hydrophobic contacts are formed when the client is in the least-folded conformation, while the fewest contacts remain when the folding is near completion. With the completion of folding, the hydrophobic contacts almost disappear, as a result of which the complex is destabilized, and the client is released in the native folded state. The electrostatic interaction, however, keeps the client bound during the time it folds and prevents aggregation as well as prenative client release.

Effectively, the client folding, not the chaperone, regulates client-binding and release. Nonetheless, Spy facilitates the folding process by functioning as an entropy sink—the client protein becomes more rigid in the folding process by transferring entropy to the chaperone which, consequently, becomes more flexible.

7.5.3 Hsp60

The Hsp60 family of chaperones are also called chaperonins. Found in all three domains of life, the chaperonins are oligomeric complexes, each about 1 MDa in size, forming stacked double-ring structures. Each of the rings encloses a central cavity. Chaperonins are ATPases; they exploit the energy of ATP-binding and hydrolysis to encapsulate misfolded proteins in their central chamber allowing the "client" proteins to fold in isolation.

Members of the chaperonin family have been divided into two groups based on sequence homology. Group I chaperonins include those from bacteria as well as mitochondria and chloroplasts. Group II chaperonins promote protein folding in eukaryotic cytoplasm and archaea.

The double-ring-like Hsp60 complex is constituted of 14 homogeneous subunits, each of which is ~57 kDa in size (Fig. 7.13). The double-ring is associated with another smaller accessory protein Hsp10 (~10 kDa). A ring-like structure formed by seven Hsp10 subunits functions as a "lid" on the cylindrical cavity of the Hsp60 complex (Fig. 7.14).

Bacterial GroEL has been the most widely studied chaperonin so far. The GroE chaperonin system of *E. coli* consists of GroEL subunits (each 548 aa) and GroES subunits (each 97 aa). Each GroEL subunit is made of three distinct domains—equatorial, intermediate, and apical (Fig. 7.15). Each domain has a specific functional role. The equatorial domain provides the inter-ring interface and contains the ATPase active site. The apical domain interacts with the unfolded client as well as GroES. The intermediate domain links the apical and equatorial domains via two "hinges." The three domains are highly dynamic and this dynamicity, in response to ATP binding to GroEL, forms the structural basis of the chaperonin functional mechanism.

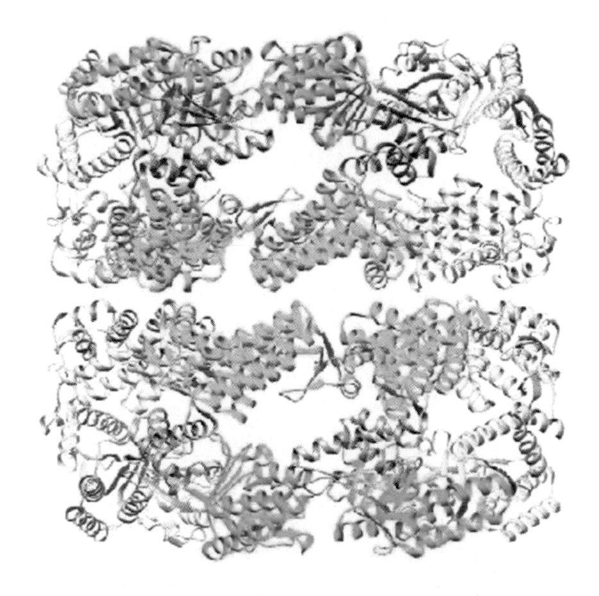

FIG. 7.13

Double-ring structure of GroEL. Domains of two subunits in each of the rings are colored as in Fig. 7.15. The other subunits are colored *light blue*.

Source: Chaudhry, C., et al., 2004. J. Mol. Biol. 342, 229.

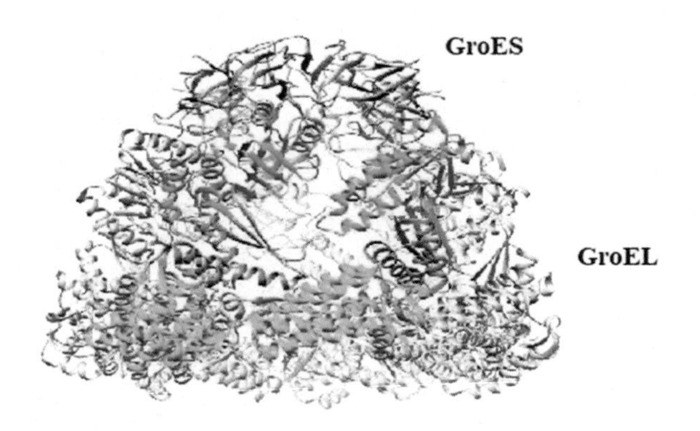

FIG. 7.14

A GroEL ring with a GroES "lid." Only one half of a GroEL-GroES complex is shown.

Source: Chen, D.-H., et al., 2013. Cell 153, 1354.

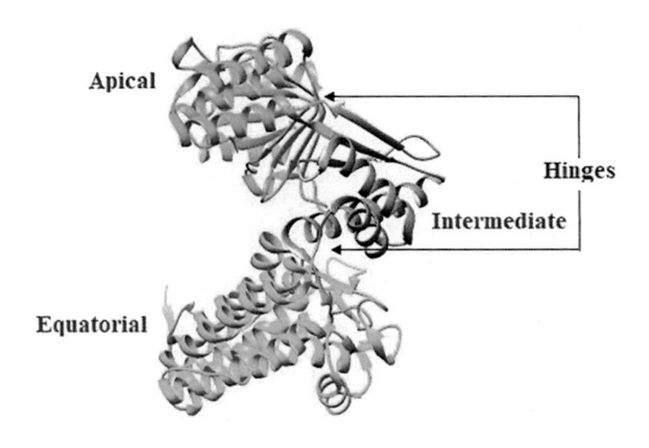

FIG. 7.15

Domains of each GroEL subunit. Color: apical—*green*; intermediate—*coral*; equatorial—*cyan*. Two hinges are indicated.

Source: Chaudhry, C., et al., 2004. J. Mol. Biol. 342, 229.

X-ray crystallography has found two major conformations of GroEL—the closed conformation in the absence of bound nucleotide and the open conformation formed in complex with GroES (Fig. 7.16). ATP-binding and hydrolysis cause transition from the closed to open conformation and triggers GroES-binding. The transition involves a rearrangement of the GroEL domains (Fig. 7.17).

GroEL-mediated protein folding is carried out in three distinct steps. In the first step, GroEL recognizes and binds a protein molecule with "nonnative" structure. Next, the bound protein is steered into the capsule-like quaternary structure formed

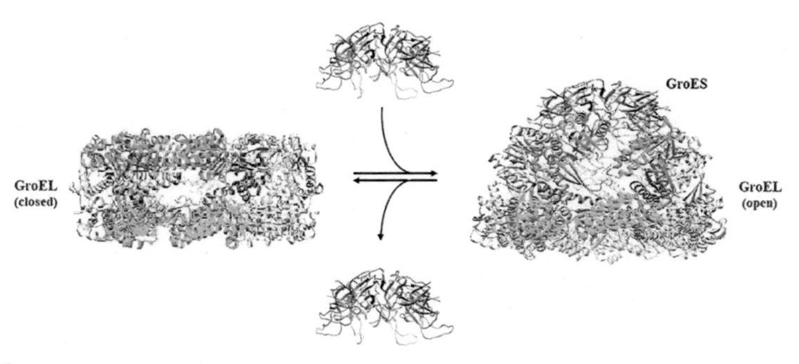

FIG. 7.16

Closed and open conformations of GroEL.

Source: Chaudhry, C., et al., 2004. J. Mol. Biol. 342, 229; Chen, D.-H., et al., 2013. Cell 153, 1354.

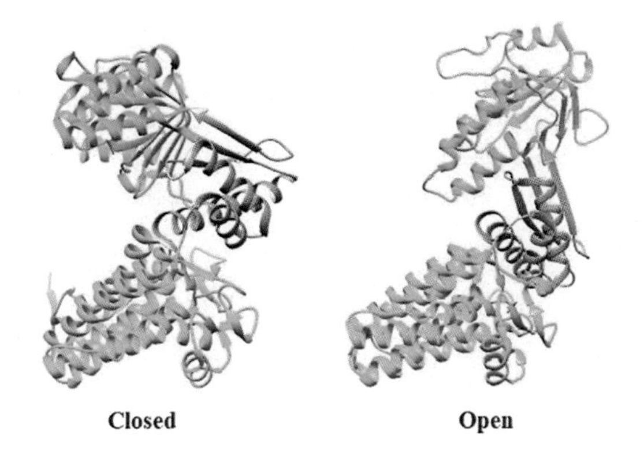

FIG. 7.17

Rearrangement of subunit domains in closed and open conformation of GroEL.

Source: Chaudhry, C., et al., 2004. J. Mol. Biol. 342, 229; Chen, D.-H., et al., 2013. Cell 153, 1354.

by the GroEL ring and GroES lid. Here, the protein is shielded from undesirable intermolecular interactions with other cellular proteins and provided an opportunity to fold in isolation. Finally, the lid dissociates and the folded protein molecule is released into the cytosol.

Nonnative proteins bind to the apical domains of GroEL, essentially through interactions with patches of hydrophobic residues (Fig. 7.18). In some cases, electrostatic interactions play a supportive role. Multiple segments of the nonnative client may bind to different apical domains of the chaperonin simultaneously and cause partial unfolding of the client protein. Client unfolding may also be caused by conformational changes in GroEL induced by ATP-binding. This unfolding rescues the client protein from kinetically trapped misfolded states.

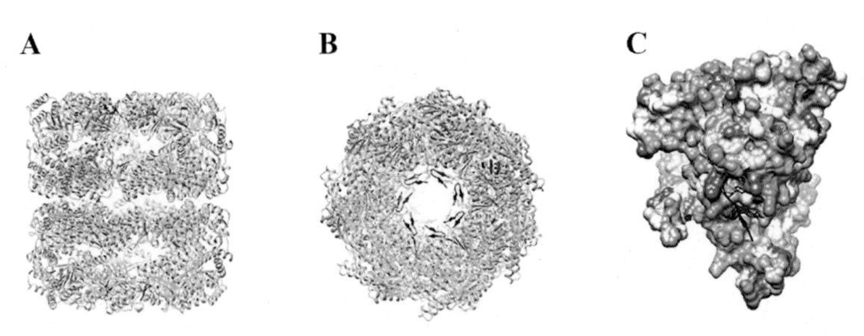

FIG. 7.18

Binding of a client peptide (brown ribbon) to the apical domains of GroEL. (A) Side view of the GroEL double-ring. (B) Top-down view of the GroEL cavity. (C) Surface of an apical domain with a client bound to a hydrophobic patch. Surface color: *dodger blue*—most hydrophilic; *white*—no hydrophilicity/hydrophobicity; *orange*—most hydrophobic.

Source: Wang, J., Chen, L., 2003. J. Mol. Biol. 334, 489.

The interior lining of the GroEL cavity in the closed state is hydrophilic (displaying an overall negative charge) with which the client interacts and folds (Fig. 7.19). The conformational changes associated with the cavity closure through GroES-binding also bury most of the hydrophobic residues in the apical domain responsible for initial client-binding.

It has been proposed that GroEL can influence client folding by different mechanisms. The first one is a passive mechanism of isolating the client protein and preventing its aggregation while folding. According to the second mechanism, the electrostatic repulsion due to the charge on the interior surface of GroEL may enhance the rate of folding. The third mechanism relates to the iterative binding and release of the client at the apical domains. This action, by unfolding the protein, may pull it out of the misfolded state and facilitate proper folding.

7.5.4 Chaperonin containing TCP1

Chaperonin containing TCP1 (or CCT) is a group II chaperonin present in the eukaryotic cytosol. It assists in the folding and assembly of several proteins, such as actin and tubulin, essential for cellular processes and viability. Although CCT folds only a small subset of cellular proteins, it shows a broad range of substrate specificity.

Like the bacterial chaperonin GroEL, CCT has a double-ring structure. Nevertheless, CCT is hetero-oligomeric, containing eight different paralogous subunits in each ring (Fig. 7.20). The mutual sequence identity among the subunits is about 30%. Driven by ATP-binding and hydrolysis, the subunits rotate to open and close a central folding chamber where nonnative proteins fold or substrates are assembled with their respective partners.

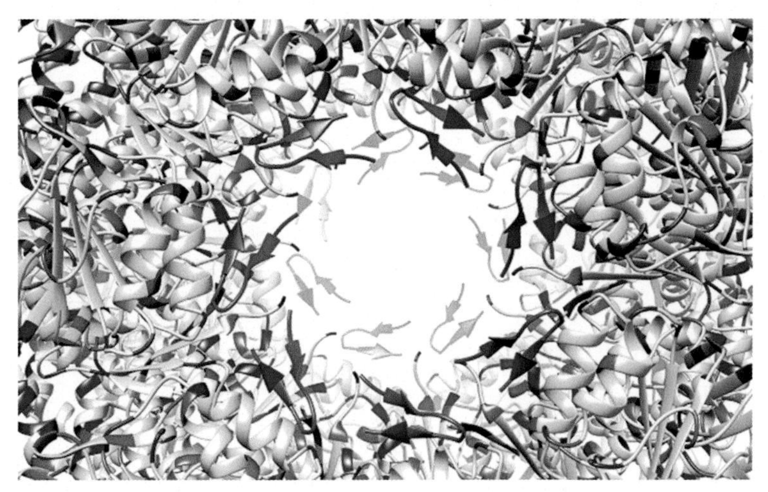

FIG. 7.19

Charged residues on GroEL cavity interior lining. Positive charges are shown in *dark blue*, negative charges in *orange*.

Source: Wang, J., Chen, L., 2003. J. Mol. Biol. 334, 489.

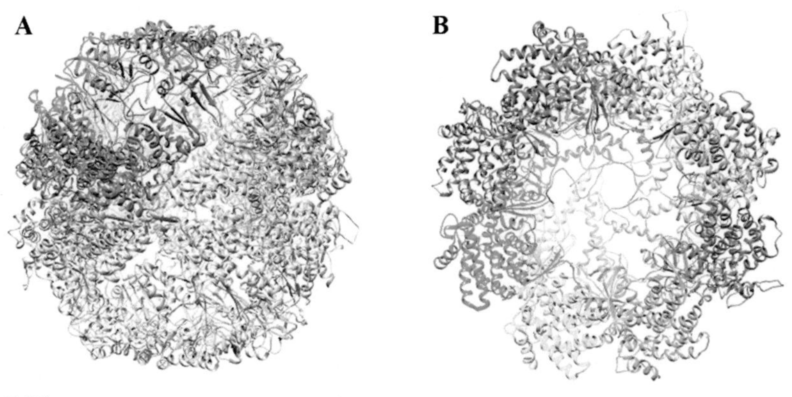

FIG. 7.20

Structure of yeast CCT. (A) Double-ring structure. All eight subunits of one ring have been colored differently. All subunits of the other ring have been colored *gray*. (B) Top-down view of one ring with the subunits differently colored.

Source: Dekker, C., et al., 2011. The crystal structure of yeast CCT reveals intrinsic asymmetry of eukaryotic cytosolic chaperonins. EMBO J. 30, 3078–3090.

It is interesting to note that, unlike its bacterial counterpart GroEL, CCT is functional without a GroES-like cofactor. Each CCT subunit does have a three-domain structure—equatorial, intermediate, and apical (Fig. 7.21). The apical substrate-binding domain contains a helical protrusion that forms a "lid" to the central folding

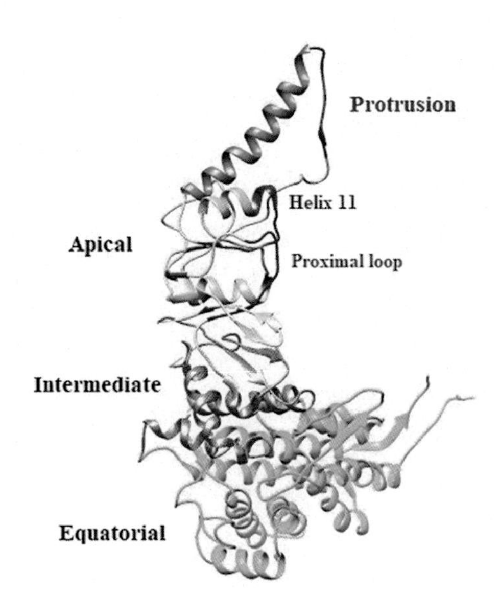

FIG. 7.21

Domains of each CCT subunit.

Source: Dekker, C., et al., 2011. The crystal structure of yeast CCT reveals intrinsic asymmetry of eukaryotic cytosolic chaperonins. EMBO J. 30, 3078–3090.

chamber where the substrate is confined. ATP-binding and hydrolysis, not binding alone, energize closure of the lid.

The apical domain of CCT contains the substrate-binding site. It is defined by a shallow groove between an α-helix (denoted Helix11 or H11) and a loop called the Proximal Loop (PL) (Fig. 7.22). The site contains hydrophobic patches surrounded by acidic residues. Hence, it is likely that, for many substrates, binding to CCT is initiated by electrostatic forces, but stabilized by hydrophobic interactions.

Needless to say, a chaperone cannot be considered a typical enzyme. One primary difference between the two is the degree of specificity as regards substrate/client-binding. Enzymes tend to be highly specific in their action, whereas, for chaperones, lack of specificity seems to be more useful.

Chaperones need to interact with a wide range of clients having different folding properties and/or multiple conformational states of the same protein. Evolutionarily, they have further widened their substrate/client range in two ways: (a) By cooperating with each other in networks, the chaperones have diversified their modes of action much beyond the capability of a single member. (b) The subunit heterogeneity of eukaryotic chaperones, such as CCT, may be due to the fact that they are required to handle a much larger proteome than their prokaryotic counterparts.

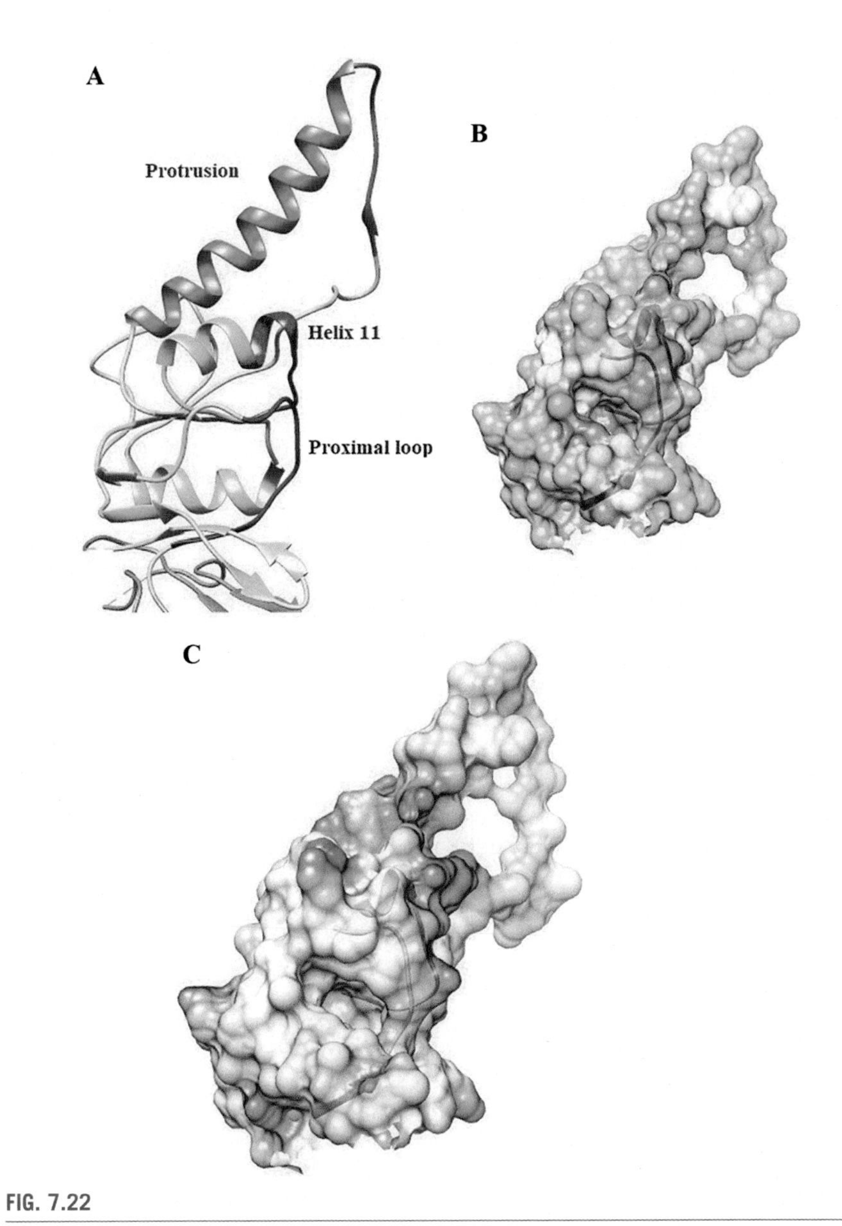

FIG. 7.22

CCT apical domain. (A) Ribbon representation. (B) Hydrophobicity surface (*blue*—most hydrophilic; *white*—no hydrophilicity/hydrophobicity; and orange—most hydrophobic. (C) Electrostatic surface (*blue*—most positive; *while*—neutral; and *red*—most negative).

Source: Dekker, C., et al., 2011. The crystal structure of yeast CCT reveals intrinsic asymmetry of eukaryotic cytosolic chaperonins. EMBO J. 30, 3078–3090.

Sample questions

1. What are the physical forces responsible for protein folding? How do they act on the atoms of a protein molecule?
2. How is protein folding a transition from disorder to order?
3. Does the descent of a protein molecule into a funnel-shaped energy landscape violate the entropy law?
4. Why are proteins dependent on chaperones for folding inside the cell?
5. Why does Trigger Factor bind the ribosome as a monomer?
6. What is the nature of interactions that enable a client protein bind to the chaperone Spy, remain bound while folding, and dissociate when folding is complete?
7. What are the functions of the GroEL cavity in the folding of a client?
8. In contrast to its bacterial counterpart GroEL, which is homo-oligomeric, the yeast chaperonin CCT is hetero-oligomeric. What is the possible implication of this difference?

References and further reading

Dekker, C., et al., 2011. The crystal structure of yeast CCT reveals intrinsic asymmetry of eukaryotic cytosolic chaperonins. EMBO J. 30, 3078–3090.

Dill, K.A., et al., 2008. The protein folding problem. Annu. Rev. Biophys. 37, 289–316.

Dill, K.A., et al., 2007. The protein folding problem: when will be solved? Curr. Opin. Struct. Biol. 17, 342–346.

Durell, S.R., Ben-Naim, A., 2017. Hydrophilic-hydrophobic forces in protein folding. Biopolymers 107(8). https://doi.org/10.1002/bip.23020.

Guo, W., Shea, J.-E., Berry, S., 2005. The physics of the interactions governing folding and association of proteins. Ann. N. Y. Acad. Sci. 1066, 34–53.

Hoffmann, A., et al., 2012. Concerted action of the ribosome and the associated chaperone Trigger Factor confines nascent polypeptide folding. Mol. Cell 48, 63–74.

Hoffmann, A., Bakau, B., Kramer, G., 2010. Structure and function of the molecular chaperone Trigger Factor. Biochim. Biophys. Acta 1803, 650–661.

Horowitz, S., et al., 2016. Visualizing chaperone-assisted protein folding. Nat. Struct. Mol. Biol. 23, 691–697.

Joachimiak, L.A., et al., 2014. The structural basis of substrate recognition by the eukaryotic chaperonin TRiC/CCT. Cell 159 (5), 1042–1055.

Kalisman, N., Schröder, G.F., Levitt, M., 2013. The crystal structures of the eukaryotic chaperonin CCT reveal its functional partitioning. Structure 21, 540–549.

Koldewey, P., et al., 2016. Forces driving chaperone action. Cell 166 (2), 369–379.

Koldewey, P., Horowitz, S., Bardwell, J.C., 2017. Chaperone-client interactions: non-specificity engenders multifunctionality. J. Biol. Chem. 292 (29), 12010–12017.

Mizobata, T., Kawata, Y., 2018. The versatile mutational "repertoire" of *Escherichia coli* GroEL, a multidomain chaperone chaperonin nanomachine. Biophys. Rev. 10 (2), 631–640.

Onuchic, J.N., Wolynes, P.G., 2004. Theory of protein folding. Curr. Opin. Struct. Biol. 14, 70–75.

Plotkin, S.S., Onuchic, J.N., 2002. Understanding protein folding problem with energy landscape theory. Q. Rev. Biophys. 35, 111–167.

Quan, S., et al., 2011. Genetic selection designed to stabilize proteins uncovers a chaperone called Spy. Nat. Struct. Mol. Biol. 18, 262–269.

Saio, T., et al., 2018. Oligomerization of a molecular chaperone modulates its activity. eLife 7, e35731. https://doi.org/10.7554/eLife.35731.

Saio, T., et al., 2014. Structural basis for protein anti-aggregation activity of the Trigger Factor chaperone. Science 344 (6184), 1250494.

CHAPTER

DNA synthesis: Replication 8

Earlier in Chapter 5, we have seen that deoxyribonucleotides (dNTPs) are the building blocks of a polynucleotide chain. Two polynucleotide chains, laid in an antiparallel format, twist around each other to form the DNA double helix. We should, however, appreciate that the nucleotides do not assemble to form a polynucleotide chain spontaneously or randomly, but need a molecular (enzymatic) machinery and a molecular directive to be joined together in an efficient and orderly manner. Here, in this chapter, we shall look into the structural features of the major macromolecular components of the molecular machinery to understand the dynamics of DNA synthesis outside and, more importantly, inside the cell.

8.1 Thermodynamics of DNA synthesis

Basically, DNA synthesis involves the hydrolysis of the inner phosphoric anhydride bond in a deoxynucleoside triphosphate (dNTP)—where N is usually either adenine (A), cytosine (C), guanine (G) or thymine (T)—and the formation of a new phosphodiester bond at the 3′-end of a DNA strand with the deoxynucleoside monophosphate (dNMP). The reaction can be represented chemically as

$$(\mathrm{NMP})_n + (\mathrm{NTP}) \rightarrow (\mathrm{NMP})_{n+1} + (\mathrm{P}) \sim (\mathrm{P}) \tag{8.1}$$

and structurally as shown in Fig. 8.1.

However, the addition of nucleotide can take place only under specific electrostatic and steric conditions in accordance with the thermodynamic principles. Chemically, the 3′-OH of $(\mathrm{NMP})_n$ should be able to disrupt the anhydride bond between the α- and β-phosphates of (NTP) and link itself to the (NMP) moiety, throwing out in the process the β- and γ-phosphates in the form of a pyrophosphate molecule (PP_i). This is possible if pK_a of the 3′-OH is lowered and a proton is transferred from O3′ to either $O2_\alpha$ (P_α), a water molecule or an aspartate unit. Whichever is the proton acceptor, it has to be brought within the optimum proton tunneling range. The activated O3′ ($O3'^-$) will then be competent for nucleophilic attack on the α-phosphate of the nucleotide. As we shall see later, a divalent metal ion (usually Mg^{2+}) coordinates the proton transfer and nucleophilic attack (S_N2 reaction).

Considering the thermodynamic possibility of the above nucleotide addition reaction, ΔG for hydrolysis of NTP is about -8 kcal/mol. On the other hand, for the endergonic phosphodiester bond formation, ΔG has been reported to

Fundamentals of Molecular Structural Biology. https://doi.org/10.1016/B978-0-12-814855-6.00008-0

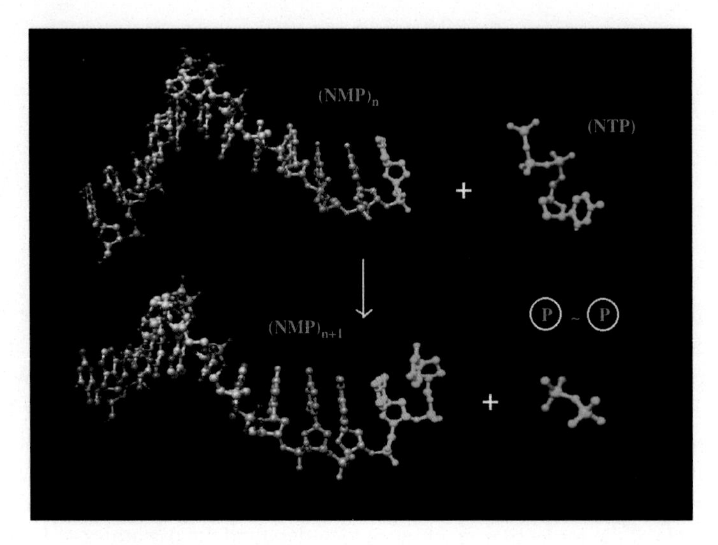

FIG. 8.1

Phosphoryl transfer reaction.

be ≈+5.3 kcal/mol. Therefore, the net free energy change for the above reaction is greater than −3 kcal/mol, which is inadequate to drive the reaction. However, PP_i is broken down into two inorganic phosphate molecules by a pyrophosphatase activity with $\Delta G \approx -5$ kcal/mol.

$$(\mathrm{P}) \sim (\mathrm{P}) \rightarrow 2(\mathrm{P})_\mathrm{i} \quad \Delta G \sim -5\,\mathrm{kcal/mol} \tag{8.2}$$

If the two reactions be coupled as

$$(\mathrm{NMP})_n + \mathrm{NTP} \rightarrow (\mathrm{NMP})_{n+1} + 2(\mathrm{P})_\mathrm{i} \tag{8.3}$$

ΔG from the overall reaction would be close to −8 kcal/mol, which is adequately favorable for the nucleotide addition process. Additionally, significant energy contributions come from two kinds of noncovalent interaction. Firstly, the base of the incoming nucleotide is engaged in stacking interaction with the nth base of the elongating strand (defined as the primer). Secondly, it also forms hydrogen bond with the base of complementary DNA strand (defined as the template).

The template strand has a more basic function than just in the matter of energy contribution—it is in the selection of the nucleotide to be added to the primer strand. If the nucleotide addition reaction (Eq. 8.3) takes place in presence of all four nucleotides (dNTPs), then in the absence of any specific direction the addition would be completely random. The specific direction is provided by the template. For duplication of genetic information of the cell with high fidelity, it is required that DNA synthesis be template-directed. The template and the primer need to come together in a specific structural arrangement called a primer:template (p/t) junction with a duplex region, where the primer is annealed to the template on one side and a single-stranded region of the template extending beyond the exposed 3′-OH terminus of the primer on the other side.

To sum up, the components of a (DNA) template-directed nucleotide addition or polymerization reaction are the two essential substrates: (a) the p/t junction and (b) four deoxynucleoside triphosphates—dATP, dCTP, dGTP, and dTTP—and Mg^{2+}. For a productive output, these components need to be assembled in a molecular machinery—an enzyme called the DNA polymerase. With the help of the individual components, the enzyme ensures a high-fidelity polymerization reaction in a 5′–3′ direction.

8.2 Mechanism of DNA synthesis

8.2.1 Structural and functional diversity

Synthesis of DNA is required in any of the three processes: (i) replication—where two identical replicas are produced from an original DNA duplex; (ii) repair—where damaged or altered nucleotide residues (sugar-phosphates or/and bases) in the DNA are restored to their normal state, and (iii) translesion synthesis (TLS)—where a damage or lesion in one strand of the DNA is bypassed by an error-prone incorporation of nucleotides in the opposite strand. A large number of DNA (template)-dependent DNA polymerases, carrying out these functions, have been discovered in a variety of organisms spread across all three domains of life—Archaea, Bacteria, and Eukarya.

Based on their amino acid sequence homology and crystal structures, the DNA-dependent DNA polymerases have been grouped into five families—A, B, C, X, and Y. Fig. 8.2 illustrates the three-dimensional structures of DNA polymerases from the five families—*Escherichia coli* Pol I Klenow fragment (family A) with replicative as well as repair functions, bacteriophage RB69 replicative DNA polymerase (RB69 gp43) (family B), *Geobacillus kaustophilus* PolC (family C) required for DNA replication, human DNA polymerase β (family X) having repair function, and *Sulfolobus acedocaldarius* polymerase Dbh (family Y) involved in translesion DNA synthesis (TLS).

There are differences between or even within the families in so far as structural details are concerned. Yet, in spite of the structural diversity, they share a common architectural framework, particularly in the polymerase domain, and utilize an identical two-metal-ion catalytic mechanism. Therefore, it is expected that they would follow a common strategy for nucleotide incorporation (Scheme 8.1).

Indeed, in all cases, polymerization begins with the binding of a primer-template (p/t) to the unliganded enzyme, the polymerase (E)—an enzyme: primer/template (E:/p/t) complex is formed (step 1). Next, a dNTP, coordinated by a Mg^{2+} ion, binds to the (E:p/t) complex to form a (E:p/t:dNTP) ternary complex (step 2). The selection of the dNTP (from the four kinds) is directed by Watson-Crick hydrogen bonding with the template (coding) base just beyond the 3′-end of the primer. The fidelity of correct nucleotide selection, as will be discussed below, is polymerase-dependent.

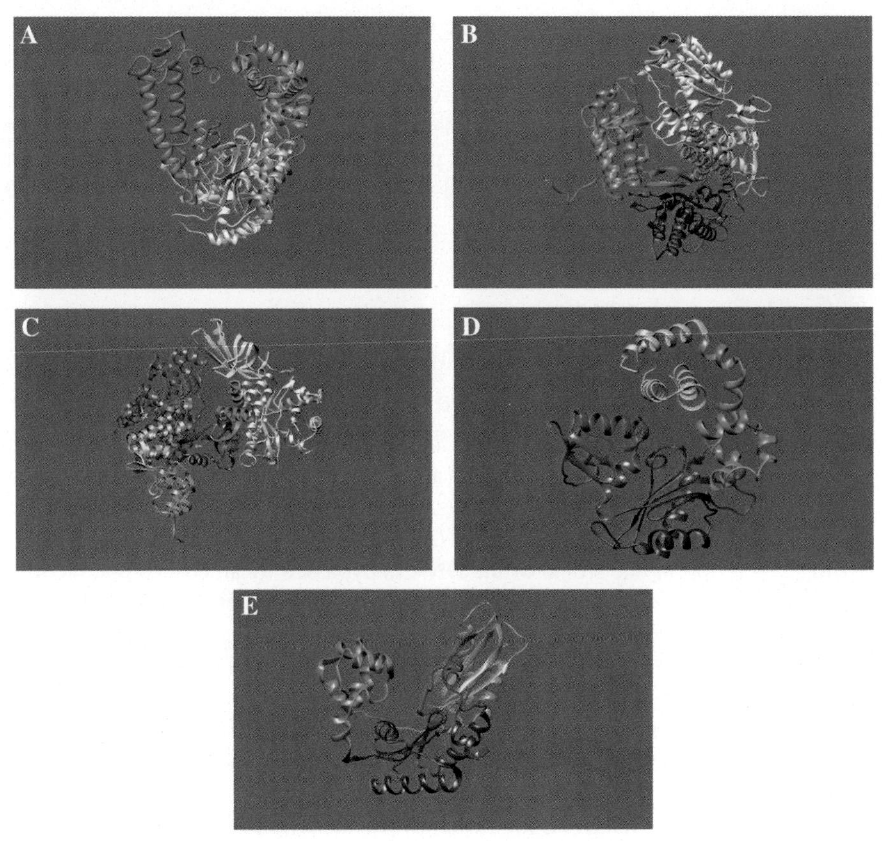

FIG. 8.2

Representative structures of five families of DNA polymerases. (A) *E. coli* Pol I Klenow fragment, (B) bacteriophage RB69 gp43, (C) *Geobacillus* PolC, (D) eukaryotic Pol β, and (E) *Sulfolobus* Dbh. Common subdomains: thumb *(green)*, palm *(brown)*, fingers *(blue)*. Specific subdomains: Pol β NH_2-terminal lyase *(golden)*; Dbh wrist *(coral)*.

Source: Brautigam, C.A., Steitz, T.A., 1998. J. Mol. Biol. 277, 363; Wang, M., et al., 2011. Biochemistry 50, 581; Evans, R.J., et al., 2008. Proc. Natl. Acad. Sci. U. S. A. 105, 20695; Sawaya, M.R., et al., 1997. Biochemistry 36, 11215; Silvian, L.F., et al., 2001. Nat. Struct. Mol. Biol. 8, 984.

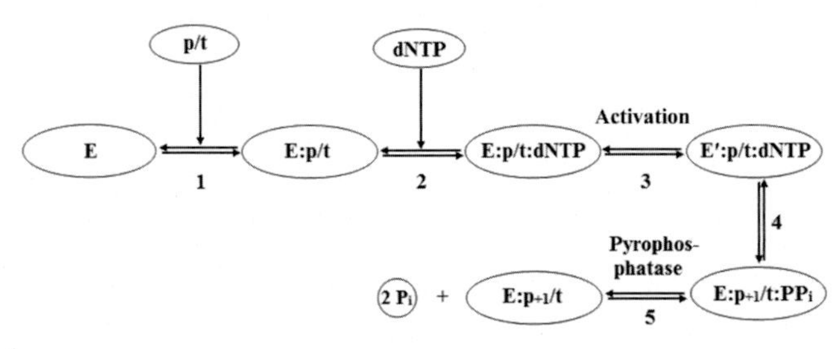

SCHEME 8.1

Nucleotide incorporation pathway.

The ternary complex undergoes a conformational change (activated) (step 3). The activated complex thus formed, (E′:p/t:dNTP), is competent to undertake the nucleotide addition chemistry (Eq. 8.3). The 3′-OH terminus of the primer, activated by the interaction with a second Mg^{2+} ion, attacks the α phosphate of the dNTP and forms a phosphodiester bond (step 4). Rapidly, the pyrophosphate is hydrolyzed into two phosphate groups by a pyrophosphatase activity (step 5).

Correct base-pairing by the formation of specific hydrogen bonds contributes only partially to the fidelity of DNA synthesis. The accuracy of nucleotide selection also depends on other factors which are polymerase-dependent. One way the DNA polymerase discriminates between correct and incorrect base-pairs is by excluding water from the active site. This increases the enthalpy difference while reducing the entropy difference, thus increasing the overall free energy difference, between the two kinds of base-paired states. Another differentiating factor is the geometries of correct and incorrect base-pairs which differ from each other. Incompatible geometry leads to steric conflicts in and around the active site. Additionally, the binding of dNTP causes conformational changes in the DNA polymerase (as we shall see below) as well as the (p/t) DNA. These conformational changes facilitate or deter the incorporation of dNTP by a mechanism known as "induced fit".

8.2.2 Common architecture of DNA polymerases

Let us try to correlate the nucleotide incorporation steps as outlined above with the structure and dynamics of the key enzyme involved, namely, the DNA polymerase. For productive interaction of the molecular components participating in DNA synthesis, the DNA polymerase, whether replicative, repair, or translesion synthesizing, is required to stably hold together the double-stranded DNA, including the p/t junction, nucleoside triphosphates, and Mg^{2+} ions. All the polymerases, therefore, share a cleft-like architecture as can be seen in Fig. 8.2 where a representative structure from each of the five families has been displayed. The shape of the DNA polymerases of the A- (*E. coli* DNA polymerase I Klenow fragment), B- (bacteriophage RB69 replicative DNA polymerase, RB69 gp43), and Y- (*S. acedocaldarius* DNA polymerase Dbh) families has been compared with that of a right hand consisting of "thumb," "palm," and "fingers." The palm subdomain is at the base with fingers and thumb subdomains forming the walls on two sides (Fig. 8.2A, B, E).

In bacterial DNA polymerase PolC (C-family) (Fig. 8.2C) and eukaryotic DNA polymerase Polβ (X-family) (Fig. 8.2D), the base of the cleft is structurally similar to the palm subdomain of other DNA polymerases, but topologically different. Therefore, for these two families of DNA polymerase, a function-based nomenclature refers to the subdomains as C- (catalytic), D- (duplex DNA-binding), and N- (nucleotide-binding/selection) corresponding to the palm, thumb, and fingers, respectively, of the other polymerases with a right-hand-like architecture.

For all the DNA polymerases, the thumb (or D-) subdomain has a role in properly positioning the duplex DNA and facilitates processivity and translocation. On the other hand, the fingers (or N-) subdomain is involved in the interactions with the incoming nucleoside triphosphate and the template base to which it is paired. Thus, the dNTP is precisely positioned in the active site. The palm (or C-) subdomain is formed by α-helices stacked against a β sheet. The strands of the β sheet are antiparallel in the A-, B-, and Y-families, but mixed (parallel and antiparallel) in C- and X-families. This subdomain carries the active site acidic residues required for the catalysis of phosphoryl transfer reaction (Fig. 8.3).

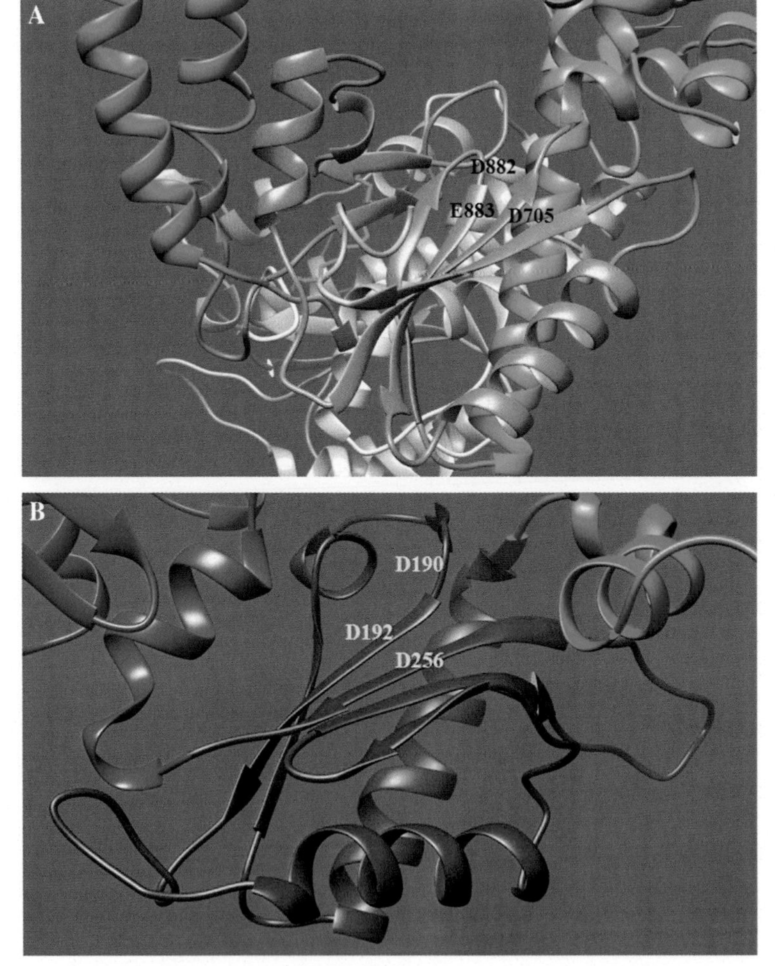

FIG. 8.3

β sheet of palm subdomain. (A) *E. coli* Klenow and (B) eukaryotic Pol β.

Source: Brautigam, C.A., Steitz, T.A., 1998. J. Mol. Biol. 277, 363; Sawaya, M.R., et al., 1997. Biochemistry 36, 11215.

8.2.3 Primer:template DNA-binding to enzyme

The first step in the polymerization process, namely, binding of the primer:template (p/t) to the enzyme (E), requires geometric and electrostatic compatibility between the two components. In the case of *E. coli* Klenow fragment, the active site cleft is about 25 Å wide and 35–40 Å deep. It can comfortably accommodate the B-form of DNA duplex whose diameter is 23.7 Å (geometric compatibility) (Fig. 8.4A). Further, we can see from the electrostatic surface of the Klenow fragment that positive charge is distributed in the cleft facilitating interactions with the negatively charged DNA backbone (electrostatic compatibility) (Fig. 8.4B).

In contrast, Dbh palm domain lacks accessible polar side chains which are known to interact with DNA bases. This may be the reason for high misincorporation frequency in Dbh-catalyzed DNA synthesis. One unique feature of Pol Y polymerases is that they contain a C-terminal domain which has been named as the wrist (Fig. 8.2E). The wrist contributes to the DNA-binding activity.

Another special feature of the TLS polymerases is that the active site is much more spacious and solvent-exposed than the high-fidelity replicases. This enables the TLS polymerases such as Dbh to accommodate large bulky DNA lesions and also water molecules to compete with nucleobases in forming H-bond. Such DNA polymerases can, therefore, discriminate between Watson-Crick (WC) and non-WC base-pairs with less efficiency than the high-fidelity polymerases.

The effect of (p/t) binding to E can be illustrated with available crystal structures of the Klenow fragment of the *Taq* polymerase, Klentaq1. When two structures—Klentaq1 unbound and bound to p/t—are superposed (using Chimera), structural changes are mostly found to be in the thumb subdomain (Fig. 8.5). The movement of the thumb as revealed by the structures has been interpreted as the thumb opening up initially, but quickly rotating in the opposite direction to bring residues in a helix-loop-helix motif at the tip of the thumb closer to the DNA. Thus, a cylindrical formation wraps the DNA and the 3′ primer terminus is brought near the polymerase active site in the palm subdomain.

Additionally, Klentaq1 holds the first base-pair of the p/t duplex in the active site by inserting Y671 (at the C-terminus of the O-helix in the fingers subdomain) into the stacking arrangement of the template bases (Fig. 8.6). In the case of *E. coli* Klenow fragment, a similar action is executed by Y766.

The eukaryotic Pol β utilizes the 8-kDa (90 residues) amino-terminal lyase domain (Fig. 8.2D), which we shall see later in the Chapter 12 to be involved in base excision repair (BER), to initiate binding of the gapped DNA (Fig. 8.7A). Unlike the replicative polymerases that are targeted to the growing 3′-terminus at the (p/t) junction, Pol β shows affinity to the 5′-margin of the gap. The lyase domain displays a predominantly positive electrostatic potential to draw in the negatively charged phosphate. The 5′-phosphate is hydrogen-bonded to K35 and K68 (Fig. 8.7B). The binding of 5′-phosphate to the Pol β is strengthened by adjacent single-stranded DNA whose backbone phosphates interact with positive electrostatic surface potential of the lyase domain. Additionally, two helix-hairpin-helix (HhH) motifs in the lyase domain interact with the backbone at two ends of the incised DNA strand to bend the DNA by ~90° and expose the terminal base-pairs.

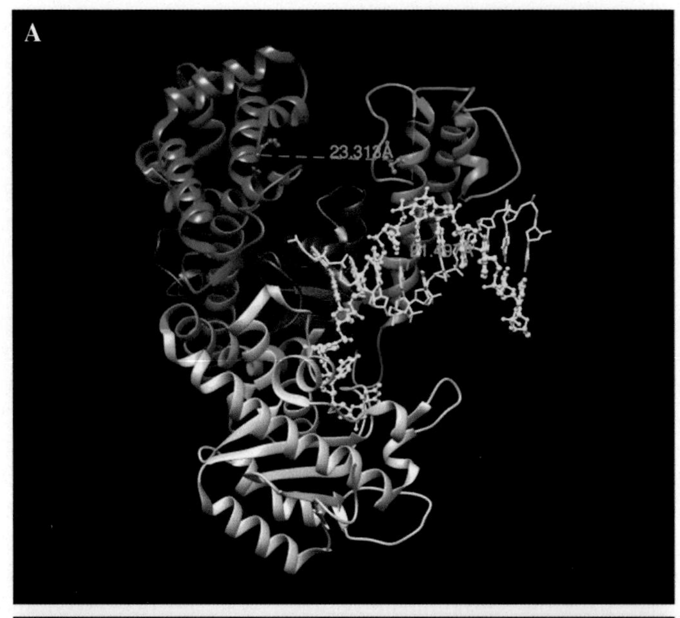

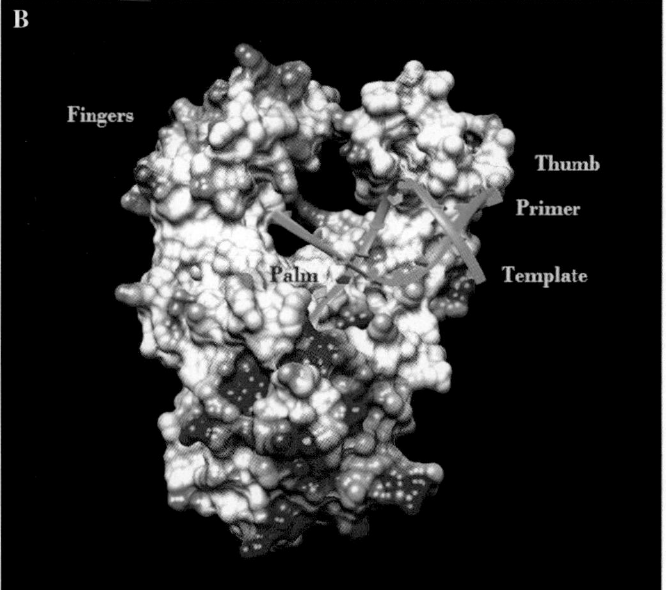

FIG. 8.4

Primer:template (p/t) duplex-binding to Klenow fragment. (A) geometric compatibility and (B) solvent-accessible surface displaying electrostatic potential: *blue*—most positive; *white*—neutral; *red*—most negative.

Source: Beese, L.S., Derbyshire, V., Steitz, T.A., 1993. Science 260, 352.

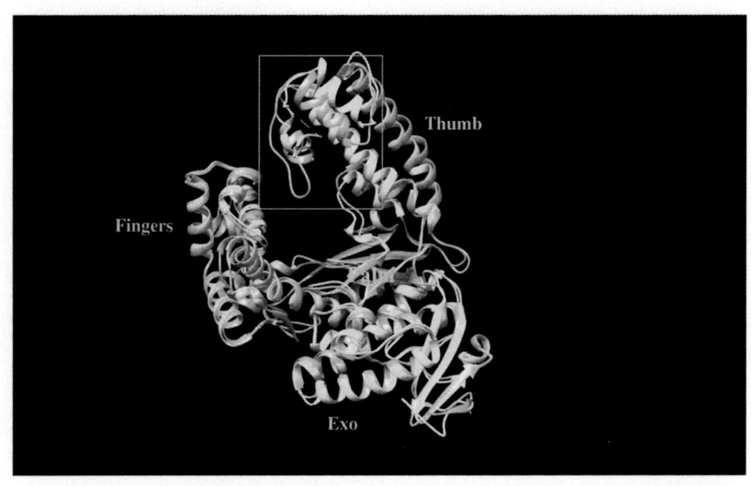

FIG. 8.5

Structural changes in the thumb domain (marked by box) of Klentaq1 bound to p/t (colored *cyan*) compared with apo form of Klentaq1 (colored *yellow*).

Source: Korolev, S., et al., 1995. Proc. Natl. Acad. Sci. U. S. A. 92, 9264; Li, Y., Korolev, S., Walksman, G., 1996. EMBO J. 17, 7514.

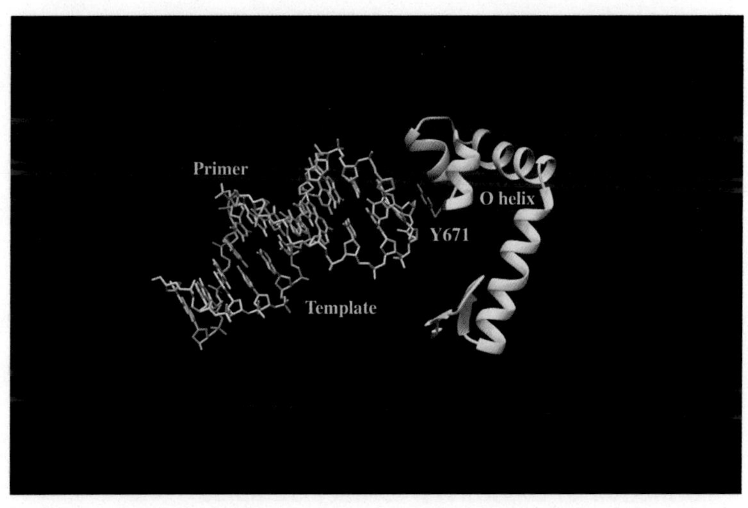

FIG. 8.6

Y671 residue *(red)* in stacking arrangement with template bases.

Source: Li, Y., Korolev, S., Walksman, G., 1996. EMBO J. 17, 7514.

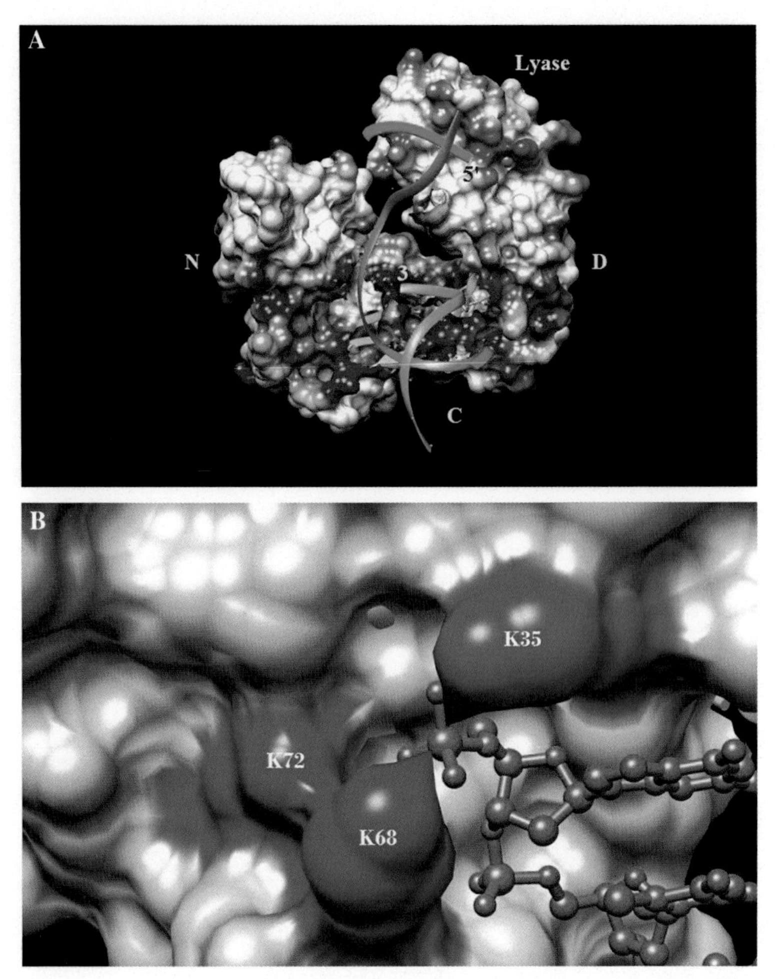

FIG. 8.7

DNA-binding to pol β. (A) Solvent-accessible surface of the enzyme displaying electrostatic potential: *Blue*—positive; *white*—neutral; *red*—negative. (B) 5′-Phosphate-binding pocket in the lyase domain.

Source: Beard, et al., 2009. J. Biol. Chem. 284, 31680; Sawaya, M.R., et al., 1997. Biochemistry 36, 11215.

8.2.4 Formation of E:p/t:dNTP ternary complex

In the next step, a dNTP binds to the E:p/t complex to form the E:p/t:dNTP ternary complex (Fig. 8.8). In the case of Klentaq1, the dNTP binds to the N-terminal end of the O-helix in the fingers subdomain. The binding is helped by the electrostatic interaction of the triphosphate group of the dNTP with the positively charged

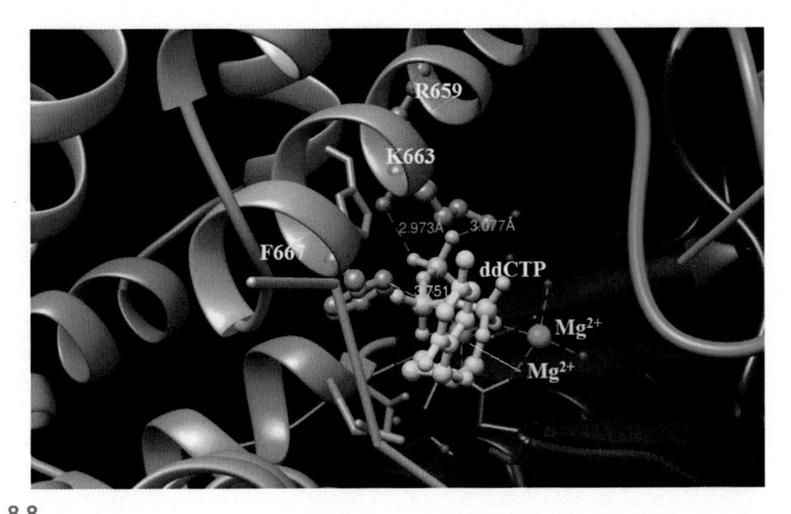

FIG. 8.8

Binding of dNTP to Klentaq1 by electrostatic and stacking interaction with O-helix residues.
Source: Li, Y., Korolev, S., Walksman, G., 1996. EMBO J. 17, 7514.

side chain of R659 and K663 and stacking interaction with F667, all three residues being present in the O-helix. The triphosphate moiety runs parallel to the O-helix. In this position, the nucleotide is at a distant (10–15 Å) from the catalytic site located in the palm and has to await a large conformational change in the fingers subdomain to arrive at the active site.

The E:p/t:dNTP (open) complex is immediately converted to the activated E′:p/t:dNTP (closed) complex. As a result of this transition, now there is a considerable reorientation of the fingers subdomain (Fig. 8.9A) and the O-helix moves closer to the active site constituted by the three carboxylate residues, D610, D785, and E786, by 46°. Y671 of the O-helix, whose aromatic ring was earlier stacked onto the first base-pair of the p/t duplex, moves away from the stacking arrangement (Fig. 8.9B) and the template base positions itself to base-pair with the correct incoming nucleotide. The transition organizes all the components of the active site topologically and geometrically to enable the polymerase to proceed to the chemical step.

Also in the case of Pol β, we can see that nucleotide-binding and ternary complex formation cause reorientation mainly in the N-subdomain (fingers) (Fig. 8.10). However, there is some movement in the lyase domain as well. Displacement of the residues is up to 13 Å. The coding template base helps binding of the incoming correct nucleotide to form the open complex. The N-subdomain closes around the nascent base-pair to activate the complex. A significant aspect of this activation is the release of D192 from salt bridge interaction with R258 to enable it to coordinate both the catalytic and nucleotide binding metal ions.

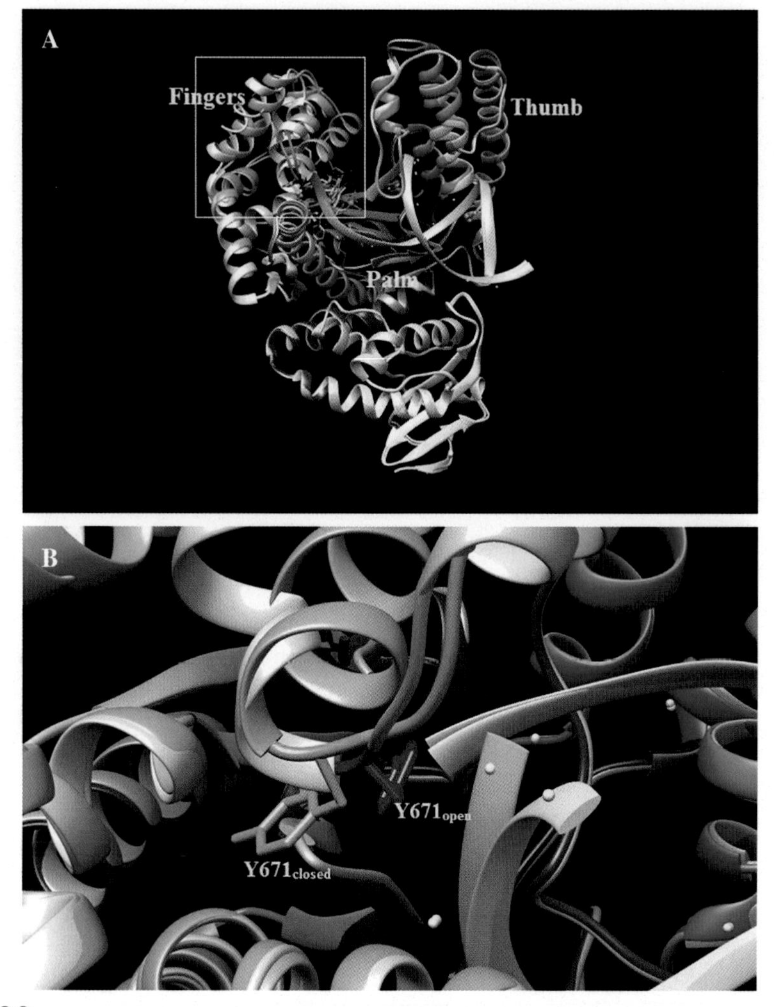

FIG. 8.9

Structural changes associated with activation of Klentaq1 ternary complex. (A) Reorientation in the fingers subdomain (marked by box). (B) Shift in the position of O-helix Tyr671.

Source: Li, Y., Korolev, S., Walksman, G., 1996. EMBO J. 17, 7514.

8.2.5 Phosphoryl transfer reaction

In all DNA polymerases, the phosphoryl transfer reaction takes place by a two-metal-ion catalytic mechanism and in most cases the metal ion is Mg^{2+} (Fig. 8.11). Two Mg^{2+} are octahedrally coordinated by the triphosphate of an approaching nucleotide and carboxylate side chains in the active site. In Klentaq1, for example, one Mg^{2+} (metal B) is ligated in the basal octahedral plane by four oxygen atoms contributed by the β and γ phosphates and carboxylates of D610 and D785. Further, on one side

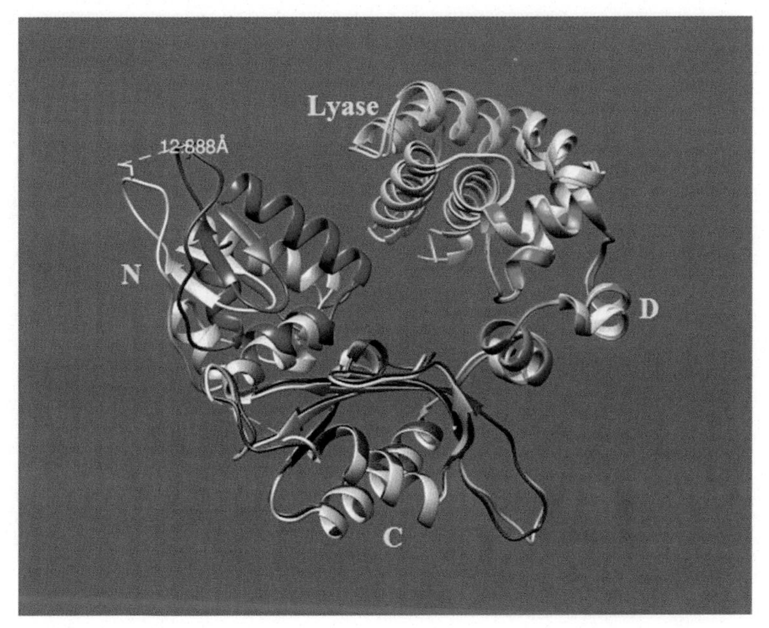

FIG. 8.10

Displacement of residues in the N-subdomain of pol β associated with activation of ternary complex.

Source: Beard, et al., 2009. J. Biol. Chem. 284, 31680; Batra, V.K., et al., 2006. Structure 14, 757.

of the octahedral plane, it is ligated to an oxygen atom of the α-phosphate and on the other side the carboxylate of Y611. The other Mg^{2+} (metal A) is pentacoordinated by oxygen atoms from the carboxylates of D785 and D610, the α-phosphate of the approaching nucleotide, and two water molecules. The vacancy in the sixth coordinating position enables the metal ion to interact with a ribose 3′-OH at the 3′ end of the primer strand and lower the affinity of 3′-OH for hydrogen (that is, lower the pK_a). Thus, activated $3'$-O^- is now competent for a nucleophilic attack on the α-phosphate, resulting in the formation of a phosphodiester bond. Metal B then assists the departure of the pyrophosphate.

Similarly, in the case of the X-family of polymerase, Pol β, one Mg^{2+} (metal B/ nucleotide-binding metal) is coordinated to two aspartate residues (D190 and D192), the triphosphate moiety (α-, β-, and γ-) of the approaching nucleotide, and a water molecule. The catalytic Mg^{2+} (metal A), on the other hand, is ligated to all three active site aspartates (D190, D192, and D256), the α-phosphate, and a water molecule, the sixth coordinating position being left vacant to interact with O3′ of the primer terminus and activate it for nucleophilic action (Fig. 8.12).

The formation of the phosphodiester bond modifies the coordination state of the catalytic Mg^{2+}, prompting it to leave the complex. However, the nucleotide-binding Mg^{2+} remains at the site of action and coordinates aspartates and PP_i. Subsequently, a

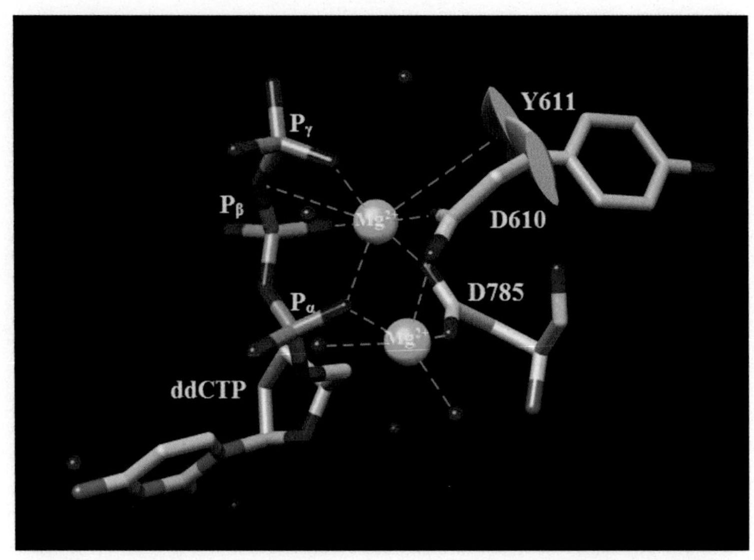

FIG. 8.11

Active site Mg^{2+} coordination in Klentaq1.

Source: Li, Y., Korolev, S., Walksman, G., 1996. EMBO J. 17, 7514.

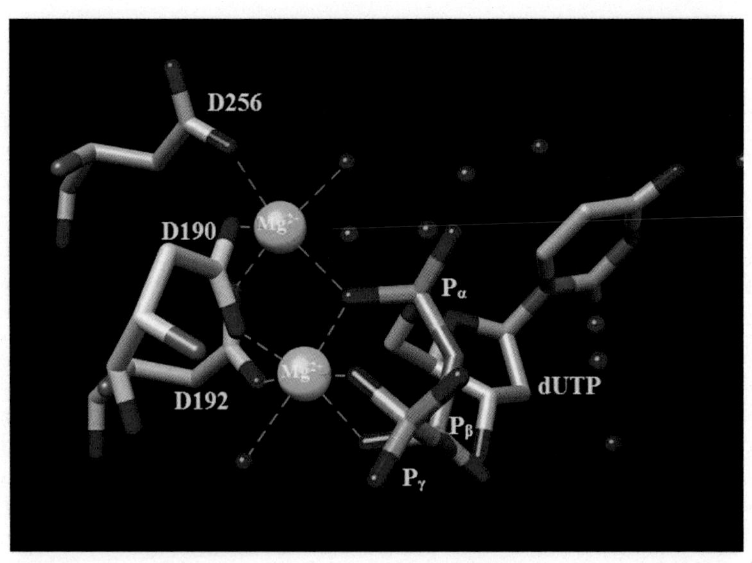

FIG. 8.12

Nucleotide binding to Pol β by active site Mg^{2+} coordination.

Batra, V.K., et al., 2006. Structure 14, 757.

competing water molecule displaces the γ-phosphate from PP_i and affects a change in the coordination state of the second metal as well.

To summarize, nucleotide incorporation in a growing polynucleotide chain is a multicomponent dynamics managed with sufficient efficiency and accuracy by the DNA polymerase. The three-dimensional geometry and electrostatic nature of the enzyme allow it to bind to the primer:template DNA with the help of the thumb (and, in some cases, an additional subdomain such as lyase or wrist) and the dNTP primarily by the fingers. The two reactants, namely, the primer:template DNA and dNTP, are juxtaposed with each other at the catalytic site of the palm where the coordination of two divalent metal ions affects the phosphoryl transfer.

8.3 Cellular context

Let us now consider a simple living organism—a bacterium. We would like to understand how a replisome travels through its genome, from start to finish, carrying out DNA synthesis. What are the requirements the replisome needs to satisfy in order to move forward at a particular instant. In contrast to the in vitro situation where we had a single-stranded template and a primer whose 3′-end was extended by the DNA polymerase, bacterial DNA is mostly a covalently closed circular duplex molecule. The two strands of the duplex, both of which must be replicated, are antiparallel to each other. Further, the replication has to take place in the environment of all kinds of molecules—small and macro—including catabolic enzymes.

We have to consider the following requirements for DNA synthesis inside the cell. First of all, at any time point during the synthesis, the duplex has to be opened in the form of a fork, known as the replication fork, at the site of synthesis to make both the strands available as templates. Obviously, on the DNA there must be a site where the duplex is initially melted—called the origin of replication—and factors recognizing and working on it. Thereafter, continuous unwinding of the duplex with concomitant movement of the replication fork is necessary. The single-stranded region that arises from duplex unwinding and fork movement has to be protected from damaging agents such as the nucleases. Also, the single-stranded nature of each DNA strand has to be maintained by preventing secondary structure formation so that both the strands can act as templates for DNA synthesis.

Most crucial is the enzyme complex that carries out the synthesis—the replicase, which is a multimeric holoenzyme containing the DNA polymerase. It should be noted here that synthesis on both the templates needs to be concurrent. However, the antiparallel layout of the two strands of duplex DNA and the 5′–3′ directionality or polarity of the synthetic activity of DNA polymerases pose an apparently paradoxical problem for DNA replication. As the DNA is continuously unwound and the fork progresses, uninterrupted synthesis of DNA in the 5′–3′ direction is easily conceivable only on one strand (leading strand). The cell resolves the paradox by discontinuous synthesis on the other strand (lagging strand). On this strand, the synthesis occurs in a series of short fragments (Okazaki fragments) strictly following the 5′–3′ polarity rule, but overall, the strand grows in a 3′–5′ direction.

We have seen earlier that DNA polymerase cannot start synthesis de novo, but can only extend an existing strand (DNA or RNA). This requires that the DNA synthesis against each template has to be "primed" by a special RNA polymerase, called the primase, with a short (10–12 bases) oligoribonucleotide (RNA) primer. Ideally for the leading strand, only one primer is required to be synthesized at the origin of replication. For the lagging strand, the primase has to synthesize a primer at the beginning of each Okazaki fragment.

Once the primer is synthesized, the DNA polymerase must take over and extend the primer by the addition of deoxyribonucleotides (dNTPs). The DNA polymerase must associate with other factors and form a replicase to fulfill two basic requirements of replication—high fidelity and processivity. High fidelity can be achieved with the help of a proofreading exonuclease. Processivity requires a mechanism to hold the polymerase onto the DNA during both leading and lagging strand synthesis.

How does *E. coli* fulfill the requirements of chromosomal replication? In its genome, which is about 5×10^6 base-pairs in size, there is a site called *oriC* where replication begins. Its initiator protein, DnaA, about which we shall discuss in further detail later in the chapter, recognizes specific features of *oriC* and melts the DNA at this site. Before DnaA dissociates from *oriC*, it recruits the homohexameric replicative helicase, DnaB. The helicase encircles the single-stranded DNA (ssDNA) on the lagging strand and moves in $5' \rightarrow 3'$ direction to unwind the parental duplex. The ssDNA generated on the lagging strand is immediately bound by single-strand DNA-binding protein (SSB) to protect ssDNA from nuclease action and prevent secondary structure formation. The *E. coli* primase, DnaG, transiently interacts with DnaB helicase to synthesize RNA primers. At this point, the task of extension of the primers is taken over by *E. coli replicase*, DNA polymerase III (Pol III). *E. coli* Pol III consist of a catalytic subunit α, a proofreading exonuclease subunit ε, and an apparently nonessential subunit θ. The ε subunit provides the 3′–5′ exonuclease (proofreading) activity, which will be discussed in a later section on bacterial replicase.

For processivity, Pol III is held onto the DNA by the β-sliding clamp. The pentameric $\tau_3\delta\delta'$ clamp loader complex assembles the sliding clamp around the DNA. The subunits of the polymerase, sliding clamp, and clamp loader together form the DNA polymerase III holoenzyme, which is the replicase in *E. coli*. The double-stranded DNA needs at least two Pol III complexes—one for each strand. Further, to ensure a mechanism whereby both strands can be synthesized concurrently, the leading- and lagging-strand Pol III complexes are coupled to the replicative helicase. The helicase, at least two replicases (and possibly a third one as a stand-by), and the primase together constitute the replication machinery of the cell at the replication fork, the replisome (Fig. 8.13).

Now, the two replicases (polymerases) on the leading and lagging strands must travel in opposite directions, yet remain coupled in the replisome. The replisome resolves this topological problem by accumulation of ssDNA loops on the lagging

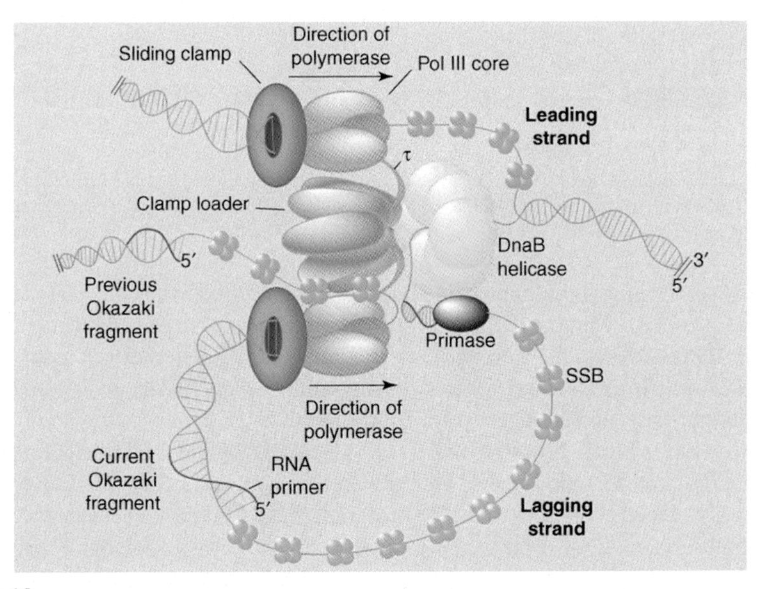

FIG. 8.13

Escherichia coli replisome with coupled DNA polymerases.

Courtesy: Pomerantz, R.T., O'Donnell, M., Replisome mechanics: insights into a twin DNA polymerase machine, Trends Microbiol., April 2007 with permission from Elsevier.

strand. As an Okazaki fragment is synthesized, the loop grows; on completion of the fragment, the loop collapses and the polymerase hops back to the clamp behind to begin the next Okazaki fragment (Fig. 8.13).

The replication process is complete only on removal of the RNA primers—one or at the most very few on the leading strand and in the Okazaki fragments on the lagging strand. Ribonuclease H, which cleaves the bonds between ribonucleotides in an RNA:DNA hybrid, removes the entire primer excepting the last ribonucleotide. A 5′–3′ exonuclease activity of DNA polymerase I removes the last one.

8.4 Initiation of bacterial DNA replication

8.4.1 Melting of the DNA—Replicator and initiator

Let us now query where on the chromosome and how replication begins and investigate the process of initiation in detail. According to the replicon model proposed by Jacob, Monod, and Cuzin, at least two components are necessary—the replicator, which is the site on the chromosome where replication begins, and the trans-acting initiator protein. In *E. coli*, they are *oriC* and DnaA, respectively.

E. coli oriC contains five 9-bp repeat elements—R1, R2, R3, R4, and R5(M)—with consensus sequence 5′-TTATCCACA-3′ (Fig. 8.14). Being sequence-specific

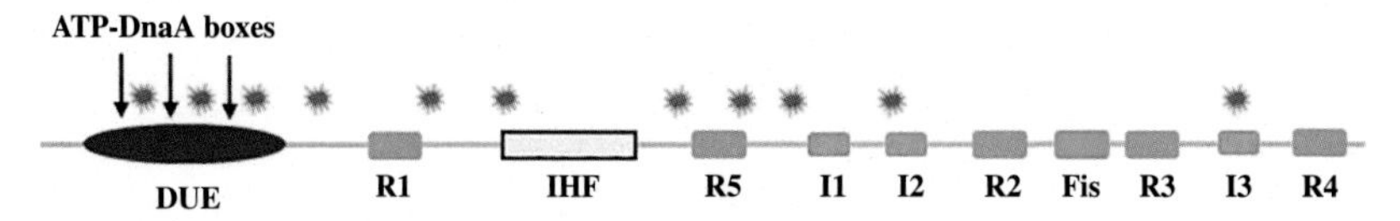

FIG. 8.14

DNA sequence motifs in *Escherichia coli oriC*. GATC sequences are marked by orange-colored stars.

binding sites for the initiator protein DnaA, these are called DnaA boxes. DnaA binds to R1, R2, and R4 with the highest affinity throughout most of the cell cycle. Weaker affinity DnaA boxes, R3 and M, and another class of sequence elements, called "I sites," which differ from the consensus sequence only marginally, bind to the ATP-bound form of DnaA just prior to initiation.

Close to the DnaA boxes in *oriC* is an AT-rich DNA unwinding element (DUE) containing three 13-mer repeats. This region also contains boxes with a 6-mer consensus 5′-AGatct-3′ binding to ATP-DnaA (and hence called ATP-DnaA boxes). As ATP binds to DnaA, the protein homo-oligomerizes to form a nucleoprotein complex containing up to 20 monomers. Consequently, with the help of negative superhelicity of the DNA and nucleotide architectural factors, IHF and HU, the nucleoprotein complex causes melting of the DNA within DUE. The unwound region offers itself as the substrate for helicase loading and subsequent assembly of the replisome.

This warrants a look at the domain structure of DnaA (Fig. 8.15) in order to find a mechanistic explanation of the multiple functions it performs. First of all, as we have already seen, it has a DNA-binding function. Secondly, it has to bind ATP for homo-oligomerization. Additionally, it has to interact with helicase and other accessory proteins to initiate replisome assembly.

In *E. coli*, the amino-terminal domain (domain I, residues 1–90) of DnaA interacts with the DnaB helicase and also supports self-association of DnaA molecules for oligomeric assembly. Domain II (residues 90–130) is a poorly conserved element acting as a flexible linker between domain I and rest of the DnaA molecule. Domain III (residues 131–347) is responsible for ATP-binding and hydrolysis and, therefore, the primary determinant of oligomerization.

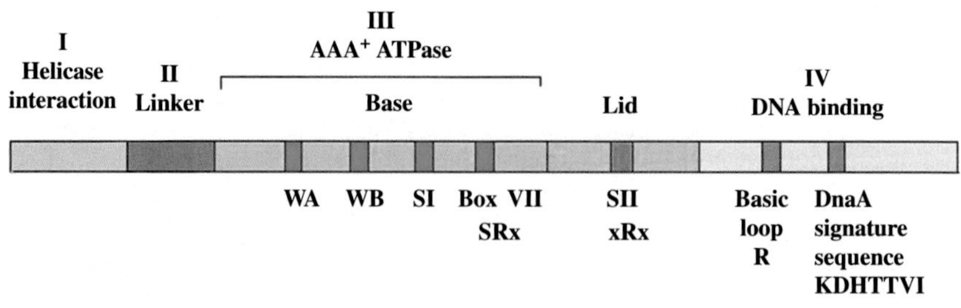

FIG. 8.15

Domain structure of DnaA. *WA*—Walker A; *WB*—Walker B; *SI*—sensor I; *SII*—sensor II.

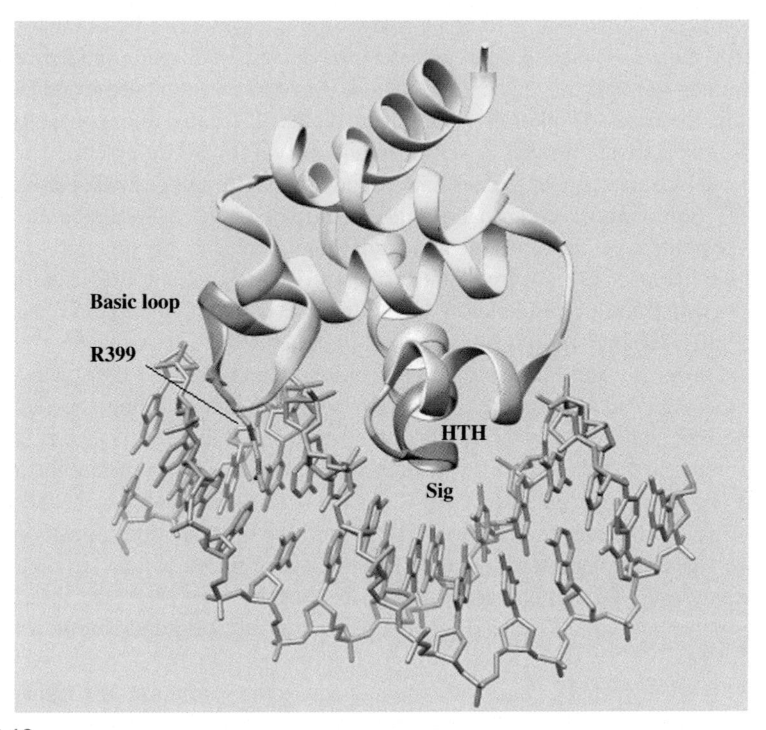

FIG. 8.16

DnaA carboxy-terminal domain *(yellow)* bound to a DnaA box *(blue)*.

Source: Fujikawa, N., et al., 2003. Nucleic Acid Res. 31, 2077.

The carboy-terminal domain (domain IV, residues 347–467) is the DNA-binding domain responsible for sequence-specific binding to the DnaA boxes in *oriC* (Fig. 8.16). Determination of crystal structure of domain IV complexed with a DnaA box has identified a helix-turn-helix (HTH) motif for DNA-binding. Residues of a "signature sequence" RDHTTVL (sig) present in one helix and the loop of the HTH motif are inserted into the major groove of the DNA and interact with five base-pairs and the phosphate backbone through hydrogen bonds, van der Waal contacts, and salt bridges. Additionally, a basic loop, containing R399 and located adjacent to the HTH motif, interacts with three more base-pairs in the minor groove through hydrogen bonds (Fig. 8.16).

DnaA belongs to the (AAA^+) superfamily of proteins. Like other members of the family, it utilizes the energy of ATP-binding and hydrolysis for its activity. Structural studies with *Aquifex aeolicus* DnaA, which has 40% primary sequence and very significant three-dimensional structural similarity to the *E. coli* homolog, have also provided useful information on the ATPase activity.

Domain III of DnaA, which is the (AAA^+) module, can be further subdivided into two subdomains—IIIa and IIIb. Subdomain IIIa (residues 130–295 in *E. coli*; 75–240 in *A. aeolicus*) consists of a compact αβα-nucleotide-binding fold—a five-stranded parallel β sheet sandwiched between helices α1 and α2 on one side and helices α3-α8 on the other—and forms the core or the base. Most of the signature sequence motifs lie within the base—Walker A (p-loop) and Walker B located at the tip of β1 and β3 strands, respectively, as well as sensor I and box VII (or SRx element). The base is connected by a short linker to the smaller subdomain IIIb, a three-helix bundle, called the "lid." Sensor II, which is an arginine at the tip of a helix, is located in the lid. All these elements together form the nucleotide-binding cleft at the junction of the base and the lid (Fig. 8.17). An invariant lysine (K178 in *E. coli*; K125 in *A. aeolicus*) interacts with β and γ phosphates of ATP. Walker B contains two conserved aspatates (D235 and D236 in *E. coli*; D180 and D181 in *A. aeolicus*). These negatively charged residues prime a water molecule for nucleophilic attack on the γ phosphate of ATP.

In the ADP-bound structure of DnaA, with the lid guarding the nucleotide-binding site, self-assembly of the monomer is strictly hindered. In the ATP-bound state, interactions between the γ phosphate of the bound nucleotide and the invariant sensor II arginine (R334 in *E. coli*; R277 in *A. aeolicus*) and also between E337 (*E. coli*)/E280 (*A. aeolicus*) and 2′-OH of the bound nucleotide reposition the lid to open the cleft. The rearrangement enables the arginine residue in the box VII SRx motif from a neighboring protomer to dock into the cleft and interact with the γ phosphate of ATP, subsequently leading to oligomerization of DnaA (Fig. 8.18).

Self-assembly of the DnaA protomers leads to the formation of a right-handed, spiral filament wrapping the duplex DNA and generating positive supercoils in it. This action results in compensatory negative writhe in the DNA as the system is topologically closed. The resulting stress causes unwinding of the DNA and the unwound state is further stabilized by the binding of ATP-bound DnaA to the single-stranded ATP-DnaA boxes. Besides, the central axis of the filament is lined with positive residues and could, therefore, act as a secondary DNA-binding region.

8.4.2 Unwinding of the DNA—Helicase and helicase loader

Once DnaA has melted the DUE, its next task is to recruit the helicase. In *E. coli*, DnaB is the replicative helicase. It has a toroidal structure with six identical subunits oriented in the same direction. Each subunit contains 471 amino acid residues. Two regions of DnaA, residues 24–86 in the N-terminal domain and 130–148 in the AAA^+ domain, and two regions of DnaB, residues 154–210 and 1–156, are involved in the protein-protein interaction. DnaB translocates along one strand of the DNA in $5' \rightarrow 3'$ direction.

Although DnaB can directly interact with DnaA, in *A. aeolicus* it requires the assistance of the helicase loader to be loaded onto *oriC* in a functionally active form. For this purpose, the C-terminal domain of DnaB interacts with the helicase loader DnaC. On the other hand, DnaC consists of two domains—a smaller N-terminal DnaB-binding domain and a larger C-terminal AAA^+ ATPase domain (Fig. 8.19). The nucleotide-binding site of DnaC is similar to that of DnaA, the only exception

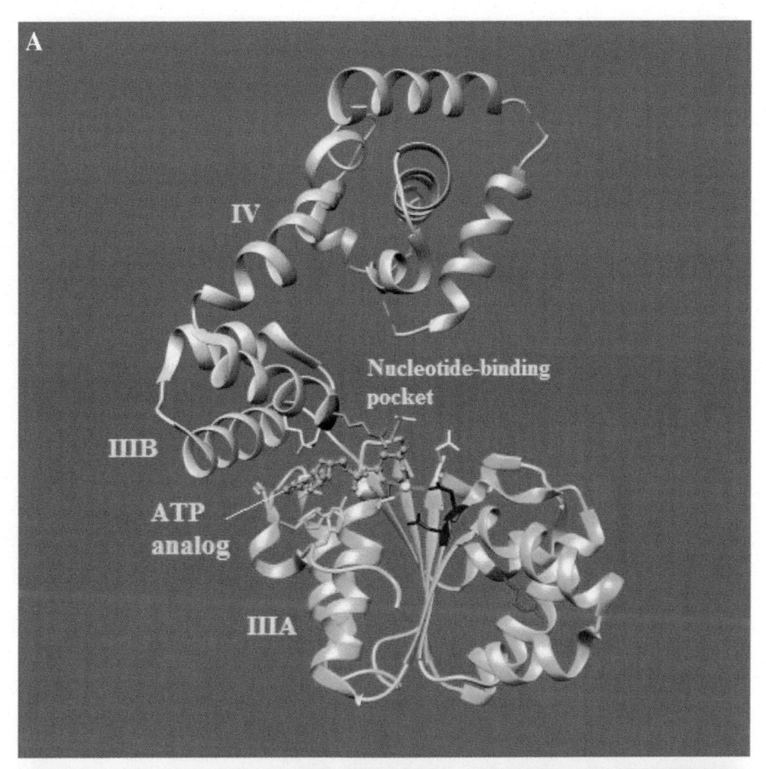

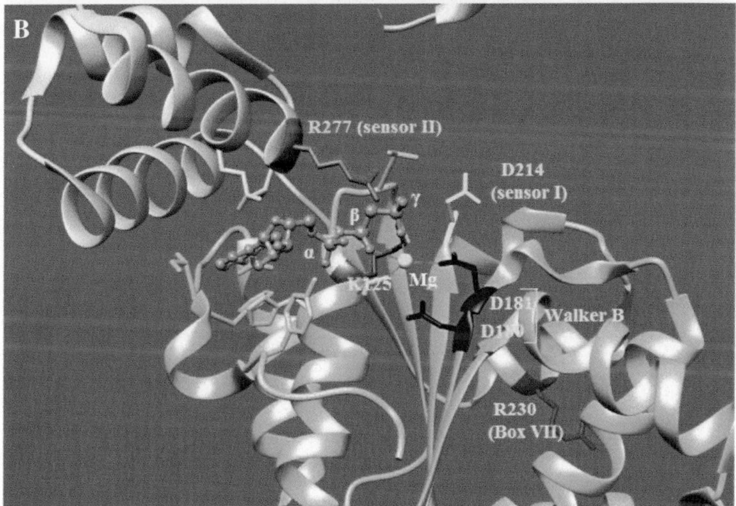

FIG. 8.17

(A) The structure of DnaA domains III and IV showing the nucleotide-binding cleft.
(B) Close-up view of the nucleotide-binding cleft. *Light green*: domain IIIa; *light pink*: domain IIIb; *tan*: domain IV.

Source: Erzberger, J.P., Mott, M.L., Berger, J.M., 2006. Nat. Struct. Biol. 13, 676.

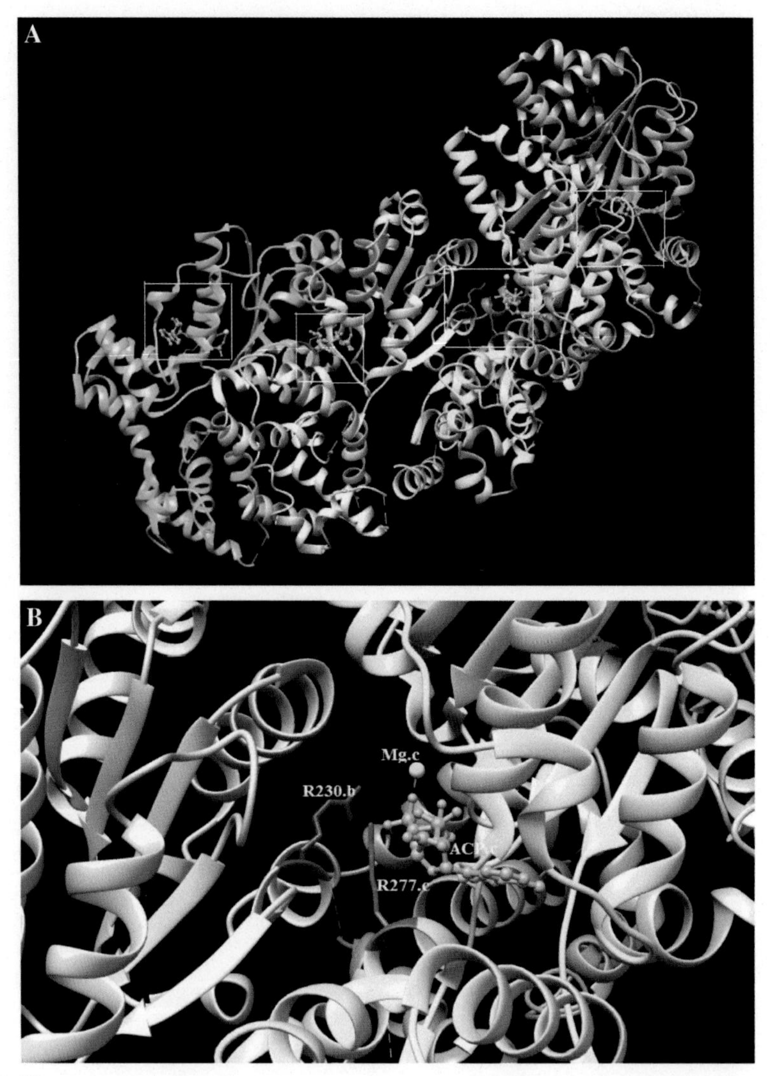

FIG. 8.18

Nucleotide-binding-mediated oligomerization of *Aquifex* DnaA. (A) Docking sites of adjacent protomers marked by boxes. (B) Close-up view of the docking site.

Source: Erzberger, J.P., Mott, M.L., Berger, J.M., 2006. Nat. Struct. Biol. 13, 676.

being the absence of the C-terminal lid. Therefore, as expected, it can also oligomerize and form a right-handed mini-filament. The free arginine in box VIII at one end of the minifilament can dock into the open ATP-binding cleft of the DnaA filament and form a heterofilament.

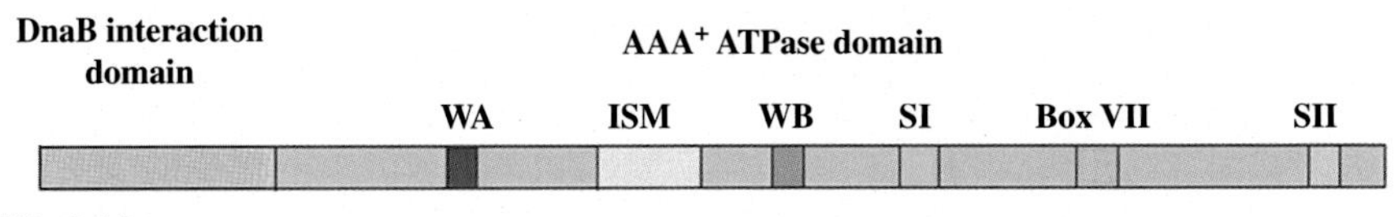

FIG. 8.19

Domain structure of DnaC. *WA*—Walker A; *ISM*—initiator-specific motif; *WB*—Walker B; *SI*—sensor I; *SII*—sensor II.

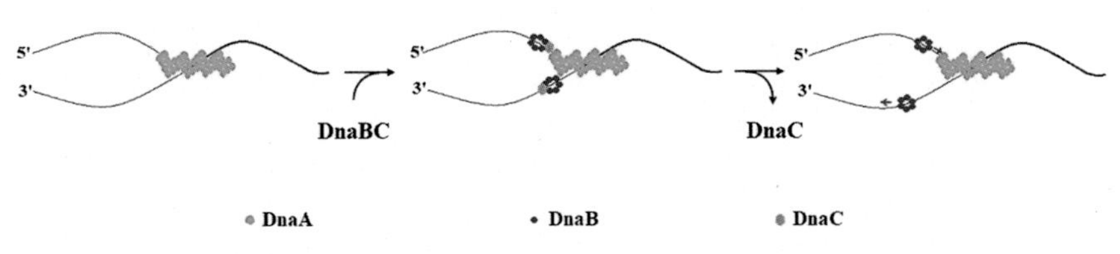

FIG. 8.20

Loading of two replicative helicases at *oriC*.

As bacterial replication is bidirectional, the basic requirement is that two copies of DnaB need to be placed in opposite directions on the two strands of the unwound DUE where the homo-oligomerized DnaA is associated with only one strand. However, different faces of DnaB are recognized by the initiator and the helicase loader. Also, in *A. aeolicus*, DnaA can efficiently bind both DnaB and DnaC. Hence, DnaA should be able to interact with the DnaBC complex through either DnaB or DnaC. This gives credence to the following model for helicase loading in *A. aeolicus*. As shown in Fig. 8.20, DnaA can deposit a single BC complex in one orientation on one strand of the origin through direct initiatior:helicase interaction. In contrast, interaction between DnaC and oligomerized DnaA should be able to place DnaBC complex on the other strand in opposite orientation.

In *E. coli*, however, the interaction between DnaA and DnaC is relatively weaker. Here, it is conceived that IHF may have a role in helicase loading. The binding of IHF adjacent to the DUE results in a sharp 160° bend of the DNA helical axis. This enables the DNA filament, which is formed on one strand of oriC, to be extended into the other strand. The helicase can then be loaded on both the strands, though not exactly opposite to each other, by protein-protein interaction with the flexible N-terminal domain of DnaA.

8.4.3 Priming DNA synthesis—Primase and SSB

Once the helicase is loaded, its task is to recruit the primase through a direct protein-protein interaction in order to complete the priming step of DNA replication. The *E. coli* primase, DnaG, is composed of three domains—the N-terminal zinc-binding domain for template recognition, the central polymerase domain for primer synthesis, and the C-terminal helicase-binding domain which couples replication fork

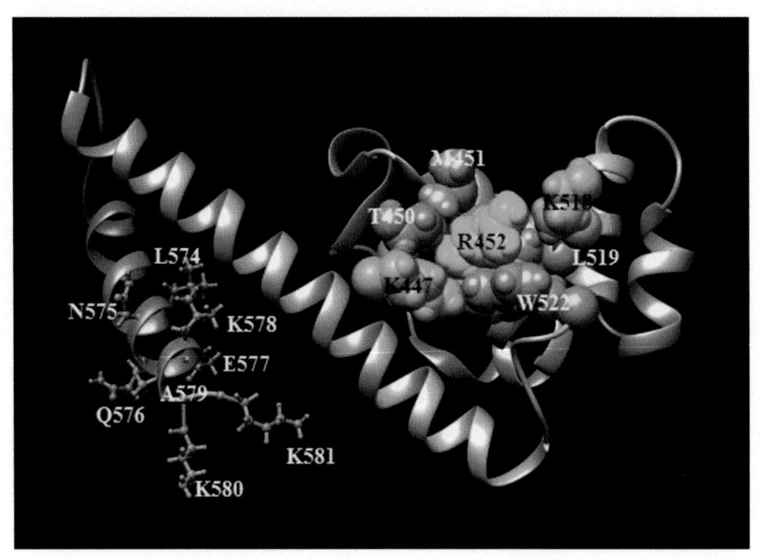

FIG. 8.21

Carboxy-terminal helicase-binding domain of DnaG. The residues interacting with DnaB are shown as pink ball and sticks. The hydrophobic pocket shown with spheres *(orange)* surrounded by basic residues *(cyan)* is for interaction with the SSB.

Source: Su, X.C., et al., 2006. FEBS J. 273, 4997.

progression with primer synthesis. The interaction with the replicative helicase involves the last eight amino acids, LNQELAKK, in the C-terminal helix hairpin of DnaG (Fig. 8.21). We shall see shortly that the C-terminal domain of DnaG also interacts with single-stranded DNA-binding protein (SSB). However, DnaG- and SSB-binding sites do not overlap (Fig. 8.21).

The helicase DnaB, on the other hand, is also able to interact with the helicase loader DnaC. The *E. coli* DnaB consists of an N-terminal domain (NTD) and a larger C-terminal domain (CTD) connected by a linker. The CTD is involved in nucleotide-binding and hydrolysis and interaction with DnaC, whereas two isoleucine residues, I135 and I141, in the linker and helical hairpins in the NTD mediate the interaction of DnaB with DnaG. However, the primase is unable to interact with DnaB when the latter is bound to DnaC.

DnaB has a dynamic structure—its hexametric ring opens and closes. By the interaction of specific residues in the N-terminal domain of DnaC and C-terminal domain of DnaB, DnaC traps DnaB as an open ring. This interaction also affects the helical hairpins in the DnaB NTD. As a result, DnaB is prevented from interacting with the primase prior to the binding of the DnaBC complex at *oriC*. The loading of the BC complex onto the unwound region of *oriC* makes specific site in the DnaB NTD accessible to the primase. Further, the interaction of the primase with DnaB and primer formation transmits a signal through a conserved

R220 the box VII motif of DnaC, leading to the release of the helicase loader and activation of the helicase.

At this point, DnaB and DnaG together constitute what is known as the primosome. The CTD of DnaB is structurally similar to the ATPase domain of RecA. It couples nucleotide hydrolysis with DNA unwinding. The primase, on the other hand, synthesizes short oligoribonucleotide primers (10–12 bases). For its activity, the primase stably associates with the template through contacts with the single-stranded DNA-binding protein (SSB).

The *E. coli* single-stranded DNA-binding protein (EcoSSB) is a tetramer; each monomeric subunit consists of 178 amino acid residues. The N-terminal of the protein (residues 1–135) contains the DNA-binding domain (OB-fold). Four positively charged concavities are present on the surface of EcoSSB, containing β sheets, basic amino acids, and loops. These concavities bind to patches of single-stranded DNA (ssDNA) in a nonspecific manner and prevent nuclease action and hairpin formation, thus stabilizing the DNA template in a suitable structure for polymerase action (Fig. 8.22). The C-terminus of SSB is amphipathic containing three aspirates followed by three hydrophobic residues (DDDIPF). Interacting with this motif is a highly conserved hydrophobic pocket (containing residues T450, M451, I455, L519, and W522) surrounded by basic residues K447, K518, and R452 on the surface of the C-terminal domain of DnaG (Fig. 8.21). The basic residues interact with the aspartate, while proline and phenylalanine contact the hydrophobic pocket.

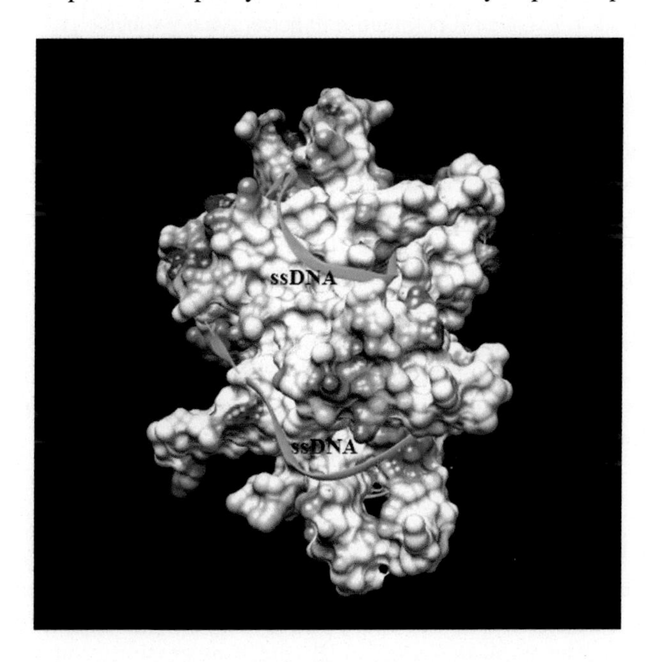

FIG. 8.22

Positively charged concavities on the surface of *E. coli* SSB protein binding to ssDNA.

Source: Raghunathan, S., 2000. Nat. Struct. Mol. Biol. 7, 648.

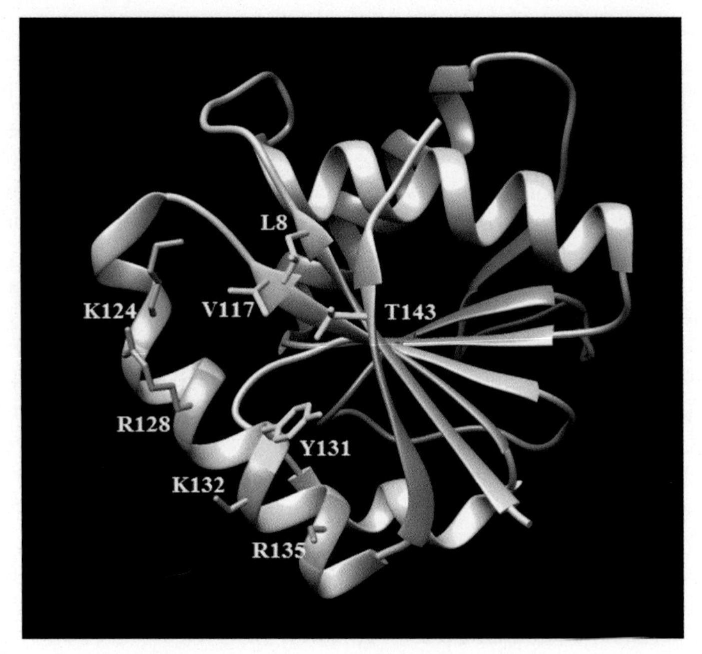

FIG. 8.23

SSB-binding site in *E. coli* DNA polymerase III holoenzyme χ subunit. Hydrophobic pocket with residues colored *yellow* are adjacent to basic residues colored *blue*.

Source: Marceau, A.H., et al., 2011. EMBO J. 30, 4236.

Once the primer has been synthesized, the primase must give way for the replicase to take over and extend the primer with deoxiribonucleotides (dNTPs). In *E. coli*, this primase-to-polymerase switch is affected by a subunit of DNA polymerase III holoenzyme, χ, which can also bind to SSB. χ has an SSB-binding site similar to that of DnaG. Its hydrophobic pocket formed by L8, V117, Y131, and T143 can accommodate the amino acid residues, P176 and P177. The adjacent basic residues, K124, R128, K132, and R135 (Fig. 8.23), interact with the negatively charged aspartates. Binding of χ to SSB destabilizes the primase-to-SSB contact, resulting in the dissociation of the primase and subsequent association of the replicase.

8.5 Bacterial replicase

In *E. coli*, DNA polymerase holoenzyme (Pol III HE) is the replicase responsible for synthesis of new DNA strands. Pol III HE comprises of three functionally distinct but interconnected subcomplexes—the Pol III core, a processivity component (clamp), and a clamp loader complex.

8.5.1 The polymerase

In the PolIII core, the 130 kDa α is the catalytic subunit primarily responsible for DNA synthesis. In order to carry out its polymerase activity, α must bind to the substrates, the RNA primer-template DNA (p/t) junction complex and deoxyribonuleotides (dNTPs).

The α polypeptide contains 1160 amino acid residues. Crystal structure of a large fragment (residues 1–917) of α has been determined at 2.3 Å resolution (Fig. 8.24) and is available in the Protein Data Bank (PDB ID: 2HQA). Fig. 8.25 shows the domain architecture of the α subunit. The Palm subdomain (residues 271–432 and 511–560) is centrally located. On the left of the Palm subdomain is the Fingers subdomain (residues 561–911) and on the right side is the Thumb subdomain (residues 433–510). Additionally, α also has a Polymerase and Histidinol Phosphatase (PHP) domain (residues 2–270).

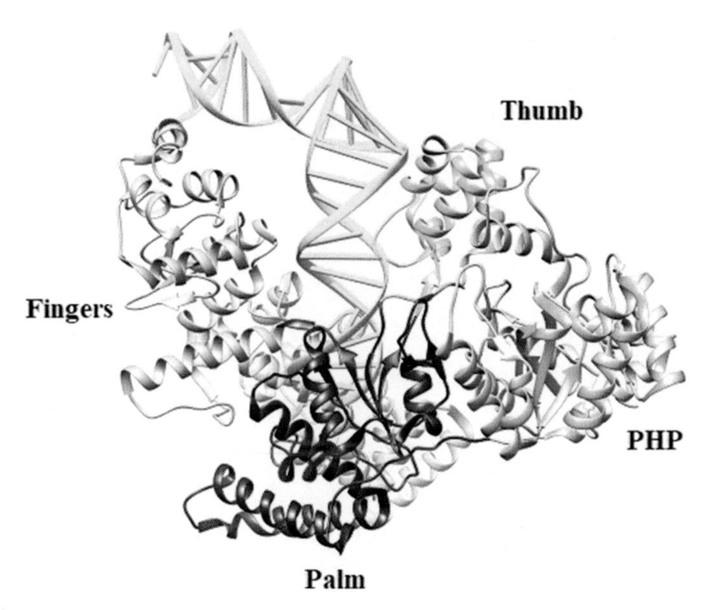

FIG. 8.24

Ribbon representation of α polypeptide with DNA.

Source: Lamer, M.H., et al., 2006. Cell 126, 381.

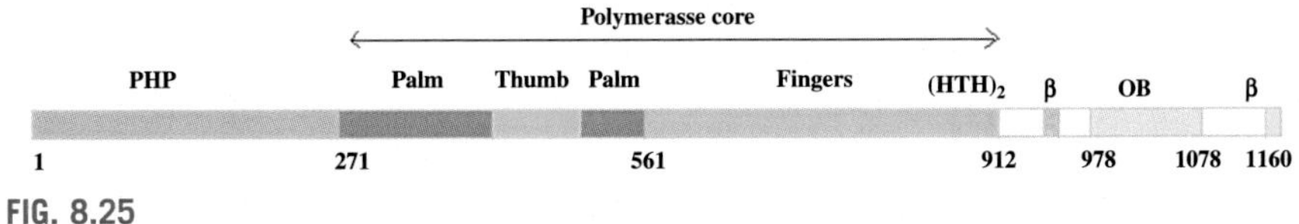

FIG. 8.25

Domain architecture of α subunit.

At least two distinct subdomains of α are necessary for its binding to 30 nucleotides of DNA in the PT complex—helix-hairpin-helix $(HhH)_2$ and oligonucleotide/oligosaccharide-binding (OB) fold (Fig. 8.25). $(HhH)_2$ is formed by two consecutive HhH motifs, which are known to bind nonspecifically to DNA by formation of hydrogen bonds between backbone nitrogen of the polypeptide and DNA phosphate oxygen. The OB-fold contains a highly conserved phenylalanine residue (F1031). The aromatic side chain of phenylalanine participates in a base-stacking interaction with single-stranded (SS) DNA. Together, Palm, Thumb, and Fingers subdomains form a deep cleft which is involved in electrostatic interactions with the sugar-phosphate backbone of the DNA along its minor groove. Several conserved positively charged residues are present in a region spanning the Thumb and Palm subdomains (Fig. 8.26).

Presence of the PHP domain in bacterial replicative polymerases is strictly conserved. However, its function in the α subunit is not fully understood. Possibly, it is involved in the hydrolysis of pyrophosphate generated in the phosphotransfer reaction and also helps in the proofreading activity of PolIII.

Three aspartate residues (D401, D403, and D555) are located in a five-stranded β-sheet of the Palm domain (Fig. 8.27). D401 and D403 coordinate two Mg^{2+} ions in the phosphotransfer reaction. D555 activates the 3′-OH of the primer strand for nucleophilic attack on the α-phosphate of the incoming nucleotide. In the active structure of α subunit, the 3′-terminal nucleotide of the primer and its template base are positioned close to a Finger domain loop (residues 753–758). The loop contains

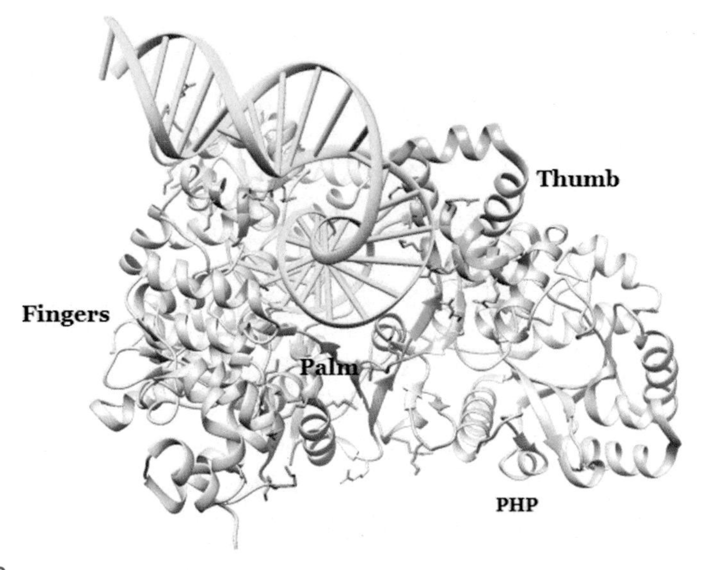

FIG. 8.26

Cross-sectional view of the DNA held within the cleft formed by Fingers, Palm, and Thumb domains. Positively charged residues are marked in *orange*.

Source: Lamer, M.H. et al., 2006. Cell 126, 381.

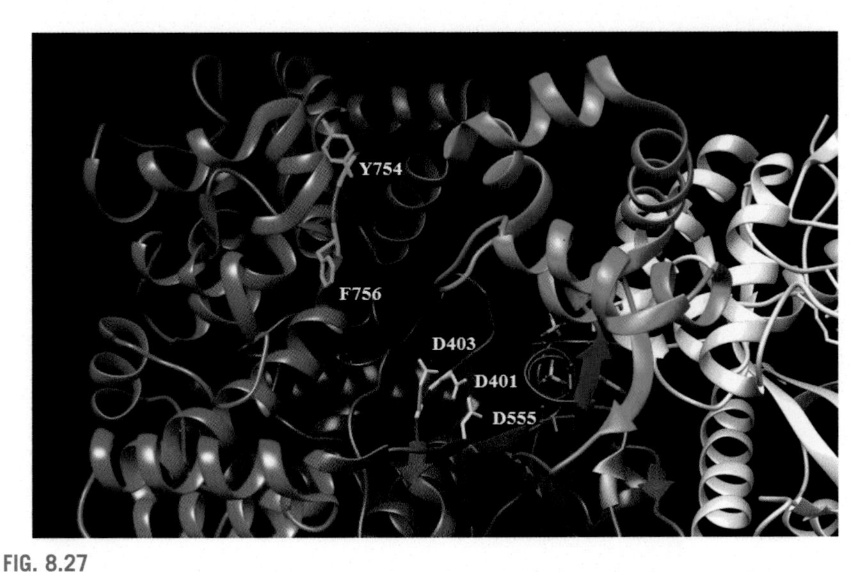

FIG. 8.27

Catalytic site of α subunit. Domains: *green*—thumb; *brown*—palm; *blue*—fingers.

Source: Lamer, M.H., et al., 2006. Cell 126, 381.

two aromatic residues (Y754 and F756) that interact with the nascent base-pair, enabling the catalytic Mg^{2+} ions to perform the phosphotransfer reaction (Fig. 8.27).

The second member of the Pol III catalytic core, the ε subunit, plays a catalytic as well as a structural role. Although known to be a 3′–5′ proofreading exonuclease, ε also influences DNA synthesis in a positive manner by stimulating the interaction of α with the β clamp. The ε subunit is a 27 kDa polypeptide with 243 amino acid residues (Fig. 8.28A). The proofreading activity resides in its N-terminal 185 residues. Residues near the C-terminal segment of ε (ε_{CTS}) are responsible for binding to α, while in the α subunit the binding site for ε is located within the first 320 residues spanning the N-terminal 270-residue PHP domain and the Palm domain. ε_{CTS} residues between K211 and the C-terminus make electrostatic and H-bond contacts with α, strengthened by a larger number of hydrophobic interactions (Fig. 8.28B). ε_{CTS} also contains a β clamp-binding motif (CBM) QTSMAF. The N-terminal proofreading domain and C-terminal α-binding site of ε are linked by a highly flexible tether between residues 183 and 201. The proofreading activity of ε will be discussed later in this section.

The third and the smallest subunit of DNA polymerase III is θ. This 8.8 kDa peptide neither has any identified enzymatic activity nor interacts with α. However, it interacts with and enhances the exonuclease activity of the ε subunit. Thus, θ appears to coordinate α-ε polymerase-exonuclease interaction.

8.5.2 The processivity clamp

α subunit is capable of DNA synthesis by itself, but with low efficiency and processivity. The processivity is increased remarkably when α is bound to β sliding clamp, a ring-shaped molecule that encircles the DNA, and clamp loader, γ. Crystal structure

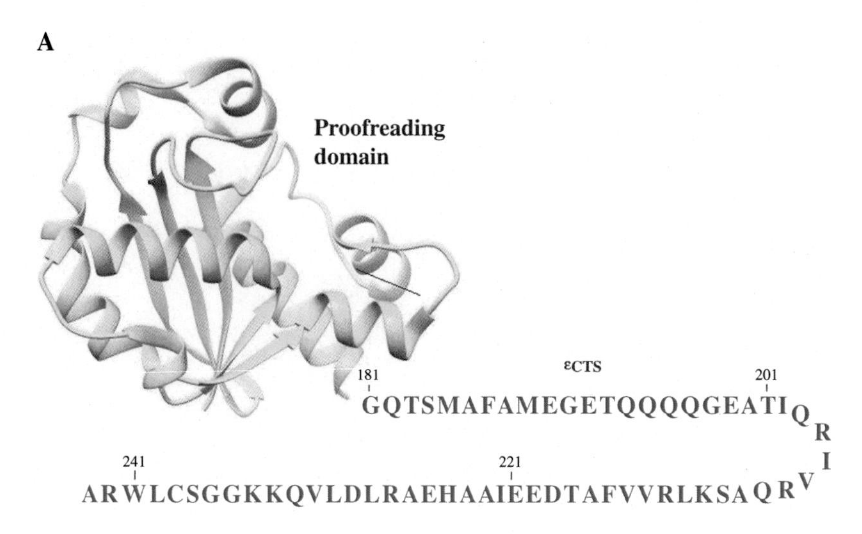

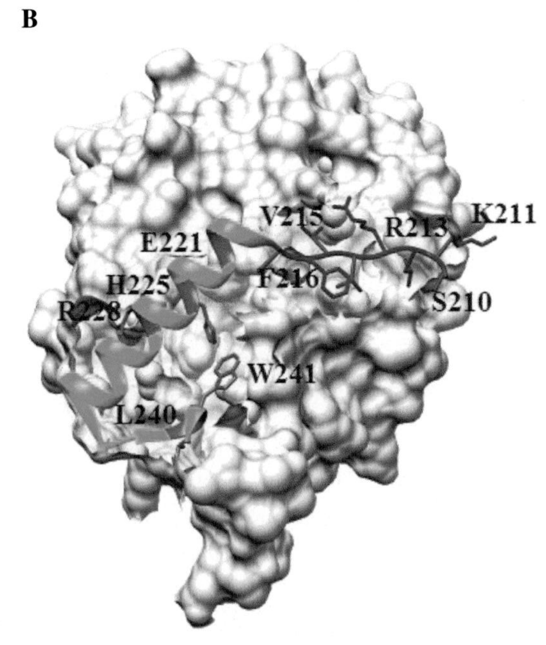

FIG. 8.28

(A) ε subunit, (B) surface representation of PolIII α subunit PHP domain and side chains of selected ε_{CTS} residues making electrostatic and H-bond contacts with α.

Source: Hamdan, S., et al., 2002. Structure 10, 535.

of the β clamp (PDB ID: 1MMI) shows that it is a head-to-tail homodimer, approximately 80 Å in diameter, with a 35 Å-central hole (Fig. 8.29). Each protomer consists of three structurally similar domains—Domain I (residues 1–125), Domain II (residues126–253), and Domain III (residues 254–366). The dimer interface is

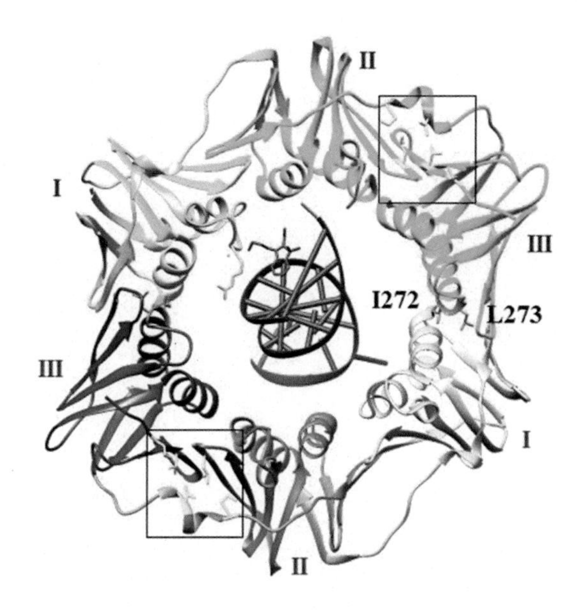

FIG. 8.29

Crystal structure of the β clamp with the dimer interface and hydrophobic cleft shown.

Source: Georgescu, R.E., et al., 2008. Cell 132, 43.

stabilized mostly by hydrophobic interactions mediated by two amino acid residues, I272 and L273 (Fig. 8.29).

The site of interaction with the α subunit as well as a few other DNA polymerases is a hydrophobic cleft in the β clamp formed by L177, P242, V247, V360, and M362 located between Domains II and III. α subunit has two clamp-binding motifs ($Qx_2[L/M]x_{0/1}F$) located in the C-terminal 243 residues—an internal motif, QADMF, (residues 920–924) and a second motif at the terminus, QVELEF, (residues 1154–1159) (Fig. 8.25). The hydrophobic residue at the fourth position of the pentapeptide motif and another hydrophobic residue (mostly aromatic) at a variable position are important for binding of the polymerase to the cleft. In addition, the OB-fold (residues 994–1073) located in the middle of the C-terminal segment helps in the clamp recognition by either interacting directly with the latter or helping the internal clamp-binding motif in doing so.

8.5.3 The clamp loader

It has been mentioned earlier that the β clamp augments the processivity of the PolIII polymerase. For this purpose, the clamp has to be loaded onto the p/t junction by a clamp loader. The *E. coli* clamp loader contains seven subunits and designated as $\delta\delta'(\gamma/\tau)_3\chi\psi$. However, it is the heteropentamer core $\delta\delta'(\gamma/\tau)_3$ that is essential for clamp loading function. γ and τ are encoded by the same gene; τ contains two additional domains IV and V that bind replicative helicase $DnaB_6$ and PolIII, respectively, thus holding all catalytic activities together. Domain IV and V are, however, not required for clamp loading.

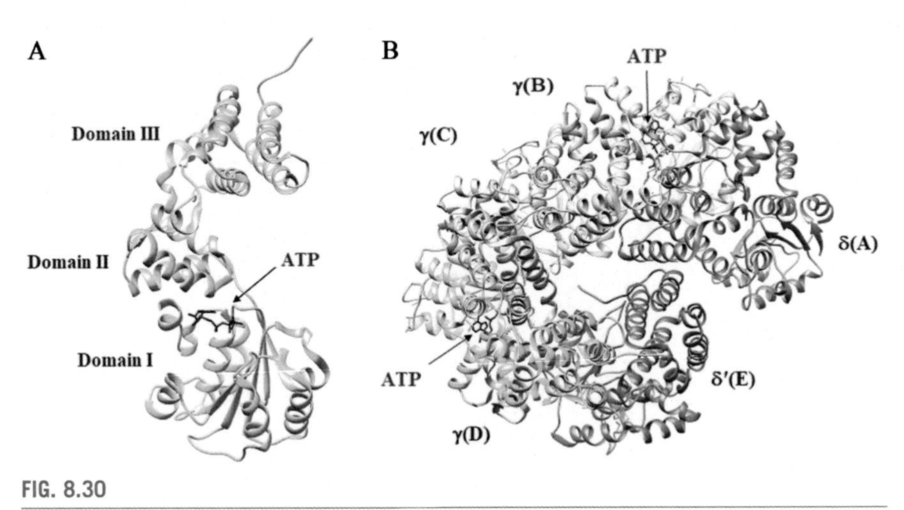

FIG. 8.30

Clamp loader structure. (A) Domain structure of each subunit and ATP-bound form of the heteropentamer core $\delta\delta'(\gamma/\tau)_3$. (B) Nucleotide-binding sites in γ-subunits.

Source: Kazmirski, S.L., et al., 2004. Proc. Natl. Acad. Sci. U. S. A. 101, 16750.

Fig. 8.30 shows the ATP-bound structure of the clamp loader complex (PDB ID: 1XXH). All five polypeptides, δ, δ′, and three copies of γ, contain the same three domains fold and bind to each other by Domain III (Fig. 8.30). The N-terminal domain (Domain I) contains Walker A (P-loop), Walker B, sensor I, and SRC motifs. Walker A and B motifs are important for nucleotide-binding and hydrolysis. Domain II contains a sensor II motif analogous to sensor I. The sensor I and II motifs "sense" the presence or absence of γ phosphate of ATP to trigger conformational changes in the clamp loader complex. The "arginine finger" of the SRC motif participates appropriately in ATP hydrolysis. Together, Domains I and II form an ATP-binding module. The C-terminal domains (Domains III) of the five subunits hold the complex together to form a ring-shaped "collar." The three γ subunits together are located in the middle, whereas δ and δ′ flank on the sides (Fig. 8.30).

In the absence of ATP, δ, which is the potential contact point with the β ring, is restricted within the clamp loader complex by its interaction with δ′. Consequently, the interaction between the clamp and the clamp loader is very weak. When ATP binds to the clamp loader, the interaction between δ and δ′ is disrupted and δ is able to interact with the C-terminal face of a closed β clamp. The N-terminal domain of δ, which resembles a triangular wedge, is involved in the interaction with β. A "hydrophobic plug" of δ squeezes into a hydrophobic cleft between Domains II and III of either β monomer, thus destabilizing the adjacent interface and forcing open the β clamp ring (Fig. 8.31). The interactions involve three hydrophobic residues of δ, namely, M71, L73, and F74, the last two residues protruding out to form the "plug." The hydrophobic pocket on the surface of β is formed by L177, P242, V247, V36,

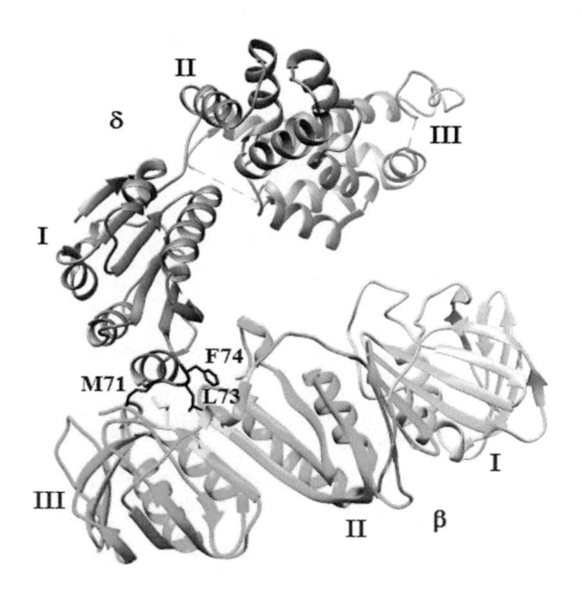

FIG. 8.31

β:δ complex. δ residues forming hydrophobic "plug" labeled. β residues forming hydrophobic pockets shown in yellow.

Source: Jeruzalmi, D., et al., 2001. Cell 106, 417.

and M362. Although δ does not directly interact with the dimer interface in β, its binding leads to a major structural distortion preventing the formation of a well-packed dimer interface. The entire process occurs in the absence of ATP hydrolysis. Thus, ATP-binding, and not hydrolysis, is required for the clamp loader binding to and opening the β clamp.

The clamp loader needs not only to bind to the β clamp but also to target the latter to the site of DNA synthesis, that is, the p/t junction. The affinity of the clamp loader for p/t junction is modulated by ATP. The ATP-bound form of the clamp loader complex has a high affinity for the p/t junction. However, the complex alone has a negligible ATPase activity. In the absence of DNA, the arginine finger (R169) from one γ subunit (γ_C) is not able to interact with the γ-phosphate of ATP in the adjacent subunit (γ_B) (Fig. 8.32A). Binding to the DNA affects a structural rearrangement in such a way that the arginine finger from one subunit is now close enough to interact with the ATP-binding site of the next and promote ATP hydrolysis (Fig. 8.32B). Consequently, the sensor I and II motifs "sensing" the absence of the γ-phosphate of ATP trigger a conformational change in the clamp loader complex converting the latter to a low-affinity state, thus causing it dissociate from the DNA.

It is not known if the open sliding clamp closes before, during, or after ATP hydrolysis. However, the closing is promoted by the electrostatic interaction between the basic interior of the β ring and negatively charged DNA.

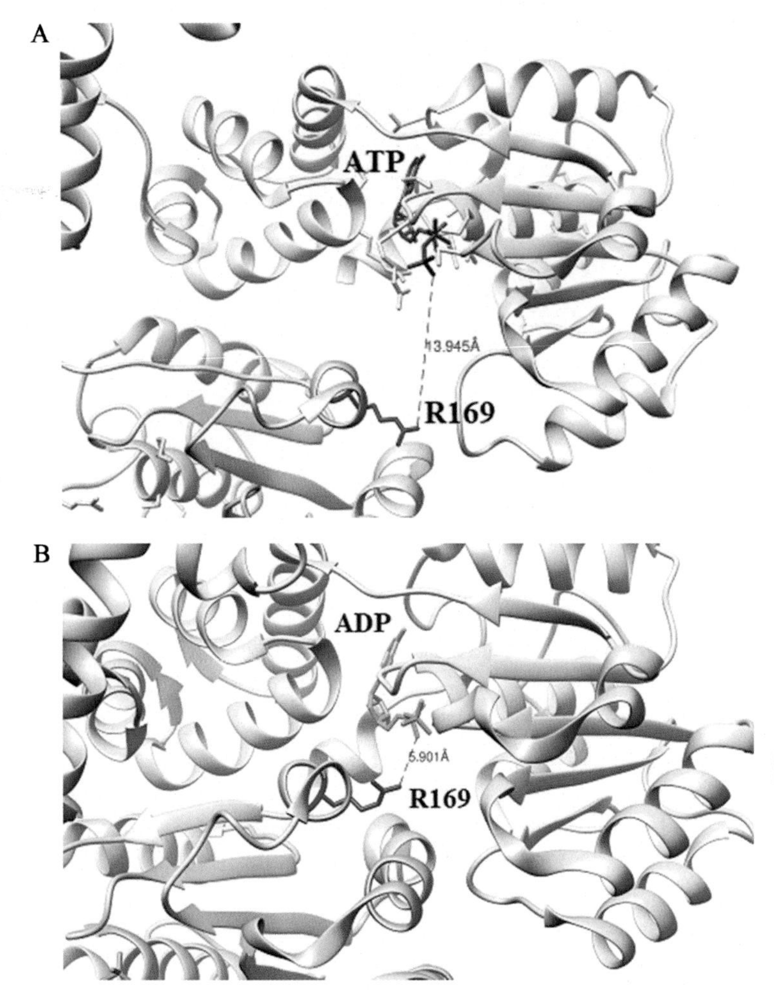

FIG. 8.32

DNA promoted ATP hydrolysis by clamp loader complex. Distance between R169 and ATP/ADP shown by dashed lines with the clamp loader is (A) not bound and (B) bound to DNA.
Source: Kazmirski, S.L., et al., 2004. Proc. Natl. Acad. Sci. U. S. A. 101, 16750; Simonetta, K.R., et al., 2009. Cell 137, 659.

8.5.4 Proofreading

We have seen that the polymerases select the correct nucleotide to be added to the 3′-end of the primer based on base-pair geometry and complementarity between the bases. But it should be recognized that base-pair geometry and complementarity between the bases can ensure only a limited degree of fidelity in DNA replication. Wrong (infrequent) tautomeric form of the bases, occurring at a frequency of 1 in 10^5, can deceive the complementarity rule and lead to the incorporation of an "incorrect" base at the primer-end. However, quick return of the "incorrect" base to its "normal" (predominantly more frequent) form generates at the primer-end a "mismatch" (with the template) that must be removed by a 3′–5′ exonuclease action for a higher level of replication fidelity. This exonuclease activity is provided by the DNA polymerase. In the case of *E. coli* Pol I, as for example, the polymerase and exonuclease active sites are located in separate domains of the same polypeptide, whereas, as mentioned earlier, in Pol III, the exonuclease activity is carried by a different subunit, ε.

The exonuclease reaction of ε is catalyzed by the two-metal mechanism, the catalytic metal and the nucleotide-binding metal, both being preferably Mn^{2+}. Two aspartate residues, D12 and D167, present in the exonuclease active site of ε coordinate the two metal ions (Fig. 8.33).

In the Klenow fragment of Pol I, the exonuclease and polymerase sites are separated by less than 30 Å that enables the primer end switch between the two sites depending on the conformational status of the p/t DNA (Fig. 8.34). When a mismatch appears at the polymerase site, the junction is destabilized and a few bases are unpaired. The exonuclease site having 10-fold higher affinity for 3′-ends, polymerization is stalled, and the 3′-end of the primer moves to the exonuclease site.

In the polymerization state of Pol III, when ε is bound to the β clamp, the distance between the polymerase and exonuclease sites is about 70 Å. However, during transition to the proofreading, the ε-β contact is snapped, α assumes a more open structure, and by the virtue of the flexible linker ε repositions its exonuclease domain close to the polymerase site to access the mismatched primer terminus. The mismatched bases are removed and the p/t junction is reformed and bound again to the polymerase active site. DNA synthesis continues.

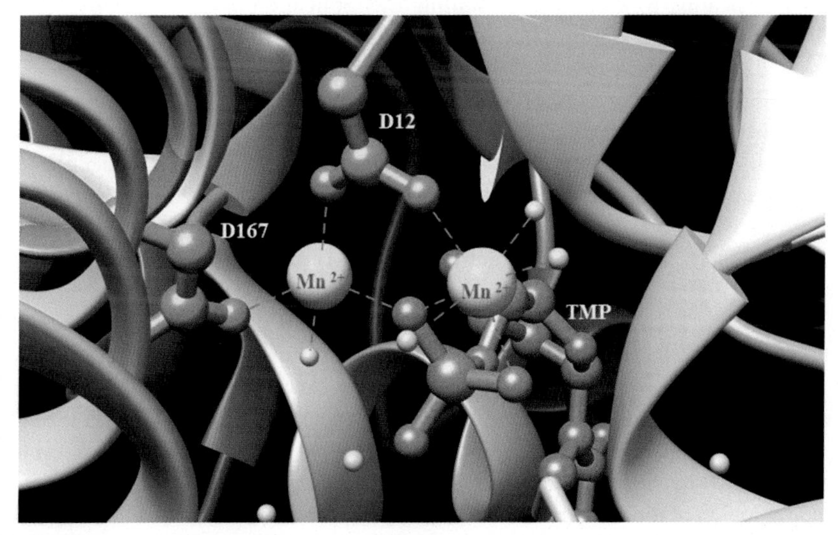

FIG. 8.33

Catalytic site in *E. coli* DNA polymerase ε subunit.

Source: Kirby, T.W., et al., 2006. J. Biol. Chem. 281, 38466.

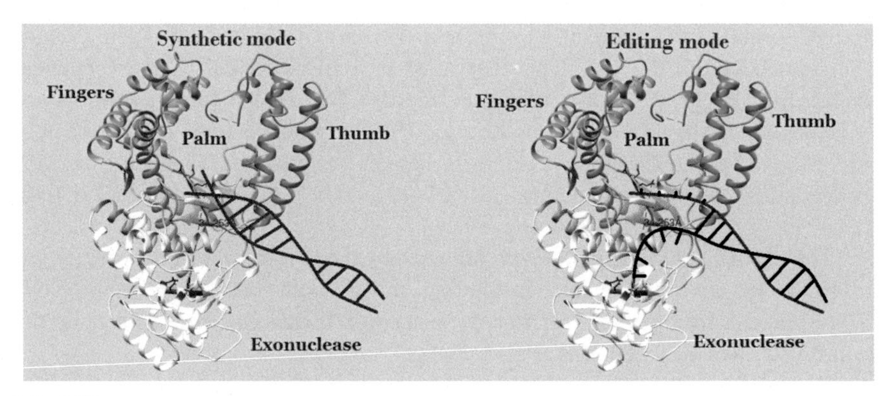

FIG. 8.34

Shuttle mechanism of proofreading.

Source: Brautigam, C.A., Steitz, T.A., 1998. J. Mol. Biol. 277, 363.

8.6 Disentanglement of replication products

We can easily realize that replication of a covalently closed circular DNA generates two topological problems. First, as the replicative helicase unwinds the DNA at the replication fork, there is a local reduction in the linking number. In such case, if the strands remain unbroken, the overall linking number of the DNA molecule would not be allowed to change. Hence, the unwinding would be compensated by positive supercoils elsewhere in the DNA. This would generate increasing strain in front of the replication fork, ultimately stalling the replication machinery. However, the cell possesses an enzymatic mechanism to relieve the strain—in *E. coli*, the enzyme involved is DNA gyrase.

The second problem is faced at the completion of replication of the circular DNA molecule. The two daughter duplexes remain intertwined as catenanes. To avoid abnormal DNA segregation at cell division, the two circular DNA molecules need to be disentangled from each other or "decatenated". Here again in *E. coli cells*, the process of decatenation is carried out by DNA gyrase.

E. coli DNA gyrase is a heterotetramer (A_2B_2) composed of two subunits, GyrA and GyrB (Fig. 8.35). Its enzymatic action involves a double-stranded break in one DNA segment, known as the gated or G-segment DNA, the passage of another DNA segment, called the transported or T-segment, which can be part of the same molecule (in case of relaxation) or a different molecule (in case of catenation or decatenation) through the gap and finally resealing the break (Fig. 8.36).

DNA-binding and cleavage are carried out by the GyrA (875 amino acid residues) subunit. Its N-terminal domain (NTD) contains a region that resembles the DNA-binding region of the *E. coli* catabolite activator protein (CAP) (Fig. 8.37). An active site tyrosine (Y122) is located in a helix-turn-helix (HTH) motif found in this region (Fig. 8.37).

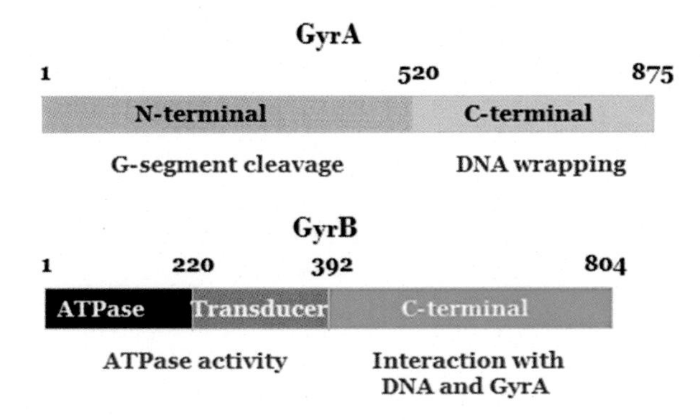

FIG. 8.35

Domain organization of *E. coli* DNA gyrase.

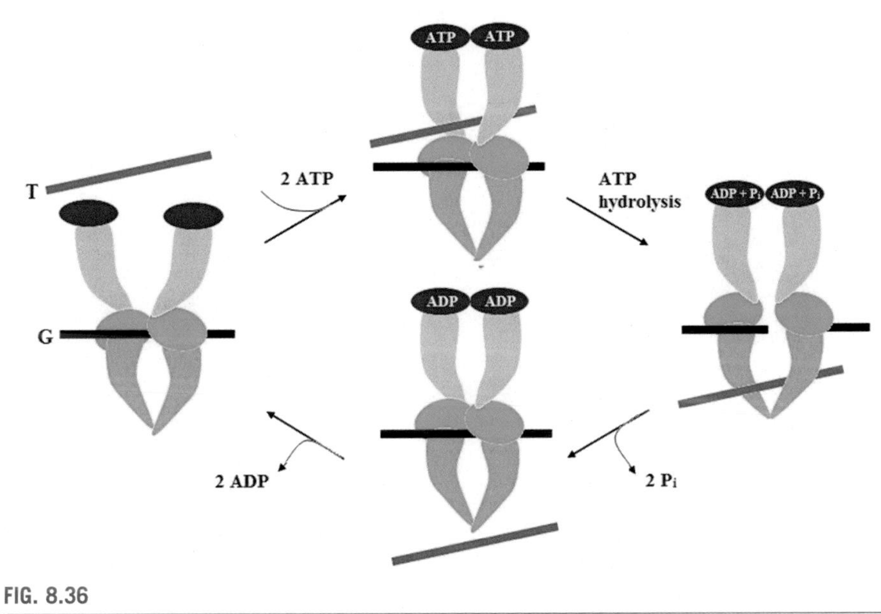

FIG. 8.36

DNA gyrase tetrameric architecture and mechanism.

The GyrB subunit contains 804 amino acid residues (Fig. 8.35). Its N-terminal fragment comprises the ATPase and transducer domains (Fig. 8.38A). The ATPase domain consists of a GHKL ATP-binding fold with an eight-stranded antiparallel β-sheet backed by five α-helices (Fig. 8.38b). The transducer domain comprises a four-stranded β-sheet backed by two α-helices (Fig. 8.38C). A critical lysine residue (K337) projects into the active site of the GHKL domain (Fig. 8.38D).

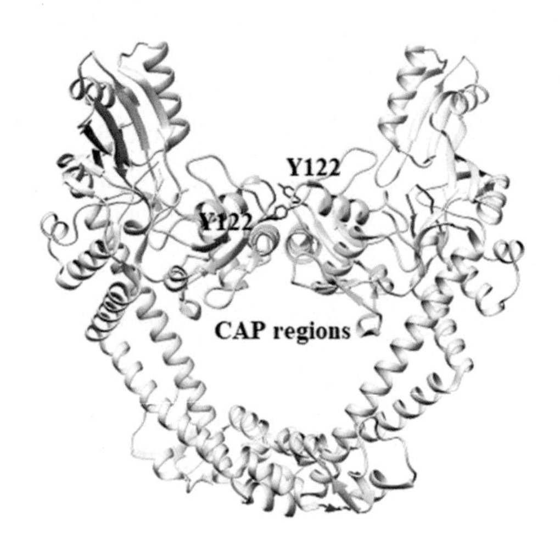

FIG. 8.37

Structure of *E. coli* DNA gyrase A subunit N-terminal domain.

Source: Hearnshaw, S.J., et al., 2014. J. Mol. Biol. 426, 2023.

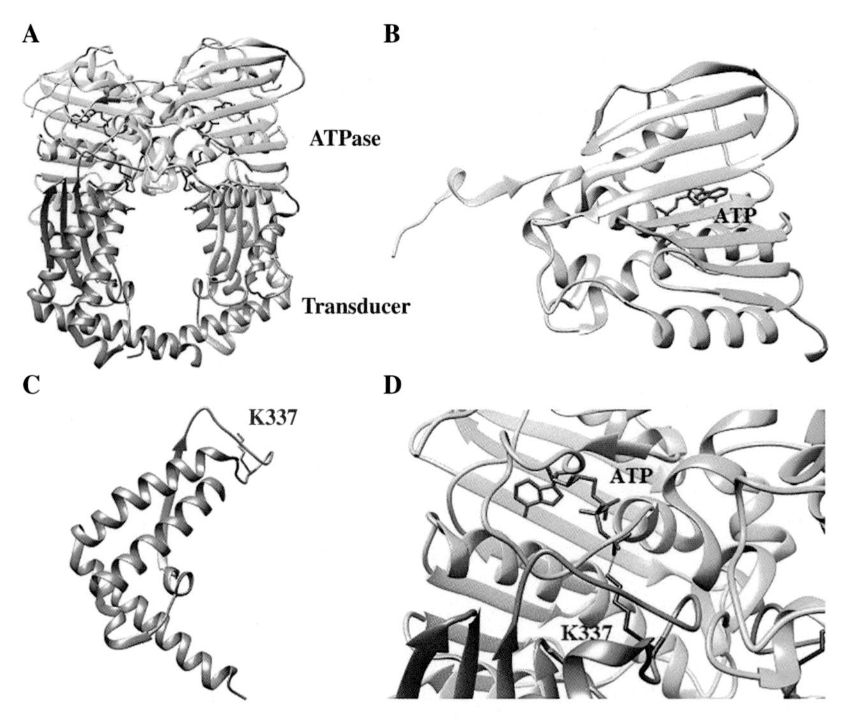

FIG. 8.38

E. coli DNA gyrase B subunit structure and inter-domain interaction. (A) N-terminal fragment, (B) ATPase domain, (C) transducer domain, and (D) lysine-mediated interaction between ATPase and transducer domains.

Source: Brino, L., et al., 2000. J. Biol. Chem. 275, 9468.

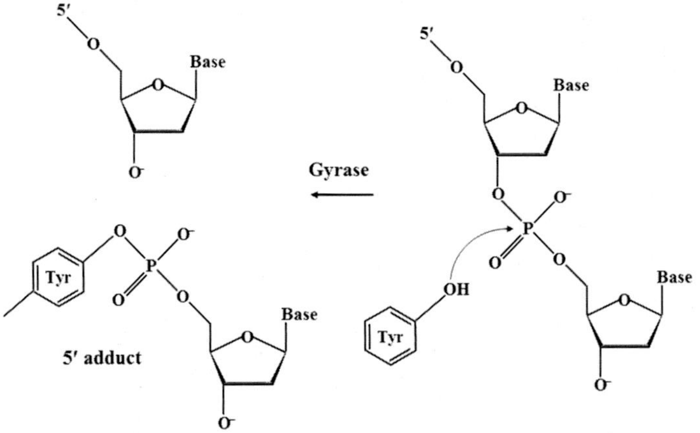

FIG. 8.39

G-segment cleavage and transesterification by DNA gyrase.

The GyrB CTDs interact with the G-segment, while the N-terminal region, consisting of the GHKL and transducer domains, self-associates to form one dimer interface or the "entrance gate" for the T-segment. The second interface or the "exit gate" is formed by the GyrA-NTDs which mediate the interaction between GyrA and GyrB.

The active site tyrosine (Y122) carries out the G-segment cleavage. The cleavage sites on the two DNA strands are staggered by four base-pairs; a 5′ overhang is thus generated in the cleaved DNA. This is followed by a reversible transesterification reaction involving the two tyrosine residues in the two GyrA subunits and two phosphoryl groups on the two DNA strands (Fig. 8.39).

The transport of one DNA duplex (T-segment) through another (cleaved G-segment) requires ATP-binding and hydrolysis. Both the GHKL and transducer domains are responsible for the ATPase activity. ATP binds to a site formed by a "floor" of β-strands and "walls" of α-helices in the GHKL domain, whereas the transducer lysine (K337) hydrogen bonds with the γ-phosphate of bound nucleotide and promotes hydrolysis.

ATP-binding powers the opening and closing of the "entrance gate" to enable the T-segment move into the cavity formed by the GyrB CTDs. This is followed by the cleavage of the G-segment forming the DNA gate. The nucleotide-binding also induces an 11-18° rotation between the GHKL and transducer domains. This transduces structural signals to other parts of the heterodimer, particularly the cleavage region. The two CAP regions are pulled apart to open the DNA gate, allowing the passage of the T-segment and exit from the enzyme (Fig. 8.36).

8.7 A glimpse of eukaryotic replication

Surely, we can expect that, compared with the bacterial system, eukaryotic replication is far more complex involving a greater number of factors. Yet, the events, right from the initiation step, are so well-regulated that any and every segment of the genome is replicated only once during a cell division cycle. Let us undertake a brief overview of eukaryotic DNA replication mechanism to appreciate its distinct features.

8.7.1 Initiation of replication

Extensive studies with the budding yeast *Saccharomyces cerevisiae* have provided structural and mechanistic insights into eukaryotic DNA replication. In contrast to the situation in bacteria, where it is the initiator binding at the replication origin that leads to unwinding of the DNA and subsequent loading of the helicase, in eukaryotic cells helicase loading is the first step in the initiation process (Fig. 8.40). Besides, in eukaryotic cells, helicase loading and helicase activation are temporally separated. Helicase loading occurs in the G1 phase of the cell cycle with the help of the origin recognition complex (ORC), which consists of Orc1/2/3/4/5/6. ORC binds to the replication origin and recruits Cdc6, Cdt1, and the hexametric helicase MCM2-7 onto the double-stranded DNA to form ORC-Cdc6-Cdt1-MCM2-7 (OCCM) complex. This step is followed by Orc1- and Cdc6-dependent ATP hydrolysis, leading to the formation of ORC-Cdc6-MCM2-7 (OCM) intermediate and recruitment of a second MCM2-7 hexamer to become ORC-Cdc6-MCM2-7-MCM2-7 (OCMM). The resultant double-hexamer is the substrate for activation in the S-phase (Fig. 8.40).

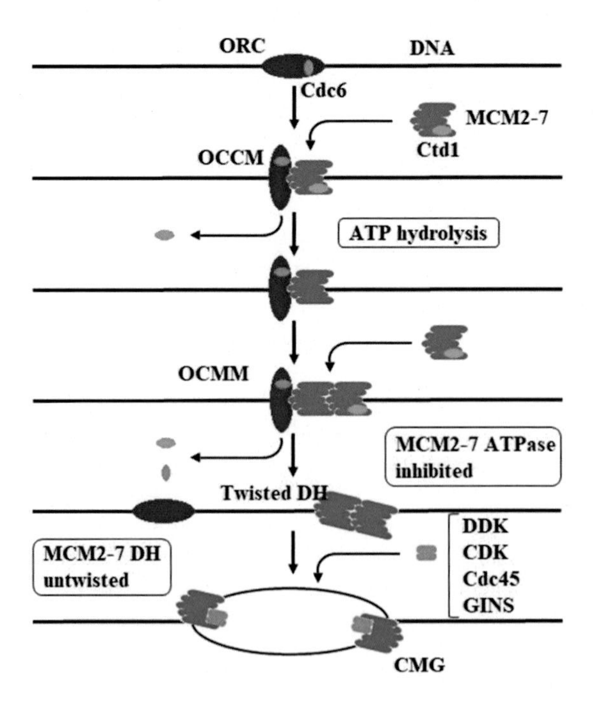

FIG. 8.40

Helicase loading and activation.

The MCM2-7-MCM2-7 double-hexamer, which is DNA-bound but inactive, is now activated in the S-phase. The Dbf4-dependent-kinase Cdc7 (DDK) interacts with Mcm2 and Mcm4 to phosphorylate the DNA-bound double-hexamer. This, together with the action of another protein kinase, cyclin-dependent kinase (CDK), promotes the binding of Cdc45 and GINS. The phosphorylation and binding of Cdc45 and GINS lead to the untwisting of the double-hexamer, which, in turn, unwinds and melts the dsDNA. Subsequently, the double-hexamer is separated and two Cdc45-MCM2-7-GINS (CMG) complexes are formed. Each CMG complex, which is the active form of MCM2-7 helicase, encircles ssDNA (Fig. 8.40).

8.7.2 Multiple polymerases at work

Replication can now be initiated with the synthesis of a primer. Three B-family of DNA polymerases—Pol α, Pol δ, and Pol ε—are involved in the replication of eukaryotic DNA. Of the three, Pol α is a dual primase-polymerase—a heterotetramer composed of two primase subunits and two polymerase subunits. In the yeast replisome, Pol α is held to the CMG helicase by Ctf4 protein which forms a trimer that can bind to both Pol α and the Sld5 subunit of GINS. The primase subunits synthesize a short RNA primer (7–12 ribonucleotides), which is extended by the polymerase subunits to yield a 30–35 nucleotide-long hybrid primer with RNA at the 5′ end and DNA at the 3′ end. The primers are then further extended by Pol ε continuously on the leading strand and Pol δ discontinuously on the lagging strand as Okazaki fragments.

As we have seen earlier, all B-family of polymerases are composed of five subdomains. The palm, thumb and fingers subdomains constitute the core of the enzyme. In addition, there are N-terminal and exonuclease (proofreading) subdomains. The exonuclease active site is usually located 40–45 Å from the polymerase active site.

The catalytic region of Pol α has the universal "right-hand" topology (Fig. 8.41A). However, unlike other DNA polymerases which make continuous contact with the minor groove of a full turn of the primer-template DNA duplex, Pol α thumb subdomain interacts almost exclusively with the RNA primer strand, in RNA/DNA A-form, through a series of hydrophobic contacts involving M1131, L1133, Y1140, P1141, and M1146 (Fig. 8.41B). As the RNA primer is extended with dNTPs, Pol α translocates beyond the RNA/DNA duplex region onto B-DNA. RNA-specific contacts are disrupted, leading to the disengagement of Pol α and take over by a replicative polymerase.

Another contrasting feature of Pol α is the absence of proofreading activity. This is due to mutation of all four conserved carboxylate residues, D114/E116/D222/D327, as found in the exonuclease active site of RB69 gp43 to S542/Q544/Y644/N757 in Pol α. Further, the β-hairpin motif, which is present in the exonuclease domain of other polymerases and required for switching of the primer between the polymerase and exonuclease active sites, is not found in Pol α (Fig. 8.41A). However, the errors generated by Pol α during initiation of the leading strand or Okazaki fragments are corrected by Pol δ.

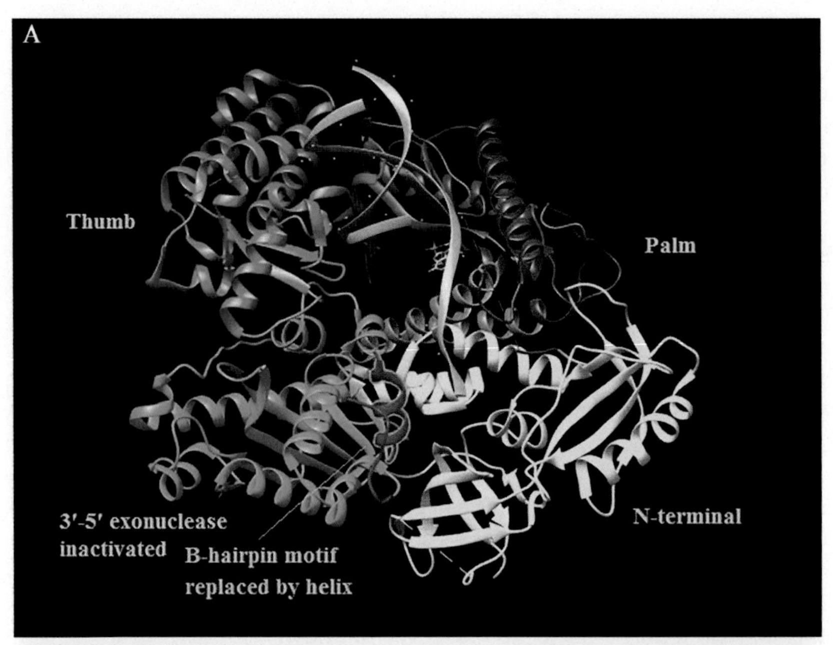

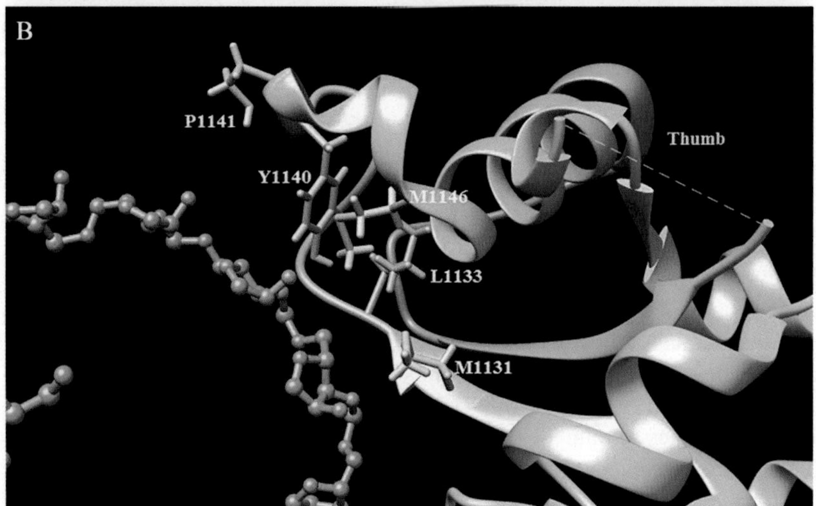

FIG. 8.41

Polymerase α. (A) Domain structure with inactive 3′–5′ exonuclease domain. A helix in the inactive exonuclease domain has substituted the β-hairpin motif. (B) Residues (labeled) of thumb domain interacting with RNA primer strand *(orange)*.

Source: Perera, R.L., et al., 2013. eLife 2, e00482.

S. cerevisiae Pol δ is composed of three subunits. The catalytic subunit (POL3) contains 1097 residues. Like other DNA polymerases, Pol δ has three conserved carboxylates, D608, D762, and D764, in its palm domain. However, these carboxylates

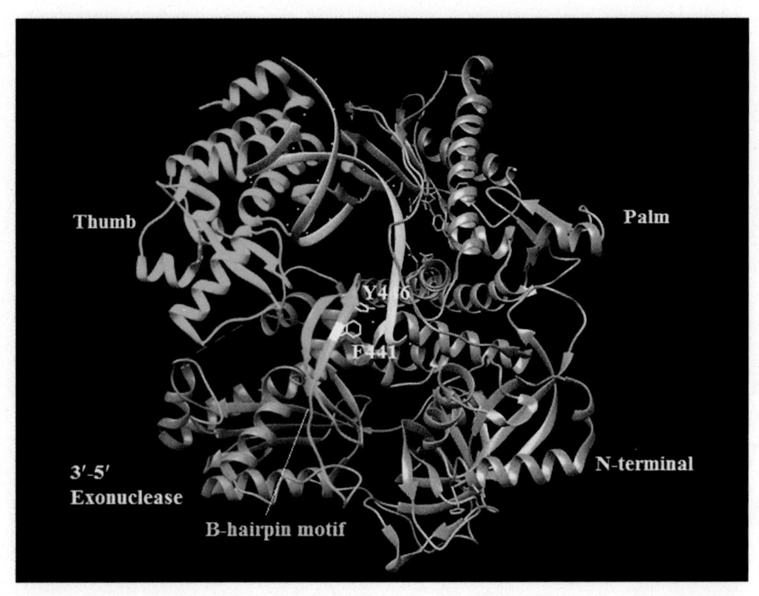

FIG. 8.42

Domain structure of Pol δ with the long β-hairpin motif involved in primer strand active site switching indicated by stick.

Source: Swan, M.K., et al., 2009. Nat. Struct. Mol. Biol. 16, 979.

coordinate Ca^{2+}, and not Mg^{2+}. The nonessential subunit Pol32 of Pol δ is able to bind Pol α. Through this interaction, Pol δ is recruited to the core replisome for proofreading of Pol α-generated error or synthesis of Okazaki fragments.

As expected, Pol δ contains an exonuclease active site. Its β hairpin protrudes into the major groove of the DNA and acts as a wedge between the duplex region and single-stranded 5′-end of the template strand (Fig. 8.42). The interaction is stabilized by two aromatic residues, F441 and Y446 (Fig. 8.42). The tip of the thumb (hinge region) is also shifted upwards towards the 5′ end of the template. These two features in Pol δ are involved in active site shuttling.

Like other mammalian B-family DNA polymerases, all three replicative polymerases of yeast contain two cysteine-rich metal-binding motifs in their C-terminal domain—CysA and CysB. The two motifs help processive DNA synthesis by Pol δ—CysA directly and CysB indirectly. CyA coordinates a Zn and promotes direct interaction of Pol δ with proliferating cell nucleus antigen (PCNA). CysB coordinates a [4Fe-4S] cluster and promotes interaction of Pol31 and Pol32 subunits. The Pol32 subunit contains a PCNA-interacting protein (PIP) motif, which, in turn, helps Pol δ interact with PCNA.

For the processivity of DNA synthesis by Pol ε, PCNA is not an absolute requirement. Remarkably, its palm domain is considerably larger than that of either Pol α or Pol δ. It contains a new domain consisting of a three-stranded β sheet and two α helices (residues 533–555 and 682–760). A region in the domain lined with positively charged residues, R686, R744, H748, R749, and K751, is in the

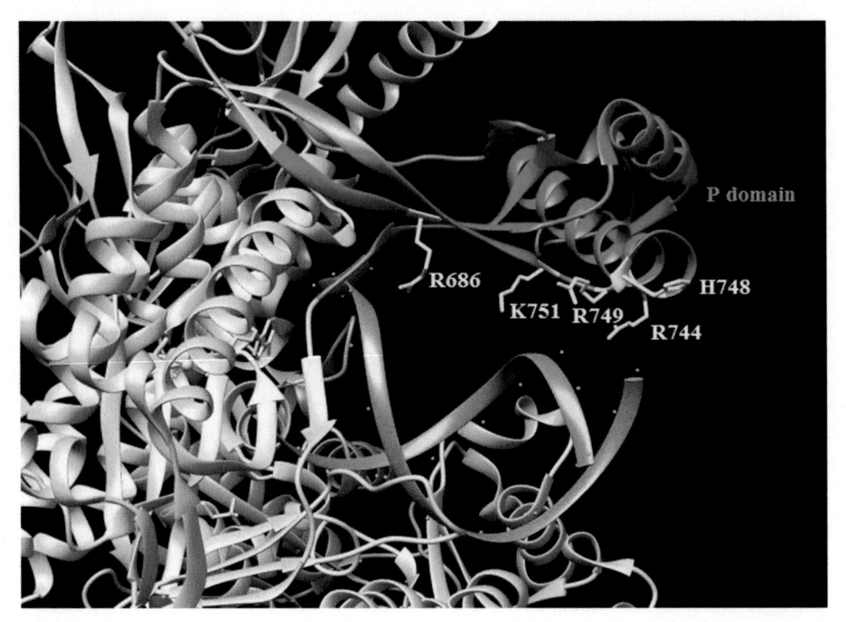

FIG. 8.43

Residues in P-domain of Pol ε encircling the nascent double-stranded DNA for processivity.
Source: Hogg, M., et al., 2014. Nat. Struct. Mol. Biol. 21, 49.

close proximity of the phosphate backbone of the template:primer (Fig. 8.43). The domain, called the P domain, encircles the nascent double-stranded DNA and ascertains processivity of the enzyme. Further, its Ddpb3 and Dpb4 subunits also contribute towards processivity.

Pol ε also possesses an exonuclease domain and proofreading function. However, its β-hairpin motif is too short to interact with the DNA. In this case, the P domain, which contacts both primer and template strand, could sense replication errors and facilitate active site switching.

It is now evident that DNA replication in eukaryotes is asymmetric, two different polymerases involved in DNA synthesis on the leading and lagging strand. The CMG (Cdc45-MCM-GINS) helicase, as we have seen earlier, surrounds the leading strand. The non-catalytic subunit Dpb2 of Pol ε binds to the Psf1 subunit of GINS, implying the polymerase binding to CMG. In vitro experiments have shown that CMG helicase selects Pol ε over Pol δ on the leading strand. On the lagging strand, however, PCNA strongly favors Pol δ.

In conclusion, whereas for replication of bacterial DNA a primase and a replicase containing two (or possibly three) copies of the same DNA polymerase carry out the polynucleotide synthesis, eukaryotic DNA replication is undertaken by the coordinated action of three DNA polymerases of the B-family, each composed of more than one subunit, having distinct but complementary functions.

Sample questions

1. ΔG for phosphodiester bond formation has been reported to be ~+5.3 kcal/mol. In such case what are the factors that make the nucleotide reaction energetically favorable?
2. What is the nature of primer:template (p/t) junction in DNA synthesis?
3. The mismatch error frequency, as predicted by free energy difference between matched and mismatched DNA termini, is ~0.4. However, experimentally, the error frequency is found to be between 10^{-3} and 10^{-6}. How does the DNA polymerase enhance the fidelity of nucleotide incorporation by such a large factor?
4. Polymerization by DNA polymerase begins by the binding of the enzyme to the primer: template (p/t) DNA. How does the *Escherichia coli* DNA polymerase I Klenow fragment satisfy the geometric and electrostatic compatibility for binding?
5. *Sulfolobus acedocaldarius* DNA polymerase Dbh palm domain lack accessible polar side chains—what is the consequence of this on DNA synthesis by Dbh?
6. Superposition of the crystal structures of the Klenow fragment of *Taq* DNA polymerase Klentaq1 unbound and bound to the primer:template (p/t) DNA indicates how the enzyme holds the p/t junction. What is the mechanism thus revealed
7. In the course of DNA base excision repair, DNA polymerase β (Polβ) binds to the gapped DNA created by the removal of damaged bases from one strand. What are the features in Polβ involved in this binding?
8. Which domain of DnaA binds to the DnaA boxes? Illustrate with sketches the nature of interactions of this domain with the DNA at the residue-base level.
9. What are the features responsible for and how do they participate in the ATP-induced oligomerization of DnaA
10. Once the helicase is loaded at oriC, it has to recruit the primase by protein–protein interaction. What is the basis of *Escherichia coli* helicase DnaC and primase DnaG interaction?
11. In *Escherichia coli*, DNA polymerase III β subunit residues L177, P242, V247, V36, and M362 structurally form a "pocket" to interact with the clamp loader subunit δ. What is the nature of the "pocket"? What is the effect of this interaction?
12. The exonuclease site of the Klenow fragment of *Escherichia coli* DNA polymerase I has 10-fold higher affinity for 3′-ends of DNA strands. How does it help in the proofreading activity of the enzyme?
13. Explain how in *Saccharomyces cerevisiae* the Dbf4-dependent kinase (DDK) and cyclin-dependent kinase (CDK) play a role in activation of the helicase.
14. In *Saccharomyces cerevisiae* Pol α is a dual primase-polymerase. What does it mean in terms of its action? What is the structural basis of this dual function?
15. How do the two cysteine-rich metal-binding motifs in yeast DNA polymerases help in the processivity of DNA synthesis?

References and further reading

Batra, V.K., et al., 2013. Amino acid substitution in the active site of DNA polymerase β explains the energy barrier of the nucleotidyl transfer reaction. J. Am. Chem. Soc. 134 (21), 8078–8088.

Beard, W.A., Wilson, S.H., 2014. Structure and mechanism of DNA polymerase β. Biochemist 53, 2768–2780.

Bell, S.P., Kaguni, J.M., 2013. Helicase loading at chromosomal origins of replication. Cold Spring Harb. Perspect. Biol. 201 (5), 1–20.

Doublié, S., Zahn, K.E., 2014. Structural insights into eukaryotic DNA replication. Front. Microbiol. 5 (444), 1–8.

Doublié, S., Sawaya, M.R., Ellenberger, T., 1999. An open and closed case for all polymerases. Structure 7, R31–R35.

Georgescu, R.E., et al., 2008. Structure of a sliding clamp on DNA. Cell 132, 43–54.

Hedglin, M., Kumar, R., Benkovic, S.J., 2013. Replication clamps and clamp loaders. Cold Spring Harb. Perspect. Biol. 201 (3), 1–19.

Hübscher, U., Maga, G., Spadari, S., 2002. Eukaryotic DNA polymerases. Annu. Rev. Biochem. 71, 133–163.

Jeruzalmi, D., et al., 2001. Mechanism of processivity clamp opening by the delta subunit wrench of the clamp loader complex of *E. coli* DNA polymerase III. Cell 106, 417–428.

Kaguni, J.M., 2006. DnaA: controlling the initiation of bacterial DNA replication and more. Annu. Rev. Microbiol. 60, 351–371.

Lamers, M.H., et al., 2006. Crystal structure of the catalytic α subunit of *E. coli* replicative DNA polymerase III. Cell 126, 881–892.

MaCauley, M.J., et al., 2008. Distinct double- and single-stranded DNA binding of *E. coli* replicative DNA polymerase III α subunit. ACS Chem. Biol. 3 (9), 577–587.

McHenry, C.S., 2011. DNA polymerases from a bacterial perspective. Annu. Rev. Biochem. 80, 403–436.

Mott, M.L., Berger, J.M., 2007. DNA replication initiation: mechanisms and regulation in bacteria. Nat. Rev. Microbiol. 5, 343–354.

Pomerantz, R.T., O'Donnell, M., 2007. Replisome mechanics: insights into a twin DNA polymerase machine. Trends Microbiol. 15 (4), 156–164.

Rothwell, P.J., Waksman, G., 2005. Structure and mechanism of DNA polymerases. Adv. Protein Chem. 71, 401–440.

Soultanas, P., 2012. Loading mechanisms of ring helicases at replication origins. Mol. Microbiol. 84 (1), 6–16.

Wu, S., Beard, W.A., Pedersen, L.G., Wilson, S.H., 2014. Structure comparison of DNA polymerase architecture suggest a nucleotide gateway to the polymerase active site. Chem. Rev. 114 (5), 2759–2774.

CHAPTER

Transcription

9

Chapter 5 has introduced the macromolecules of a cell—their structures and functions. Further, we have seen in the previous chapter how one of the macromolecules, DNA, that carries the genetic information of the cell is replicated. This chapter will address the question as to how this genetic information is transferred from DNA to another macromolecule, RNA, in a process called transcription.

9.1 Basic requirements for transcription

In a sense transcription is similar to DNA replication—both involve the synthesis of a new strand of nucleic acid. However, in the case of transcription, the new strand is synthesized by the addition of ribonucleotides, and not deoxyribonucleotides.

Further, like DNA replication, the assembly of ribonucleotides in transcription is not random; it is also directed by a DNA template strand. Nevertheless, in contrast to replication where the entire genome is copied once, and only once, in every cell division, transcription copies only specific segments of the genome (genes) and can generate one to several thousand copies from a given segment. Even in such case, the DNA segment needs to be unwound and the bases unpaired so that nucleotide incorporation can proceed in accordance with the principle of base complementarity. One more distinguishing feature is that, unlike replication, in transcription the RNA product does not remain base-paired with the template.

In transcription, each nucleotide addition step involves the breakage of an α, β-phosphoanhydride bond in an NTP molecule, and generation of a pyrophosphate (PP_i). This is an energetically favorable reaction with free energy change $\Delta G \approx -$ 5.6 kcal/mol under standard conditions. However, this chemical energy must be coupled to the movement of the transcription machinery along the DNA template.

Therefore, it implies that the transcription process can be undertaken only by an enzymatic machinery that fulfills the following requirements: (a) that it recognizes "transcribable" regions of the genome and interacts with the DNA in an effective manner including opening up the DNA duplex and making a single strand available as a template; (b) that it conveys ribonucleotides opposite to the template bases in the active site in accordance with the principle of complementarity; (c) that it joins two ribonucleotides by a phosphodiester bond either at the end of a preformed primer or ab initio; (d) that it extends the RNA chain by continuous addition of ribonucleotides and peels off the growing chain from the template when the site of nucleotide

Fundamentals of Molecular Structural Biology. https://doi.org/10.1016/B978-0-12-814855-6.00009-2

addition has moved just a "little" ahead; and (e) that it terminates the transcription process at specific sites and dissociates from the DNA.

In all three domains of life—bacteria, archaea, and eukaryotes—the key constituent of this enzymatic machinery is the multisubunit RNA polymerase (RNAP). It comprises five subunits in the common core that are conserved in all the three domains. Bacterial RNAP, being the simplest form of this family, has been the most widely studied to understand the phenomena and mechanisms of transcription. Let us, therefore, first consider bacterial transcription and look at the structures and functions of bacterial RNA polymerases before moving on to the more complex ones from eukaryotes.

9.2 Bacterial transcription

9.2.1 Bacterial RNA polymerase

We know that *Escherichia coli* has been the most widely used organism in the studies on gene expression and regulation at the transcriptional level. Yet, due to the difficulty in crystallization of *E. coli* RNA polymerase (EcoRNAP), it was not the first among the bacterial RNA polymerases to have its structure determined at the molecular level. The structures of RNAPs from the *Thermus* genus (*T. aquaticus* and *T. thermophilus*), first as a core and then as a holoenzyme, were determined much earlier. Nevertheless, since a high degree of sequence conservation was found to exist among the RNAPs from all bacterial species, the mechanistic understanding of the transcription apparatus as derived from the *Thermus* RNAP has been generalized also to the other bacterial RNAPs. Therefore, in this chapter, the three RNAPs have been interchangeably used to explain the transcription mechanisms.

The bacterial RNAP core enzyme is formed by the sequential assembly of five subunits (Fig. 9.1). Its overall structure resembles a crab claw with two "pincers"

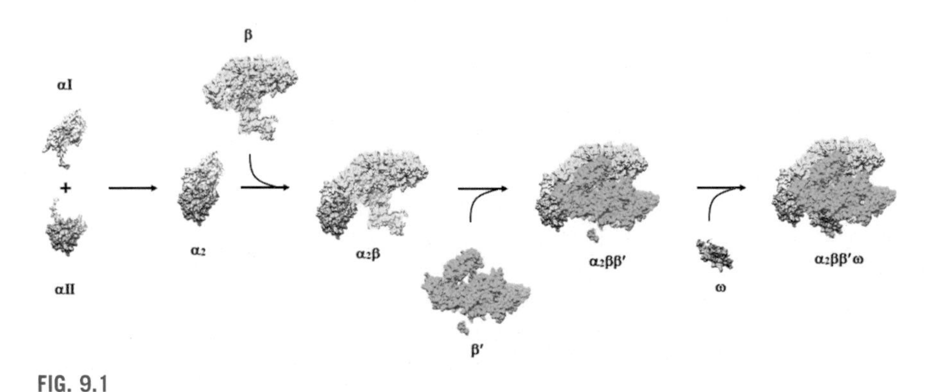

FIG. 9.1

Taq RNAP core assembly prepared by using the surface structures of the subunits.
Source: Minakhin, L., et al., Proc. Natl. Acad. Sci. U. S. A. 98, 892.

forming a cleft that binds DNA and contains the active center. β′, the largest subunit forming one pincer called the "clamp," is mobile, while the second-largest subunit β forms the other pincer. The clamp swings between open and closed states about a hinge, called the "switch region" (Sw1-5), located at its base. The difference between the open and closed states is $\geq 20°$.

The bases of the pincers are joined to the N-terminal domains of α-subunit dimer that serves as a platform for RNAP assembly (Fig. 9.1). Each N-terminal domain of the α-dimer (αNTD) is tethered to a C-terminal domain (αCTD) by a linker. αCTDs project out from the upstream DNA-facing side of RNAP. They are DNA-binding elements, but can also interact independently with transcription factors.

A ribbon representation of Taq RNAP core in Fig. 9.2 illustrates its functional components. The cleft between β and β′ is partitioned into three channels: (a) the main "primary channel" accommodates downstream dsDNA and RNA-DNA hybrid; (b) the "secondary channel" is for the entry of NTP; and (c) the "RNA exit channel" separates the RNA-DNA hybrid strand and interacts with RNA hairpins during pausing and termination (Fig. 9.2). The ω subunit, bound near the C-terminus of β′, is thought to serve as a β′ chaperone.

The active center is located on the back wall of the primary channel. It consists of a long α-helical "bridge" (bridge helix, BH) connecting the β and β′ subunits, two flexible α-helices of a "trigger" loop (TL), an extended loop (called the F-loop, FL), and a catalytic loop with three aspartates for coordinating essential Mg^{2+} ions (Figs. 9.2A, B and 9.3).

Structurally, the BH, TL, FL, and catalytic loop are all part of the β′ subunit. The BH is a distinct feature of all multisubunit RNAPs and, as we shall see later, it interacts with other elements involved in the nucleotide addition cycle (NAC) and DNA-binding. The TL, which is also involved in the elongation process, contains a dynamic domain that undergoes a folding-unfolding transition during the NAC. It contains another domain that remains helical throughout the cycle. The adjacent FL modulates the structural changes in the TL.

The core enzyme, by itself, possesses the basic RNA polymerization activity. However, it cannot specifically bind to the DNA sequence at the beginning of a gene that signals normal initiation of transcription. It needs the association of a transcription factor called σ to specifically initiate transcription from a defined site. The σ-bound form of the enzyme is called the RNA polymerase holoenzyme. Like many other bacteria, *E. coli* contains multiple σ factors, but the predominant σ factor in *E. coli* is σ^{70}, which is a member of a homologous family of proteins. The principal σ factors in *T. thermophilus* and *T. aquaticus* (designated as σ^{A}) belong to the same family.

σ^{70} has four regions of highly conserved amino acid sequences (σ1–σ4). The regions are further divided into subregions as shown in Fig. 9.4 The entire polypeptide folds into four independent structural domains σ1.1, σ2, σ3, and σ4 (Fig. 9.4). The domains are separated by separated flexible linkers. The linkers, σ1.1–2 and σ3–4, allow significant movement to domains, σ1.1 and σ4, respectively.

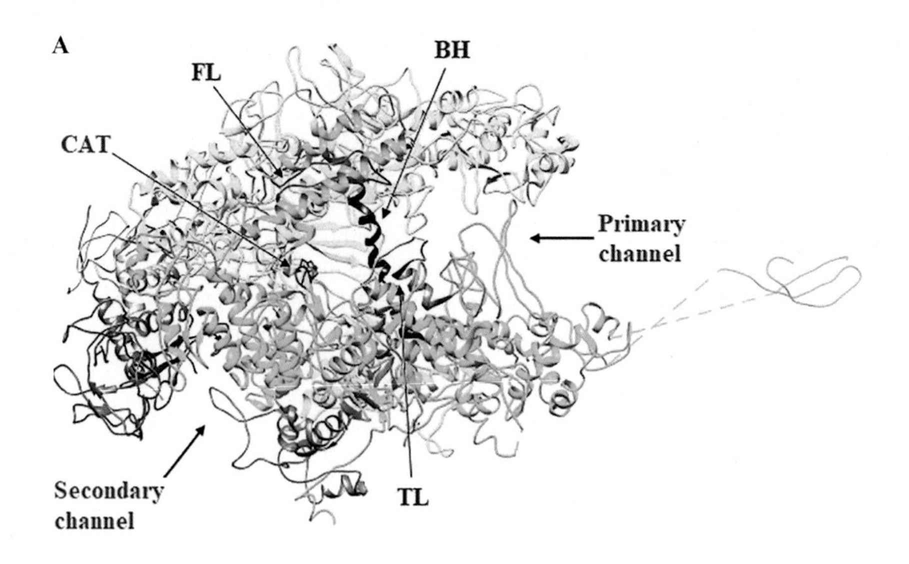

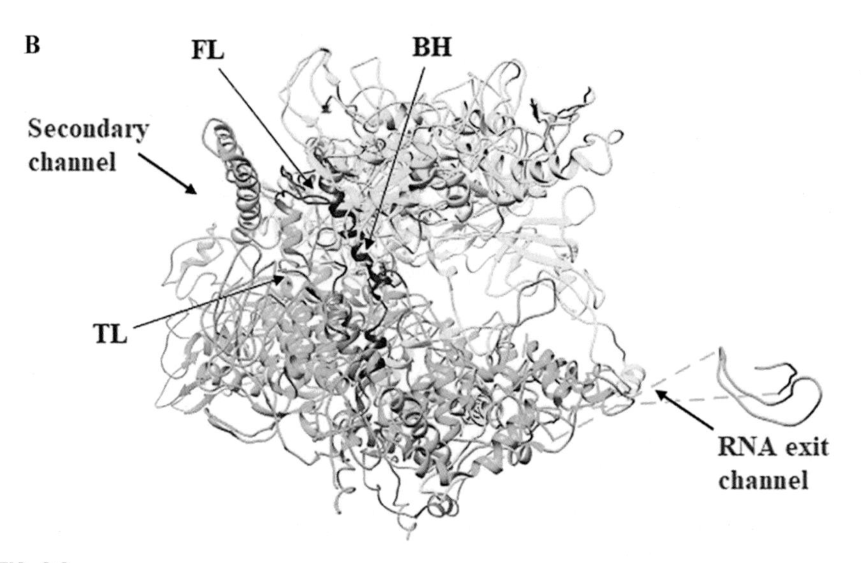

FIG. 9.2

Ribbon representation of *Taq* RNAP core in two rotational views (A and B). The subunits are colored as follows: α_{I}—*light gray*; α_{II}—*dark gray*; β—*yellow*; β'—*cyan*; ω—*blue*. β' features: catalytic loop (CAT—*red*), F-loop (FL—*green*), bridge helix (BH—*brown*), and trigger loop (TL—*coral*).

Source: Minakhin, L., et al., Proc. Natl. Acad. Sci. U. S. A. 98, 892.

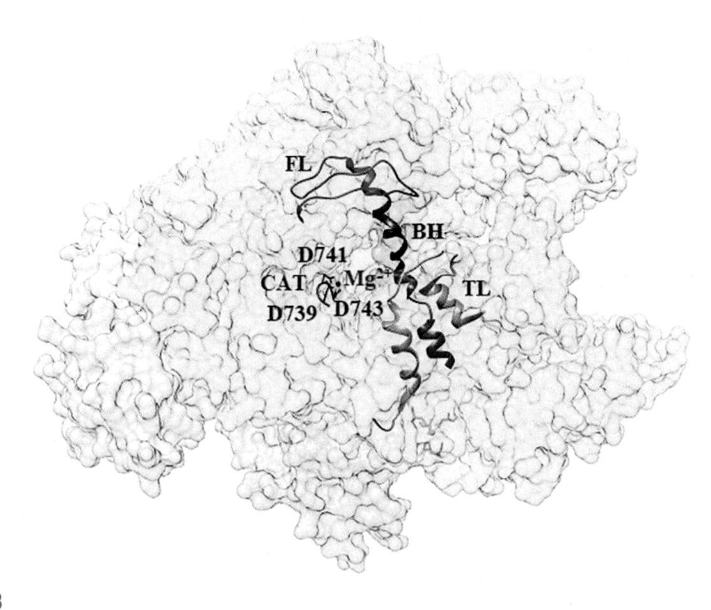

FIG. 9.3

β′ features (as in Fig. 9.2) represented by ribbons against surface background of remaining core structure.

Source: Minakhin, L., et al., Proc. Natl. Acad. Sci. USA. 98, 892.

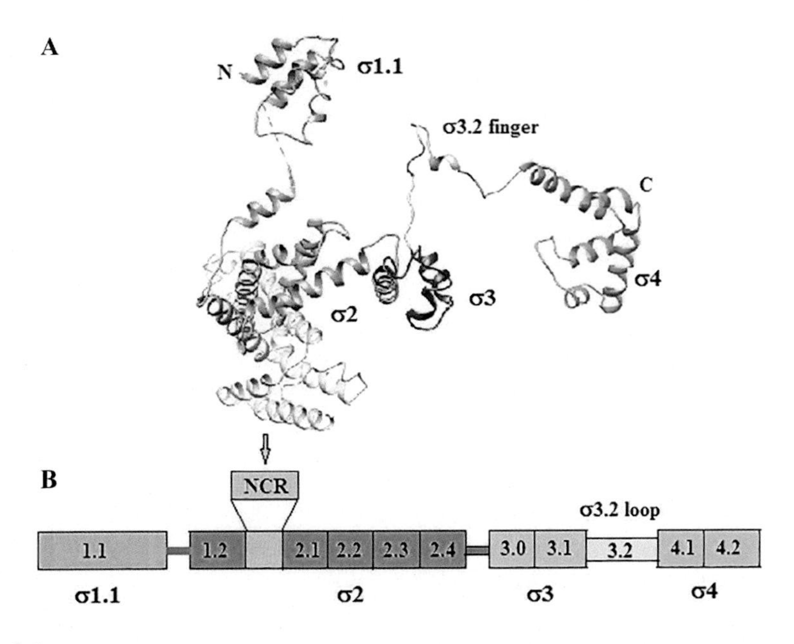

FIG. 9.4

E. coli σ structure: (A) three-dimensional and (B) domain. NCR—nonconserved region.

Source: Murakami, K.S., 2013. J. Biol. Chem. 288, 9126.

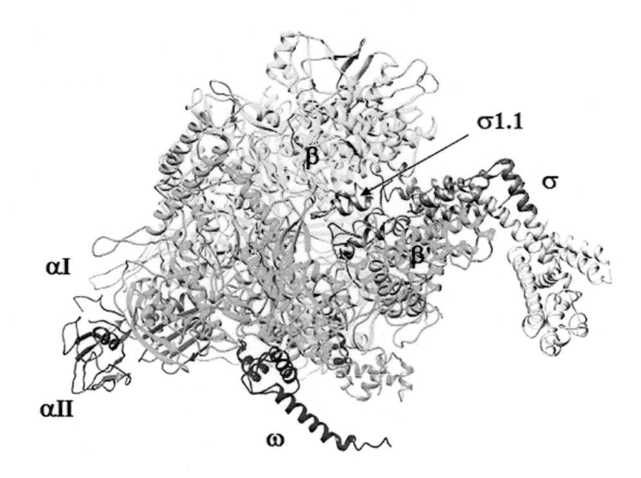

FIG. 9.5

E. coli RNAPσ^{70} structure. Core subunits: α1—*light gray*; αII—*dark gray*; β—*yellow*; β'—*cyan*; ω—*dark blue*. σ^{70} domains: σ1—*pink*; NCR—*light blue*; σ2—*coral*; σ3—*firebrick*; σ4—*orange*.

Source: Murakami, K.S., 2013. J. Biol. Chem. 288, 9126.

As σ binds to the core to form the holoenzyme, an extensive interface (~10 Å) is created between the two. Spread over the interface are >100 contacts involving amino acid residues present in all the σ domains and linkers on one side and those in the β and β' subunits of the core on the other. These contacts include salt bridges, direct hydrogen bonds, water-mediated hydrogen bonds, and van der Waals (hydrophobic) contacts. The structure of *E. coli* RNAPσ^{70} (E σ^{70}) is shown in Fig. 9.5.

Regions σ2, σ3, and σ4 render the holoenzyme the capacity to bind to specific sites in the DNA. The major part of the σ subunit (regions σ1–σ3) is bound to the core surface at the entrance of the cleft. The subregion σ1.1 is particularly interesting. It contains four α helices and is surrounded by σ2, the β-lobe, the β' clamp, and the cleft. Its acidic residues shield the basic residues of the β lobe and β' clamp (Fig. 9.6), thereby preventing either nonspecific double- or single-stranded DNA from accessing the RNAP active site. Thus, it gives the holoenzyme an additional specificity of binding. Only specific binding by other regions of σ triggers the displacement of σ1.1 from the DNA-binding main channel of the holoenzyme.

9.2.2 Promoter

The defined site which the RNA polymerase binds, along with initiation factors if required, to initiate transcription is known as a promoter. Sequence alignment of a large number of *E. coli* promoters identified two conserved hexameric sequences TTGACA and TATAAT, respectively, at positions −35 and −10, where +1 is the transcription start site (Fig. 9.7). They are respectively called the −35 and −10 regions or elements and the primary determinant of the basal strength of a promoter.

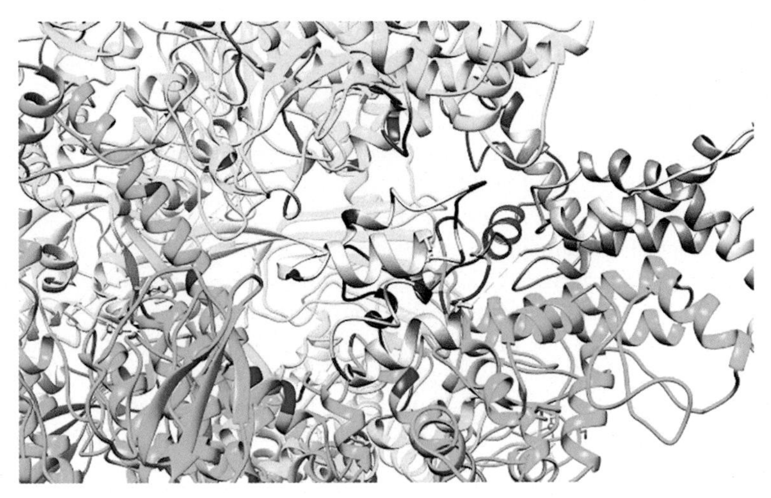

FIG. 9.6

Shielding of basic residues *(dark blue)* of β and β′ by acidic residues *(red)* of σ1.1.

Source: Murakami, K.S., 2013. J. Biol. Chem. 288, 9126.

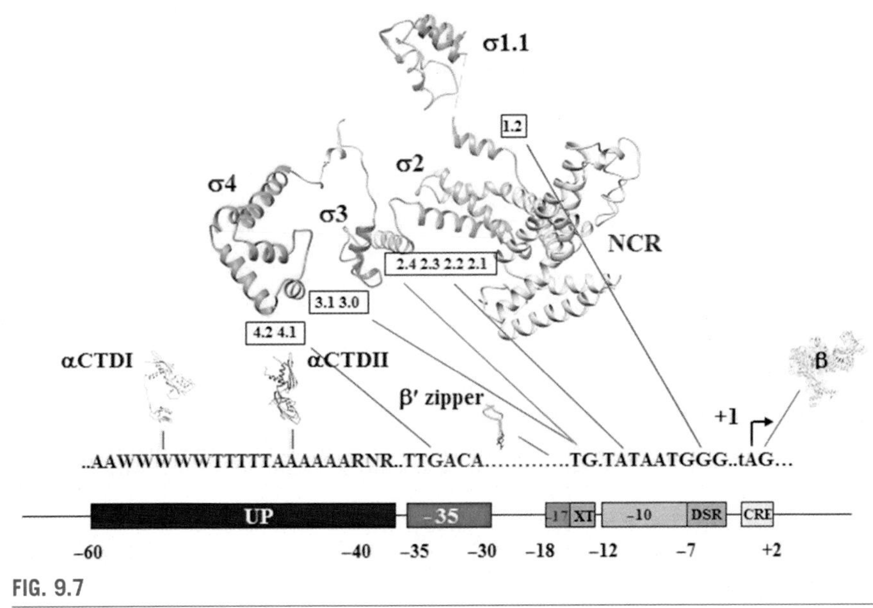

FIG. 9.7

E. coli RNAPσ^{70} subunits—promoter elements interaction.

Source: Murakami, K.S., 2013. J. Biol. Chem. 288, 9126.

However, the −35 and −10 sequences are not exactly identical among different σ^{70}-dependent promoters. Variation of these sequences in individual promoters affects their specific interactions with the RNAP subunits σ and α and, in turn, basal strengths.

Subsequently, it has been found that sequence elements immediately upstream or downstream of the −35 and −10 elements also contribute to the specific recognition of the promoter by $E\sigma^{70}$ (Fig. 9.7). A G+C-rich region between the −10 hexameter and the transcription start site in some bacterial promoters (such as rRNA and tRNA promoters) has been designated as "discriminator" (DSR). A guanine at the −5 position of the nontemplate strand in the discriminator is the most favorable for interaction with *E. coli* σ^{70}—one or both of a tyrosine residue (Y101) and a methionine residue (M102) in σ1.2 is/are likely to be involved in the interaction. Another region in the promoter between position −3 and +2 has been designated as "core recognition element" (CRE). This element, as we shall see below, interacts with residues of RNAP β subunit.

9.2.3 Initiation: Holoenzyme-DNA interactions

Bacterial transcription initiation process consists of five steps: (a) association of the RNAP core enzyme $\alpha_2\beta\beta'\omega$ (E) with the specificity factor, such as σ^{70}, to form the holoenzyme ($E\sigma^{70}$); (b) recognition and binding of the promoter DNA by $E\sigma^{70}$ to form a closed promoter complex (RP_c); (c) a series of conformational changes (isomerization) occurring in the RP_c to form an open promoter complex (RP_o); (d) conversion of RP_o to an initial transcription complex (RP_{init}) in the presence of rNTPs and formation of the first phosphodiester bond between the rNTPs located at the +1 and +2 sites (dinucleotide synthesis); and (e) synthesis of an RNA of a critical length (typically 11–15 nt with ~9 nt as RNA-DNA hybrid) by the RNAP, loss of RNAP-promoter DNA contacts, gradual dissociation of σ and escape of the core polymerase from the promoter, and formation of a highly stable protective ternary elongation complex (EC) (Fig. 9.8). In all these steps, σ plays the key role with its association and dissociation; however, the other components of the holoenzyme aptly contribute. Let us consider each step in further detail together with the underlying molecular interactions.

Promoter recognition and binding

The initial contacts leading to the formation of RP_c are established with the −10, extended −10, and −35 elements of the promoter by σ2, σ3, and σ4 (subregions 2.2–2.4, 3.0, and 4.2, respectively) (Fig. 9.7). These contacts are essentially polar and van der Waals in nature. RP_c formation is also aided by the interaction of α subunit with the A/T-rich sequences of the UP element. At the same time, certain residues of σ3 and β′ zipper contribute to promoter binding by interacting with the −17/−18 "Z-element" (phosphates) (Fig. 9.7).

The specificity of α binding to the A-tract of the UP element arises partly from the base contacts in the minor groove. As we have seen above, each of the two αCTD

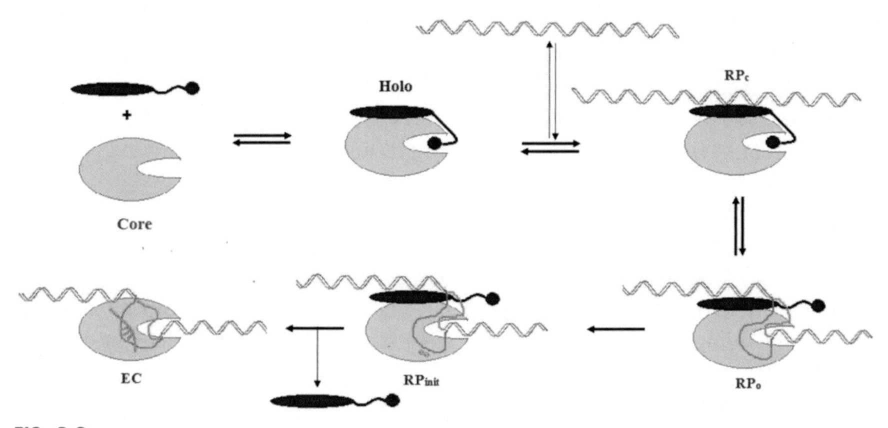

FIG. 9.8

Essential steps in transcription initiation. *Brown* circle represents σ1.1; *brown* oval represents the rest of σ. The DNA template (t) strand is colored *green*, the nontemplate (nt) strand colored *orange*.

Source: Murakami, K.S., 2013. J. Biol. Chem. 288, 9126.

subunits of the RNAP contains two helix-turn-helix (HTH) motifs. Four residues are required for binding to the UP element—R265 in HTH1 and G296, K298, and S299 in HTH2 (Fig. 9.9). The contacts are essentially between N3A and O2T and the basic side chains of R265 and K298 through hydrogen bonds. Besides, two residues D259 and E261 in HTH1 most likely interact with σ.

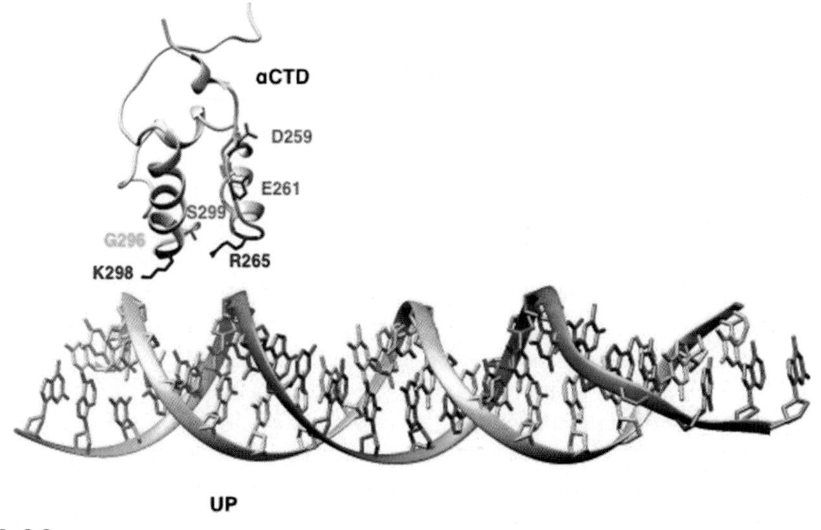

FIG. 9.9

αCTD binding to UP element.

Source: Murakami, K.S., 2013. J. Biol. Chem. 288, 9126.

In the case of σ3-Z-element interaction, the residues Eco R451 and H455 in the subregion 3.0 are involved, while the β′ zipper contacts through its residues Eco Y46, R47, T48, and F49. Most of these interactions other than those involving σ2 and σ4 are weak. Nonetheless, they increase promoter occupancy by RNAP and induce local distortion in DNA structure, such as a bent or kink at three places around −25, thus influencing subsequent steps in the initiation process.

At this stage, the negatively charged face of σ1.1, which mimics DNA, is bound to the basic surface of the β lobe and β′ clamp, thereby blocking the access to downstream DNA (Fig. 9.6). In fact, its opposite positively charged face interacts with the downstream DNA to further stabilize RP_c. Subsequently, β′ clamp opening and DNA-unwinding in the open promoter complex lead to the displacement of σ1.1.

Open promoter complex

As we have seen, RP_c is intrinsically unstable. Rather, the specific binding of RNAP holoenzyme to promoter DNA causes a series of conformational changes, called "isomerization," in both the molecules. The promoter DNA is in an "open" state for ~13 bp from −11 position to just beyond the transcription start site (+1). Thus, an initiation "bubble" is created. During this melting, the contacts between σ2 and duplex form of the promoter in RPc, which were polar and van der Waals in nature, are mostly replaced by interactions between conserved aromatic residues in σ2 and bases from −11 to −7 in the nontemplate (nt) DNA strand.

The melting starts with the −11A (nt), in the −10 element, flipping out of the duplex DNA into a recognition pocket in σ2. A conserved tryptophan (aromatic)-dyad (σ^A: W256 and W257; σ^{70}: W433 and W434) forms a chair-like structure and interacts with base-paired −12T (nt) at the upstream edge of the bubble substituting the flipped-out −11A (nt) (Fig. 9.10A). The first tryptophan of the dyad forms a π-stack with the face of −12T (nt). The pocket which accommodates −11A comprises three aromatic residues (σ^A: Y253, F242, and F248; σ^{70}: Y430, F419, and Y425) and two polar residues (σ^A: R246 and E243; σ^{70}: R423 and E420) (Fig. 9.10B). Further, −11T (t), left unpaired by the flipping-out of −11A (nt), may be stabilized by stacking interaction with another conserved aromatic residue (σ^A: Y217; σ^{70}: Y394). Similarly, another canonical nucleotide of the −10 element, −7T (nt), flips out of the duplex and is held in a pocket comprising five σ2 and σ1.2 residues which, excepting one, are either charged or polar (σ^A: E114, N206, L209, K249, and S251; σ^{70}: E116, N383, L386, K426, and S428) (Fig. 9.10C).

Interactions of RNAP with other promoter elements also contribute to RP_o stabilization. The DSR region of the nt-strand interacts with eight σ1.2 residues, of which M102 and R103 (in σ^{70}) are functionally the most important. A purine-rich DSR contributes to the high stability of RP_o. On the other hand, the CRE, comprising nucleotide positions −3 to +2 on the nt-strand, interacts specifically with 10 β-subunit residues. Of these 10 residues, 6 form a pocket that accommodates the flipped-out +2G at the downstream edge of the bubble. An aspartate residue D326 (D446 in σ^{70}) forms hydrogen bonds with +2G, while a tryptophan residue W171 (W183 in σ^{70}) unstacks +1T from +2G (Fig. 9.11). All these interactions stabilize the transcription bubble in RP_o.

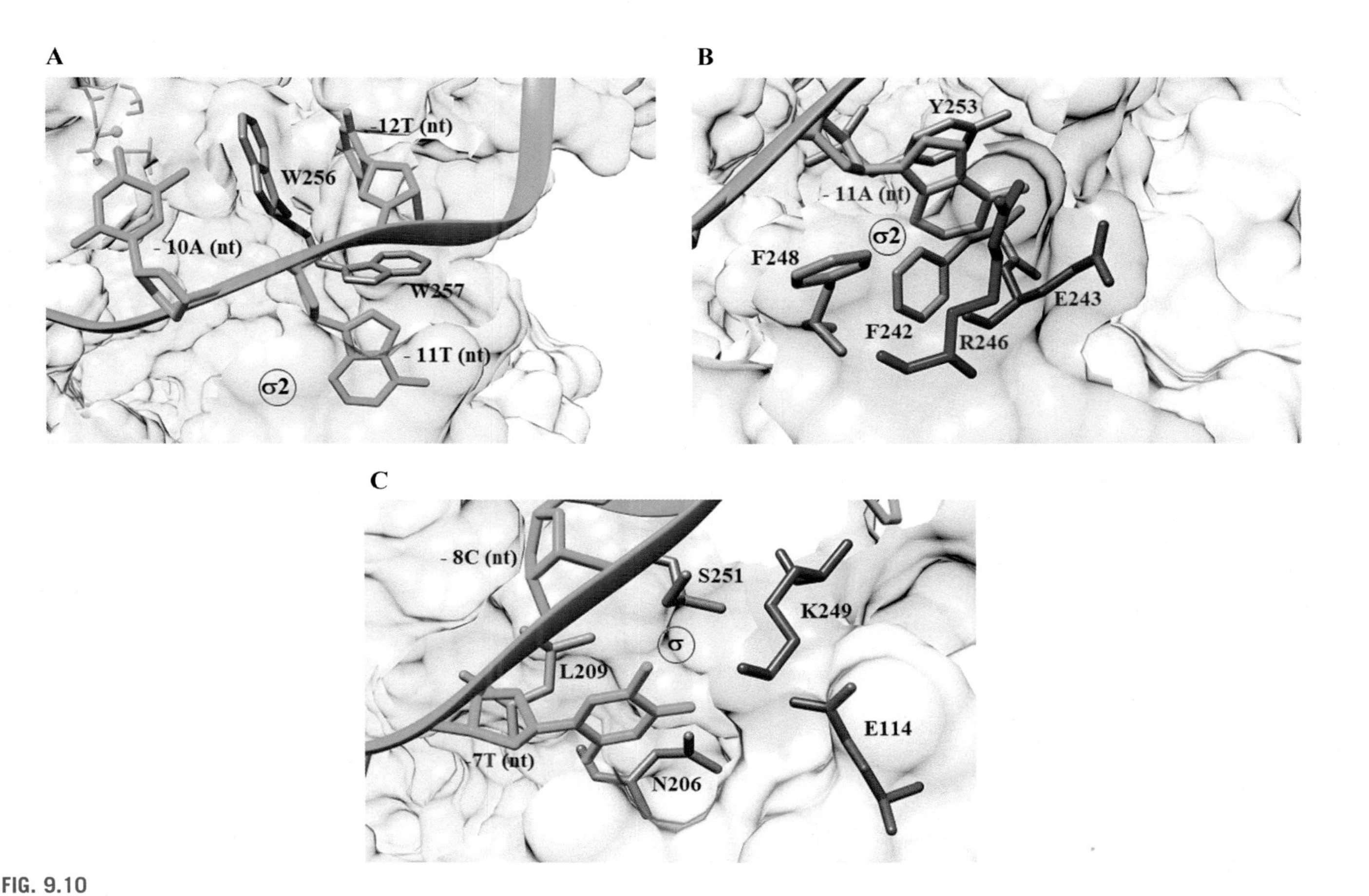

FIG. 9.10

Interactions of σ^A residues with promoter bases. Interacting residues are represented by stick on σ surface: (A) tryptophan dyad vis-à-vis −12T, (B) pocket accommodating −11A, and (C) pocket holding −7T.

Source: Bae, B., et al., 2015. eLife. https://doi.org/10.7554/eLife/08504.001.

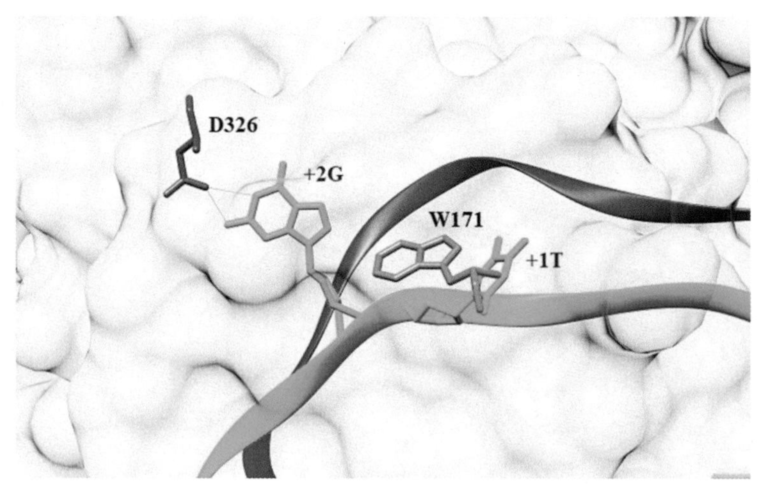

FIG. 9.11

Interaction of σ^A residues with CRE. H-bonds between D326 and +2G are shown by *gray* lines.
Source: Bae, B., et al., 2015. eLife. https://doi.org/10.7554/eLife/08504.001.

Initiation

Next, a cluster of basic residues of σ2.4 and σ3.0 (σ^A: R288, R291, and R220; σ^{70}: R465, R468, and R397), interacting with the phosphate backbone of the t-strand from −13 to −10, bends it by 90° and pulls it into the main channel (Fig. 9.12). This action puts DNA +1 position adjacent to the catalytic center.

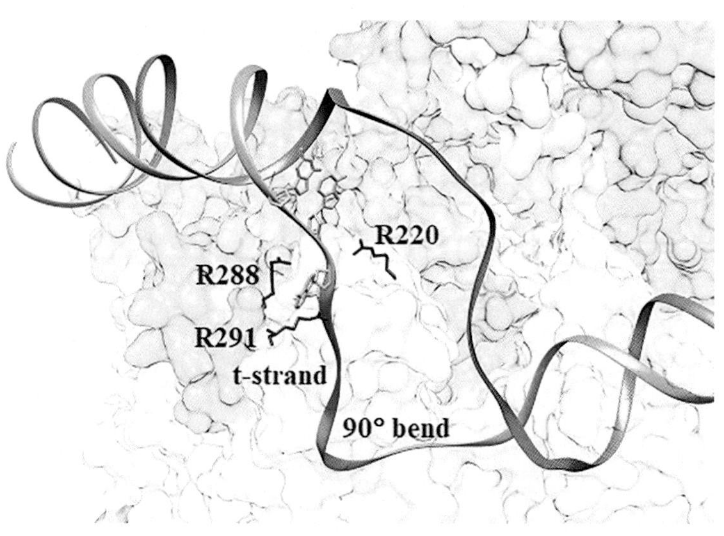

FIG. 9.12

Electrostatic interaction of a cluster of basic residues *(dark blue)* of σ^A with t-strand.
Source: Bae, B., et al., 2015. eLife. https://doi.org/10.7554/eLife/08504.001.

Now that a 12–15 nucleotide-long transcription bubble between −11 and +2/+4 is formed, and the start site is in an appropriate position in the active center, RP_o is converted to an initial transcription complex (RP_{init}). In this complex (fourth step in transcription initiation), the first phosphodiester bond is formed between rNTPs positioned at +1 and +2 sites and RNA synthesis begins (no requirement of a primer).

Two approaches have been taken to determine the structure of RP_{init}. In one, the RP_o crystal structure included RNA oligonucleotides complementary to the t-strand around the start site. In the other, a de novo RP_{init} complex was generated by soaking the holoenzyme-promoter DNA crystals with two NTPs complementary to the template bases at positions 1 and 2. Although neither of the two would reflect the natural conditions, some basic features of the molecular interactions in the transcription initiation complex have been revealed from these structures.

It has been observed that the two NTPs interact with a number of β- and β′-residues in the active site (Fig. 9.13). As for example, a glutamine residue, Q567 (σ^{70}: Q688), and a histidine residue, H999 (σ^{70}: H1237), of the β-subunit form salt bridges with a nonbridging oxygen of γ-phosphate of the first nucleotide, while a lysine residue K838 (σ^{70}: K1065) interacts with the nonbridging oxygen of α-phosphate. Additionally, an arginine R704 (σ^{70}: R425) of the β′ subunit forms salt

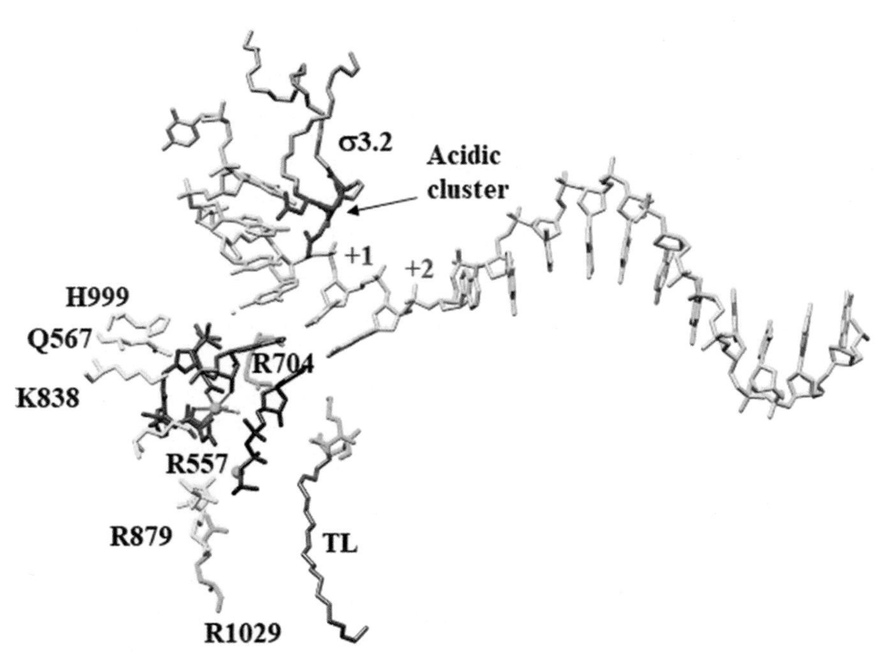

FIG. 9.13

Residues of RNAP interacting with NTPs and nascent RNA. The NTP's at positions +1 and +2 are colored *coral* and *brown*, respectively. β and β′ residues are marked in *yellow* and *cyan*, respectively. Charged residues of σ3.2 tip shown in *magenta*.

Source: Basu, R.S., et al., 2014. J. Biol. Chem. 289, 24549.

bridges with both 2′-OH and 3′-OH of the same nucleotide and can, therefore, discriminate between NTP and dNTP. Similarly, the second nucleotide, which is positioned through base-pairing with the template DNA at the +2 position, interacts with some basic residues (such as R557 and R879 of β and R1029 of β′ as marked in Fig. 9.13) and also the β′ subunit trigger loop (TL). Further, placement of a purine nucleotide at the +1 position is favored by base-stacking interaction if there is a purine nucleotide in the template strand at the −1 position. Base-stacking may play a contributory role even in the binding of the nucleotide at the +2 position.

As the formation of the dinucleotide is completed, RNAP tends to move further downstream to continue with the transcription process. Yet, at this stage, the upstream DNA-RNAP contacts that have been established in RP_o are stable enough to hold back the enzyme. In such case, possibly with the help of several positively charged residues lining the walls of the active site and certain degree of rotational freedom the downstream dsDNA possesses, RNAP pulls in the downstream DNA (from +2 to +15) while unwinding the latter. This is a phenomenon called "DNA scrunching." In the process, the transcription bubble expands from an initial size of ~13 nt to over 20 nt and the nascent RNA is extended. However, at the same time, DNA-unwinding stress and DNA-compaction stress build up in the RP_{init} complex.

Promoter escape

As the nascent RNA attempts to grow beyond 4–5 nt, another conflicting situation arises in RP_{init}. The nascent RNA encounters a steric hindrance by the σ3.2 loop (also called σ3.2 finger), which intrudes the RNAP active center and lies in its path. The tip of σ3.2 contains a cluster of acidic residues (Fig. 9.13). So there is charge repulsion between this cluster and the 5′-phosphate of the RNA. This, together with the stress associated with DNA scrunching, is possibly the reason for RNAP repeatedly synthesizing short pieces of RNA and not being able to escape from the promoter. This is called abortive initiation.

Alternatively, there also exists a finite probability that the nascent RNA, instead of falling-off itself, may cause disorderliness in the σ3.2 tip and push its way towards σ4.1. Therefore, in the fifth and final step of transcription initiation, when RNAP has synthesized an RNA of a critical length of about 15 nt (~9 nt being in the transcription bubble), σ is released stochastically. RNAP undergoes a conformational change, escapes from the promoter, and forms a highly stable protective ternary elongation complex (EC).

9.2.4 Elongation

Formation of the EC does not involve major structural changes in the RNAP core. Rather, with the exit of the σ factor, its contact sites (as already described) become available for interaction with the nucleic acids and establishment of a stable processive EC. As a consequence, there are specific adjustments in the core enzyme conformation that meet the requirements of stability and processivity. Some of the rearrangements in the β and β′ subunits are the following: (a) in the β′-pincer, an α-helical coiled-coil (CC) (*T. thermophilus* β′ 540–581), a β′ loop called the "lid"

(*T. thermophilus*: β′ 525–539), and another loop, the "rudder" (*T. thermophilus*: β′ 582–602), all move away from their earlier positions in the holoenzyme to put the "lid" into a stack with the upstream edge of the RNA/DNA hybrid and the "rudder" into a position between the downstream DNA and the hybrid; (b) the β′ N-terminal domain (NTD) (*T. thermophilus* β′ 51–499) moves closer to the downstream edge of the double-stranded DNA and helps form the RNA exit channel; and (c) the β domain (*T. thermophilus* β 1–130, β 335–396) in the pincer, which had earlier contacts with σ, moves closer to the RNA/DNA hybrid. With all these movements (Fig. 9.14), the

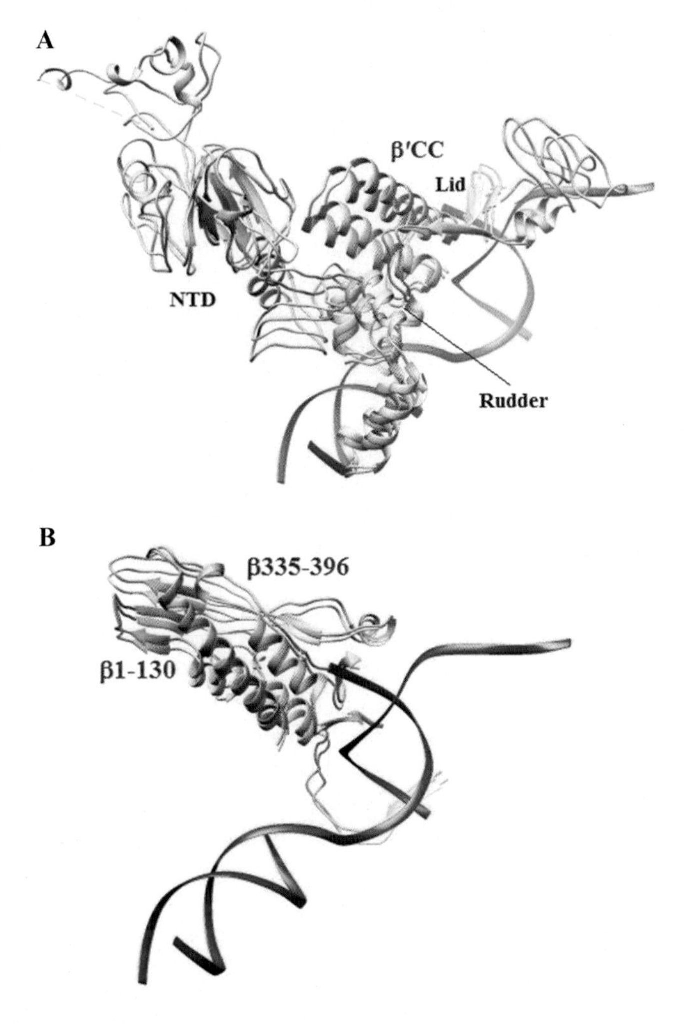

FIG. 9.14

Some rearrangements in (A) β′ subunit and (B) β subunit with the formation of ternary elongation complex (TEC). Coloring: before TEC formation *(light blue)*; after TEC formation *(golden)*.

Source: Vassylyev, D.G., 2007. Nature 448, 163; Artsimovitch, I., 2005. Cell 122, 351.

RNAP claws close-in to reduce the dimension of the main channel. In this manner, the surface compatibility between the protein and the nucleic acid is established and, thus, stability and processivity of the EC are ensured.

The EC, at any instant of time, consists of an 8- to 9 bp RNA-DNA hybrid within an 11- to 12-bp melted DNA bubble. On the upstream side of the hybrid is ssRNA moving through the exit channel and on the downstream side of the bubble is dsDNA waiting to be unwound. The duplex conformation downstream is maintained by self-stabilizing base-pairing as well as long distance electrostatic and van der Waals interactions of the DNA with RNAP. A β' subunit three-helix bundle (*Thermus thermophilus* β' 101–131) is involved in several polar and van der Waals interactions with the dsDNA phosphates.

The major sources of stability of the EC are, however, polar and van der Waals interactions between the RNAP and the RNA-DNA hybrid backbone (phosphates). These interactions are supplemented by hydrogen bonds between the polymerase and ssRNA in the exit channel.

The closed-clamp structure, which holds the EC, is stabilized by three (Sw1-3) of the five switches—Sw1 contacts the template strand in the RNA-DNA hybrid, Sw2 stabilizes the downstream edge of the hybrid, and Sw3 provides a hydrophobic binding pocket for the first ssRNA base. Stacking interactions with the lid is another stabilizing factor for the upstream edge of the hybrid.

Nucleotide addition cycle

During elongation, processivity of the RNAP enables it to undertake thousands of nucleotide addition cycles (NACs). However, the enzyme adds only one nucleotide to the nascent RNA in each cycle. For every nucleotide added, there is a forward translocation of the enzyme relative to the template by just one nucleotide. This movement results in the separation of the downstream DNA by one bp and the displacement of one nucleotide of the nascent RNA from the DNA template at the upstream end of the RNA/DNA hybrid. Concomitantly, one bp of the upstream DNA duplex is annealed.

The EC can accommodate nine bp of the RNA/DNA hybrid. Several residues from the lid stack onto the upstream base-pair of the hybrid. Therefore, further growth of the hybrid is sterically hindered, and strand separation is constrained.

A concerted motion of the bridge helix (BH) and the trigger loop (TL), functioning as a BH/TL unit, has an important role in the NAC (Fig. 9.15). In the posttranslocated position, when the NTP-binding site (denoted as position +1) is available for the next NTP, RNAP is in an open conformation, TL is unfolded, and BH is kinked. As the NTP binds, TL folds into a "double-helical" structure, BH adopts a straighter conformation, and RNAP goes to a closed conformation. The incoming NTP is correctly positioned when the incorporation is complete and a pyrophosphate PP_i is released. The TL unfolds, the BH is kinked, and the polymerase readopts an open conformation.

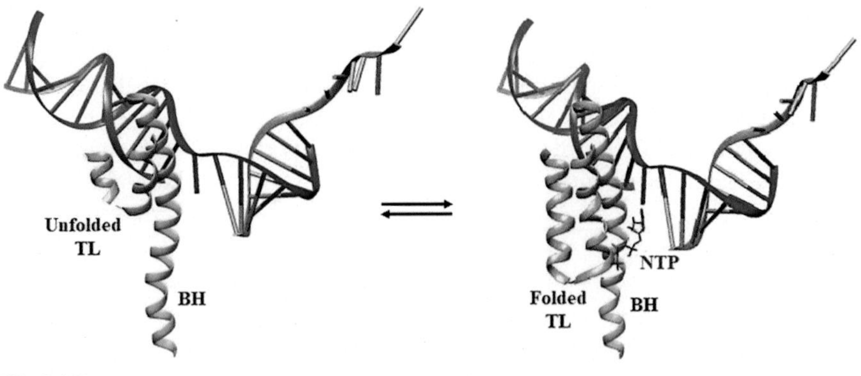

FIG. 9.15

Dynamics of BH/TL in NAC. DNA strands are shown in *magenta* and *blue*.

Source: Vassylyev, D.G., 2007. Nature 448, 163.

RNAP fidelity

The fidelity of the transcription process depends on three activities of the RNAP complex: (a) selection of the correct NTP substrate (error prevention); (b) detection of a misincorporation; and (c) error removal by polymerase backtracking and RNA cleavage.

As we have already seen, the selection of NTP involves two steps. The correct NTP binds to position +1 of an open active center of the enzyme by Watson-Crick interactions with a DNA template base. The mobile TL, which is involved in the subsequent closure of the active center, and some of its surrounding residues ensure the correctness of the NTP substrate.

In case a noncomplementary NTP is incorporated, position +1 is "frayed away" from the template and further closure of the active center is prevented. This causes an "elemental" pausing of RNAP with no further extension of the nascent RNA, that is, an "off-line" state of the elongation complex is induced. The misincorporated nucleotide is now prone to be removed by RNAP's intrinsic nucleolytic RNA cleavage activity. However, for the cleavage to occur, the target phosphodiester bond must be aligned with the catalytic site. This is achieved by what is known as RNAP backtracking.

In such a circumstance, the RNAP moves backward (backtracked) along the DNA, while one or a few end nucleotide(s) of RNA is(are) threaded into the secondary channel. The backtracked complex, by its intrinsic nucleolytic activity, cleaves the extruded part of the RNA 3′-end. In bacteria, the cleavage activity is enhanced by stimulating factors such as GreA and GreB proteins. A new RNA 3′-OH and a vacant NTP-binding site are generated. The complex returns to an "online" state and transcription resumes.

9.2.5 Termination

Earlier, we have discussed how stability of the EC is important for the RNAP to undertake an efficient transcription process. It is, therefore, obvious that transcription termination necessitates destabilization of the EC. In bacteria, two mechanisms that destabilize the EC, leading to termination of transcription, are (a) intrinsic and (b) factor (Rho)-dependent.

Intrinsic

The canonical intrinsic terminator comprises two sequence elements—(a) a short GC-rich inverted repeat of ~20 bp immediately followed by (b) a stretch of 7–8 AT-base-pairs. When the sequences are transcribed by the RNAP, the synthesized RNA has the potential to form a stem-loop structure or "hairpin" T_{hp}, which is followed by a 7–8 nt U-rich tract (Fig. 9.16A). In fact, these elements in the RNA are responsible for the termination process.

As the U-tract is transcribed, the RNAP is compelled to pause at the end of the tract. The duration of the pause is sufficiently long for the formation of T_{hp} in the exit channel (Fig. 9.16B). The pairing is aided by the flap domain tip. Complete formation of the T_{hp} is accompanied by the displacement of −10 RNA base (with respect to possible NTP-binding position) from the Sw3 binding pocket, unstacking of the lid from the RNA–DNA bp at −9 position, and melting of 3–4 bp of the upstream RNA-DNA hybrid. Subsequently, the EC disintegrates and the RNA, DNA, and RNAP are all released.

Rho-dependent

In contract to intrinsic termination, the second mechanism of termination in bacteria is dependent on the RNA translocase activity of a homohexameric helicase known as Rho (Fig. 9.17). To execute its transcription termination action, the enzyme proceeds through the following steps: (a) binds to substrate RNA—primary binding and secondary binding; (b) translocates RNA by its motor activity; and (c) taking advantage of the paused state of the EC, disengages the RNAP from it.

The binding of Rho to the nascent RNA requires certain features in both the enzyme and the RNA sequences. In the nascent RNA transcript, Rho binds preferentially to C-rich, G-poor sequences lacking extensive secondary structures. These primary binding regions are called Rho-utilization (*rut*) sites. Each of these *rut* sites is approximately 80–90 nt long containing pyrimidine dinucleotides interspersed by loops at least 7–8 nt long.

rut binds the N-terminal domain (NTD) of Rho in an open-ring state. Rho-NTD contains an oligonucleotide-oligosaccharide binding fold (OB) and a helix bundle (NHB). A narrow cleft in the OB can accommodate only a pyrimidine dinucleotide, preferentially a consensus 5′-YC-3′ (Fig. 9.18A). In *E. coli* Rho, the first nucleotide, "Y," is held within a pocket formed by Y80, E108, and Y110. On the other hand, the base of the second nucleotide, "C," stacks on F64 aromatic side chain and also interacts with R66 and D78. Positive charges in OB as well as adjacent NHB help in RNA

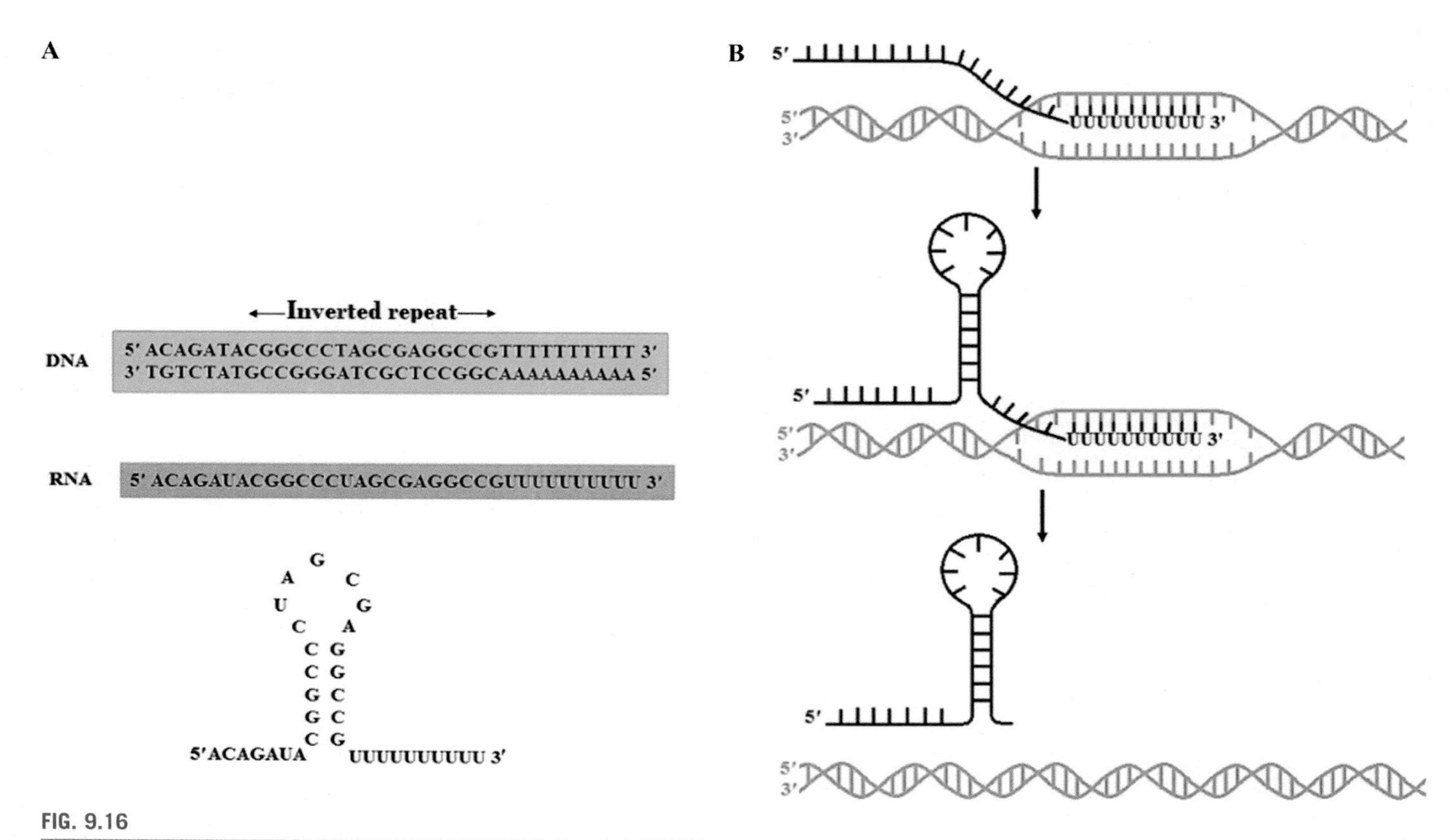

FIG. 9.16

Intrinsic terminator and transcription termination. (A) Top, sequences in the DNA; middle, corresponding RNA sequence; bottom, terminator hairpin, T_{hp}. (B) Top, U-tract transcribed; middle, terminator hairpin formed; bottom, transcript released.

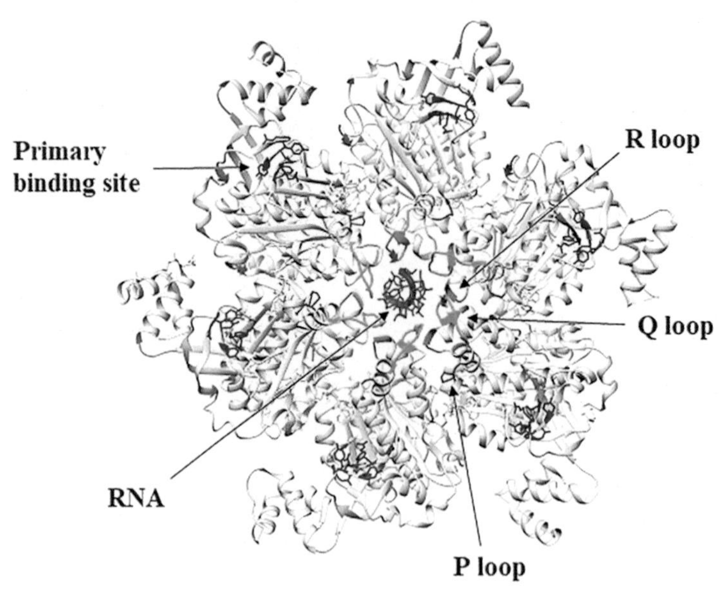

FIG. 9.17

Closed-ring structure of Rho homohexamer bound by RNA in the secondary binding site. In each protomer, primary binding site in OB-fold *(deep blue)*, Q loop *(cyan)*, R loop *(coral)*, P loop *(red)* are shown.

Source: Thomsen, N.D., 2009. Cell 139, 523.

binding to the enzyme by electrostatic interactions with the negatively charged phosphate backbone (Fig. 9.18B).

The RNA secondary binding site in Rho-CTD consists of the conserved Q- and R-loops (Fig. 9.19A), which contact the ribonucleotides in the RNA in a "spiral" staircase conformation. Compared with primary binding, RNA-binding of the secondary site is more nonspecific. However, the carbonyl group of Rho V284 (*E. coli*) in the Q-loop can H-bond with only a 2′-OH of RNA ribose (Fig. 9.19B). Therefore, unlike the primary binding site which can bind to either ssDNA or ssRNA, secondary binding is specific to RNA. Rho CTD also contains the Walker A and B motifs (P-loop) that create the ATP-binding pocket, while *E. coli* residues E211, E212, and R366 form the active (ATP hydrolysis) site.

As *rut* binds the NTD, it steers RNA into the exposed central pore of the CTD. In addition to *rut*, when ATP also binds Rho, ring closure is induced and the enzyme is activated for catalysis. Driven by ATP hydrolysis, Rho translocates the nascent transcript until it reaches the RNA exit channel. Continued ATP hydrolysis then forces the shearing of the RNA/DNA hybrid, extraction of the nascent transcript from the RNA exit channel, collapse of the EC bubble, and release of the template DNA.

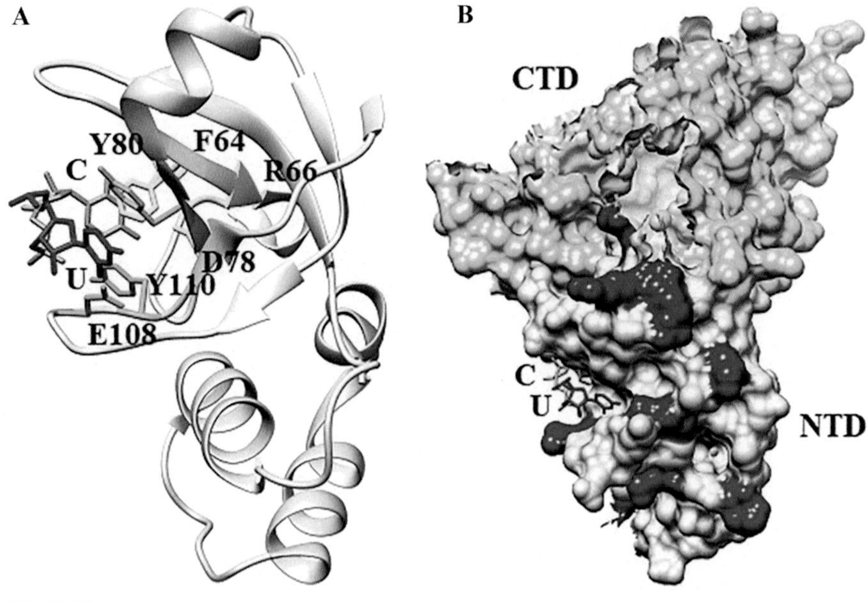

FIG. 9.18

Primary RNA binding by Rho-NTD. (A) Pyrimidine dinucleotide interacting with amino acid residues in OB cleft. (B) Surface of Rho protomer showing dinucleotide-binding cleft. NTD—*cyan*; CTD—*light green*. Positively charged amino acid residues in NTD colored *deep blue*.

Source: Skordalakes, E., 2003. Cell 114, 135.

9.3 Eukaryotic transcription

In bacteria, all the genes are transcribed by a single RNA polymerase. In contrast, eukaryotic genes are transcribed by three RNA polymerases, Pol I-III. Pol I transcribes large rRNA precursor genes, while Pol III transcribes genes encoding tRNAs, small nuclear RNA, and 5S rRNA. However, most of the cellular genes, including all protein-coding genes, are transcribed by Pol II. Here, we shall focus exclusively on Pol II.

9.3.1 RNA polymerase II

Since RNA polymerase, whether bacterial or eukaryotic, carries out the same chemical reaction (as mentioned earlier), it is but expected that the transcribing enzymes from different organisms would have some common features, particularly in and around the catalytic site. At the same time, since the eukaryotic RNA polymerase has to function in a more complex environment than the bacterial polymerase, it is also expected that the eukaryotic enzyme would possess additional features to address the challenges of higher complexity.

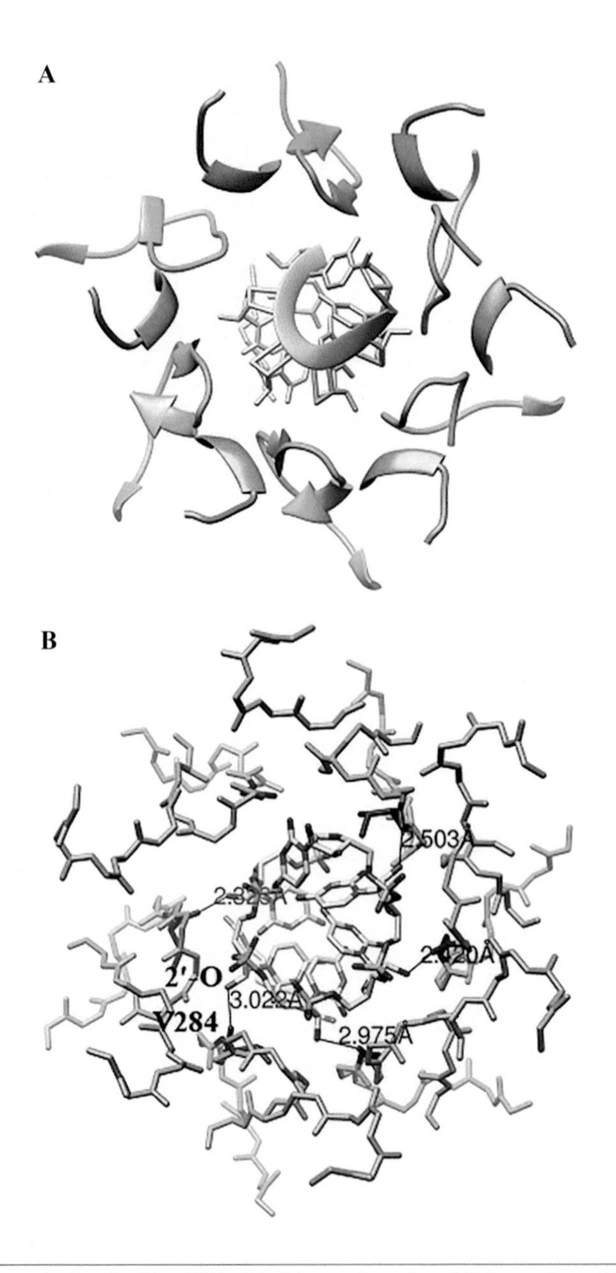

FIG. 9.19

RNA secondary binding by Rho-CTD. (A) Q-loops *(cyan)* and R-loops *(coral)* in a spiral staircase conformation around RNA. (b) H-bonds *(deep red)* between V284 carbonyl and 2′-OH shown with length.

Source: Thomsen, N.D., 2009. Cell 139, 523.

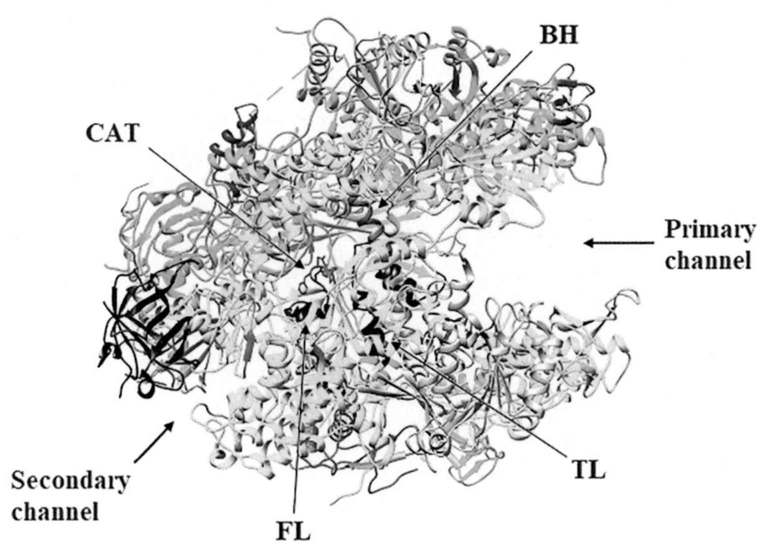

FIG. 9.20

Structure of *Saccharomyces cerevisiae* Pol II. The subunits are colored as follows: Rpb1—*yellow*; Rpb2—*cyan*; Rpb3—*green*; Rpb5—*pink*; Rpb6—*dark green*; Rpb8—*brown*; Rpb9—*light brown*; Rpb10—*medium blue*; Rpb11—*dark gray*; Rpb12—*magenta*. Rpb1 features: catalytic loop (CAT, *purple*), F-loop (FL, *red*), bridge helix (BH, *orange*), and trigger loop (TL, *brick red*).

Source: Cramer, P., et al., 2001. Science 292, 1863.

Saccharomyces cerevisiae Pol II has been the most widely studied eukaryotic RNA polymerase. Its core contains 12 subunits, two of which are required only for initiation and can, therefore, dissociate from the core. The crystal structure of the 10-subunit core is shown in Fig. 9.20. The two largest subunits, Rpb1 and Rpb2, form the opposite sides of a central cleft. The other subunits are arranged around the periphery of the enzyme. The subunits are stably held together by a large number of salt bridges, hydrogen bonds, and structural water molecules.

The N- and C-terminal regions of Rpb1 and the C-terminal region of Rpb2 form mobile "clamp" on one side of the cleft. Five protein "switch" regions connect the clamp to the rest of the enzyme and confer mobility to it. The cleft contains the active center of the polymerase. It is restricted at one end by a protein wall.

The surface charge of the cleft floor and inward sides of the cleft including the clamp is essentially positive. This facilitates placement of the DNA in the active site for transcription (Fig. 9.21).

Like the bacterial polymerase, the largest subunit of *S. cerevisiae* Pol II contains structural features such as the bridge helix (BH), trigger loop (TL), fork loops (FLs), and catalytic loop, all of which are crucial in the transcription process (Fig. 9.20). The BL ensures processivity of the polymerase, the TL participates in correct NTP selection, and the FLs help maintain the transcription bubble.

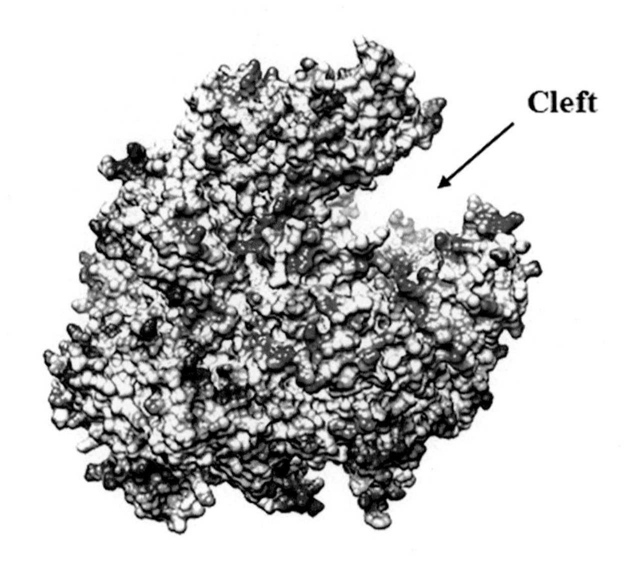

FIG. 9.21

Solvent-accessible surface of yeast Pol II displaying electrostatic potential: *blue*—most positive; *white*—neutral; *red*—most negative.

Source: Cramer, P., et al., 2001. Science 292, 1863.

The active site of Pol II contains in the catalytic loop three conserved aspartate residues, D481, D483, and D485, to which a Mg^{2+} ion is stably bound (metal A). A second Mg^{2+} ion (metal B) is brought to the active site by the substrate nucleotide.

It can be seen from Fig. 9.20 that the overall shape of *S. cerevisiae* Pol II is similar to that of the *Taq* RNA polymerase core (earlier shown in Fig. 9.2A). When the three-dimensional structures are superposed, similarity is evident between the two enzymes particularly around the active center (Fig. 9.22). However, differences exist in the peripheral and surface structures as the two enzymes must interact with different sets of factors in the entire transcription process.

9.3.2 Promoter

Like the bacterial RNAP, Pol II also requires a specific region (promoter) of the DNA for accurate transcription initiation. The eukaryotic core promoter, usually 40–60 nucleotides, contains a number of sequence elements located both upstream and downstream from the transcription start site (Fig. 9.23). The most prominent of these element are the TFIIB recognition element (BRE), the TATA box (TATA), the initiator (Inr), and the downstream promoter element (DPE). However, a particular promoter may contain all, some, or none of these elements. There are some other elements, such as the downstream core element (DCE), found in the region

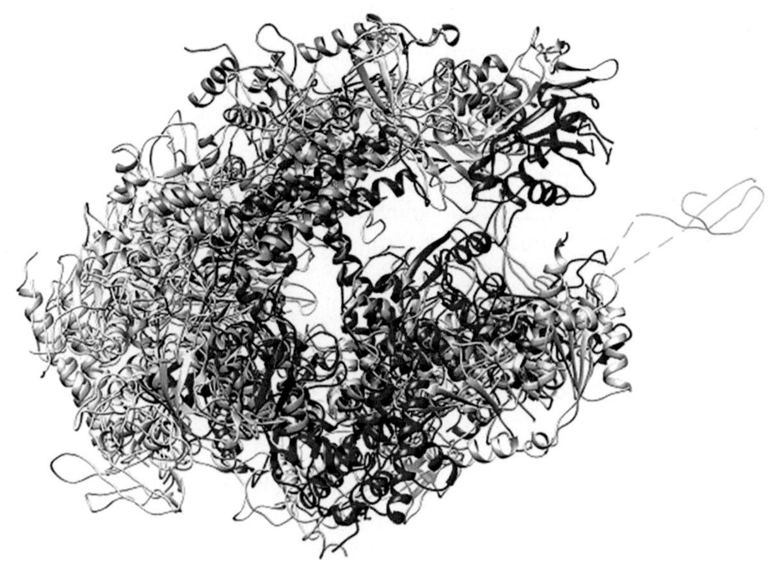

FIG. 9.22

Superposition of three-dimensional structures of *Taq* RNAP and Yeast Pol II. Coloring scheme—*Taq* RNAP: β′—*orange* and β—*light pink*; yeast Pol II: Rpb1—*red* and Rpb2—*pink*. All other Taq RNAP subunits *light gray* and yeast Pol II subunits *dark gray*.
Source: Minakhin, L., et al., Proc. Natl. Acad. Sci. U. S. A. 98, 892; Cramer, P., et al., 2001. Science 292, 1863.

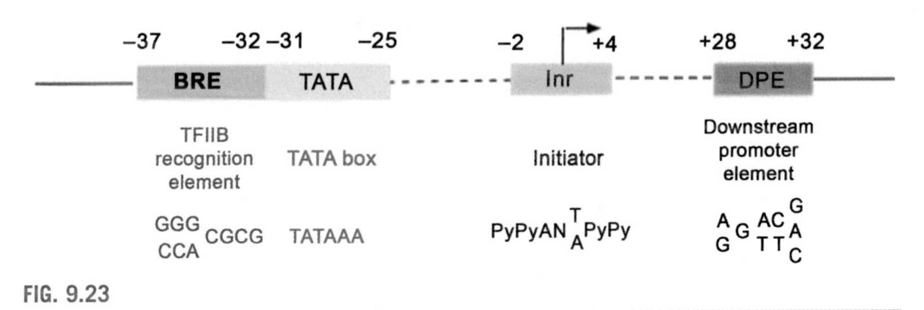

FIG. 9.23

Eukaryotic Pol II core promoter elements.

between +10 and +45, which also have been found to promote transcription initiation in certain genes.

9.3.3 General transcription factors

All multisubunit RNA polymerases, whether bacterial or eukaryotic, need accessory factors to specifically initiate transcription at promoter sequences. We have already seen that the bacterial RNAP core requires only one additional factor (σ factor).

In eukaryotic cells, on the other hand, a complex set of general transcription factors (GTFs) is involved in the transcription initiation process—promoter recognition, polymerase recruitment, interaction with regulatory factors, DNA-unwinding, and transcription start site recognition. In the following discussion, we shall focus on some of the crucial GTFs.

The GTFs help RNA Pol II bind the core promoter in a state defined as the pre-initiation complex (PIC). This is analogous to the bacterial closed complex. In TATA-box containing promoters, the formation of PIC begins at TATA.

TFIID

At first, the GTF TFIID recognizes the TATA element. TFIID contains the TATA-binding protein (TBP) and 14 TBP-associated factors (TAFs). Some of the TAFs bind the other core elements such as the Inr, DPE, and DCE—these bindings are, however, weaker than TBP-TATA-binding.

The C-terminal region of TBP consists of two subunits, each of about 90 amino acids, having only 30% sequence similarity. Yet their structural identity gives the TBP region a twofold pseudo-symmetry. TBP forms a saddle-shaped structure with each of the subunits consisting of a five-stranded curved antiparallel β sheet and two α helices. The central eight strands of the curved β sheet make the underside of the saddle concave (Fig. 9.24).

In a conspicuous departure from the characteristics of transcription factors to bind the chemically rich major groove of DNA, the extensive region of β sheet in TBP

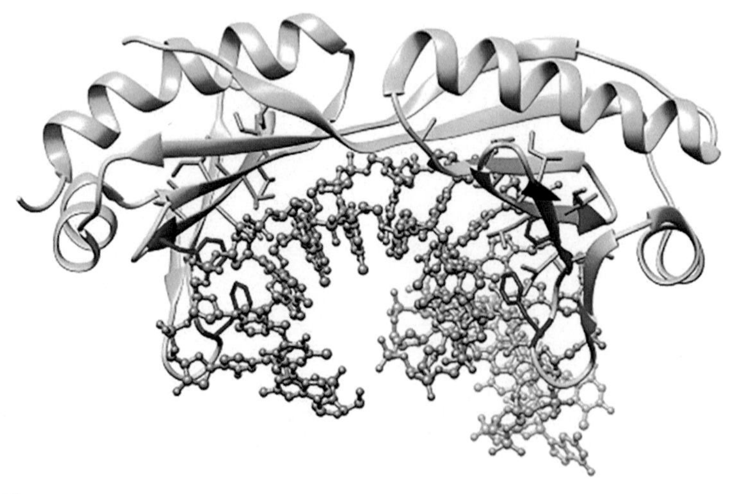

FIG. 9.24

TBP/TATA-box complex. TBP—*light green*; DNA—*pink*. Residues in β sheet interacting with minor groove colored *orange*; phenylalanine residues intercalating between base-pairs colored *deep blue*.

Source: Kim, Y., et al., 1993. Nature 365, 512.

interacts with the minor groove in the TATA element (Fig. 9.24). The protein-DNA interface is formed predominantly by hydrophobic or van der Waals contacts, although some hydrogen bonds between the side chains of TBP and the base edges also exist.

When bound to the DNA, TBP causes a major distortion in the TATA element which now deviates significantly from the canonical B-DNA and appears more like the A-form. The minor groove in the center of TATA becomes extremely wide and shallow, the width exceeding 9 Å as compared to ~4 Å for B-DNA. The DNA is bent ~80° towards the major groove. Two pairs of phenylalanine side chains intercalating between base-pairs at either end of the TATA box also contribute to the bending (Fig. 9.24). It can be understood why A:T base-pairs (making up TATA) are favored as they are more amenable to distortions. Overall, TBP-binding to TATA produces an asymmetric platform for subsequent PIC assembly.

Some of the TAFs have structural similarity with histones—proteins which are known to bind and package DNA into nucleosomes. Whether TAFs bind DNA is a matter of speculation. However, TAF1 N-terminal domain (TAND) binds the DNA-binding surface of TBP and prevents it from binding the DNA. The transcription factor TFIIA displaces TAND from TBP to enable the latter bind the DNA. In this respect, TFIIA functions as a cofactor that binds TFIID and stabilizes the TFIID-DNA complex.

TFIIA

The crystal structure of TFIIA has shown that it possesses a two-domain folding pattern (Fig. 9.25A). It is a heterodimer whose large and small subunits associate to form both the domains—a six-stranded β-sandwich domain and a four-helix bundle domain. The β-sandwich domain interacts with the N-terminal end of TBP and also with the DNA (Fig. 9.25B).

TFIIB

The next GTF to enter the PIC is TFIIB. It contains two conserved domains—an N-terminal zinc ribbon domain and a C-terminal core domain. The N-terminal domain contains a putative Zn^{2+} ion-binding motif $CX_2CX_{17}CX_2C$ that is essential for recruitment of TFIIF/Pol II to the PIC (Fig. 9.26). The C-terminal core domain interacts with the C-terminal stirrup of TBP (Fig. 9.27) making, at the same time, base-specific contacts with the major groove upstream (BRE) and minor groove downstream of TATA. At this stage, the PIC is a TBP-TFIIA-TFIIB-DNA complex as shown in Fig. 9.28. The asymmetric nature of the complex gives directionality to the subsequent transcription process.

Once the PIC-intermediate containing promoter DNA, TBP, TFIIA, and TFIIB is formed, TFIIF enters the complex along with Pol II. The entry is assisted by a protein-protein interaction between TFIIF and TFIIB.

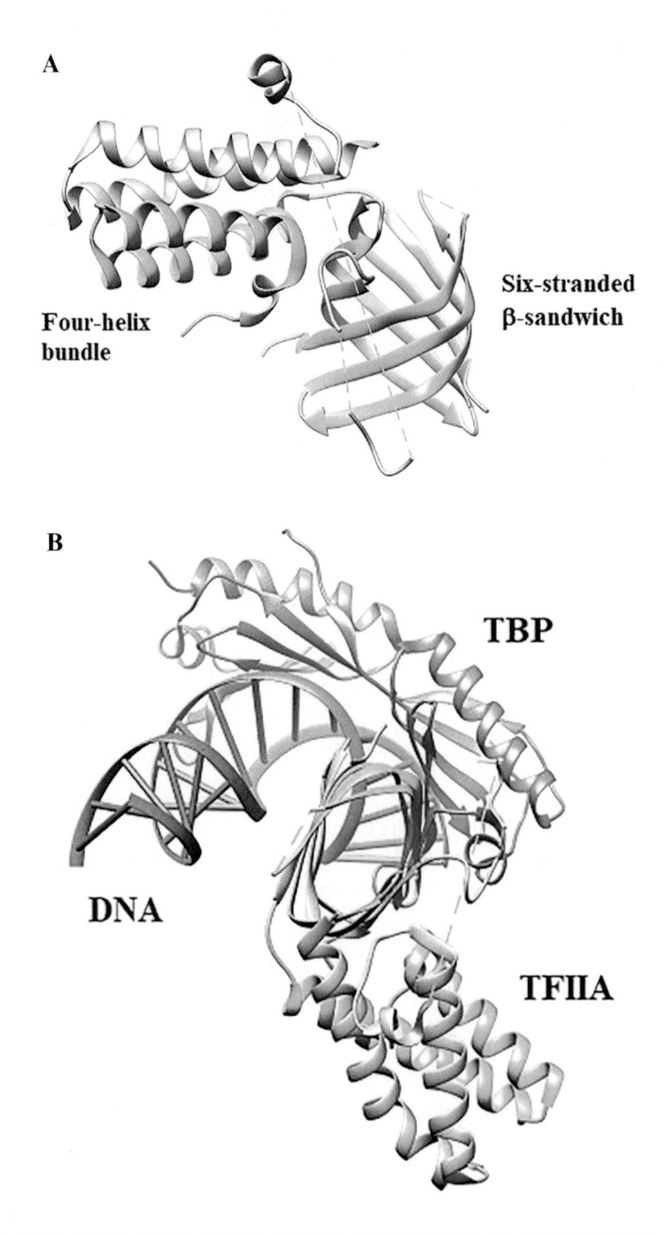

FIG. 9.25

(A) Domain structure of TFIIA. Large and small subunits are colored *light pink* and *light green*, respectively. (B) TATA/TBP/TFIIA complex.

Source: Jin, X., et al., PDB deposit 2003-11-26; unpublished.

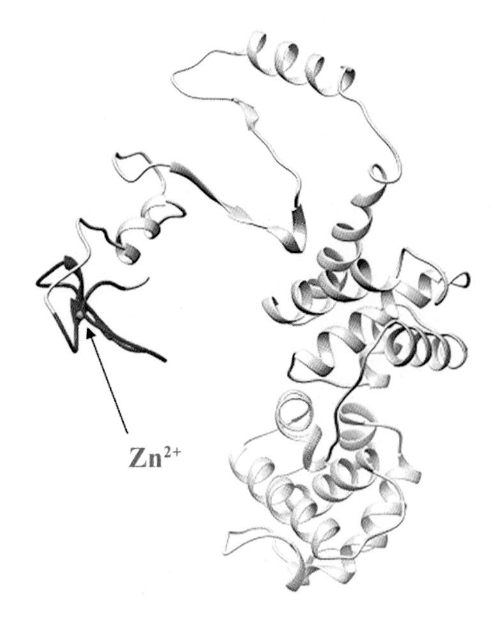

FIG. 9.26

Zn^{++}-binding motif in the TFIIB N-terminal domain.

Source: Sainsbury, S., et al., 2013. Nature 493, 437.

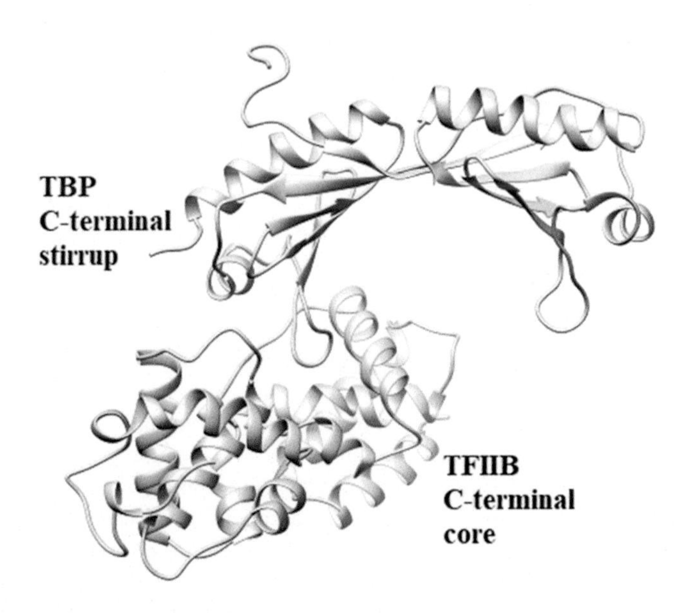

FIG. 9.27

TFIIB C-terminal core domain interacting with TBP C-terminal stirrup.

Source; Nikolov, D., et al., 1995. Nature 377, 119.

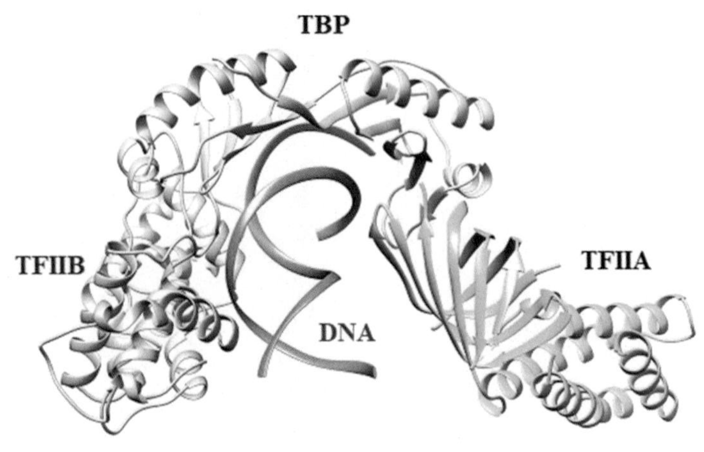

FIG. 9.28

Structure of TBP-TFIIA-TFIIB-DNA complex.

Source: Tan, S., et al., 1996. Nature 381, 127.

TFIIF

TFIIF is a heteromer containing two conserved subunits Tfg1 and Tfg2 in yeast (Rap74 and Rap30 in homosapiens). Each of the subunits contains an N-terminal dimerization domain and a C-terminal winged helix (WH) domain.

Pol II-TFIIF complex has not been very amenable to X-ray crystallographic analysis. However, protein cross-linking and cryo-EM have revealed useful information on Pol II-TFIIF interactions. TFII has been found to interact with PolII over an extended surface. Tfg1 and Tfg2 form a dimerization domain and bind Pol II Rpb2 lobe on the edge of the active center cleft near the downstream DNA. Possibly, Tfg2 extends along the PolII cleft connected by a flexible linker, while Tfg1 interacts with Rpb4/7 subunits. Tfg2 WH domain binds Pol II protrusion near the promoter DNA upstream of TATA.

TFIIF is essential for transcription initiation at both TATA-containing and TATA-less promoters. Interaction of Tfg2 WH domain with DNA upstream of the transcription start site helps position the DNA in the Pol II active site. Earlier in Chapter 5, we have seen that WH domain has a DNA-binding property.

At the same time, TFIIF prevents nonspecific DNA-binding to Pol II due to an extremely positively charged downstream cleft. Tfg1 places an unstructured negatively charged linker region in the cleft, thus applying a repulsive force on the DNA. This characteristic is similar to that of the bacterial σ^{70} region 1.1.

The assembly of TFIIA, TBP, TFIIB, and TFIIF at the double-stranded promoter DNA is a PIC-intermediate in a closed state. Transition to an open complex (OC) state is necessary for transcription initiation. The transcription factors TFIIE and TFIIH step in at this stage.

TFIIE

At first, TFIIE binds to the complex containing TFIIA, TBP, TFIIB, TFIIF, and PolII. TFIIE is a heterodimer consisting of two subunits Tfa1 and Tfa2 in yeast (TFIIEα and TFIIEβ in mammals). TFIIE interacts with both the promoter and Pol II.

Tfa1 contains an N-terminal winged helix (WH) domain which presumably binds DNA. In the case of Tfa1, the predicted DNA-binding groove is negatively charged which is unsuitable for binding DNA. Nonetheless, the Tfa2 subunit contains a central tandem WH domain in which the N-terminal WH (WH1) has been found to bind DNA in vitro. TFIIE binds DNA in the region between −10 and +2. On the other hand, the Tf1 WH dimerizes with the Tf2 tandem WH domain, stabilizing the latter's binding to the promoter. The Tfa1 WH also anchors TFIIE to the Pol II clamp. It spans the Pol II cleft from the clamp (in Rpb1) to the protrusion (in Rpb2).

The PIC formation is complete when TFIIE recruits TFIIH. For this, the two factors need to interact. The interaction has been biochemically and X-ray crystallographically studied using human GTF's hTFIIE and hTFIIH.

hTFIIE consists of two subunits—hTFIIEα and hTFIIEβ. hTFIIEα contains a C-terminal acidic amino acid-rich domain (AC-D) which is responsible for binding to hTFIIH. On the other hand, the p62 subunit of hTFIIH contains an N-terminal pleckstrin homology domain (PH-D). p62 PH-D has an array of positively charged residues on its surface and a couple of hydrophobic pockets (Fig. 9.29). A flexible acidic stretch of hTFIIEα AC-D, which forms a β-stand upon binding, wraps around the PH-D. The complex is stabilized by a combination of electrostatic and hydrophobic interactions.

In yeast, the interaction between TFIIE and TFIIH is more of a speculation. The yeast homolog of p62, Tfb1, contains a PH-D (residues 1–115). Despite a low level

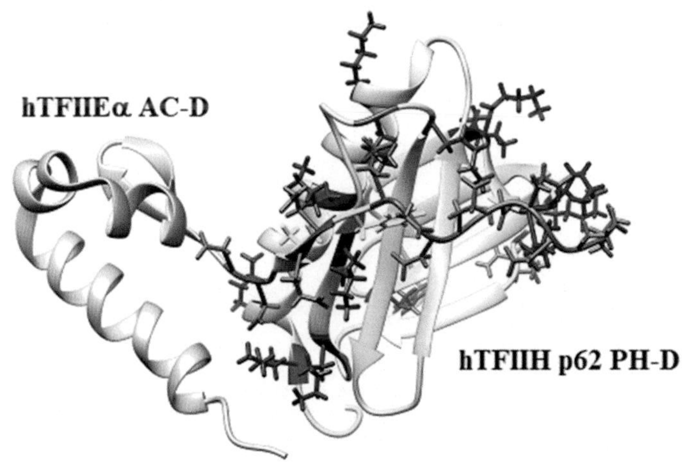

FIG. 9.29

hTFIIE α AC-D—hTFIIH p62 PH-D complex. Coloring of residues: acidic—*coral*; basic—*blue*; hydrophobic—*yellow*.

Source: Okuda, M., et al., 2008. EMBO J. 27, 1161.

of sequence similarity between the two PH-D's, their three-dimensional similarity is remarkable (Fig. 9.30). Like p62, the Tfb1 PH-D surface is also arrayed with positively charged residues (Fig. 9.31). Since the yeast TFIIE Tfa2 (small subunit) possesses a stretch of acidic residues (86-DDDDDDED-93), interaction between this acidic motif and the Tfb1 PH-D is a distinct possibility.

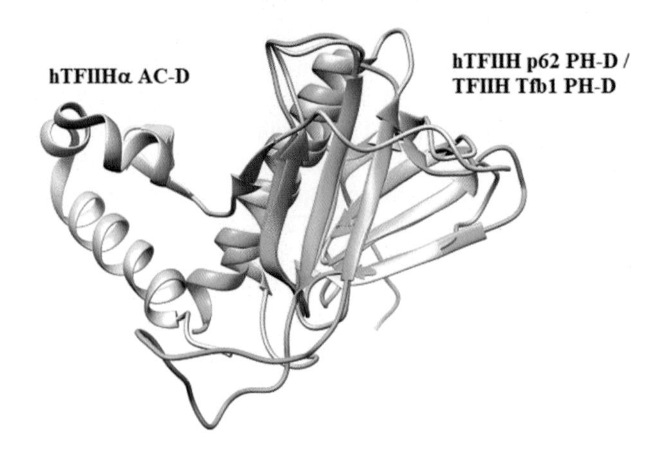

FIG. 9.30

Superposition of human hTFIIH p62 PH-D *(tan)* and yeast TFIIH Tfb1 PH-D *(cyan)* three-dimensional structures.

Source: Di Lello, P., et al., 2005. Biochemistry 44, 7676; Okuda, M., et al., 2008. EMBO J. 27, 1161.

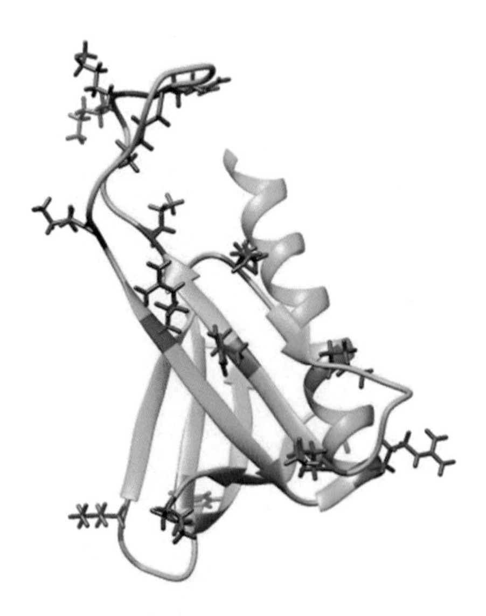

FIG. 9.31

Positively charged residue on TFIIH Tfb1 PH-D surface shown in *blue*.

Source: Di Lello, P., et al., 2005. Biochemistry 44, 7676.

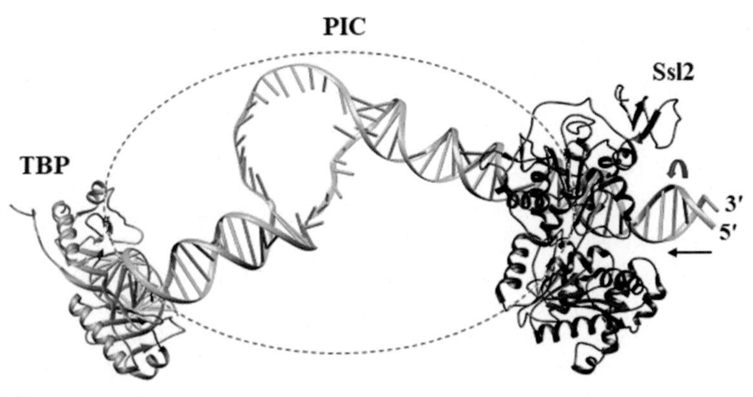

FIG. 9.32

DNA translocase activity of yeast THIIH Ssl2 subunit. Arrows indicate the movements of dsDNA. PIC, pre-initiation complex.

Source: He, Y., et al., 2016. Nature 533, 359.

TFIIH

TFIIH comprises 10 subunits—a seven subunit core complex and a three subunit Cdk activating kinase complex. Its Ssl2 subunit (human homolog XPB) possesses a DNA-dependent ATPase activity and functions as a double-stranded DNA translocase. This subunit is required for DNA opening and transition to the OC.

Ssl2 interacts with promoter DNA downstream from the initial site of DNA-unwinding. It uses the energy of ATP hydrolysis to move DNA into the Pol II cleft and tracks along the nontemplate DNA strand in the 5′–3′ direction. DNA is inserted into the cleft by a right-handed threading of downstream DNA through Ssl2 binding groove (Fig. 9.32). However, the movement of TFIIH is constrained by its association with other PIC components such as TFIIE and Med11 subunit of the Mediator complex. All these together lead to DNA-unwinding and subsequent separation of DNA strands in the OC (Fig. 9.32).

9.4 Transcriptional regulation

Till now, we have seen how a gene is transcribed by the transcription machinery to synthesize a messenger RNA (mRNA) molecule in the cell. In a later chapter, we shall consider how the mRNA is translated into a protein product. Some of the genes in a cell are expressed at a more or less constant level at all times (constitutive expression). Most of the genes are, however, regulated, positively or negatively, according to the requirement of the cell. Gene expression can be regulated at every step from the gene to its product; however, the most fundamental regulation of gene expression is at the level of transcription initiation.

Transcriptional regulation is mediated by regulatory molecules which can be proteins or RNAs. Here we shall try to understand the mechanisms of regulation mediated by protein molecules. Regulation can be positive or negative—the positive regulators, or activators, increase the level of transcription, whereas the negative regulators, or repressors, decrease the level of transcription.

9.4.1 Prokaryotic regulation

At first, let us talk about transcriptional regulation in prokaryotic organisms. In bacteria, there are genes whose promoters are bound by RNA polymerase only weakly. This may be due to the fact that either some of the promoter elements, as described earlier, are either missing or the promoter sequence is effectively deviated from the consensus. In such case, the polymerase binds the promoter with a low frequency and leads to a low "constitutive" or "basal" level of transcription. An activator helps the polymerase bind the promoter with a greater efficiency. Typically, the activator has a DNA-binding domain which helps it bind the DNA close to the promoter and an activating domain that simultaneously interacts with the RNA polymerase to bring it to the promoter, the mechanism being called "recruitment." As a result, the level of transcription is increased.

On the other hand, the constitutive level of transcription of a gene may be undesirable for the health of a cell and need to be repressed. A repressor molecule makes that happen by binding the DNA at a site, called the operator, which overlaps the promoter. In that case, the RNA polymerase is prevented from binding the promoter and the level of transcription goes down. The mechanisms of both repression and activation can be explained with the help of *E. coli lac* operon.

The lac *operon*

In bacteria, functionally related genes are clustered together so that they can be regulated from a single promoter. Such a genetic unit where a cluster of contiguous genes are coordinately regulated is known as an operon. In the *lac* operon of *E. coli*, three genes (structural genes) coding for enzymes that carry out lactose metabolism, *lacZ*, *lacY*, and *lacA*, are grouped together under the control of the *lac* promoter (p_{lac}). Transcription from p_{lac} produces a single messenger RNA (polycistronic message). This

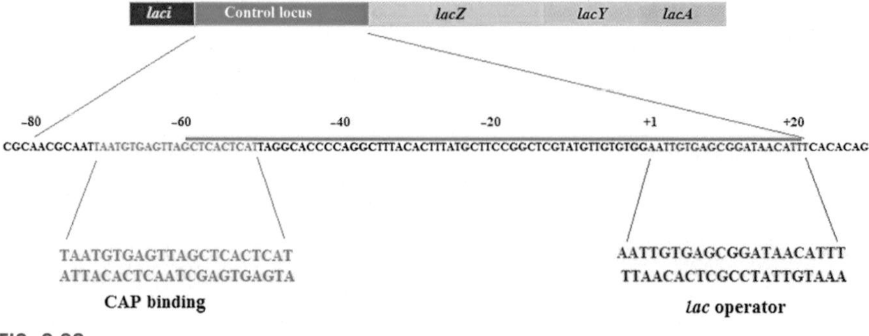

FIG. 9.33

Regulatory sequences of the *lac* operon. The sequence of only one strand of the control locus has been shown. *Blue* line above the sequence indicates the RNA polymerase-binding region. The central G:C base-pair in the *lac* operator is colored *dark blue*.

mRNA is subsequently translated to yield three protein products. Both types of control, negative and positive, operate in the *lac* operon (Fig. 9.33).

Negative control (repression)

The negative control on the *lac* operon is exerted by a regulatory protein molecule called the Lac repressor. It is encoded by the *laci* gene located close to structural genes. However, *laci* comprises an independent transcription unit with its own constitutively expressed promoter and terminator. In the absence of lactose, the Lac repressor has been found to actively repress transcription of the *lac* operon, whereas in the presence of lactose, it is inactive and the lac genes are derepressed. Hence, the structure of the Lac repressor should be such that (i) it is able to bind to a site in the DNA overlapping p_{lac}, (ii) it responds to the presence of lactose in the environment (the lactose signal), and (iii) it exerts additional control, if possible. Let us examine its structure in order to understand how it carries out these functions.

The structure of Lac repressor, as determined by X-ray crystallography, shows that it is a homotetrameric protein (Fig. 9.34). Each monomer of 360 amino acids residues consists of five parts (Fig. 9.35). A three-helix bundle "headpiece" (residues 1–49) and a "hinge helix" (residues 50–58) together constitute the DNA-binding domain (DBD; residues 1–62). A core domain (residues 62–333) is divided into N- and C-subdomains. An inducer binding pocket is present at the junction of the two subdomains. The C-terminal region contains an α helix (residues 340–357).

Now, where in the DNA does the Lac repressor bind? It binds to a regulatory element called an operator. The 21-bp lac operator (lacO) sequence, which extends from +1 to +21 with respect to the transcription start site, is pseudo-symmetric; the approximate

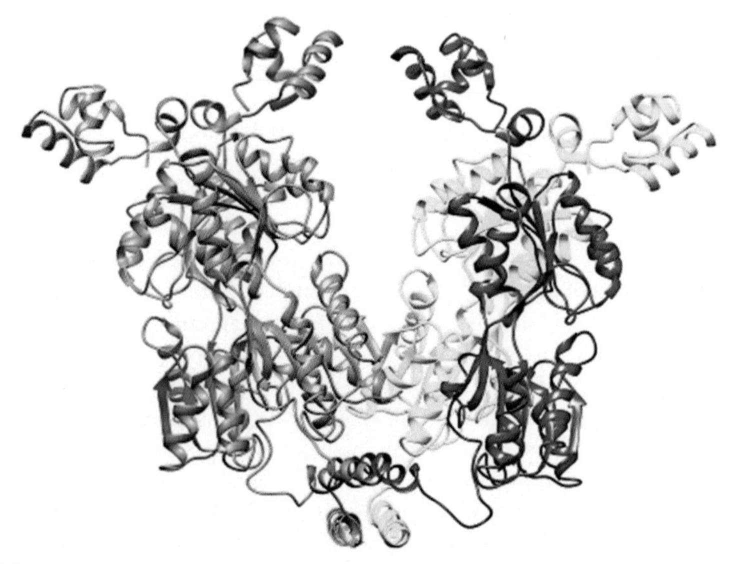

FIG. 9.34

Tetrameric structure of the Lac repressor. Each subunit has been colored differently.

Source: Lewis, M., et al., 1996. Science 271, 1247.

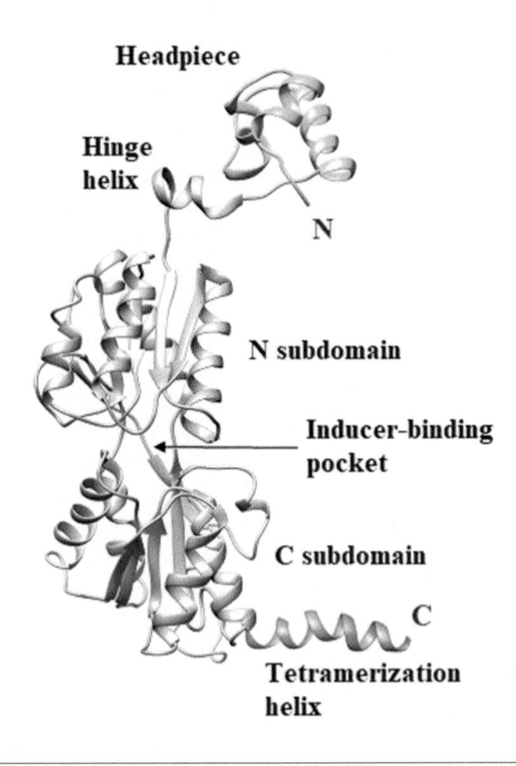

FIG. 9.35

Structure of a Lac repressor monomer showing individual units.

Source: Lewis, M., et al., 1996. Science 271, 1247.

dyad axis passes through the central G:C base-pair (Fig. 9.33). We have seen earlier that the promoter extends up to +2. Moreover, the binding region of the RNA polymerase extends even beyond (up to +20). Hence, the polymerase and repressor-binding regions are overlapping (Fig. 9.33). The repressor monomers self-associate to form stable dimers. An extensive interface created through conserved interactions of the C-subdomains and variable interactions of the N-subdomains hold together the two monomeric subunits. (Fig. 9.36). The repressor binds the DNA as a dimer with the DBD HTH motifs of the two subunits entering two consecutive major grooves (Fig. 9.37). It can be said that the dimer is the functional repressor.

We have seen in Chapter 5 that the helix-turn-helix (HTH) of a protein is a common DNA-binding motif. So it is not surprising that the HTH motif of the repressor headpiece binds the major groove of the operator. What is unusual here, and for that reason more interesting, is that apolar amino acid residues L56, A53, and A57 of the hinge helix are projected into minor groove of the operator. Here, they are involved in hydrophobic interactions with the deoxyribose and base portions of the central base-pairs. Further, polar amino acids N50 and Q54 of the hinge helix form hydrogen-bonds with the phosphate backbone (Fig. 9.37). Insertion of the hinge helix

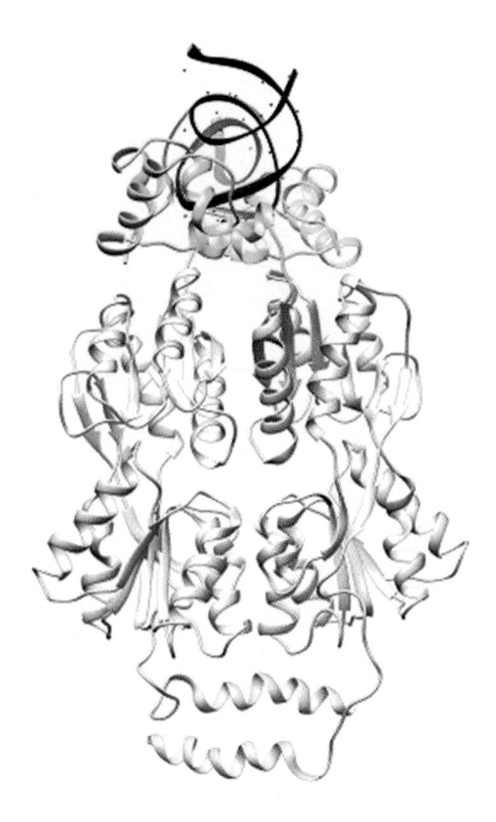

FIG. 9.36

Dimerization of Lac repressor monomers.

Source: Lewis, M., et al., 1996. Science 271, 1247.

leads to opening of the minor groove and conformational distortion of the operator (bending ~40° away from the repressor).

Now, since binding of the repressor to the DNA (operator) "switches off" the transcription of the lac operon, an "inducer" should be able to reduce this binding affinity to "switch on" the transcription. In the natural system, the inducer is allolactose, the product of a side reaction of β-galactosidase with lactose. However, in most of the biochemical and structural studies on the regulation of lac operon, isopropylthiogalactoside (IPTG), which is a lactose analog containing the galactoside ring, has been used as an inducer. One way the inducer could reduce the repressor-DNA-binding affinity is by titrating out the DNA-binding site of the repressor. However, as seen in the structure of the repressor, the inducer binding site is different from the DNA-binding site.

In fact, the switching off and on of *lac* operon transcription by the Lac repressor follows the conformational selection mechanism that has been explained in Chapter 5. As an allosteric protein molecule, the repressor can adopt at least two

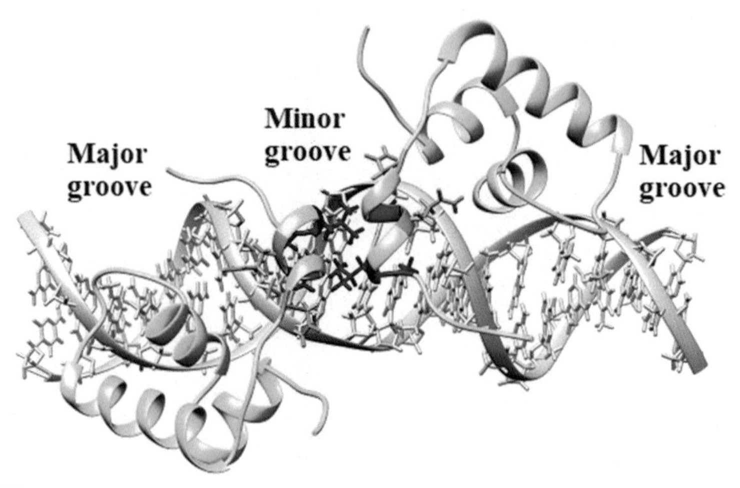

FIG. 9.37

Lac repressor-operator interactions. HTH motifs in the major grooves; apolar amino acids *(blue)* and polar amino acids *(orange)* of hinge helix in the minor groove.

Source: Spronk, C.A., et al., 1999. Structure Fold. Des. 7, 1483.

distinct conformations—one conformation (R) has a high affinity for the operator while the other (R*) has a high affinity for the inducer. Essentially, the inducer shifts the equilibrium towards R*, as a consequence of which the concentration of the active repressor is reduced.

X-ray crystallography has revealed several alterations in the inducer-bound conformation as compared with the operator-bound conformation. Here are a couple of examples. In the operator-bound conformation, K84 residues belonging to the two subunits of the dimer are buried at the center of the interface, whereas in the inducer-bound conformation they are out on the surface with an increased separation between them (Fig. 9.38). On the other hand, there is no interaction between His74 of one subunit and Asp278 of the other in the operator-bound conformation as they are >9 Å apart. In the inducer-bound conformation, they are closer together ($\sim$3.5 Å) and able to form an ion-pair (Fig. 9.39). Similarly, a network of interactions between the N-subdomain and the DBD, crucial for stabilizing the operator-bound conformation, is also altered. In all, the alignment of the two structures shows that the deviation is <0.7 Å for the C-subdomains, but >1.5 Å for the N-subdomains. Therefore, we can see that the allosteric signal from the inducer-binding pocket is transmitted through the N-subdomains to the headpieces and the repressor-operator binding affinity is reduced.

The lac repressor-operator system also possesses a built-in mechanism for additional repression on the lac operon. Besides the operator (let us now call it the principal operator O_1) which we have been discussing so far, there are two auxiliary operators, O_2 and O_3, that are also involved in the regulation of lac operon transcription in *E. coli*. O_2 and O_3 are located, respectively, 401 bp downstream and 92 bp upstream of O_1. Binding of the Lac repressor to either of the two operators

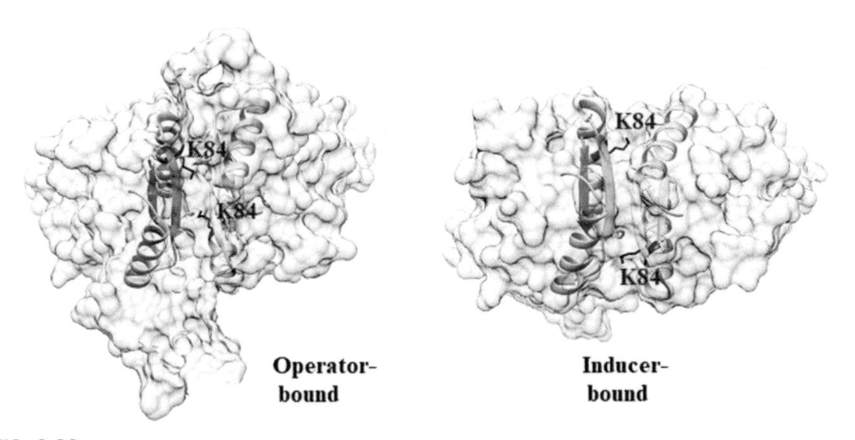

FIG. 9.38

Altered locations of K84 residues in the repressor monomer-monomer interface.

Source: Daber, R., et al., 2007. J. Mol. Biol. 370, 609; Bell, C.A., Lewis, M., 2000. Nat. Struct. Mol. Biol., 7, 209.

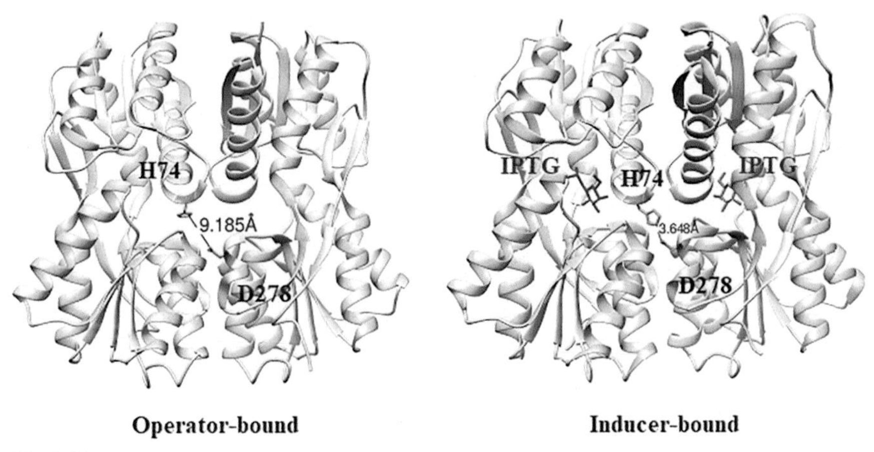

FIG. 9.39

Interaction involving H74 and D278 at the monomer-monomer interface.

Source: Daber, R., et al., 2007. J. Mol. Biol. 370, 609; Bell, C.A., Lewis, M., 2000. Nat. Struct. Mol. Biol. 7, 209.

is weak; nonetheless, they are required for the maximal repression of transcription. The bidentate repressor can simultaneously bind two operators, O_1 and either O_2 or O_3, holding the intervening DNA segment in a loop. Such looping reinforces repression by increasing the occupancy of the principal operator by repressor molecules.

Now, we have seen that the Lac repressor is a homotetramer. Each of the monomers contains a C-terminal α-helix (residues 340–357) and a four-helix bundle tethers the two dimers forming a tetramer. We have also seen that each dimer is capable of binding the two halves of the (pseudo)-symmetric operator. So, conceptually,

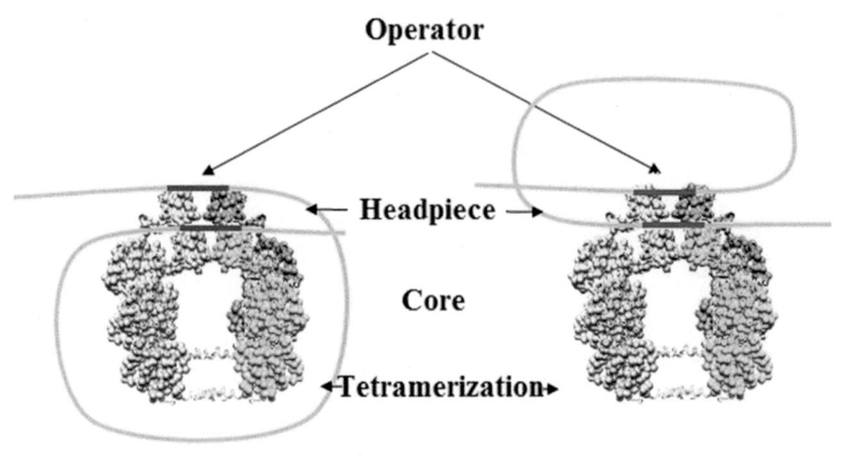

FIG. 9.40

Possible modes of binding of the two dimmers of the Lac repressor tetramer to different *lac* operators separated by a distance.

it should be possible for the two dimers of the tetramer to bind simultaneously two different operators in a manner as shown in Fig. 9.40. Crystal structure has shown that the two dimers bind two different operator DNA fragments on the same side of the tetramer forming a deep V-shaped cleft in between (Fig. 9.41). The relative

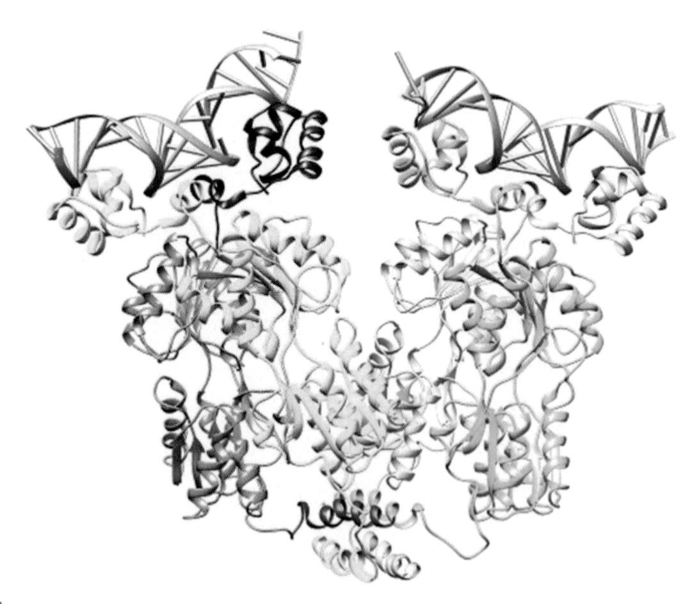

FIG. 9.41

Crystal structure of binding of a V-shaped Lac repressor tetramer to different *lac* operators.
Source: Lewis, M., et al., 1996. Science 271, 1247.

orientation of the dimers is flexible enough to accommodate varied spacing between the operators.

Positive control (activation)

The −35 region of p_{lac} is deviated from the consensus sequence and, therefore, not optimal for binding of the RNAP. Further, the promoter also lacks an UP element which the α subunits of RNAP, and with them the entire holoenzyme, can bind. Hence, if unaided by any positive regulator (that is, an activator), the efficiency of transcription from the lac promoter is relatively low even in the absence of any functional repressor. The regulatory protein that fulfills the requirement of activation is called catabolite activator protein (CAP).

For activation of the *lac* operon transcription, CAP needs to bind a specific DNA sequence at or near the lac promoter and recruit the RNAP to the promoter. Like Lac repressor, CAP also responds to an environmental signal—Lac repressor functions in the absence of lactose, whereas CAP responds to the absence of glucose. The glucose signal is transmitted to CAP by cyclic AMP (cAMP). Hence, for its activity, CAP is required to possess the following structural features: (i) separate DNA-binding and activating surfaces and (ii) a cAMP-binding pocket.

CAP is a homodimer with 209 amino acid residues in each subunit (Fig. 9.42). Two distinct domains constitute each subunit—an N-terminal cAMP-binding domain or CBD (residues 1–136) and a C-terminal DNA-binding domain or

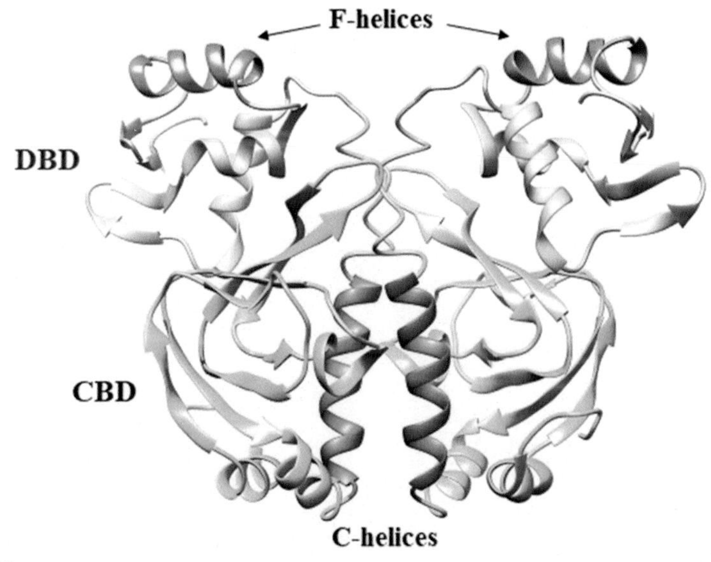

FIG. 9.42

Dimeric structure of CAP. Coloring scheme: C-helices—*dark green*, rest of CBD—*light green*; F-helices—*coral*, rest of DBD—*light blue*.

Source: Popovych, N., et al., 2009. Proc. Natl. Acad. Sci. U. S. A. 106, 6927.

DBD (residues 139–209). A short hinge (residues 137–138) connects the two domains. The CBD contains a cyclic nucleotide-binding pocket and a long α-helix (called C-helix). The C-helices of two subunits form an inter-subunit coiled-coil as the dimer interface. The DBD contains a helix-turn-helix (HTH) motif, called F-helix, for binding to DNA.

Fig. 9.42 shows the structure of CAP in its apo- (unliganded) form. In this form, which is the "off" or inactive state, CAP binds DNA weakly and nonspecifically. Its F-helices (DBDs) are not in register with successive major grooves and, therefore, cannot bind DNA specifically.

cAMP binds to CAP through hydrogen-bond interactions of its adenine base (at positions N1, N6, and N7) with the side chain residues of T127 and S128 located in the unstructured segment (Fig. 9.43). The binding of cAMP causes an allosteric transition of CAP from the inactive conformation (apo-CAP) to an active conformation (CAP-cAMP$_2$). A prominent change, coil-to-helix transition, occurs in the C-helix. In the apo-CAP, the C-helix, and the inter-subunit C-helix/C′-helix coiled-coil, extends from P110 to Q125. In CAP-cAMP$_2$, the C-helix is further extended by another three turns up to F136 (Fig. 9.44).

Besides the interaction of cAMP with T127 and S128, there are possibly two more contributors to the coil-to-helix transition—(i) interactions between the sugar phosphate moiety of cAMP and a phosphate-binding cassette (residues G71 to A84) in CAP; in the apo-CAP R82 of one chain is interacting with E129′ of the other, whereas in the cAMP-bound form of CAP R82 enters the cAMP-binding pocket and forms a salt bridge with the cAMP phosphate (Fig. 9.45); and (ii) the forced ejection (indicated by red arrow) of an aromatic residue W85 from the hydrophobic cAMP-binding pocket by cAMP (Fig. 9.46).

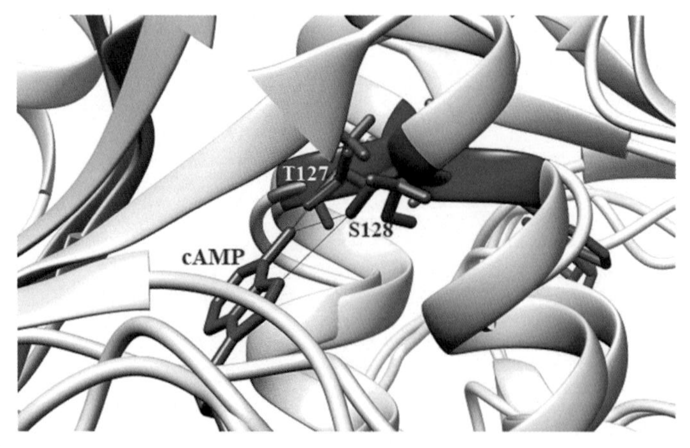

FIG. 9.43

cAMP binding to CAP through H-bonds (*red* lines).

Source: Popovych, N., et al., 2009. Proc. Natl. Acad. Sci. U. S. A. 106, 6927; Kumarevel, T.S., et al., PDB deposit 2006-05-12, unpublished.

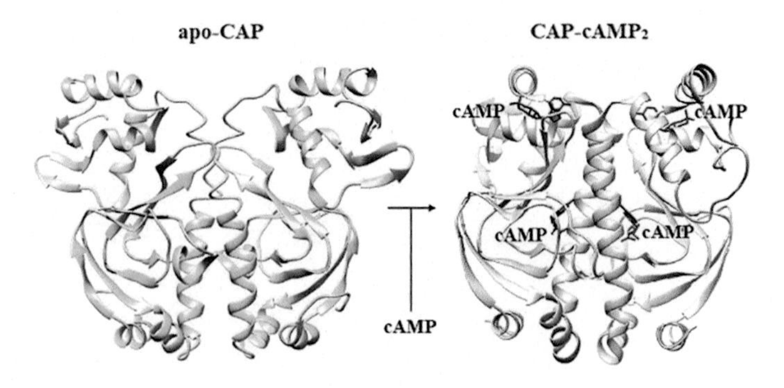

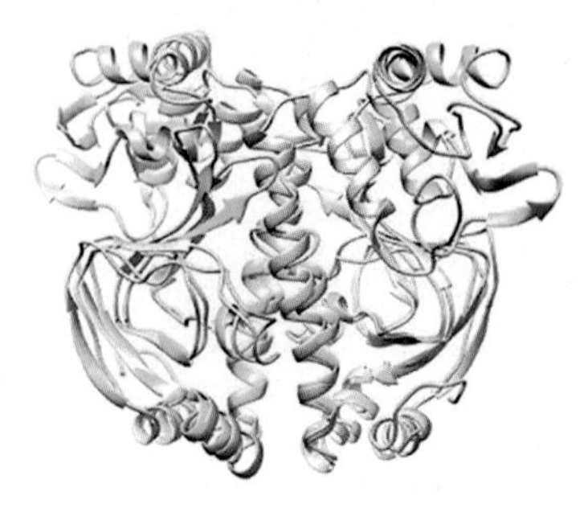

FIG. 9.44

cAMP-mediated allosteric transition of CAP.

Source: Popovych, N., et al., 2009. Proc. Natl. Acad. Sci. U. S. A. 106, 6927; Kumarevel, T.S., et al., PDB deposit 2006-05-12, unpublished.

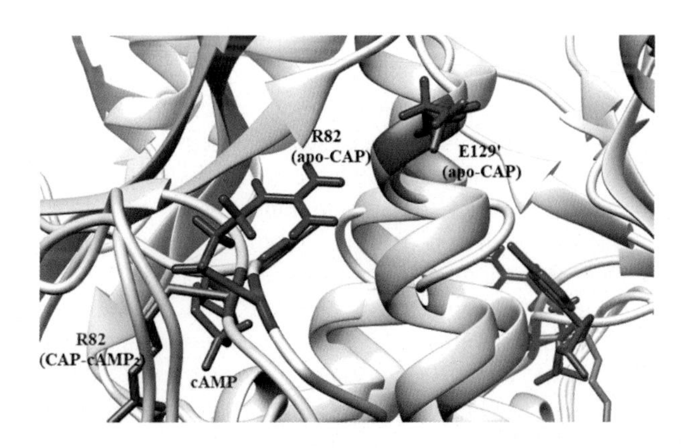

FIG. 9.45

Relocation of R82 upon cAMP-binding. Coloring: apo-CAP—*gray*; CAP-cAMP$_2$—*light blue*; R82 (apo-CAP)—*dark blue*; R82 (CAP-cAMP$_2$)—*green*.

Source: Popovych, N., et al., 2009. Proc. Natl. Acad. Sci. U. S. A. 106, 6927; Kumarevel, T.S., et al., PDB deposit 2006-05-12, unpublished.

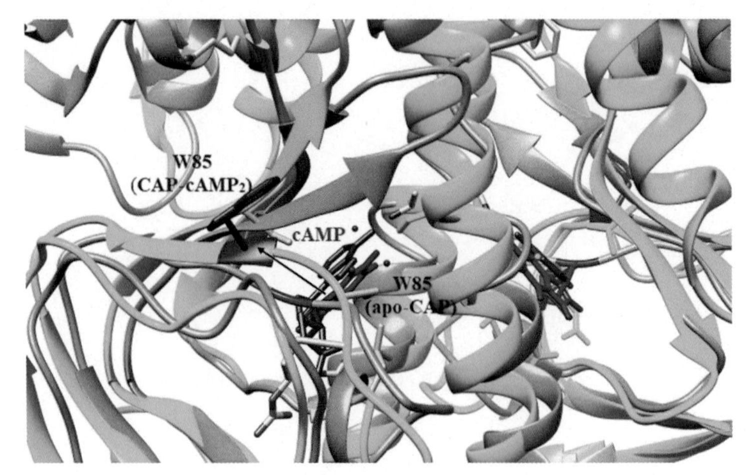

FIG. 9.46

Ejection of W85 from hydrophobic cAMP-binding pocket. Coloring: apo-CAP—*coral*; CAP-cAMP$_2$—*cyan*; W85 (apo-CAP)—*dark blue*; W85 (CAP-cAMP$_2$)—*red*; *black arrow*—ejection.

Source: Popovych, N., et al., 2009. Proc. Natl. Acad. Sci. U. S. A. 106, 6927; Kumarevel, T.S., et al., PDB deposit 2006-05-12, unpublished.

The allosteric transition of CAP, like Lac repressor, is also due to conformational selection mechanism. CAP exists in equilibrium between at least two conformational states in the absence of cAMP—inactive and active. Binding of cAMP to a preexisting small active population shifts the equilibrium towards the active form.

In the inactive or apo-form of CAP dimer, the distance between the two F-helices is 41 Å. As a result of coil-to-helix transition, the F-helices of the two subunits are brought closer to each other by ~7 Å and rotated by ~60°. This movement of the F-helices places them in the right position and orientation to be inserted into successive major grooves of the DNA (Fig. 9.47).

CAP binds as a dimer to a 22 bp consensus DNA sequence about 60 bp upstream of the transcription start site (Fig. 9.33). The sequence displays a two-fold symmetry. Now the question is how does CAP activate transcription? Once CAP has bound to the DNA, it interacts through its activating region 1 (AR1; Fig. 9.48) with the α subunit of RNAP. This interaction facilitates the binding of αCTD, and with it the entire RNAP, to promoter DNA and helps stimulate transcription initiation. αCTD possesses three distinct determinants on its surface to interact with three different molecules—CAP, DNA, and σ^{70} (Fig. 9.49).

Fig. 9.50 shows the structure of CAP-αCTD-DNA complex. Interaction between CAP and αCTD is mediated by "activating region 1" (AR1) of the promoter-proximal subunit of CAP and the "287 determinant" of αCTD involving a small interface with six residues on each side. Several atoms in the backbone as well as side chains participate in hydrogen bonding, van der Waal's interaction, and a salt bridge.

The "265 determinant" of αCTD is responsible for its binding to the DNA. αCTD residues 264, 268, 294, 296, 298, and 299 make hydrogen bonds with DNA backbone

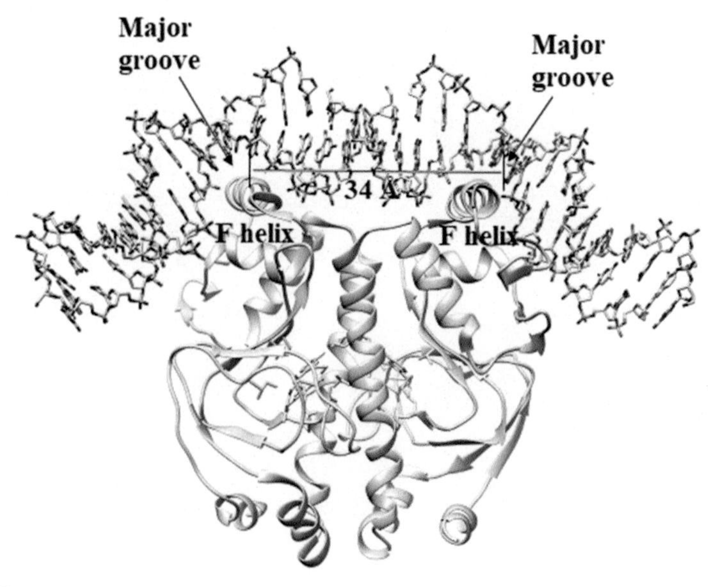

FIG. 9.47

Positioning of F-helices into successive DNA major grooves.

Source: Chen, S., et al., 2001. J. Mol. Biol. 314, 63.

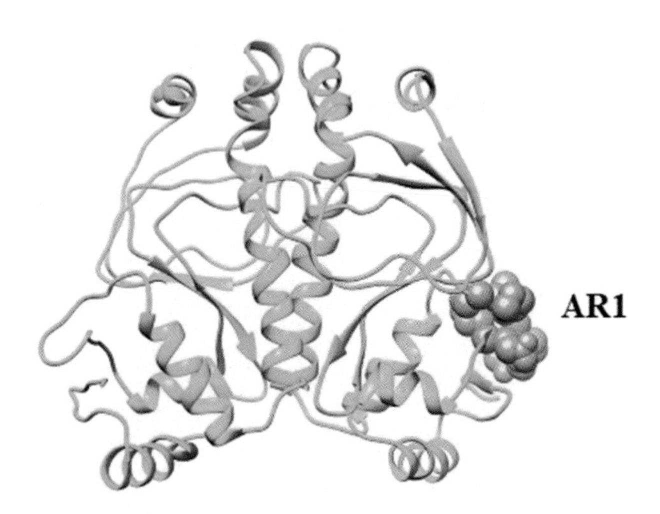

FIG. 9.48

Determinant of CAP for transcription activation.

Source: Benoff, B., et al., 2002. Science 297, 1562; Chen, S., et al., 2001. J. Mol. Biol. 314, 63.

phosphates. Arg265 is the most critical residue; it makes multiple hydrogen bonds with A:T base-pair atoms. CAP also places αCTD adjacent to σ^{70}, facilitating interaction between the "261 determinant" of αCTD and residues 573–604 in σ^{70} region 4.

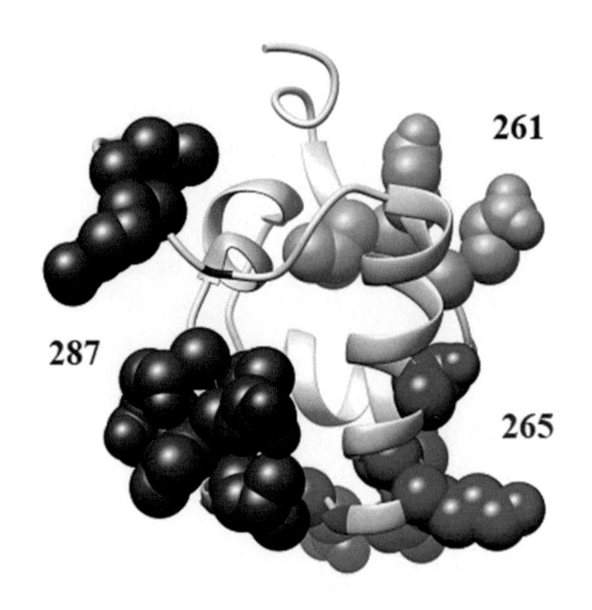

FIG. 9.49

αCTD determinants involved in CAP-dependent transcription.

Source: Benoff, B., et al., 2002. Science 297, 1562

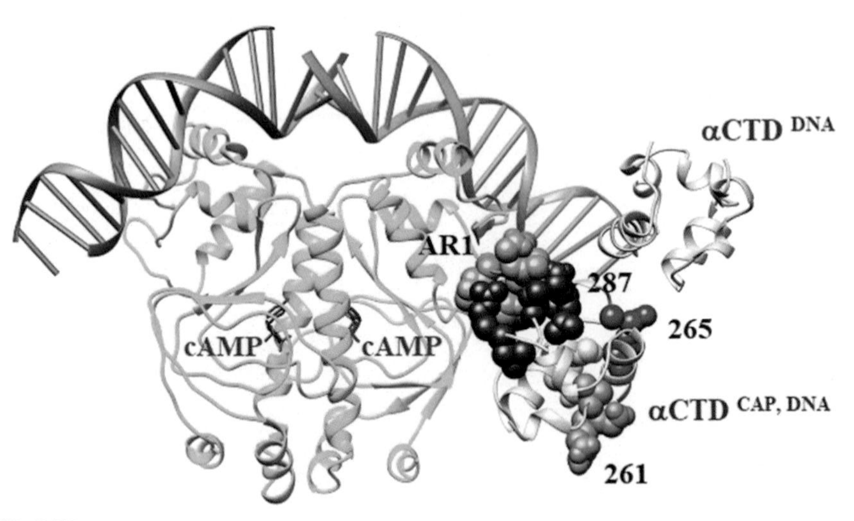

FIG. 9.50

CAP-αCTD-DNA complex.

Source: Benoff, B., et al., 2002. Science 297, 1562; Chen, S., et al., 2001. J. Mol. Biol. 314, 63.

9.4.2 Eukaryotic regulation

Like bacteria, eukaryotic gene transcription is mostly controlled at the level of initiation. Therefore, it is not at all illogical to expect that the basic principles of bacterial transcription regulation will be applicable in the eukaryotic system as well. Indeed, eukaryotic transcription is also regulated by gene regulatory proteins—activators and repressors, one augmenting transcription while the other diminishing it. However, there are two prominent differences between prokaryotic and eukaryotic cells so far as transcription initiation is concerned.

One difference is in the nature of substrate (DNA template) presented to the transcription machinery. Eukaryotic DNA is packaged, as nucleosomes, in proteins call histones. The protein-DNA complex thus formed is called chromatin. When a chromatin template, and not a free DNA, is used, the "basal" transcription even in vitro is repressed in the absence of additional necessary factors.

The second difference is in the degree of complexity in the transcription machinery itself. The difference starts right from the polymerase. We have already seen that the bacterial core RNAP contains 5 subunits, while eukaryotic RNA Pol II contains 12. Further, whereas the bacterial RNAP core needs the association of a multidomain σ factor for promoter-binding and transcription initiation, RNA Pol II is aided by an entire battery of GTFs—some of which are constituted of more than one subunit.

Moving a step backward along the pathway of gene regulation, we find that in bacteria the regulatory molecule directly interacts with the polymerase. In eukaryote, most often an intermediate lies in between. This intermediate is not a single protein molecule, but a multiprotein complex called Mediator.

The difference is also apparent in the number of regulator-binding sites. Although individual gene regulators, whether bacterial or eukaryotic, bind short DNA sequences, the binding sites in eukaryotes are, in majority of the cases, greater in number and located farther away from the transcription start site.

Another *cis*-acting DNA element (less frequently found in prokaryotes but widespread in eukaryotes) that increases the transcription of genes is the enhancer. Enhancers function independently of orientation and at variable distances from their target promoter(s). Each of the enhancers contains one or more binding sites for activator protein(s).

Mediator

Using the mass spectrometric technique, distinct forms of Mediator have been isolated as stable complexes. The consensus Mediator "core" in *S. cerevisiae* contains 21 subunits (26 subunits in human). The core is organized in three modules—Head, Middle, and Tail, while a Cdk8 kinase module (CKM) reversibly associates with the core (Fig. 9.51).

The Head module comprises seven subunits—Med6, Med8, Med11, Med17, Med18, Med20, and Med22—and can be divided into three structural domains—neck, fixed jaw, and variable jaw (Fig. 9.52). In the PIC, Head is responsible for interactions with transcription factors TFIID and TFIIH as well as Pol II.

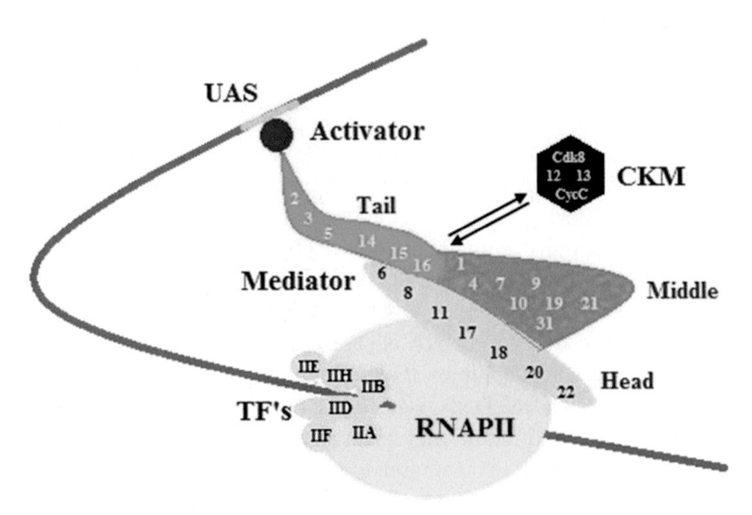

FIG. 9.51

S. cerevisiae Mediator. The numbers in the Mediator modules represent the subunits (Med's). UAS, upstream activating sequence.

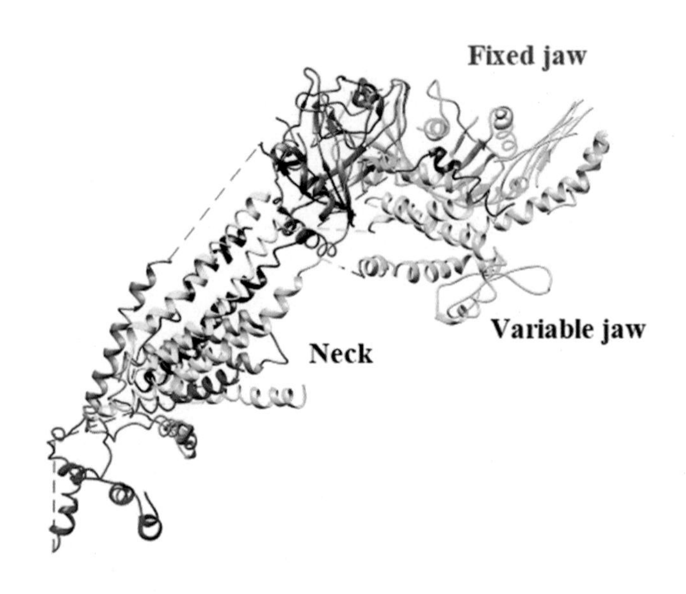

FIG. 9.52

Mediator Head module. Subunit color: Med6—*purple*; Med8—*dodger blue*; Med11—*yellow*; Med17—*light green*; Med18—*orange*; Med20—*cyan*; Med22—*brown*.

Source: Imasai, T., et al., 2011. Nature 475, 240.

Fig. 9.53 shows a Mediator-RNA Pol II PIC. Mediator subunits Med6, Med8, Med17, Med18, and Med20 are found to be interacting with Rpb1. Pol II subunits Rpb4 and Rpb7 also interact with Med8 and Med17 and so does Rpb3 with Med20. Nevertheless, the most crucial Mediator-Pol II interaction involves the

FIG. 9.53

Mediator-RNA PolII PIC. Pol II subunits Rpb3, Rpb4, and Rpb7 are colored *cyan*, *forest green*, and *brown*, respectively; the rest of PolII is colored *orange*. Coloring scheme for Mediator subunits: Med6—*spring green*; Med8—*orange red*; Med17—*deep blue*; Med 18—*pink*; Med20—*yellow*; all other Head subunits are colored *light green* and Middle subunits are *light blue*. TFs are colored *purple*.

Source: Robinson, P.J., et al., 2016. Cell 166, 1411.

C-terminal domain (CTD) of Pol II Rpb1. Mediator subunits Med6, Med8, and Med17 interact with the CTD. Serine and tyrosine residues present in a β-turn heptapeptide motif (SPTSPSY) of the CTD enter into H-bonding and hydrophobic interactions with several conserved residues of the Head module (Fig. 9.54).

In one sense, Mediator can be considered as a general transcription factor. However, the structural flexibility and variable subunit composition are its distinct features. Its subunits can adopt different orientations with respect to each other. As for example, the Med7/Med21 heterodimer, which is located at the interface of the Head and Middle modules, serves as a flexible hinge to allow reorganization of Mediator's structure upon binding to Pol II or transcriptional activators (Fig. 9.55).

Besides, the presence or absence of Med26 in the Mediator complex is a demonstration of its compositional variability. The Med26 subunit has a role in the Pol II initiation-elongation transition. Of the isolated Mediator complexes, the ones containing the four subunit CKM module are devoid of the Med26 subunit. Med26 dissociates from Mediator with the binding of CKM.

Mediator is a general requirement for regulated transcription. Therefore, it interacts with a variety of gene-specific transcription factors (TFs). This flexibility is

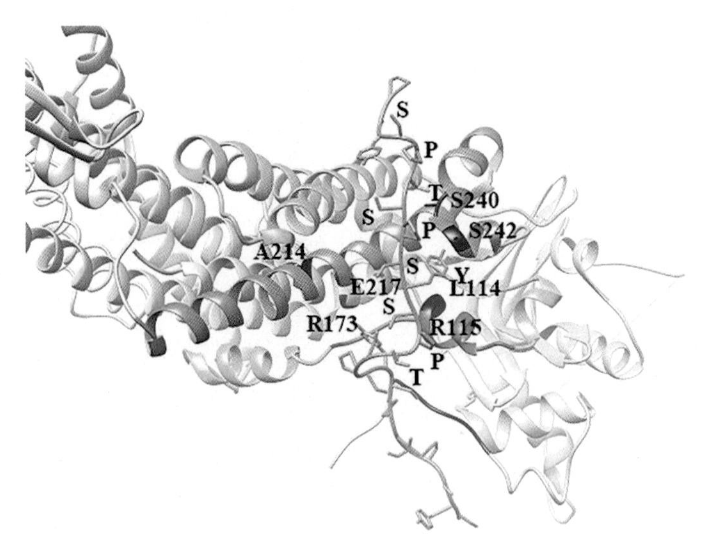

FIG. 9.54

Interaction of PolII Rpb1 CTD with Mediator Head. Rpb1 CTD shown in *orange*, while Med6, Med8, and Med17 colored *light blue*, *light green*, and *light pink*, respectively. The interacting segments of each of the three mediator subunits colored in respective darker shades; interacting residues shown in *navy blue*.

Source: Robinson, P.J., et al., 2012. Proc. Natl. Acad. Sci. U. S. A. 109, 17931.

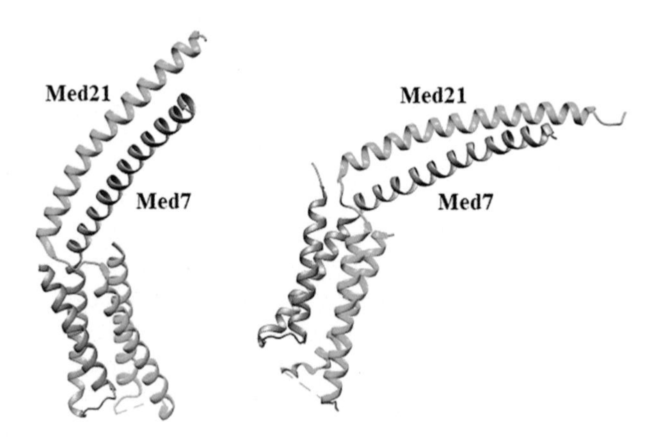

FIG. 9.55

Different conformational states of Med7/Med21 subcomplex.

Source: Baumli, S., et al., 2005. J. Biol. Chem. 280, 18171.

provided by the presence of intrinsically disordered regions (IDRs) in Mediator. Remarkably, most of these IDRs are present in the Tail subunit, which is involved in the interaction with the TFs. In fact, IDRs are utilized by Mediator for interactions even between subunits within the complex (Fig. 9.56).

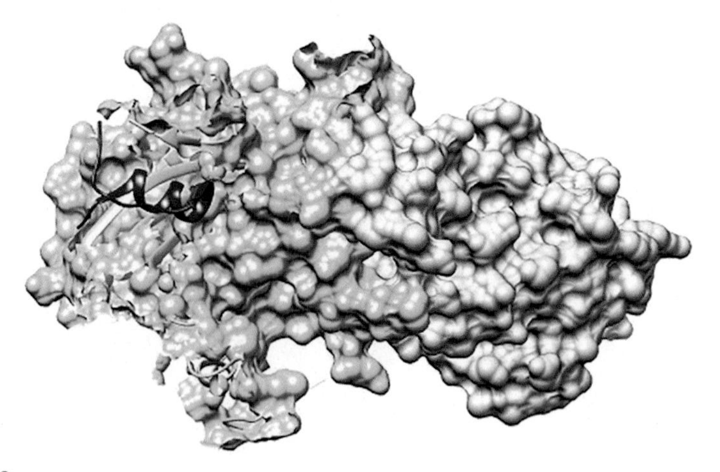

FIG. 9.56

IDR in Mediator formation. Med8 IDR *(red)* forms α-helix when bound to Med18 *(light green)* and Med20 *(light blue)*.

Source: Lariviere, L., et al., 2006. Nat. Struct. Mol. Biol. 13, 895.

Gene regulatory proteins

The gene regulatory proteins can be either activators or repressors; however, in this text we are going to focus on activators only. The activators may not act on the transcription machinery directly. The principal targets of the activators are the coactivators.

Eukaryotic activators can modulate transcription by either recruiting the transcription machinery to the DNA or chromatin remodeling. Activation by recruitment is the most pervasive mechanism and, therefore, most widely studied. In fact, chromatin remodeling can also be considered as a special case of the recruitment mechanism since here again chromatin-modifying coactivators are recruited by the activators.

Structurally, like the bacterial activators such as CAP, the eukaryotic activators are also bipartite with separate DNA-binding domain (DBD) and activating domain (AD). The DBDs of different activators contain some well-defined DNA-binding motifs which have been described in Chapter 5. In contrast, the ADs of eukaryotic activators are mostly unstructured in the absence of a binding partner. Two examples of eukaryotic activators will be given here to illustrate that, despite the differences in the degree of complexity between prokaryotic and eukaryotic transcription regulation, the basic molecular structural principles underlying the binary interactions of the activator with the DNA and the coactivator remain the same.

Gcn4

Yeast Gcn4 is a transcription activator responsible for the regulation of >70 genes involved in a variety of processes related to metabolic stress response and autophagy. It is a homodimeric protein, each subunit containing an N-terminal activating domain

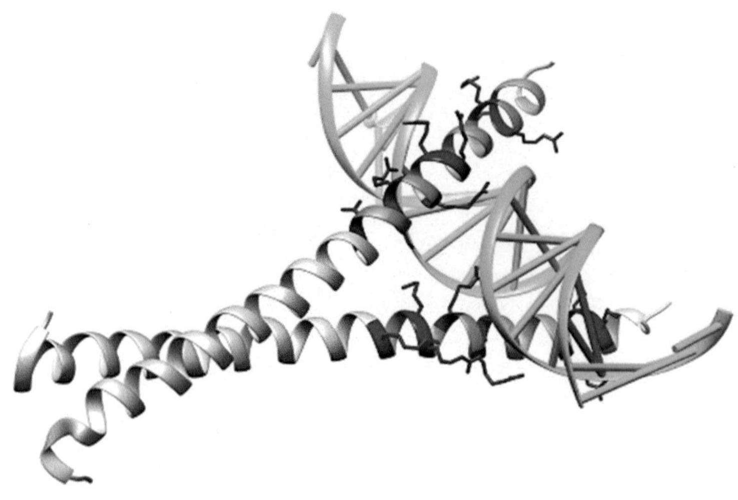

FIG. 9.57

Gcn4 DNA-binding and dimerization. Basic regions are colored *dodger blue* with basic residues shown in *deep blue*.

Source: Ellenberger, T.E., et al., 1992. Cell 71, 1223.

(N-terminal AD; residues 1–100), a central activating domain (cAD; residues 101–134), and a DNA-binding domain (DBD; residues 209–281).

Gcn4 is a member of the basic-region-leucine-zipper (bZIP) family of DNA-binding proteins (described in Chapter 5). Each of the two subunits of Gcn4 contains in the C-terminal DBD a basic region, which interacts with the DNA, and a leucine-zipper motif, which forms a coiled-coil and mediates dimerization (Fig. 9.57). The protein recognizes a near-palindromic sequence 5′-ATGACTCAT-3′. The bZIP dimer is a pair of continuous α-helices. The dimerization interface is almost perpendicular to the axis of the DNA. Each protein chain passes through a major groove and residues in the basic region interact with DNA bases and phosphate oxygen.

One of the targets of Gcn4 in the preinitiation complex (PIC) is the Gal11/Med15 mediator. Gal11 has three conserved Gcn4-binding domains (ABD1-3). Interaction of cAD with ABD1 (residues 158–238) was studied by NMR spectroscopy. ABD1 consists of four α-helices of varying lengths, α1–4. α1, α3, and α4 create a cleft with a floor essentially formed by hydrophobic residues V170, Y220, A216, M213, L208, A203, T200, and W196 (Fig. 9.58A). The hydrophobic cleft is surrounded by a positively charged surface as can be seen from Fig. 9.58B, which shows the surface electrostatic potential of ABD1.

Gcn4 AD assumes an α-helical conformation when bound to Gal11. The helix is amphipathic in nature. Gcn4-Gal11-binding is driven primarily by multiple low-affinity hydrophobic interactions, rather than high-affinity high-specificity ones. For this reason, the two proteins are able to bind each other in multiple orientations. Three different orientations of Gcn4 with respect to Gal11 are shown in Fig. 9.59.

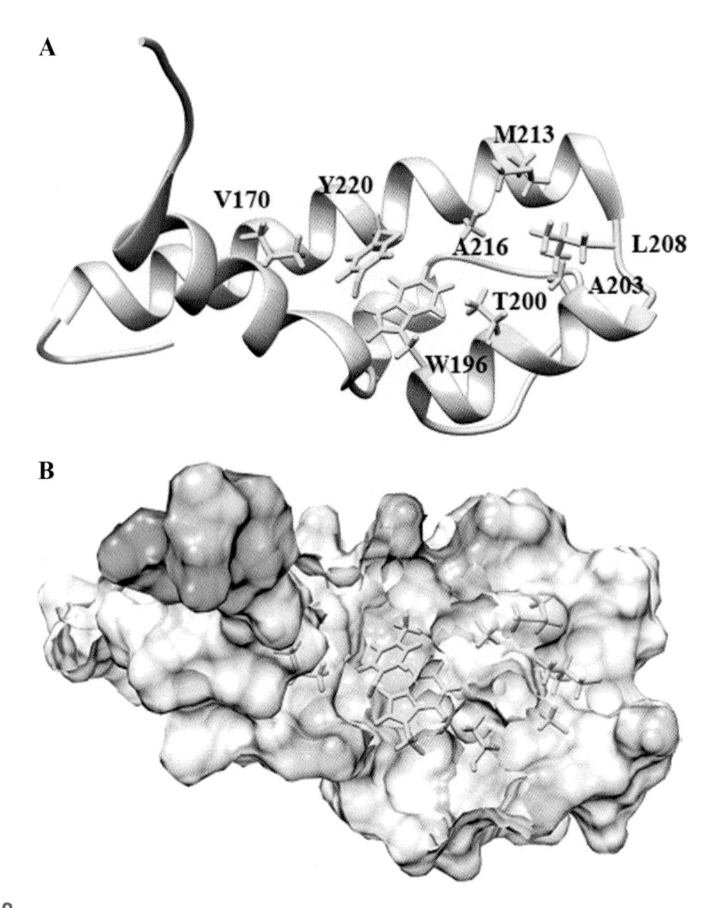

FIG. 9.58

Gal11 ABD1. (A) Amino acid residues of hydrophobic cleft. (B) Surface electrostatic potential: *blue*—most positive; *red*—most negative; *white*—neutral.

Source: Brzovic, P.S., 2011. Mol. Cell 44, 942.

In each of the orientations, the hydrophobic face of Gcn4 formed by residues W120, T121, L123, and F124 binds to the hydrophobic cleft of ABD1(Fig. 9.60; side chains are not shown for the sake of clarity). Although one surface of the Gcn4 amphipathic helix and the periphery have opposite electrostatic potentials, structural and mutational analyses have shown that there are no specific polar or ionic interactions between cAD and ABD1. This may also be a reason for cAD to bind ABD1 in multiple orientations showing the characteristics of a "fuzzy complex."

p53

Next, we consider the role of tumor suppressor protein p53 as a potent transcriptional activator. p53 is known to induce the expression of several target genes whose products are responsible for cell-growth arrest and apoptosis.

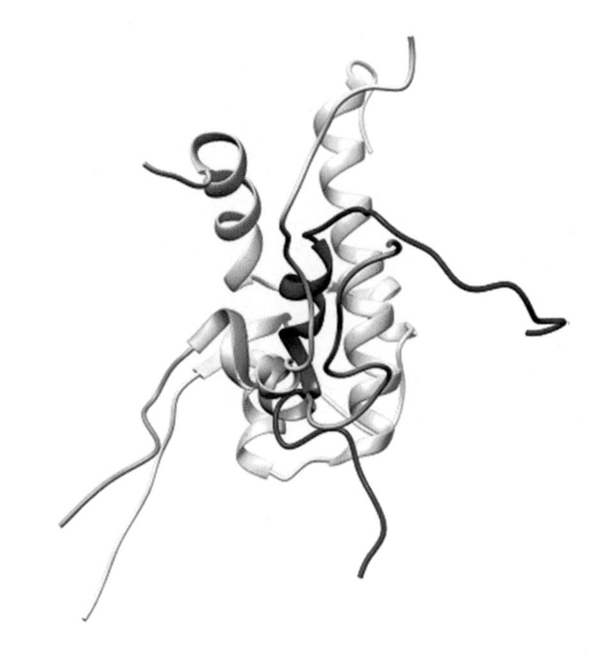

FIG. 9.59

Three different orientations of Gcn4 AD (different shades of *blue*) bound to Gal11 ABD1 *(pink)*.

Source: Brzovic, P.S., 2011. Mol. Cell 44, 942.

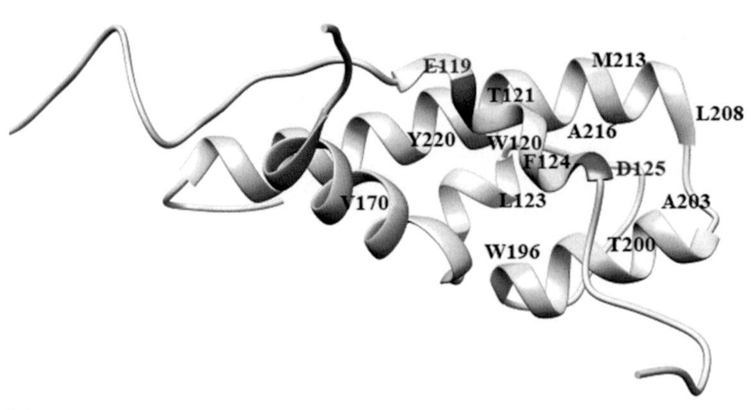

FIG. 9.60

Binding of Gcn4 AD to Gal11 ABD1 through hydrophobic interactions. All hydrophobic residues colored *yellow*: Gal11 residues labeled *brown*; Gcn4 residues labeled *deep blue*. Gcn4 charged residues colored and labeled *red*.

Source: Brzovic, P.S., 2011. Mol. Cell 44, 942.

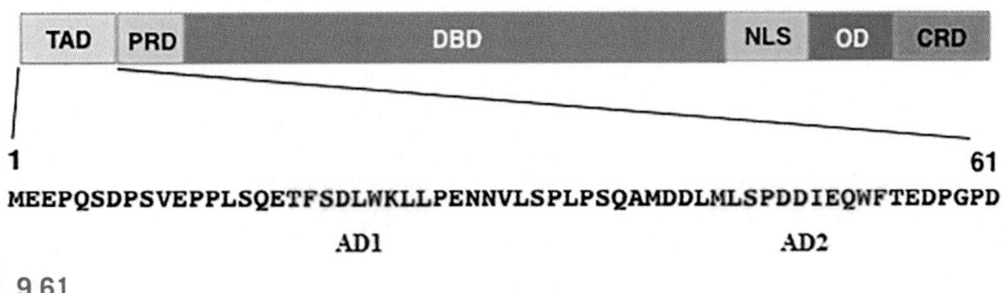

FIG. 9.61

Domain structure of p53.

The functional unit of p53 is a tetramer. Each polypeptide chain of the tetramer consists of multiple domains: an N-terminal transactivation domain (TAD), a proline-rich domain (PRD), a DNA-binding domain (DBD), a nuclear localization signal (NLS), an oligomerization domain (OD), and a C-terminal regulatory domain (CRD) (Fig. 9.61). It will be worth looking at the respective structures and interactions of the DBD and TAD to understand the role of p53 as an activator.

The DNA-binding sites of p53 consist of two decameric motifs, or half-sites, with consensus sequence 5′-RRRCWWGYYY-3′ (where R = purine (A or G), Y = pyrimidine (C or T), and W = A or T). The separation between the half-sites ranges anywhere between 0 and 20 base-pairs. p53 binds to the two half-sites as a tetramer or "a dimer of dimers" (Fig. 9.62). Side chains of amino acid residues R280, K120, A276, and C277 of each monomer make direct contacts with DNA bases in the major groove (Fig. 9.63). R280 plays an anchoring role by forming H-bonds with the conserved guanine base, G8.

The TAD of p53 is intrinsically disordered in the absence of a binding partner. It can be divided into two activation subdomains containing short amphipathic interaction motifs AD1 (residues 18–26) and AD2 (residues 44–54), respectively (Fig. 9.61). Each of the motifs folds into stable amphipathic helix when bound to its target.

One of the targets of p53 is the transcriptional coactivator cyclic AMP response element-binding protein (CBP). On the other hand, CBP is known to interact with several other transcription factors. The interaction between p53 and CBP activates transcription of p53-regulated stress response genes.

CBP contains two folded transcriptional adapter zinc finger (TAZ) domains—TAZ1 and TAZ2. These two domains of CBP, and two other, interact with p53TAD. Binding of p53TAD to TAZ1 and TAZ2 has been studied by NMR spectroscopy. When bound, AD1 and AD2 regions fold into amphipathic helices.

Each of the two TAZ domains consists of a bundle of four α-helices (α1–4). In TAZ2, a hydrophobic patch is formed by the residues I1773, M1799, V1802, V1819, L1823, A1825, L1826, C1828, and Y1829 at the interface between α1, α2, and α3 (Fig. 9.64A). The AD2 helix, with its hydrophobic residues I50, W53, and F54, and two hydrophobic residues M44 and L45 preceding the helix bind to the patch (Fig. 9.64B).

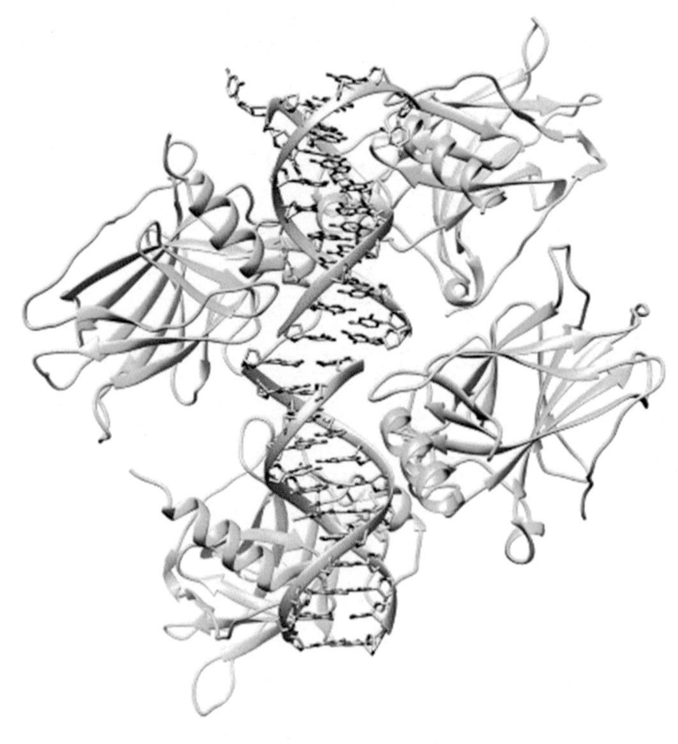

FIG. 9.62

p53 tetramer/DNA complex.

Source: Kitayner, M., et al., 2006. Mol. Cell 22, 741.

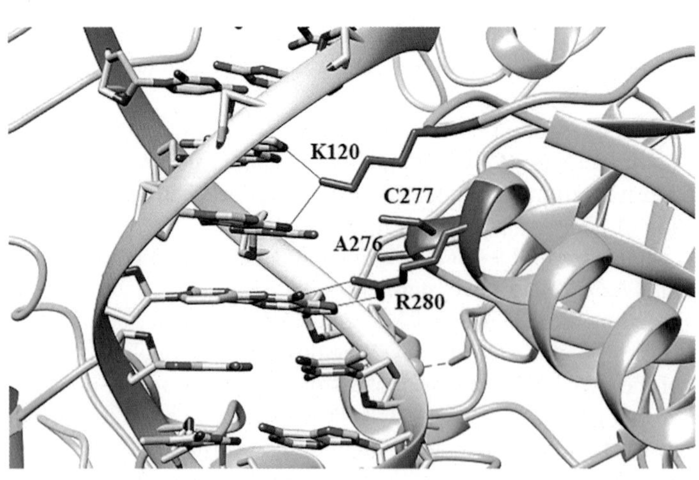

FIG. 9.63

Contacts of p53 amino acid residues with DNA bases in major groove. *Red* lines represent H-bonds.

Source: Kitayner, M., et al., 2006. Mol. Cell 22, 741.

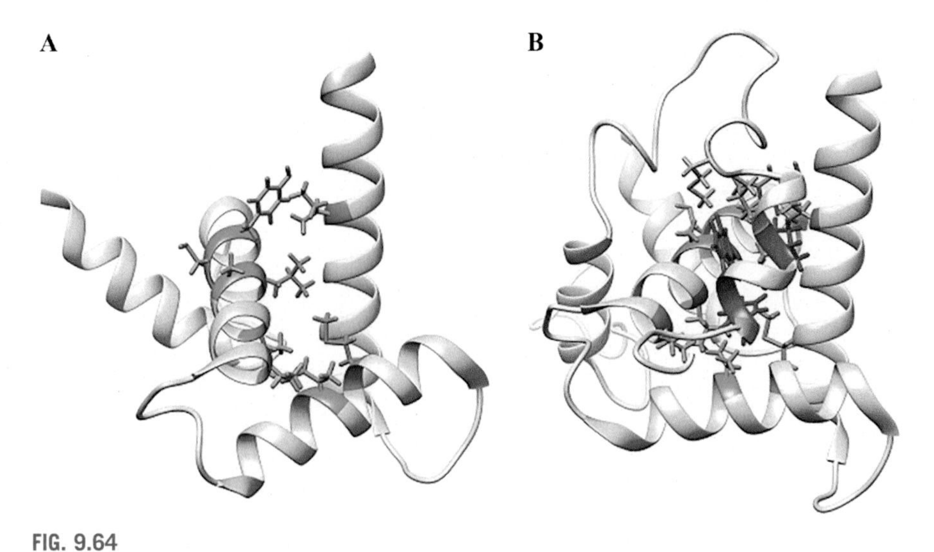

FIG. 9.64

Hydrophobic interactions of p53 AD2 *(light pink)* with CBP TAZ2 *(light blue)*. (A) TAZ2 hydrophobic residues colored *dodger blue*. (B) Hydrophobic residues of AD2 helix colored *purple* and preceding the helix colored *orange*.

Source: Krois, A.S., 2016. Proc. Natl. Acad. Sci. U. S. A. 113, E1852.

Similarly, a deep hydrophobic groove formed by the residues T1812, L1823, I1824, A1825, F1843, I1847, and L1851 between α3 and α4 helices and the α2–α 3 loop is the secondary binding site for p53TAD. Ad1 projects its hydrophobic side chains of F19, L22, W23, L25, and L26 into this groove. There is an additional pi-cation interaction between TAZ2 K1850 side chain and the indole ring of AD1 W23. AD1 and AD2 also bind to TAZ1 and in this case, too, the binding is mediated by hydrophobic interactions.

Having understood the structural basis for binding of p53TAD to CBP TAZ domains, it is not difficult to appreciate that CBP-mediated transactivation of the stress response genes by p53 would depend on the strength of interaction between the two proteins. In response to stress, several sites on p53 are phosphorylated. Nine of these sites (serines and tyrosines) are located within the TAD. In the AD-TAZ complex, some of these phosphorylated residues can electrostatically interact with the basic residues of TAZ due to their positional proximity. As a result, the binding affinity is increased several-fold facilitating transcription activation.

Thus, we have seen that, compared to the regulation in bacteria, eukaryotic transcription initiation is controlled by a more complex network of factors and interactions. Nevertheless, each of the interactions in a regulatory network, bacterial or eukaryotic and simple or complex, can be explained on the basis of the structures of the interacting molecules.

Sample questions

1. What are the specific interactions of σ^{70} with the DNA sequence at the beginning of a gene? How does σ^{70} prevent nonspecific binding of RNA polymerase to DNA?
2. What is the structural basis for abortive initiation?
3. How are the stability and processivity of the bacterial elongation complex ensured?
4. Describe the dynamics of the bridge helix (BH) and trigger loop (TL) of the RNA polymerase in the nucleotide addition cycle (NAC)?
5. What are the interactions of the Rho-factor with RNA that destabilize the transcription elongation complex (EC) leading to termination of transcription?
6. Which factor tends to preclude TATA-binding protein from binding to DNA? How is this inhibition overcome and the TFIID-DNA complex stabilized?
7. The formation of RNA polymerase II-dependent preinitiation complex (PIC) is complete when TFIIE recruits TFIIH. How do the acidic amino acid-rich domain (AC-D) in TFIIE and the pleckstrin homology domain (PH-D) in TFIIH help this recruitment?
8. How does the homotetrameric nature of the Lac repressor ensure the maximal repression of *lac* operon transcription?
9. What are the structural changes in the catabolite activator protein (CAP) brought about by cAMP binding that enable the protein to bind to a specific site in the DNA?
10. How does the "intrinsically disordered" nature of p53 transactivation domain (TAD) help its interactions with the targets?

References and further reading

Anandapadmanaban, M., et al., 2013. High-resolution structure of TBP with TAF1 reveals anchoring patterns in transcription regulation. Nat. Struct. Mol. Biol. 20 (8), 1008–1014.

Armache, K.-J., Kettenberger, H., Cramer, P., 2003. Architecture of initiation-competent 12-subunit RNA polymerase II. Proc. Natl. Acad. Sci. U. S. A. 100 (12), 6964–6968.

Bae, B., et al., 2015. Structure of a bacterial RNA polymerase holoenzyme open promoter complex. elife. 4e08504. https://doi.org/10.7554/eLife.08504.

Basu, R., et al., 2014. Structural basis of transcription initiation by bacterial RNA polymerase holoenzyme. J. Biol. Chem. 289 (35), 24549–24559.

Bell, C.E., Lewis, M., 2001. The lac repressor: a second generation of structural and functional studies. Curr. Opt. Struct. Biol. 11, 19–25.

Benoff, B., et al., 2002. Structural basis of transcription activation: the CAP-αCTD-DNA complex. Science 297, 1562–1566.

Brzovic, P.S., et al., 2011. The acidic transcription activator Gcn4 binds the Mediator subunit Gal11/Med15 using a simple protein interface forming a fuzzy complex. Mol. Cell 44, 942–953.

Busby, S., Ebright, R.H., 1999. Transcription activation by catabolite activator protein (CAP). J. Mol. Biol. 293, 199–213.

Cramer, P., Bushnell, D.A., Kornberg, R.D., 2001. Structural basis of transcription: RNA polymerase II at 2.8 Ångstorm resolution. Science 292, 1863–1877.

Daber, R., et al., 2007. Structural analysis of lac repressor bound to allosteric effectors. J. Mol. Biol. 370 (4), 609–619.

Feklistov, A., Darst, S.A., 2011. Structural basis for promoter −10 element recognition by the bacterial RNA polymerase σ subunit. Cell 147, 1257–1269.

Feklistov, A., et al., 2017. RNA polymerase motions during promoter melting. Science 356, 863–866.

Fishburn, J., et al., 2015. Double-stranded DNA translocase activity of transcription factor TFIIH and the mechanism of RNA polymerase II open complex formation. Proc. Natl. Acad. Sci. U. S. A. 112 (13), 3961–3966.

Friedman, A.M., Fischmann, T.O., Steitz, T.A., 1995. Crystal structure of lac repressor core tetramer and its implications for DNA looping. Science 268, 1721–1727.

Fuxreiter, M., et al., 2014. Disordered proteinaceous machines. Chem. Rev. 114, 6806–6843.

Gnatt, A.L., et al., 2001. Structural basis of transcription: an RNA polymerase II elongation complex at 3.3 Å resolution. Science 292, 1876–1882.

Grünberg, S., Hahn, S., 2013. Structural insights into transcription initiation by RNA polymerase II. Trends Biochem. Sci. 38 (12), 603–611.

Hahn, S., 2004. Structure and mechanism of the RNA polymerase II transcription machinery. Nat. Struct. Mol. Biol. 11 (5), 394–403.

Jeronimo, C., Robert, F., 2017. The Mediator complex: at the nexus of RNA polymerase II transcription. Trends Cell Biol. 27 (10), 765–783.

Kitayner, M., et al., 2006. Structural basis of DNA recognition by p53 tetramers. Mol. Cell 22, 741–753.

Krois, A.S., et al., 2016. Recognition of the disordered p53 transactivation domain by the transcriptional adapter zinc finger domains of CREB-binding protein. Proc. Natl. Acad. Sci. U. S. A. 113 (13), E1853–E1862.

Lawson, C.L., et al., 2004. Catabolite activator protein: DNA binding and transcription activation. Curr. Opt. Struct. Biol. 14, 1–11.

Lee, J., Borukhov, S., 2016. Bacterial RNA polymerase-DNA interaction – the driving force of gene expression and the target for drug action. Front. Mol. Biosci. 373. https://doi.org/10.3389/fmolb.2016.00073.

Lello, P.D., et al., 2006. Structure of the Tfb1/p53 complex: insights into the interaction between the p62/Tfb1 subunit of TFIIH and activation domain of p53. Mol. Cell 22, 731–740.

Lewis, M., 2013. Allostery and the lac operon. J. Mol. Biol. 425, 2309–2316.

Mejia, Y.X., Nudler, E., Bustamante, C., 2015. Trigger loop folding determines transcription rate of *Escherichia coli's* RNA polymerase. Proc. Natl. Acad. Sci. U. S. A. 112 (3), 743–748.

Mooney, R.A., Darst, S.A., Landick, R., 2005. Sigma and RNA polymerase: an on-again, off-again relationship? Mol. Cell 20, 335–345.

Murakami, K.S., 2013. X-ray crystal structure of *Escherichia coli* RNA polymerase σ^{70} holoenzyme. J. Biol. Chem. 288 (13), 9123–9134.

Murakami, K.S., 2015. Structural biology of bacterial RNA polymerase. Biomol. Ther. 5, 848–864.

Okuda, M., et al., 2008. Structural insight into the TFIIE-TFIIH interaction: TFIIE and p53 share the binding region on TFIIH. EMBO J. 27, 1161–1171.

Popovych, N., et al., 2009. Structural basis for cAMP-mediated allosteric control of the catabolite activator protein. Proc. Natl. Acad. Sci. U. S. A. 106 (17), 6927–6932.

Ray-Soni, A., Bellecourt, M.J., Landick, R., 2016. Mechanisms of bacterial transcription termination: all good things must end. Annu. Rev. Biochem. 85, 319–347.

Sainsbury, S., Niesser, J., Cramer, P., 2013. Structure and function of the initially transcribing RNA polymerase II-TFIIB complex. Nature 493, 437–441.

Sydow, J.F., Cramer, P., 2009. RNA polymerase fidelity and transcriptional proofreading. Curr. Opt. Struct. Biol. 19, 732–739.

Tóth-Petróczy, Á., et al., 2008. malleable machines in transcription regulation: the Mediator complex. PLoS Comput. Biol. 4 (12), e1000243. https://doi.org/10.1371/journal.pcbi.1000243.

Vassylyev, D.G., et al., 2002. Crystal structure of a bacterial RNA polymerase holoenzyme at 2.6 Å resolution. Nature 417, 712–719.

Vassylyev, D.G., et al., 2007. Structural basis for transcription elongation by bacterial polymerase. Nature 448, 157–164.

Zuo, Y., Steitz, T.A., 2017. A structure-based kinetic model of transcription. Transcription 8 (1), 1–8.

CHAPTER

RNA processing

10

The previous chapter has described how a gene is transcribed to produce RNA and the manner in which the process of transcription is regulated. We have also seen that, in eukaryotic cells, the protein-coding genes are transcribed by RNA polymerase II (RNAP II). However, before RNAP II-generated transcripts could be translated into protein products (a phenomenon to be described in the next chapter), these transcripts need to be suitably "formatted" or processed to acquire a form known as messenger RNA (mRNA). The primary transcription product of the protein-coding genes is called the precursor messenger RNA (pre-mRNA).

Three major events constitute pre-mRNA processing: (a) 5′-end capping, (b) splicing, and (c) 3′-end polyadenylation. In 5′-capping, the 5′-triphosphate of the nascent transcript is hydrolyzed to a diphosphate and a guanosine monophosphate is added in a reverse 5′–5′ orientation. Subsequently, the GpppN- cap is methylated to form m^7GpppN-. In splicing, the noncoding sequences (introns) that separate the coding sequences (exons) are removed and the exons are joined together. 3′-polyadenylation involves cleavage of the transcript at a specific site and addition of a poly(A) tail.

Of the three processes, 5′-capping is mostly cotranscriptional (though a few instances of posttranscriptional capping have been reported), splicing may or may not be cotranscriptional, while the 3′-end processing begins cotranscriptionally but continues posttranscriptionally. The coupling between any of the three processes and transcription is maintained by the C-terminal domain (CTD) of the largest RNAP II subunit, Rpb1.

10.1 RNA Pol II CTD

RNAP II CTD consists of tandem heptapeptide repeats with a consensus sequence $Y_1S_2P_3T_4S_5P_6S_7$. There are 26 such repeats in yeast RNAP II, while 52 in the human counterpart. The CTD forms an unstructured extension connected to the polymerase core by an 80-residue linker (Fig. 10.1). The linker binds RNAP II subunit, Rpb7, which forms a subcomplex with Rpb4.

All five hydroxylated residues of the heptapeptide, Y_1, S_2, T_4, S_5, and S_7, are potential phosphorylation sites. The dynamic plasticity and diversity of the binding surface resulting from differential phosphorylation at multiple sites enable the CTD to interact with a wide range of nuclear factors, many of which are involved in

Fundamentals of Molecular Structural Biology. https://doi.org/10.1016/B978-0-12-814855-6.00010-9

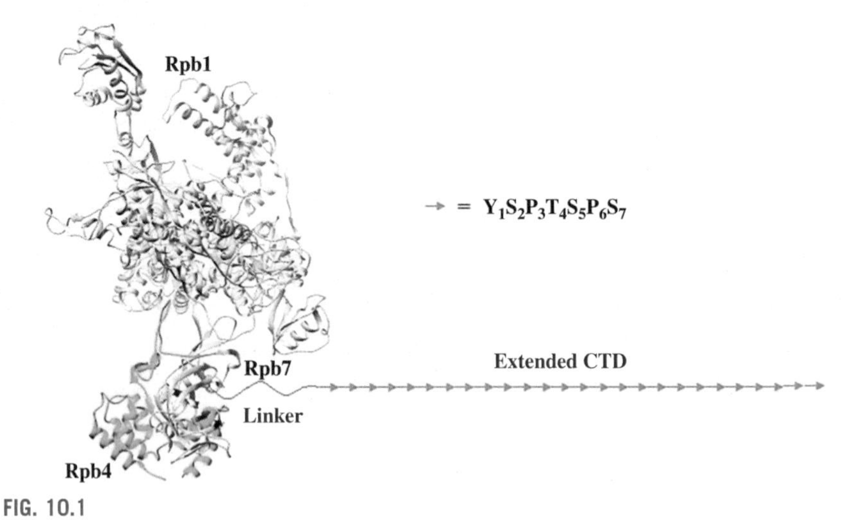

FIG. 10.1

Saccharomyces cerevisiae RNAP II CTD.

Source: Armache, K.-J., et al., 2005. J. Biol. Chem. 280, 7131.

pre-mRNA processing. The phosphorylation state of the CTD changes during the transcription cycle.

A number of kinases and phosphatases are involved in the dynamic modifications of the CTD by phosphorylation and dephosphorylation, respectively, thereby playing important roles in transcriptional control and RNA processing. An example is cyclin-dependent kinase 7 (Cdk7), which phosphorylates S_5 and S_7. On the other hand, Fcp1 is a CTD phosphatase preferentially hydrolyzing S_2-PO_4.

10.2 5′-End capping

5′-end capping is an outcome of three sequential enzymatic activities—RNA triphosphatase, RNA guanylyltransferase, and RNA guanine-N7 methyltansferase. This cap, denoted as m^7GpppXpY- (where X and Y are the first two nucleosides of the nascent transcript), is also known as Cap0 (Fig. 10.2). In certain cases, further methylation on the 2′-hydroxyl (2′-O) of the first (X) and second (Y) nucleosides by RNA 2′-O-ribose methyltransferase leads to the formation of Cap1 and Cap2, respectively.

In yeast *Saccharomyces cerevisiae*, the RNA triphosphatase, RNA guanylyltransferase, and RNA guanine-N7 methyltransferase reactions are catalyzed, respectively, by Cet1, Ceg1, and Abd1. As RNAP II switches from initiation to processive elongation, the CTD is phosphorylated, first at S_5. This recruits Ceg1 to the CTD. However, the activity of Ceg1 requires to be stimulated by an allosteric interaction with Cet1. The two enzymes, Cet1 and Ceg1, form a stable complex known as the capping enzyme (CE).

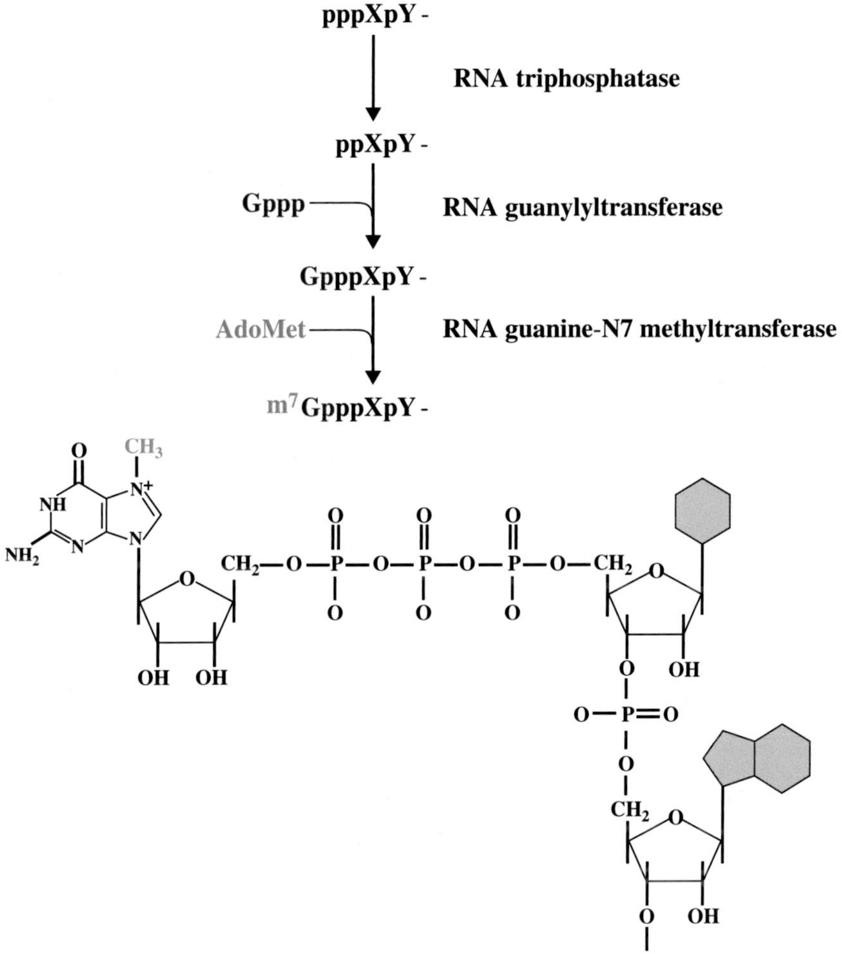

pppXpY -
RNA triphosphatase
ppXpY -
Gppp
RNA guanylyltransferase
GpppXpY -
AdoMet
RNA guanine-N7 methyltransferase
m^7GpppXpY -

FIG. 10.2

Sequential enzymatic activities leading to RNA 5′-end cap.

Although RNA triphosphatase reaction is the first step in 5′-capping, Cet1 cannot directly interact with the CTD. It has to be recruited to the site of action through interactions with guanylyltransferase, which can directly bind the CTD.

The structural basis of interactions between guanylyltransferase and the CTD has been revealed by *Candida albicans* RNA guanylyltransferase Cgt1 co-crystallized with $Y_1S_2P_3T_4S_5(\text{-}PO_4)P_6S_7$ heptad repeats. Cgt1 has 40% sequence similarity and significant structural similarity with *S. cerevisiae* Ceg1. Like Ceg1, Cgt1 also consists of two structural domains—an N-terminal nucleotidyl transferase domain (NT) and a C-terminal OB-fold domain (OB) (Fig. 10.3). The CTD peptide binds

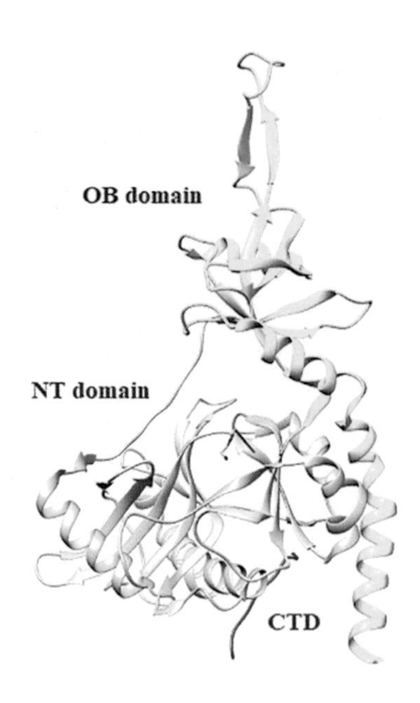

FIG. 10.3

Candida albicans RNA guanylyltransferase Cgt1-CTD complex. NT domain and OB domain of Cgt1 are colored *light blue* and *golden*, respectively; CTD is colored *purple*.

Source: Fabrega, C., et al., 2003. Mol. Cell. 11, 1549.

within a "saddle-shaped" cleft in the NT domain with S_5-PO_4 placed inside a positively charged pocket (Fig. 10.4). The CTD, which is otherwise unstructured, adopts an extended β-like conformation on binding to the Cgt1 NT domain. There are three CTD docking sites (CDS1-3) on the surface Cgt1. Fig. 10.5 illustrates the interactions between the CTD and CDS1. Residues K152, R157, and Y165 of CDS1 are involved in electrostatic and H-bonding interactions, whereas F63, L163, G164, F196, and M199 make van der Waals contacts.

CTD-bound guanylyl transferase recruits the triphosphatase to form the CE complex. In *S. cerevisiae*, CE consists of a Cet1 dimer bound to either one or two Ceg1 molecules (Fig. 10.6A). The OB domain of Ceg1 interacts with a WAQKW motif in Cet1. Next to the WAQKW motif is a flexible linker that allows Ceg1 the conformational mobility it requires for RNA capping, while maintaining its association with Cet1 as well as RNAP II (Fig. 10.6B).

CE spans the RNAP II RNA exit channel. As the nascent transcript emerges from the RNA exit channel, it first binds to the Cet1 active site and then proceeds to the Ceg1 active site before dissociating from the CE complex.

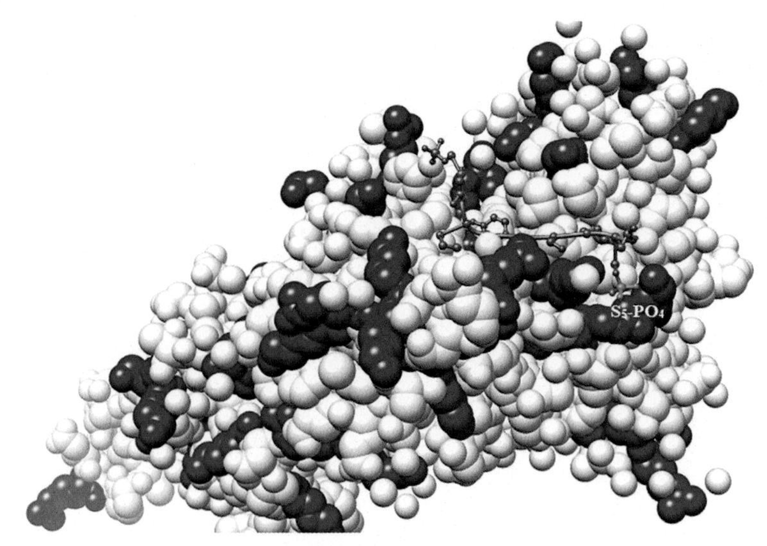

FIG. 10.4

Cgt1 surface interacting with CTD. Positive charged residues in Cgt1 are shown by *blue*-colored spheres; phosphate group in S_5 is colored *green*.

Source: Fabrega, C., et al., 2003. Mol. Cell. 11, 1549.

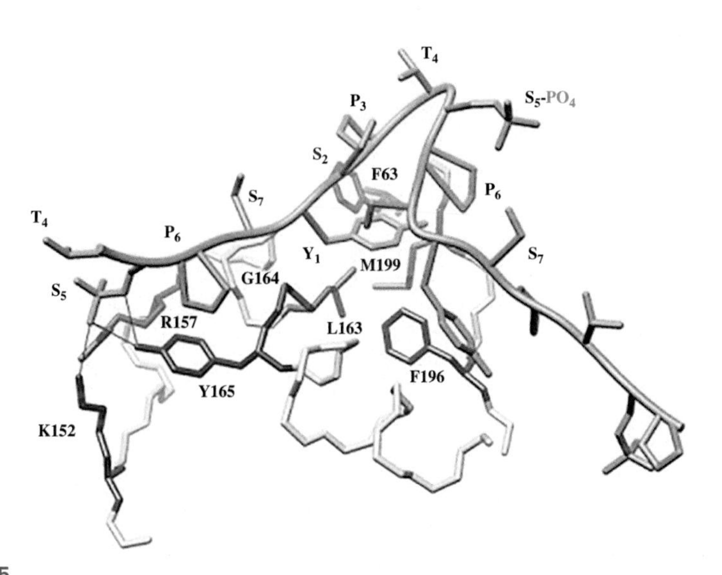

FIG. 10.5

Interacting residues in CTD and one of the CTD docking sites, CDS1. Interacting CDS1 residues are shown in *dark blue*, the other Cgt1 residues in *light blue*. CTD is colored *pink*. A few H-bonds are shown by *blue* lines.

Source: Fabrega, C., et al., 2003. Mol. Cell. 11, 1549.

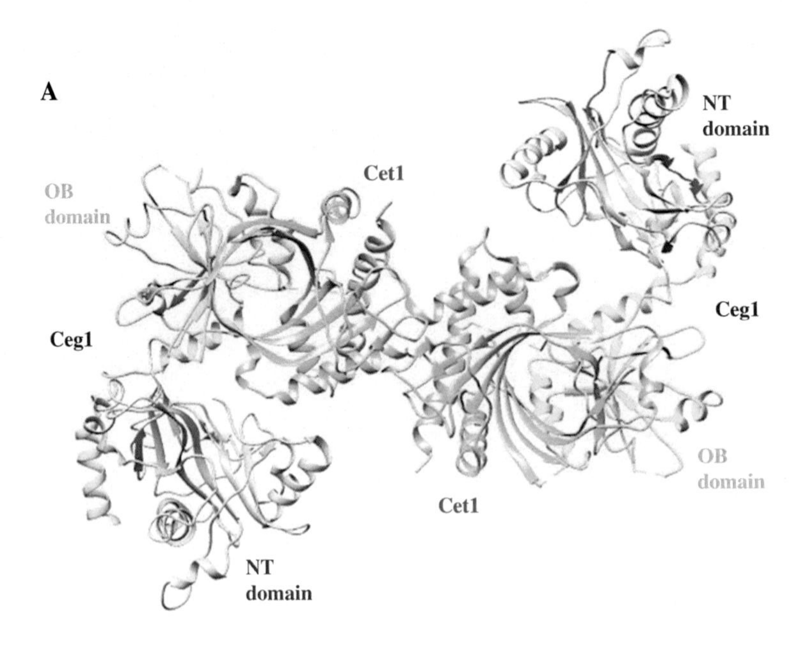

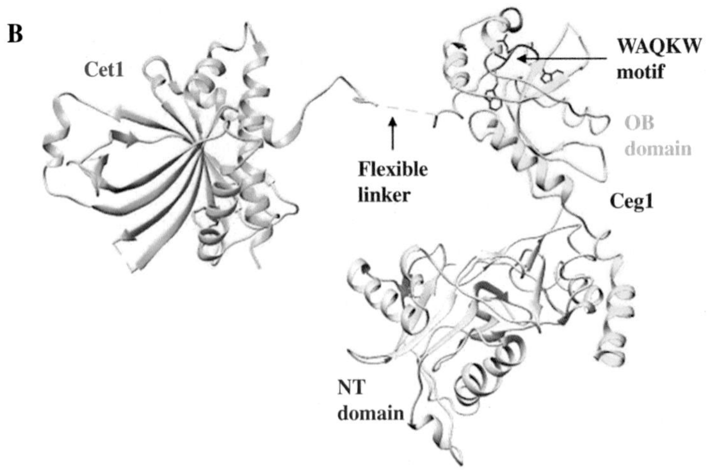

FIG. 10.6

Structure of *S. cerevisiae* CE complex. (A) Cet1 dimer *(light green)* is bound to two Ceg1 molecules, each of which is colored *light blue* (NT domain) and *golden* (OB domain). (B) Mode of Cet1-Ceg1-binding.

Source: Gu, M., Rajashankar, K.R., Lima, C.D., 2010. Structure 18, 216.

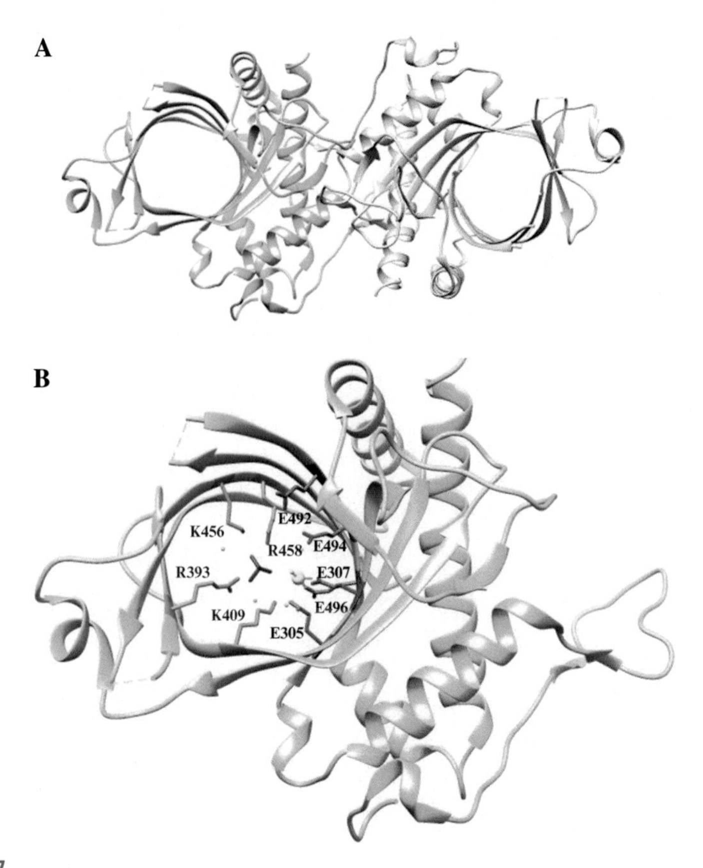

FIG. 10.7

(A) *S. cerevisiae* RNA triphosphatase dimer (each monomer colored differently). (B) Active site within a triphosphate tunnel in each monomer—acidic residues *(orange red)* coordinating a metal ion *(golden)* and basic residues *(dodger blue)* coordinating a sulfate ion *(purple)* are shown.

Source: Lima, C.D., Wang, L.K., Shuman, S., 1999. Cell 99, 533.

The Cet1 homodimer contains two independent active sites within parallel tunnels (triphosphate tunnel) (Fig. 10.7A). Each tunnel consists of an eight-stranded antiparallel β-barrel. One side of the active site is arrayed with a few acidic residues E305, E307, E492, E494, and E496, which coordinate metal ions required for the triphosphatase activity. Basic residues R393, K409, K456, and R458, present on the other side of the active site, are involved in coordinating a sulfate ion. Since sulfate is a structural analog of phosphate, the interactions of the basic residues with sulfate are presumed to reflect interactions with the γ-phosphate of pppRNA (Fig. 10.7B).

As the γ-phosphate is clipped away from the 5′-end of the nascent transcript, the resultant diphosphate end now becomes the substrate for guanylyl transferase activity. RNA guanylyl transferase transfers a GMP to the diphosphate end in a two-step ping-pong reaction. In the first step, the GMP is transferred from a GTP to a conserved lysine (K67) present in the NT domain leading to the formation of a covalent enzyme-(lysyl-N)-GMP intermediately accompanied by the release of a pyrophosphate (Fig. 10.8A). In the second step, the enzyme binds the ppRNA substrate affecting GMP transfer from the (lysyl-N)-GMP adduct to the diphosphate end and forming 5′-GpppXpY- (Fig. 10.8B).

In order to execute the two steps, the enzyme configurationally shuttles between "open" and "closed" states created by the relative orientations of the OB and NT domains. The domains open to bind GTP, close to catalyze nucleotidyl transfer to the enzyme, reopen to eject the pyrophosphate, and close again to catalyze nucleotidyl transfer to ppXpY-. The structure needs to open one more time to release 5′-GpppXpY-, which is then methylated at N7 of G by the methyltransferase enzyme.

RNA guanine-N7 methyltansferase transfers a methyl group from S-adenosylmethionine (AdoMet) to GpppXpY- to form m^7GpppXpY- and S-adenosylhomocysteine (AdoHcy). Crystal structures of a guanine-N7 methyltansferase, Ecm1, from the intracellular parasite *Encephalitozoon cuniculi*, determined without or in complex with substrates, have helped elucidate the mechanisms of methyl transfer.

The Ecm1 structure can be split into two segments. It contains a deep intersegment cleft with two ligand-binding pockets—one for the methyl donor AdoMet, whereas the other for the methyl acceptor guanosine (Fig. 10.9A). The methyl transfer occurs without any direct contact between the enzyme and the N-7 atom of guanine, the methyl carbon of AdoMet, or the sulfur of AdoHcy. The methylation is rather achieved by optimization of the proximity and relative orientations of the substrates and a favorable electrostatic environment.

Ecm1 residue Y145 is responsible for bridging interactions between the cap guanine and AdoMet ligands. The electrostatic character of AdoMet and the cap-binding site has an influence on the substrate affinity and product dissociation. The surface beneath the AdoMet-binding site is electronegative, whereas that below the guanine-binding site is close to neutral (Fig. 10.9B). The yeast guanine-N7 methyltransferase, Abd1, also adopts a similar mechanism of methyl transfer.

In general, capping reactions are universal in eukaryotes and DNA viruses. However, it should be appreciated that a remarkable diversity exists in the genetic and physical organization of the cap-forming enzymes among different organisms.

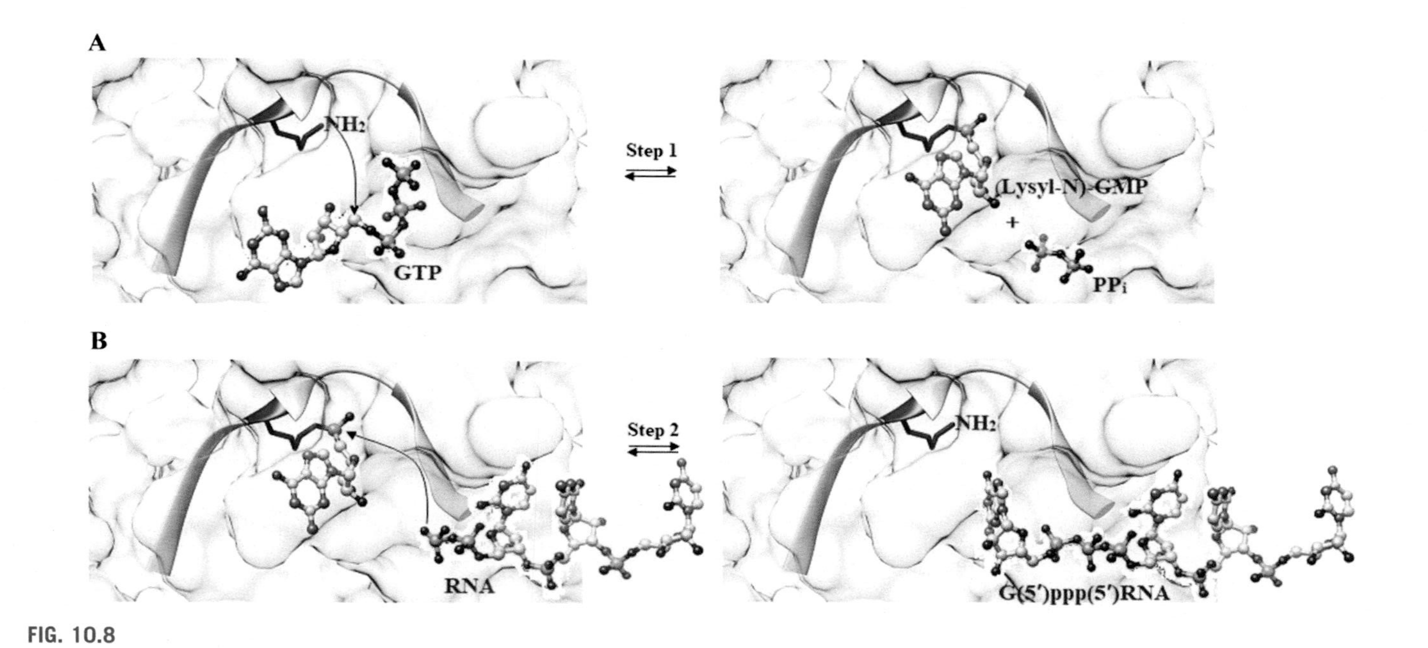

FIG. 10.8

Two-step RNA guanuylyltransferase reaction. (A) Step 1: Transfer of GMP from GTP to a conserved lysine, K67 *(dark blue)*, in the NT domain. (B) Step 2: Transfer of GMP to ppRNA and formation of GpppXpY-. Part of *C. albicans* Cgt1 *(light blue)* containing the conserved lysine shown against the background of Cgt1 surface.

Source: Fabrega, C., et al., 2003. Mol. Cell. 11, 1549.

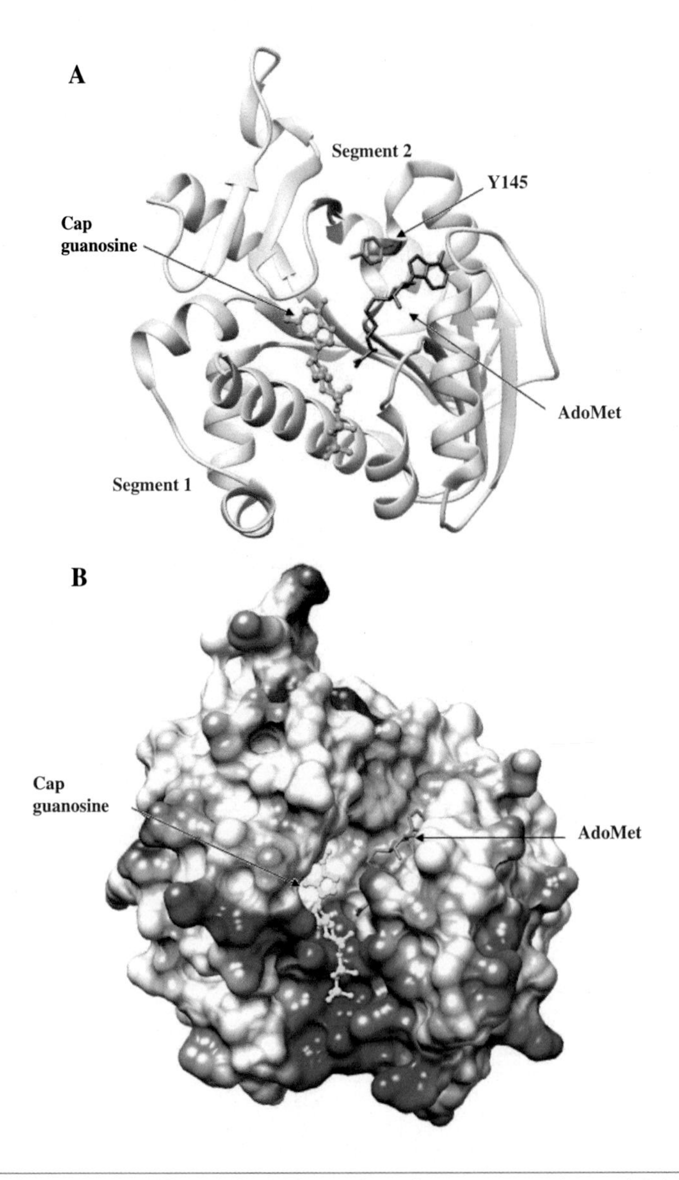

FIG. 10.9

E. cuniculi RNA guanine-N7 methyltransferase bound to cap guanosine and AdoMet. (A) Ribbon representation. (B) Electrostatic surface: *red*—negative, *white*—neutral, and *blue*—positive.

Source: Fabrega, C., et al., 2004. Mol. Cell. 13, 77.

10.3 3′-End polyadenylation

It is interesting to note that, although only two enzymatic steps are required for 3′-end processing of the primary transcript (pre-mRNA), the complete machinery involved in the execution consists of more than 20 polypeptides in yeast cells and more than 80 in human cells. Specific *cis* elements in the RNA are required to steer the protein factors through the process.

In mammals, although the site of endonucleolytic cleavage in the RNA exhibits no consensus, it is mostly preceded by a dinucleotide CA. The sequence elements which are responsible for precise positioning of the 3′-end processing complex are present both upstream and downstream of the cleavage site (Fig. 10.10). Located 10–35 nucleotides (nt) upstream of the cleavage site is a highly conserved hexamer, AAUAAA, which is known as the polyadenylation signal (PAS). On the other hand, a U/GU-rich downstream sequence element (DSE) is located within 40 nt from the cleavage site. Further, in a number of cases, multiple UGUA elements are present 40–100 nt upstream of the poly(A) site.

The gigantic 3′-end processing machinery consists of four multiunit protein subcomplexes—cleavage and polyadenylation factor (CPSF), cleavage stimulating factor (CstF), cleavage factor I (CFI_m) and cleavage factor II ($CFII_m$), and a poly(A) polymerase (PAP). All the components contribute to the cleavage reaction; for PAS-dependent poly(A) synthesis in the absence of cleavage, CPSF and PAP are sufficient. The C-terminal domain (CTD) of RNAP II Rpb1 subunit is also an important component of the 3′-end processing complex. The three-dimensional structures of many of these proteins, alone or in complex, have already been determined. With such a large number of proteins involved, the mechanistic picture of the process is far from complete. Nevertheless, we can keep ourselves restricted to one or two subunits of each of the subcomplexes and look into the special structural features underlying their functions and gain an overall understanding of the macromolecular interactions involved.

10.3.1 CFI

Human CFI_m recognizes UGUA elements and functions early in the assembly of the 3′-end processing complex. It is a heterotetrameric subcomplex consisting of a highly conserved 25 kDa subunit CFI_m25 along with two molecules of CFI_m59, CFI_m68 or CFI_m72, CFI_m68 being the most common second subunit. CFI_m25 has been found to be both necessary and sufficient for CFI_m binding to UGUA.

CFI_m68 contains an RNA recognition motif (RRM). However, it cannot bind RNA on its own. CFI_m68 interacts with CFI_m25 to facilitate RNA binding of CFI_m. The crystal structure of CFI_m25, in complex with CFI_m68 RRM and RNA, has shown that CFI_m25 homodimer is flanked on opposite sides by two CFI_m68 RRM domains, while each of the two CFI_m25 subunits binds one UGUA element specifically (Fig. 10.11A).

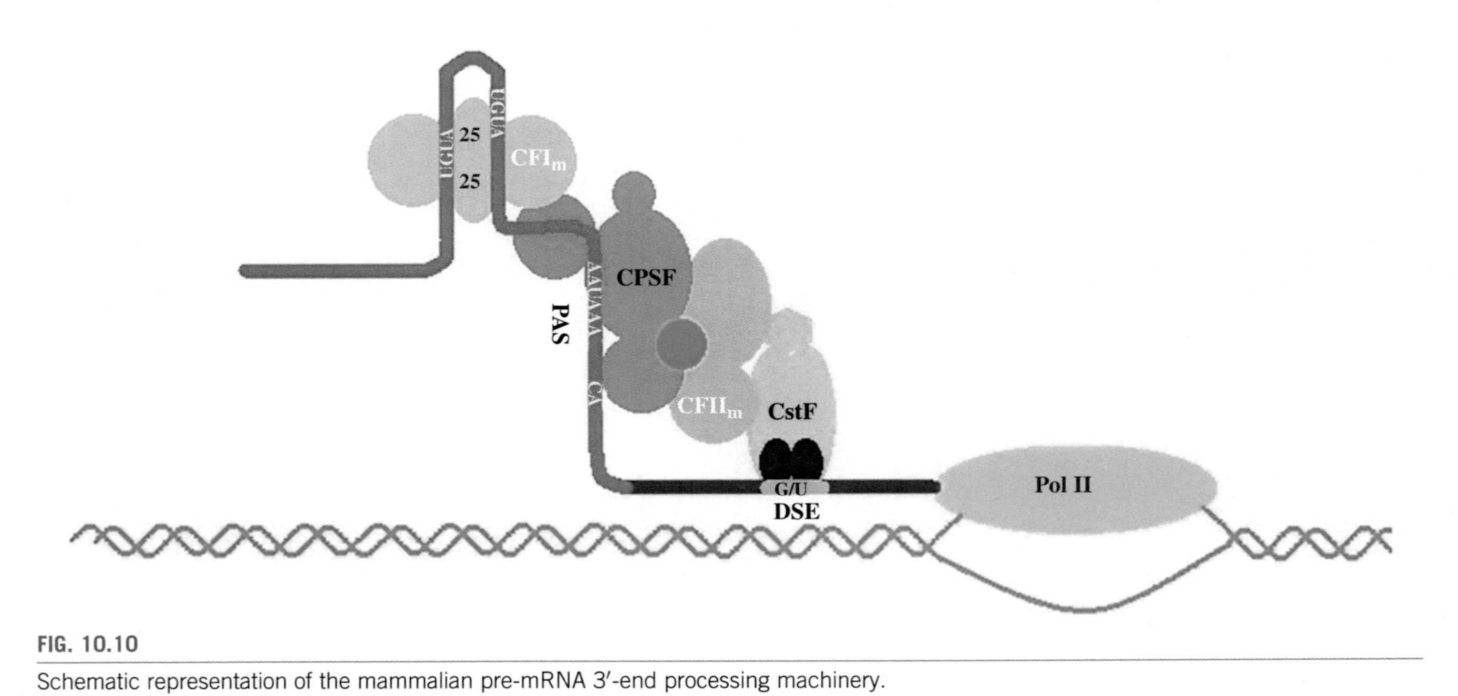

FIG. 10.10

Schematic representation of the mammalian pre-mRNA 3′-end processing machinery.

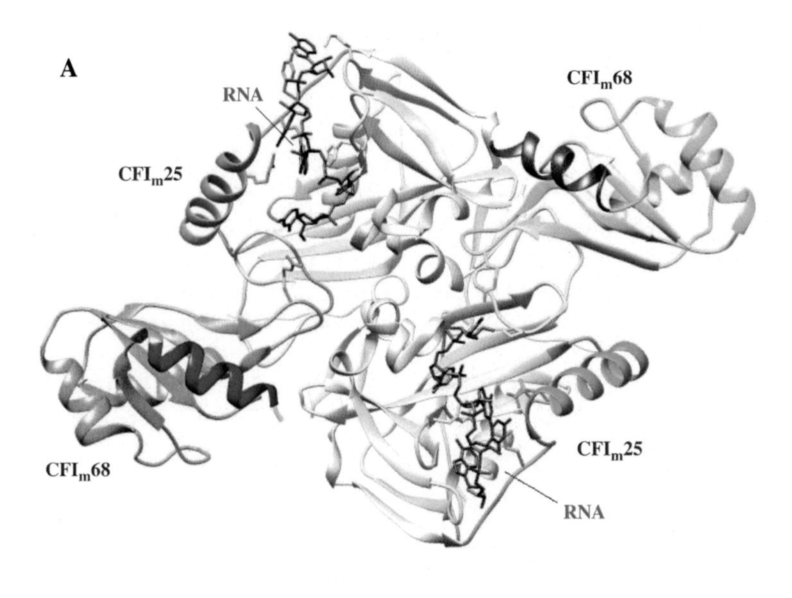

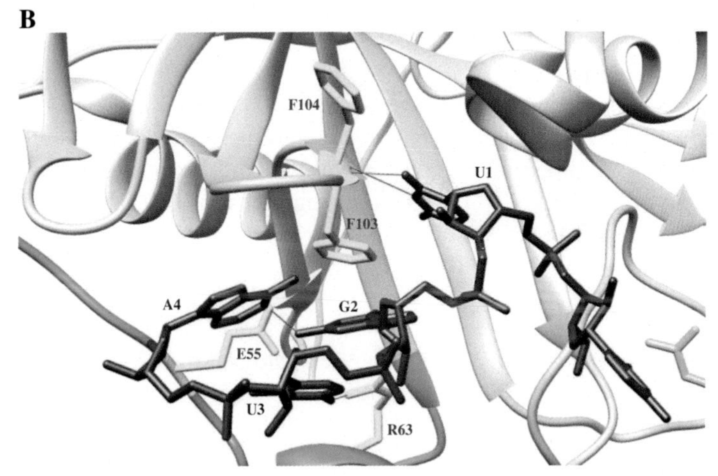

FIG. 10.11

CFI$_m$ subcomplex. (A) Structure of CFI$_m$25/CFI$_m$68 RRM/RNA. Coloring scheme for each CFI$_m$25 subunit: Nudix domain—*light blue*; loop-helix motif—*dodger blue*; and rest of subunit—*gray*. Two flanking CFI$_m$68 RRMs colored differently—*light green and coral*; additional C-terminal α-helices colored firebrick and *dark blue*. (B) CFI$_m$25-UGUA interactions. H-bonds are shown in *red* lines.

Source: Yang, Q., et al., 2011. Structure 19, 368.

The ability of CFI_m25 to bind RNA is due to a modified Nudix fold domain it possesses. Nudix fold-containing proteins hydrolyze nucleoside diphosphates linked to other moieties. The canonical Nudix fold ($\alpha\beta\alpha$) is a β-sheet with α helices on each side. However, CFI_m25 contains two modifications: (a) it lacks two of the four glutamate residues of a typical Nudix enzyme which coordinate divalent metal ions and, therefore, any catalytic activity, and (b) in CFI_m25, the conserved loop-helix motif (Fig. 10.11A) that precedes the Nudix domain blocks the conventional substrate-binding pocket of Nudix proteins, but provides essential residues for the formation of an alternative binding pocket. These two modifications together convert the hydrolase domain into a sequence-specific single-stranded RNA-binding domain (RBD).

The alternative binding pocket in CFI_m25 specifically binds UGUA elements. In particular, F104 of the Nudix domain and E55 and R63 of the preceding loop-helix motif enter into H-bonding interactions with U1, G2, and U3, while A4 is recognized by H-bonding interactions with G2 (Fig. 10.11B). F103 provides additional hydrophobic forces to stabilize CFI_m25-UGUA interactions.

The RRM of CFI_m68 displays the canonical $\beta_1\alpha_1\beta_2\beta_3\alpha_2\beta_4$ topology with two α-helices packed on one side of the four-stranded β-sheet. An additional C-terminal α-helix of the RRM lies on the top of the β-sheet to block the canonical RNA-binding surface. CFI_m68 is also missing one of the three aromatic residues that stack on to the RNA bases. These two factors make CFI_m68 incapable of directly binding RNA. It can, however, interact with CFI_m25 through the β_1/α_1 and β_2/β_3 connecting loops.

In the CFI_m25 dimer, the two Nudix domains are oriented antiparallel to each other. Therefore, in order to bind simultaneously to the dimer, two UGUA elements are required to make an 180° turn. CFI_m68 enhances the RNA-binding affinity of CFI_m by stabilizing the RNA loop connecting two UGUA elements (Fig. 10.12).

10.3.2 CPSF

CPSF recognizes the PAS and also provides a platform for a few other 3′-end processing factors including PAP. It is required for both polyadenylation and cleavage reactions. One of its subunits, CPSF73, is responsible for the endonucleolytic cleavage activity.

CPSF73 belongs to the metallo-β-lactamase (MBL) family of proteins. The canonical MBL domain, with an αβ/βα architecture, consists of two mixed β-sheets flanked by α-helices; the metal-binding site is located at the edge of the domain. Human CPSF73 N-terminal residues (aa 1–208), together with residues 395–460 at the C-terminal end, form the MBL domain (Fig. 10.13). Several members of the MBL family are known to possess Zn-dependent nuclease activity.

Another domain of human CPSF (aa 209–394) contains a completely parallel six-stranded β-sheet surrounded by α-helices on opposite faces (Fig. 10.13). This domain, which can be considered as a cassette inserted into the MBL domain, has been named as the β-CASP domain. In a sense, the β-CASP family is a subset of the larger MBL family.

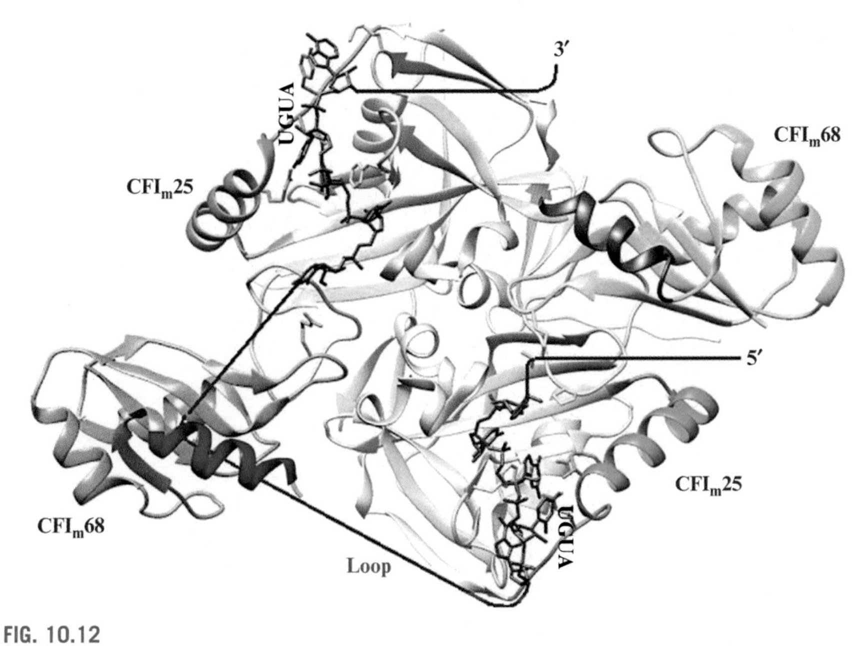

FIG. 10.12

CFI$_m$-facilitated RNA looping.

Source: Yang, Q., et al., 2011. Structure 19, 368.

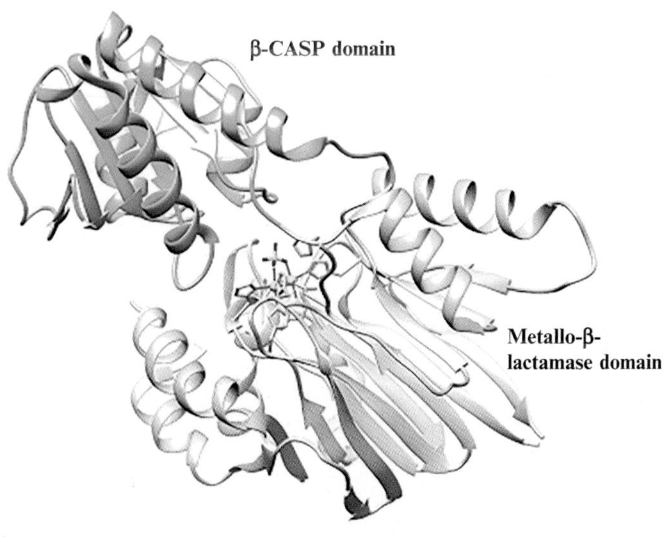

FIG. 10.13

Structure of human CPSF73. Zinc atoms in active site shown as *yellow* spheres.

Source: Mandel, C.R., et al., 2006. Nature 444, 953.

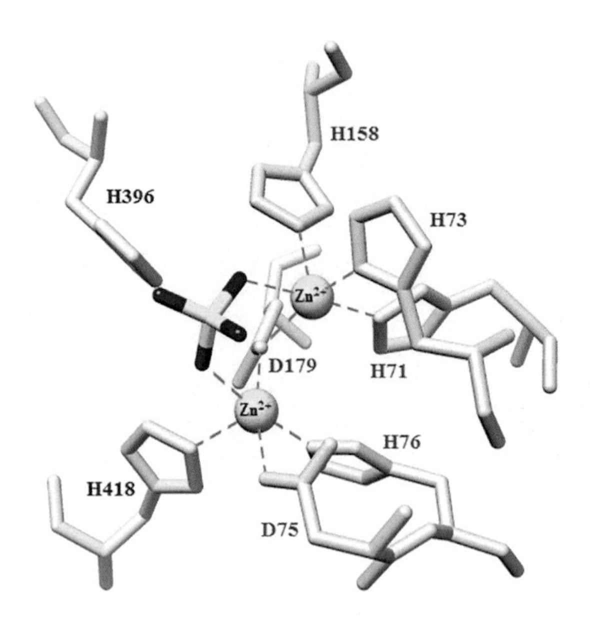

FIG. 10.14

Active site of human CPSF73. MBL domain residues coordinating Zn^{2+} ions labeled.

Source: Mandel, C.R., et al., 2006. Nature 444, 953.

The active site of CPSF73 is located at the interface of its MBL and β-CASP domains. Zn is required for the nuclease activity of CPSF73 and the crystal structure shows two Zn^{2+} ions are present in the active site. Both the ions are coordinated by conserved residues in the MBL domain (Fig. 10.14).

10.3.3 CstF

The U/GU-rich DSE is recognized by heterotrimeric CstF through its 64 kDa subunit CstF64. CstF is essential for the cleavage reaction, but not for polyadenylation. Separately, CPSF and CstF bind RNA weakly, but together they form a very stable complex with pre-mRNA and recruit the other factors required for cleavage and polyadenylation.

CstF64 RNA recognition motif (RRM) is folded in the typical $\beta_1\alpha_1\beta_2\beta_3\alpha_2\beta_4$ conformation. It is necessary and sufficient for sequence-specific recognition of RNA. The central β-sheet is the RNA recognition surface. Although CstF64 can bind to a wide range of U/GU-rich sequences, the binding is particularly strong when two consecutive Us are present in the sequence.

Structure determination and modeling have identified a binding pocket on the surface of the central β-sheet. In this pocket, while CstF64 residues S17 and R46 form H-bonds, F19 and F61 are involved in stacking interactions with UU in the DSE (Fig. 10.15).

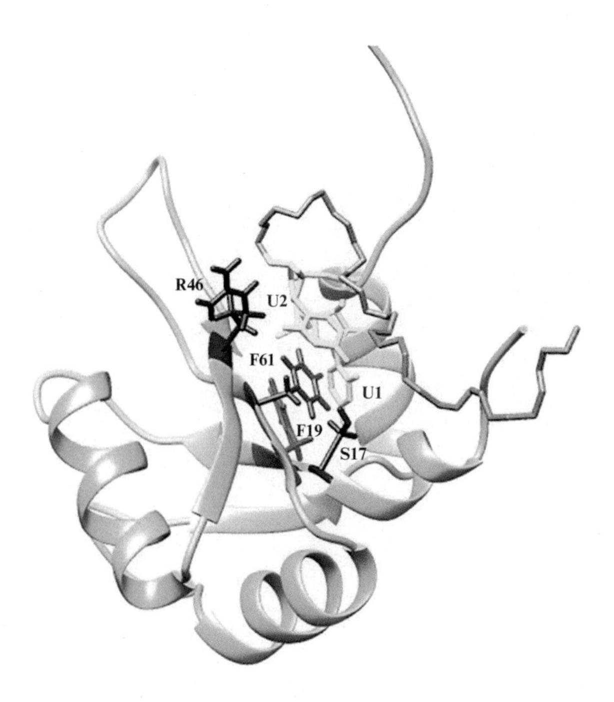

FIG. 10.15

CstF64-DSE interactions. CstF64 RRM labeled *light green*. Interacting CstF64 residues colored in *firebrick*; UU dinucleotide shown in *yellow*.

Source: Perez-Canadillas, J.M., Varani, G., 2003. EMBO J. 22, 2821; Wang, X., Tanaka Hall, T.M., 2001. Nat. Struct. Mol. Biol. 8, 141.

10.3.4 CFII

CFII is not a very well-characterized factor in the mammalian 3′-end processing complex. Nonetheless, it has been found to be a mediator of interactions of the complex with RNAP II Rpb1 CTD. The RNAP II CTD, which we have already seen to play a crucial role in 5′-end processing, is considered to be necessary for efficient 3′-end processing as well.

One of the constituent proteins of CFII, Pcf11, possesses an RNAP II CTD-interacting domain (CID) in its N-terminal region. Pcf11 has a role in coupling transcription to 3′-end processing. In the Pcf11-S_2(-PO_4) CTD crystal structure, the bound CTD is seen to be adopting a β-turn conformation, possibly through "induced fit," as a consequence of Pcf11 binding. The phosphate group of S_2 forms H-bonds within the CTD and thereby stabilizes the β-turn structure (Fig. 10.16).

It can be noted that not all Pol II-generated transcripts undergo polyadenylation. As for example, pre-mRNAs, transcribed from the replication-dependent histone genes, are typically cleaved but not polyadenylated. Similarly, a number of

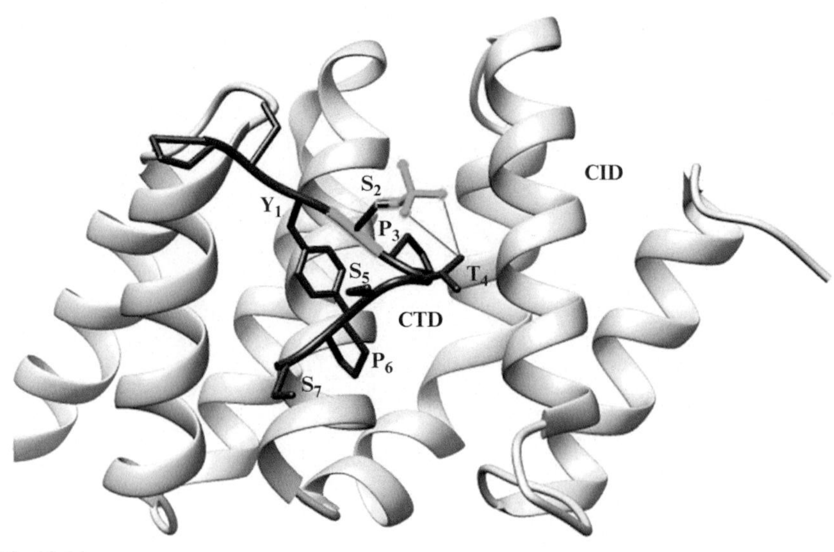

FIG. 10.16

Structure of Pcf11 CID *(light blue)* in complex with a RNAP II CTD peptide *(firebrick)*. PO_4 is colored *green* while H-bonds are shown by *orange* lines.

Source: Meinhart, A., Cramer, P., 2004. Nature 430, 223.

noncoding RNAs such as small nuclear RNAs (snRNAs), which are also products of Pol II-specific transcription, do not possess a poly(A) tail at the 3′-end.

10.4 RNA splicing

Most of the eukaryotic pre-mRNAs contain blocks of coding sequences, called exons, separated by intervening noncoding sequences known as introns. The introns are subsequently removed by splicing which is an essential step in the eukaryotic gene expression pathway.

10.4.1 Chemistry

Typically, an intron is defined by short conserved sequences in the pre-mRNA—(a) a 5′ splice site (5′ ss), (b) a branch-point sequence (BPS), (c) a polypyrimidine tract (PYT), and (d) a 3′ splice site (3′ ss) (Fig. 10.17). Chemically, the removal of an intron can be seen as two consecutive S_N2 type transesterification reactions (explained in Chapter 4). In the first reaction, the 2′ OH of a crucial adenosine residue in the BPS attacks the -PO_4 group at the 5′ ss. A cleavage occurs at this site and 5′ end of the intron is ligated to the branch adenosine. Thus, a lariat structure is formed.

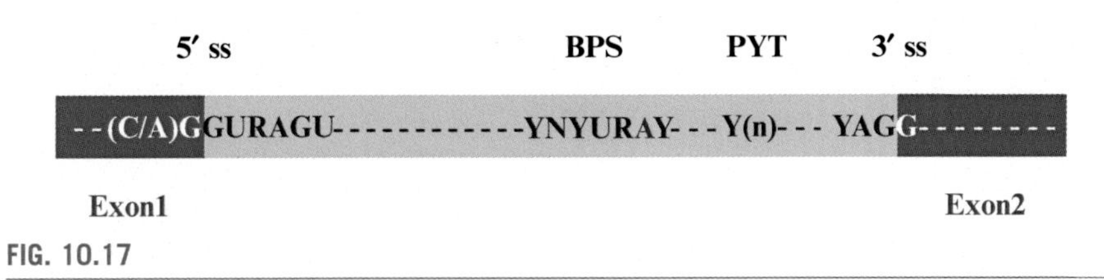

FIG. 10.17

Conserved sequences in pre-mRNA defining an intron.

Now that the 3′ OH group of the 5′ exon is free, it can attack the 3′ ss. As a consequence, the 5′ and 3′ exons are joined and the intron is released as a lariat (Fig. 10.18).

Clearly, one can see that the machinery responsible for carrying out the splicing process is required to perform three functions: (a) recognize the 5′ ss, BPS, and PYT sequences that, together with the 3′ ss, define the intron, (b) restructure the intron in such a way that these sites are brought together as required, and (c) catalyze the cleavage and joining reactions. In order to perform these tasks, the splicing machinery engages a complex network of RNA-RNA, RNA-protein, and protein-protein interactions.

It may be noted that, although the basic structure of the intron is rather simple, the enzymatic machinery which is assigned the task of splicing, the spliceosome, is a gigantic RNA/protein complex. The spliceosome consists of five small nuclear ribonucleic acid particles (snRNPs) assembled around five small nuclear RNAs (snRNAs)—U1, U2, U4, U5, and U6—and numerous non-snRNP proteins.

The pre-mRNA and snRNAs do not self-assemble into a catalytically active structure. Several proteins play indispensable roles in the formation of the spliceosome and recognition of the splice sites. They effectively participate in the dynamics of RNA-RNA, RNA-protein, and protein-protein interaction networks of the spliceosome and assist the pre-mRNA to be suitably positioned for catalysis. It should be noted that many of the RNA-RNA, RNA-protein, or protein-protein binary interactions are relatively weak. Nevertheless, multiple weak interactions together ensure the overall stability of the snRNP complexes formed.

10.4.2 Spliceosome dynamics

The spliceosome assembly occurs in an orderly stepwise manner. The steps involved are the following: (a) At first, the U1 snRNP is recruited to the 5′ ss and non-snRNP factors, SF1/BBP and U2AF, interact with the BPS and PYT, respectively (formation of the E complex). (b) U2 snRNP stably associates with the BPS replacing SF1—the A complex (also known as the prespliceosome) is formed. (c) The next to be recruited is the U4/U6·U5 tri-snRNP, which is preassembled from the U5 and U4/U6 snRNPs. This is now the precatalytic B complex. (d) The complex undergoes extensive structural rearrangements leading to the activation of the spliceosome. The interaction between U1 snRNA and the 5′ ss is disrupted; U6 begins to interact with the 5′ ss.

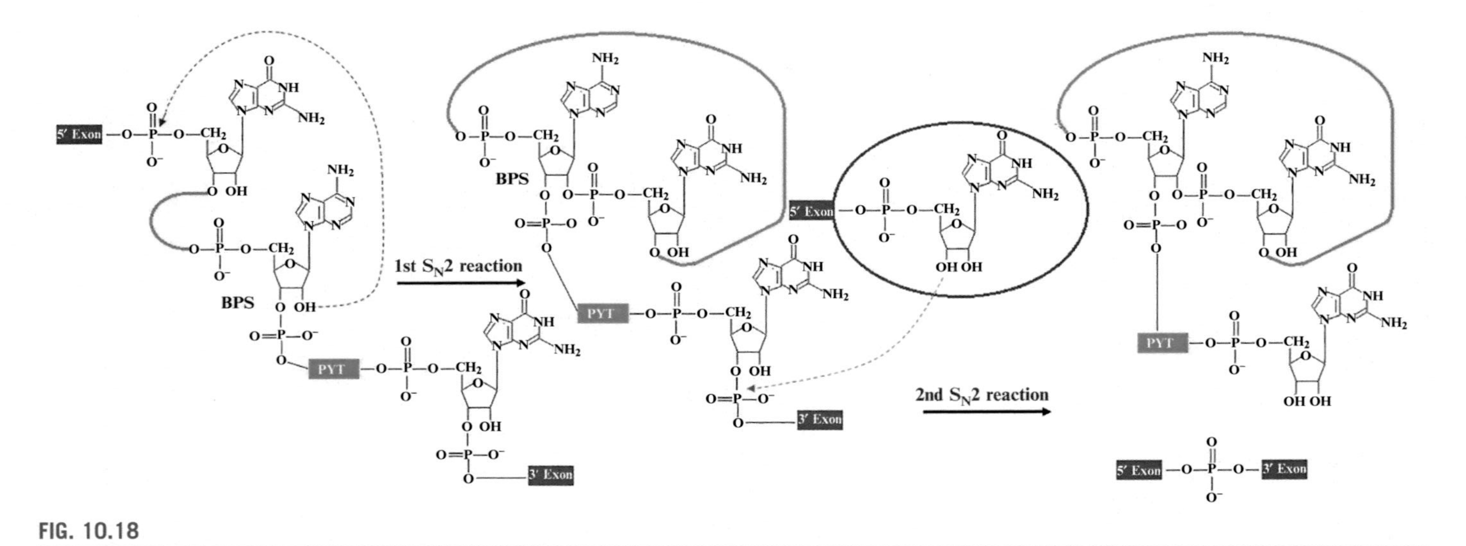

FIG. 10.18

Removal of intron by two consecutive S_N2 type transesterification reactions. Nucleophilic actions shown by *red* dashed lines.

At the same time, the interaction between U4 and U6 is also disrupted and replaced by the interaction between U2 and U6. These rearrangements facilitate the conversion of the spliceosome to an activated form (the B^{act} complex). (e) From B^{act}, the DEAH-box RNA helicase Prp2 generates the B* complex which catalyzes the first of the two splicing steps (Fig. 10.19) and becomes converted to the C complex. (f) The C complex catalyzes the second step (Fig. 10.19).

In human spliceosome, U1 snRNP is the largest subcomplex whose crystallographic structure has been determined (Fig. 10.20). It comprises U1 snRNA, seven Sm proteins (Sm B/B′, Sm D1, Sm D2, Sm D3, Sm E, Sm F, and Sm G) which are common to all snRNPs, and three U1-specific proteins (U1-A, U1-C, and U1-70K). U1 snRNA forms four stem loops (SL1-4). The seven Sm proteins form a heptameric ring on the U1 snRNA. Upstream of the Sm-binding site, U1 snRNA forms a four-way junction which is stabilized by Sm-binding.

U1-70k and U1-A contain RNA recognition motif (RRM) domains which help them bind SL1 and SL2, respectively. The N-terminal arm of U1-70k extends from its RRM domain, wraps around the core Sm heptamer, and eventually contacts U1-C. Several U1-70k N-terminal residues interact with those of Sm proteins through H-bonds and salt bridges.

Specific residues of U1-70k RRM domain interact with SL1 (Fig. 10.21). The bases of U1 snRNA C33 and G34 stack against each other while being sandwiched between the side chains of K138 and R191. Similarly, F106 and F148 are involved in stacking interactions with C31 and A32. Besides, the base of G28 flips out of the RNA helix and gets sandwiched between Y112 and R172.

U1-C does not make any contacts with the bases of either the U1 snRNA or the 5′ ss. Instead, it makes contacts with the sugar-phosphate backbones of both RNAs. Through the backbone interactions, U1-C stabilizes the duplex between U1 RNA and 5′ ss (Fig. 10.22).

Human U2 snRNP comprises U2 snRNA, seven Sm proteins, and 15 U2-specific proteins. U2 RNA folds into a structure known as the branch-point interacting stem loop (BSL). In this conformation, the branch site recognition sequence is present in the single-stranded loop making itself available for base-pairing with the pre-mRNA BPS. A number of conserved pseudouridines (Ψ_S) are present in the branch site recognition region (BSSR) (Fig. 10.23).

In the E complex, cross-intron bridging interactions exist between Prp40 in U1 snRNP at the 5′ ss and SF1/BBP at the BPS or U2AF at the PYT. In the course of prespliceosome formation, the BPS-SF1/BBP interactions or the PYT-U2AF interactions are replaced by BPS-BSSR interactions, which lead to the formation of the prespliceing complex (complex A) in an ATP-dependent manner.

Spliceosomal dynamics, that are the sequential rearrangements in the spliceosome's RNA-RNA and RNA-protein networks, require double-stranded (ds) RNA unwinding and RNP-disruptive activities. Such activities are carried out by some members of the DExD/H-box family of proteins. DExD/H-box proteins (also called DEAD-box proteins) are characterized by nine conserved motifs, one of which (motif II or Walker B motif) contains the amino acids D-E-A-D; hence, the name

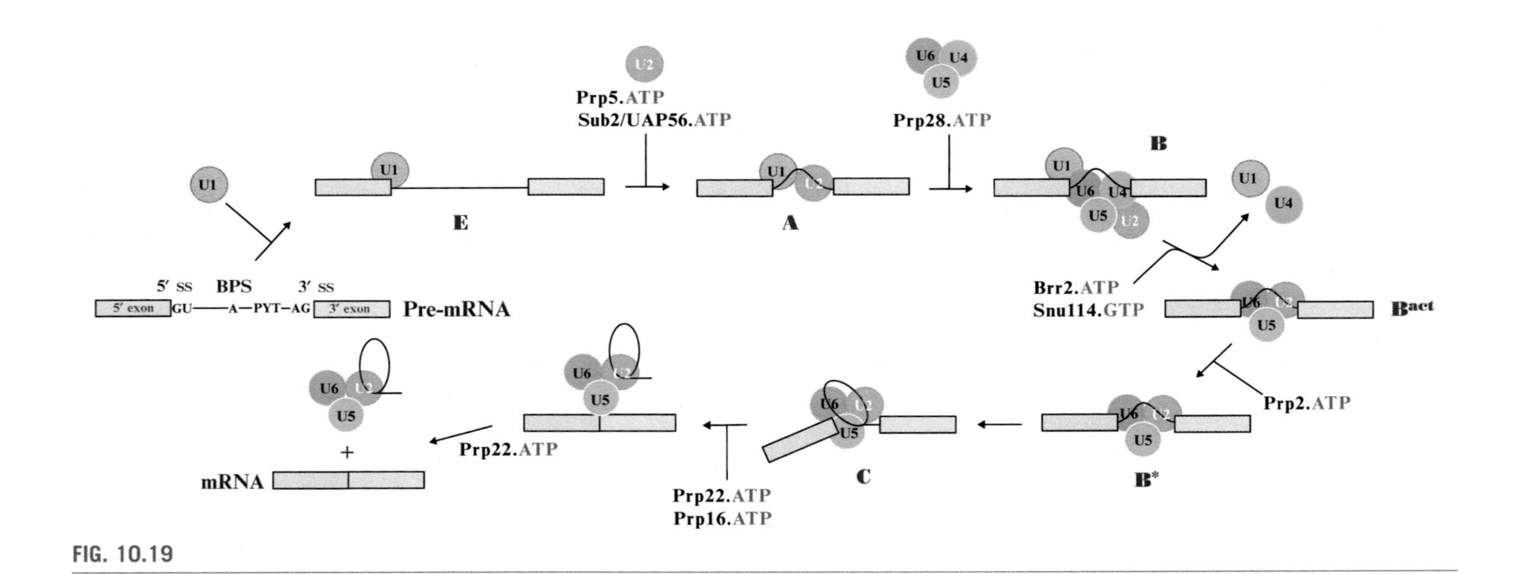

FIG. 10.19

Dynamics of spliceosome assembly and activity.

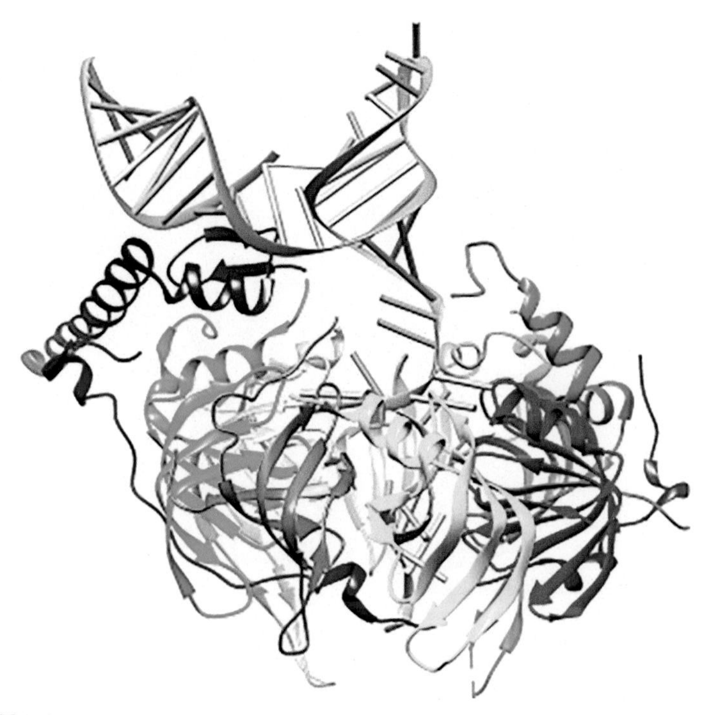

FIG. 10.20

Human U1 snRNP subcomplex structure. Coloring scheme: U1 snRNA—*light blue/gray*; Sm B—*pink*; Sm D1—*light green*; Sm D2—*forest green*; Sm D3—*cyan*; Sm E—*yellow*; Sm F—*magenta*; Sm G—*orange*; U1-70K—*dark blue*; U1-C—*firebrick*.

Source: Kondo, Y., et al., 2015. Elife 4, https://doi.org/10.7554/eLife.04986.

to the family. These proteins also contain other motifs required for ATP-binding and hydrolysis. The energy of ATP hydrolysis is utilized for their dsRNA unwinding and RNPase activities.

ATP hydrolysis by the DEAD-box protein Prp5 is required for stable U2 snRNP formation. A sequence complementary to the BPS is present near the 5′ end of the U2 snRNA. This sequence contains pseudouridines (Ψ_S) that interact with Prp5. In the U2-BPS duplex, the branch-point adenosine bulges out and its 2′ OH poses as the nucleophile for the first catalytic reaction (Fig. 10.24).

In the U4/U6.U5 tri-snRNP, the U4 and U6 snRNAs are considerably base-paired with one another (Fig. 10.25). As the tri-snRNP associates with complex A leading to the formation of complex B, U1 is replaced with U6 at the 5′ ss. One of the DExD/H-box proteins, Prp28, catalyzes this replacement and thereby promotes the transition of complex B to B^{act}. It is presumed that Prp28 disrupts the RNA-protein interaction that stabilizes the U1/5′ ss base-pairing.

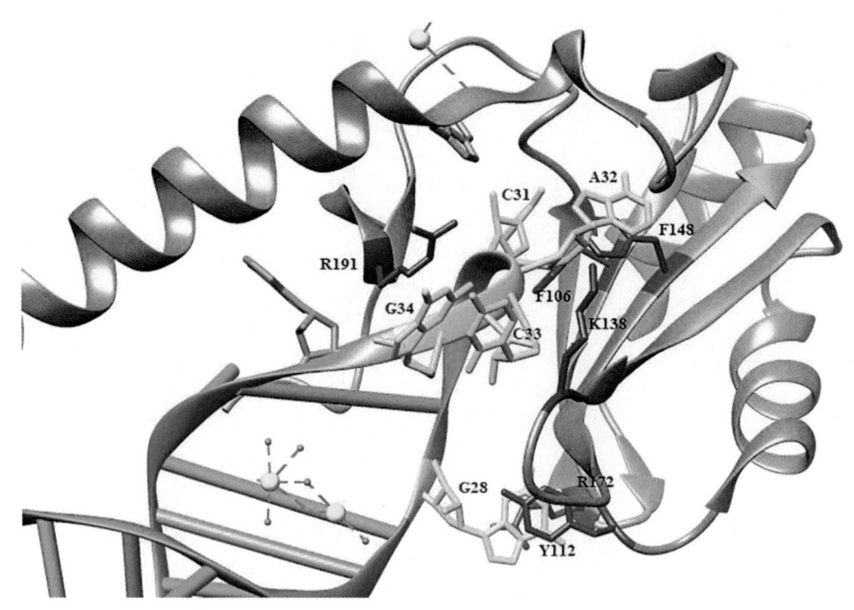

FIG. 10.21

Interaction between U1-70k RRM *(coral)* and SL1 of U1 snRNA *(dodger blue)*. Interacting bases of U1snRNA shown in *light green* and residues of U1-70k in *magenta*.

Source: Kondo, Y., et al., 2015. Elife 4, https://doi.org/10.7554/eLife.04986.

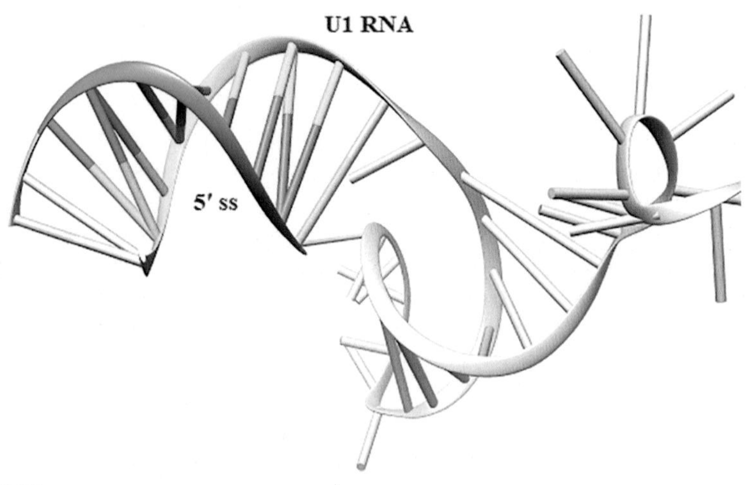

FIG. 10.22

Duplex between U1 RNA and 5′ ss stabilized by U1-C.

Source: Kondo, Y., et al., 2015. Elife 4, https://doi.org/10.7554/eLife.04986.

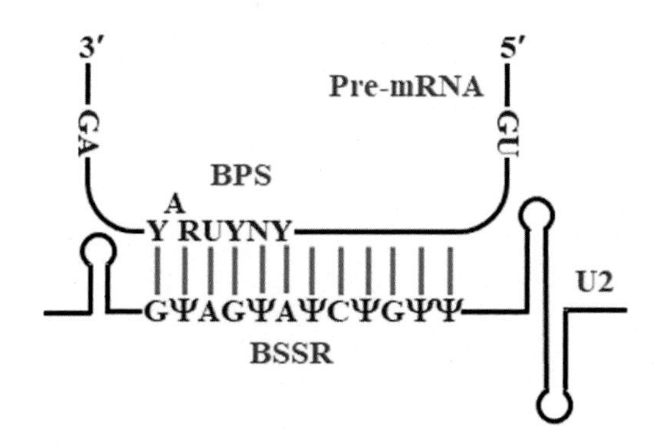

FIG. 10.23

U2—pre-mRNA interaction. Ψ—pseudouridine.

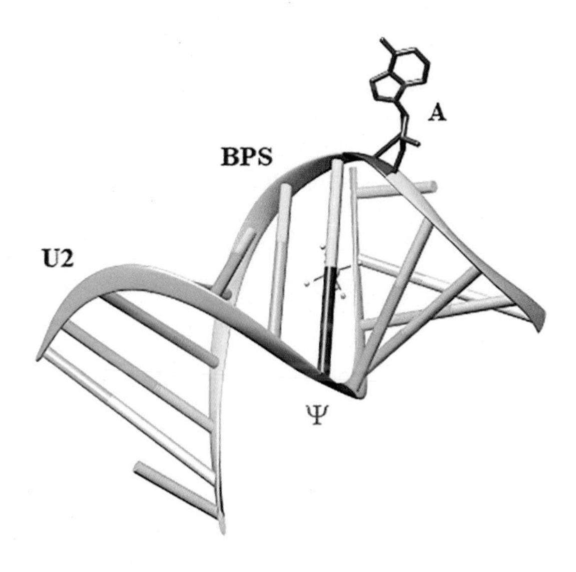

FIG. 10.24

U2 snRNA—BPS duplex.

Source: Berglund, J.A., Rosbash, M., Schulz, S.C., 2001. RNA 7, 682.

Another DExD/H-box protein, Brr2, catalyzes the unwinding of U4/U6, facilitating spliceosome activation and ejection of U4. Brr2 is an integral component of the spliceosome. However, its activity is tightly regulated by Prp8 and the GTPase Snu114. The DExD/H-box protein Prp2 converts B^{act} into the catalytically competent B* complex.

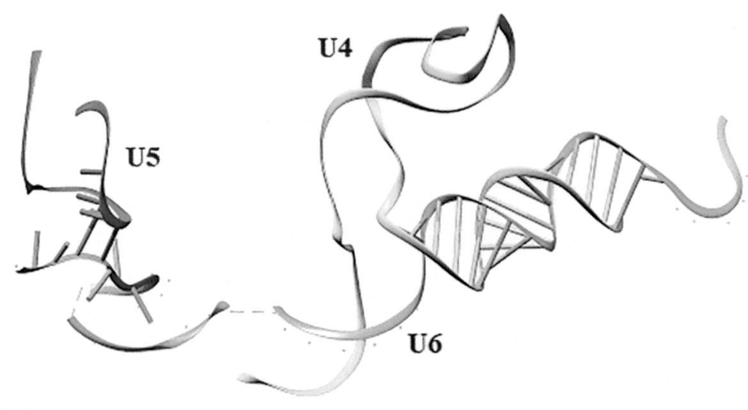

FIG. 10.25

U4/U6.U5 tri-snRNP.

Source: Agafonov, D.E., et al., 2016. Science 351, 1416.

The DExD/H-box ATPase Prp16 facilitates a conformational rearrangement in the spliceosome before the second catalytic step of splicing. The action of Prp16 is followed by that of the DExD/H-box protein Prp22, which is also responsible for mRNA release from the spliceosome. It is likely that U6/5′ ss interaction is disrupted before the second step. Further, since the spliceosome uses the same active site for both catalytic steps, the lariat intermediate is displaced from this site to accommodate the 3′ ss. U5 then contacts exon nucleotides just downstream of the 3′ ss and aligns both exons for ligation.

There are two other important aspects of splicing whose details are not discussed here—(a) alternative splicing and (b) self-splicing. In alternative splicing, the 5′ ss of a specific exon is not joined to the 3′ ss of the next exon, but to one further downstream. Alternative splicing may, therefore, be considered as the differential inclusion or exclusion of a part of the primary transcript into the mature mRNA.

We have seen how the spliceosome catalyzes RNA cleavage and ligation with high fidelity. Nevertheless, the inherent flexibility of the spliceosome makes it responsive to regulation. Certain *cis*-acting elements in the pre-mRNA, called exonic (or intronic) splicing enhancer (ESE or ISE) or silencers (ESS or ISS), interact with specific protein factors to regulate alternative splicing.

In self-splicing, the intron folds into a specific conformation within the pre-mRNA so as to catalyze its own release. These self-splicing introns have been categorized into two groups—group I and group II. Like the spliceosome, group II introns catalyze splicing by the way of two consecutive transesterification reactions. However, despite their robust autocatalytic activity, these introns rarely function alone—they are aided by intron-encoded proteins called maturases. Group I introns follow a different splicing pathway—they use a G nucleotide (instead of the branch-point A residue) for the first transesterification reaction and finally release a linear intron rather than a lariat structure.

10.5 RNA editing

Besides splicing, the sequence of RNA can also be changed posttranscriptionally by a process known as RNA editing. However, unlike splicing, editing does not remove stretches of the transcript; instead, individual bases are inserted, deleted, or altered. As a consequence, the coding information in the RNA is modified.

10.5.1 Site-specific deamination

One of the RNA editing mechanisms involves site-specific deamination of adenosine or cytosine. Deamination converts cytosine (C) to uridine (U) and adenosine (A) to inosine (I). In humans, deamination of adenosine is the most common form of RNA editing. I pairs with C instead of U. The conversion is carried out by ADAR (adenosine deaminase acting on RNA) enzymes.

The ADAR protein contains double-stranded RNA-binding domains (dsRBDs) and a C-terminal deaminase domain. Specific adenosines in the duplex region of RNA are deaminated. For the purpose of deamination, ADAR flips the reactive base out of the RNA duplex into its active site (Fig. 10.26A, B).

Several nucleic acid modifying enzymes, such as DNA glycosylases, are also known to utilize the base-flipping mechanism for their action. Nevertheless, unlike these enzymes which act on bases located in distorted duplex structures, ADAR interacts with specific adenosines within normal base-paired duplex.

Chemically, the ADAR reaction involves the formation of a hydrated intermediate which then loses ammonia to generate inosine. This hydrated intermediate is mimicked by the hydrate of nucleoside analog 8-azanebularine. Based on this fact, crystal structures of the deaminase domain of human ADAR2 bound to RNA duplex containing 8-azanebularine were analyzed to explain the basis of the base-flipping mechanism of ADAR (Fig. 10.26A, B). The structures show that the 8-azanebularine interacts with several amino acid residues such as V351, T375, K376, E396, and R455. As Fig. 10.26B shows, E396 H-bonds with N1 and O6 of the mimic nucleoside. The backbone carbonyl of T375 H-bonds with 2′-OH of 8-azanebularine and T375 side chain with its 3′-phosphodiester. R455 and K376 hold the flipped nucleoside in the active site by interacting with its flanking phosphodiester backbone. V351 provides a hydrophobic surface for interaction with the target nucleobase.

In addition, a loop present near the active site of ADAR2 also participates in the editing interaction. The residue E488 of the loop penetrates the helix and occupies the space vacated by the flipped out base. E488 H-bonds with both the complementary strand opposite the orphaned base and the nucleotide immediately 5′ to the editing site.

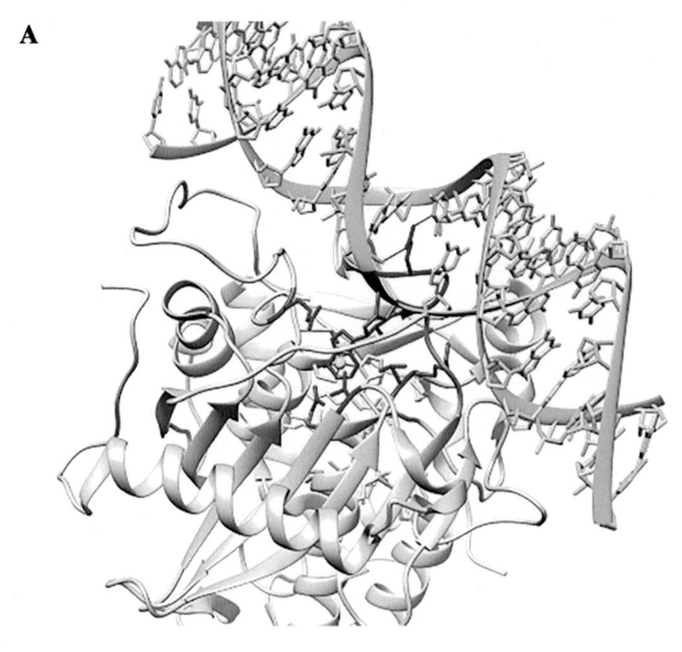

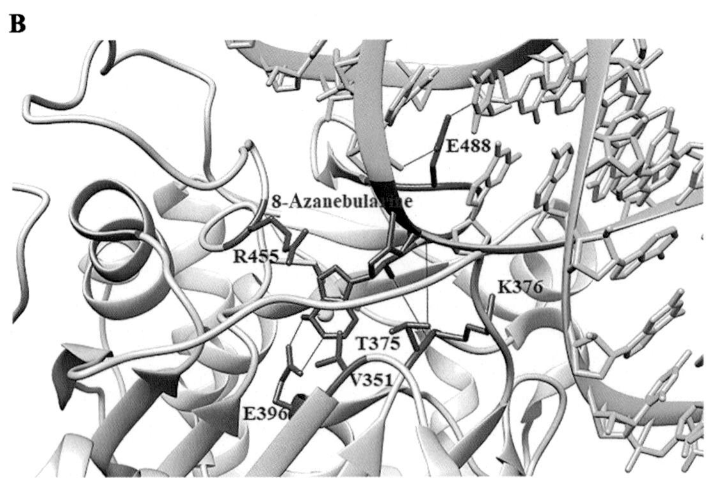

FIG. 10.26

Interactions of human ADAR2 with reactive base in RNA duplex. (A) Coloring scheme: ADAR2—*light blue*; dsRNA—*light green*; and reactive base—*orange*. (B) Region of interaction magnified. ADAR2 loop penetrating RNA duplex is shown in *dark blue*. H-bonds are shown by *red* lines.

Source: Mathews, M.M., et al., 2016. Nat. Struct. Mol. Biol. 23, 426.

10.5.2 Guide RNA-directed editing

An alternative mechanism of RNA editing has been discovered in the pathogenic protozoa *Trypanosoma brucei*. Here, uridine residues (Us) are inserted into, or deleted from, mitochondrial messenger RNA precursors to convert them into translatable mature mRNAs. The positions and numbers of inserted or deleted Us are determined by mitochondrially encoded 50–60 nucleotide (nt) RNA molecules known as guide RNAs (gRNAs). Editing reactions are catalyzed by the RNA editing core complex (RECC), also called ~ 20S editosome.

Each gRNA consists of three regions: (i) a 5′ "anchor" region that can form duplex with its cognate pre-mRNA based on complementarity, (ii) an adjacent central region that specifies the sites of insertion or deletion of Us, and (iii) a 3′ oligo (U) tail that possibly promotes the interactions among the gRNA, pre-mRNA, and RECC. Multiple gRNAs are involved in the editing of a single mRNA molecule.

RECCs contain several catalytic activities and the editing proceeds by a coordinated sequence of enzymatic steps (Fig. 10.27). Initially, an "anchor" duplex is formed between the gRNA and the pre-mRNA. The pre-mRNA is then cleaved at the first unpaired nucleotide adjacent to the duplex by an endonuclease, REN1 or REN2. Us are then either added to the pre-mRNA by a 3′-terminal uridylyl transferase (TUTase), RET2, or removed from the pre-mRNA by U-specific exoribonucleases (exoUases), REX1 and REX2. Finally, the processed RNA fragments are religated by a RNA ligase, REL1 or REL2.

Let us elaborate on the structural-functional aspects of the TUTase RET2. *T. brucei* RET2 consists of three domains—the N-terminal domain (NTD), the middle domain (MD), and the C-terminal domain (CTD)—and three conserved catalytic aspartates, D97, D99, and D267, located in a cleft of the protein molecule (Fig. 10.28A). The NTD contains a five-stranded antiparallel β-sheet flanked by three α-helices. The MD, which is an "insertion" between D99 and D267, contains six α-helices and a four-stranded antiparallel β-sheet. The CTD contains seven α-helices and four short β-strands.

Crystal structures of RET2 in complexes with UTP and divalent metal ions have shown that the ligand UTP molecule and a divalent metal ion are also located in the catalytic cleft of the protein (Fig. 10.28A). The side chain of C83 and the aromatic ring of Y319 stack on the UTP sugar ring from two sides and stabilize it through hydrophobic interactions (Fig. 10.28B).

The 2′-OH is involved in H-bonding with side chain atoms of N277 and S278, which thus discriminate an rUTP from a dUTP as the incoming nucleotide. The uridine base is involved in an extensive interaction with the CTD, particularly with the residues N277, T317, Y319, E424, and R435.

The specificity of RET2 for UTP is determined by a crucial water molecule which bridges the N3 amine group of the uridine base with two carboxylate oxygens of D421 and E424 (Fig. 10.28B).

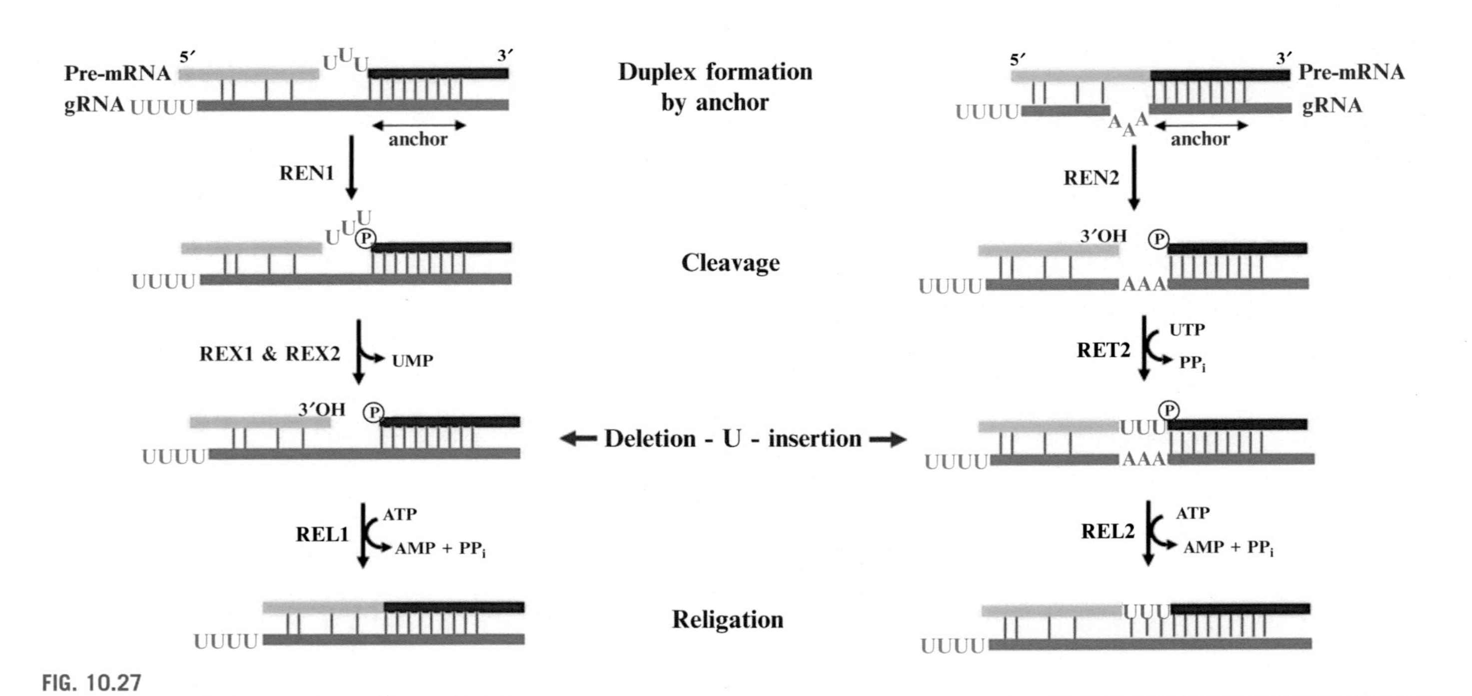

FIG. 10.27

RNA editing reactions carried out by the RECC.

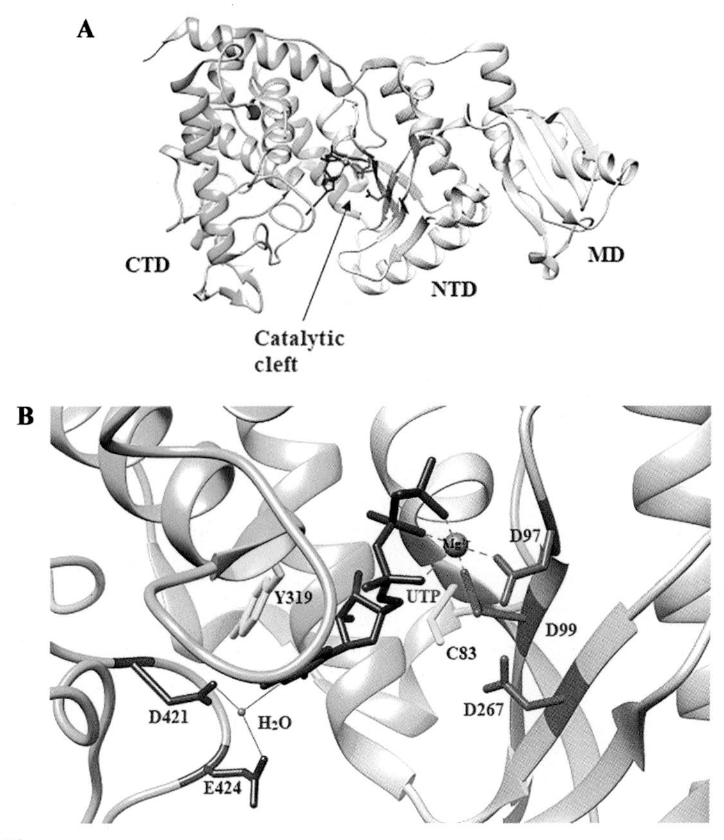

FIG. 10.28

Structure of *T. brucei* RET2 in complex with UTP and Mg^{2+}. (A) Domains: NTD *(pink)*, MD *(light blue)*, and CTD *(light green)*. (B) Catalytic cleft enlarged. H-bonds shown by *black* lines.

****

Let us conclude this chapter with the appreciation of two facts. One, the pre-mRNA processing that has been discussed here is not a set of isolated cellular phenomena. It is intimately connected to the upstream events in transcription as well as downstream events including translation. Secondly, regulation of RNA processing is capable of generating an enormous diversity in eukaryotic gene expression at the RNA level. As a consequence, an organism can synthesize a much larger repertoire of protein types than that would be expected from one gene—one polypeptide hypothesis.

Sample questions

1. How is the differential phosphorylation of the RNA polymerase II carboxy-terminal domain important for its role in pre-mRNA processing?
2. Highlight the structural basis of interactions between *Candida albicans* guanylyltransferasse Cgt1 and the RNA polymerase II carboxy-terminal domain in 5′-end capping.
3. The mammalian cleavage factor I (CFI) subunit CFI_m25 possesses a Nudix fold domain, yet it lacks the hydrolase activity of other Nudix fold-containing proteins—state reasons. What is, then, its role in the CFI complex?
4. Which factor is responsible for the cleavage activity in pre-mRNA 3′-end processing? What is the basis of its nucleolytic action?
5. RNA splicing is essentially two consecutive S_N2 type transesterification reactions—elaborate the chemistry.
6. How do U1-specific proteins stabilize the interaction between U1 snRNA and 5′ ss?
7. Spliceosomal dynamics requires RNA unwinding and RNAP-disruptive activities—which are the factors responsible for such activities?
8. What are the key interactions between human ADAR2 and an RNA duplex that lead to flipping of the target base?
9. What determines the specificity of *Trypanosoma brucei* 3′-terminal uridylyl transferase, RET2, for UTP?

References and further reading

Aphasizhev, R., Aphasizheva, I., 2014. Mitochondrial RNA editing in trypanosomes: small RNAs in control. Biochimie 100, 125–131.

Caňdillas, J.M.P., Varani, G., 2003. Recognition of GU-rich polyadenylation regulatory elements by human CstF-64 protein. EMBO J. 22 (11), 2821–2830.

Deng, J., et al., 2005. Structural basis for UTP specificity of RNA editing TUTases from *Trypanosoma brucei*. EMBO J. 24, 4007–4017.

Fabrega, C., Shen, V., Shuman, S., Lima, C.D., 2003. Structure of an mRNA capping enzyme bound to the phosphorylated carboxy-terminal domain of RNA polymerase II. Mol. Cell 11, 1549–1561.

Ghosh, A., Lima, C.D., 2010. Enzymology of RNA cap synthesis. Wiley Interdiscip. Rev. RNA 1 (1), 152–172.

Ghosh, A., Shuman, S., Lima, C.D., 2011. Structural insights to how mammalian capping enzymes reads the CTD code. Mol. Cell 43, 299–310.

Gu, M., Rajashankar, K.R., Lima, C.D., 2010. Structure of the *Saccharomyces cerevisiae* Cet1-Ceg1 mRNA capping apparatus. Structure 18, 216–227.

Hang, J., Wan, R., Yan, C., Shi, Y., 2015. Structural basis of pre-mRNA splicing. Science 349 (6263). 1191–1198.al.

House, H.E., Lynch, K.W., 2008. Regulation of alternative splicing: more than just ABCs. J. Biol. Chem. 283 (3), 1217–1221.

Hsin, J.-P., Xiang, K., Manley, J.L., 2014. Function and control of RNA polymerase II C-terminal domain phosphorylation in vertebrate transcription and RNA processing. Mol. Cell. Biol. 34 (13), 2488–2498.

Jurado, A.R., et al., 2014. Structure and function of pre-mRNA 5′-end capping quality control and 3′-end processing. Biochemistry 53, 1882–1898.

Kondo, Y., Oubridge, C., van Roon, A.M., Nagai, K., 2015. Crystal structure of human U1 snRNP, a small nuclear ribonucleoparticle reveals the mechanism of 5′ splice site recognition. eLife 4, e04986. https://doi.org/10.7554/eLife.04986.

Licatalosi, D.D., Darnell, R.B., 2010. RNA processing and its regulation: global insights into biological networks. Nat. Rev. Genet. 11, 75–87.

Linder, P., 2006. Dead-box proteins: a family affair – active and passive players in RNP remodeling. Nucleic Acids Res. 34 (15), 4168–4180.

Martinez-Rucobo, F.W., et al., 2015. Molecular basis of transcription-coupled pre-mRNA capping. Mol. Cell 58, 1079–1089.

Mathews, M.M., et al., 2016. Structure of human ADAR2 bound to dsRNA reveal base-flipping mechanism and basis for site selectivity. Nat. Struct. Mol. Biol. 23 (5), 426–433.

Meinhart, A., et al., 2005. A structural perspective of CTD function. Genes Dev. 19, 1401–1415.

Moore, M.J., Proudfoot, N.J., 2009. Pre-mRNA processing reaches back to transcription and ahead to translation. Cell 136, 688–700.

Neugebauer, K.M., 2002. On the importance of being co-transcriptional. J. Cell Sci. 115, 3865–3871.

Neve, J., et al., 2017. Cleavage and polyadenylation: ending the message expands gene regulation. RNA Biol. 14 (7), 865–890.

Nilsen, T.W., 2015. RNA splicing: an intimate view of a spliceosome component. elife 4, e06200. https://doi.org/10.7554/eLife.06200.

Perriman, R., Ares, M., 2010. Invariant U2 snRNA nucleotides form a stem loop to recognize the intron early in splicing. Mol. Cell 38, 416–427.

Shao, W., et al., 2012. A U1-U2 snRNP interaction network during intron definition. Mol. Cell. Biol. 32 (2), 470–478.

Will, C.L., Lührmann, R., 2011. Spliceosome structure and function. Cold Spring Harb. Prospect. Biol. 3, a003707.

Wu, G., et al., 2016. Pseudouridines in U2 snRNA stimulate the ATPase activity of Prp5 during spliceosome assembly. EMBO J. 35 (6), 654–677.

Xiang, K., Tong, L., Manley, J.L., 2014. Delineating the structural blueprint of the pre-mRNA 3′-end processing machinery. Mol. Cell. Biol. 34 (11), 1894–1910.

Yang, Q., Doublié, S., 2011. Structural biology of poly(A) site definition. Wiley Interdiscip. Rev. RNA2 (5), 732–747.

Yang, Q., Gilmartin, G.M., Doublié, S., 2010. Structural basis of UGUA recognition by the Nudix protein CFI_m25 and implications for a regulatory role in mRNA 3′ processing. Proc. Natl. Acad. Sci. U. S. A. 107 (22), 10062–10,067.

Zhang, l., Li, X., Zhao, R., 2013. Structural analysis of the pre-mRNA splicing machinery. Protein Sci. 22, 677–692.

CHAPTER

Translation

11

Once genetic information, carried in DNA in the form of a linear sequence of four deoxyribonucleotide bases (A, T, G, C), has been transferred to messenger RNA (mRNA) as a linear sequence of four ribonucleotide bases (A, U, G, C), the task of the cell is to "forward" the message to another macromolecule, protein, in the form of a linear sequence of 20 amino acids. Therefore, we shall discuss, in this chapter, a process in which the 4-letter language of RNA is "translated" into the 20-letter language of protein. This process is called translation.

11.1 Transfer of genetic information—From mRNA to protein

11.1.1 Protein synthesis

We have earlier seen in Chapter 5 that just as DNA and RNA are polynucleotide chains comprising deoxyribonucleotides and ribonucleotides, respectively, proteins consist of polypeptide chains formed by amino acids. The incorporation of an amino acid into a growing polypeptide chain occurs as shown in Fig. 11.1. Let us note here that, in contrast to DNA synthesis, but like RNA synthesis, a polypeptide chain can be synthesized ab initio.

As already indicated, the sequential incorporation of amino acids into a polypeptide chain is not random, nor is it spontaneous. Here again, a molecular machinery and a molecular directive ensure that the amino acids are joined together in an efficient and orderly manner. The machinery for the synthesis of proteins in a cell consists of: (a) the mRNA, (b) tRNAs, (c) aminoacyl-tRNA synthetase, and (d) the ribosome.

The mRNA provides the directive—information or "code," that needs to be interpreted or "decoded" by the translation machinery for its appropriate action. tRNAs provide the physical interface between the amino acids being incorporated into the growing polypeptide chain and the mRNA. Aminoacyl-tRNA synthetases are the enzymes that link amino acids to specific tRNAs. The ribosome is a multimegadalton complex that suitably positions and coordinates all the interacting components.

11.1.2 Messenger RNA—Genetic code

The information for protein synthesis that the mRNA carries is in the form of nucleotide triplets known as codons. It is obvious that each position of the mRNA can be occupied by any of the four nucleotides. So, the total number of possible triplets

Fundamentals of Molecular Structural Biology. https://doi.org/10.1016/B978-0-12-814855-6.00011-0

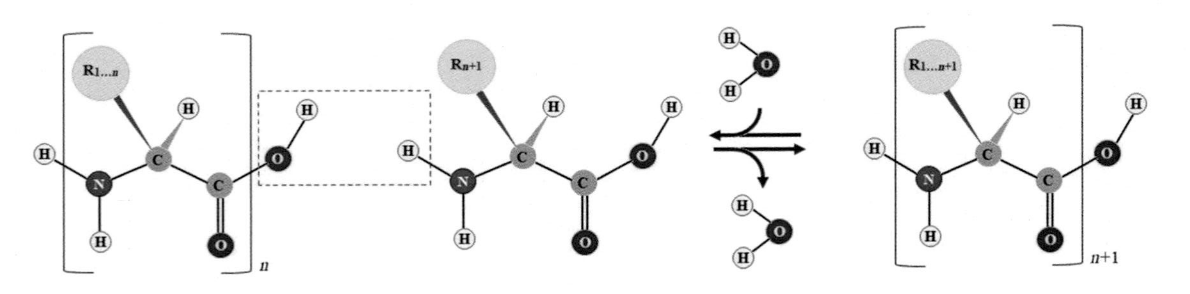

FIG. 11.1

Incorporation of amino acid into growing polypeptide chain.

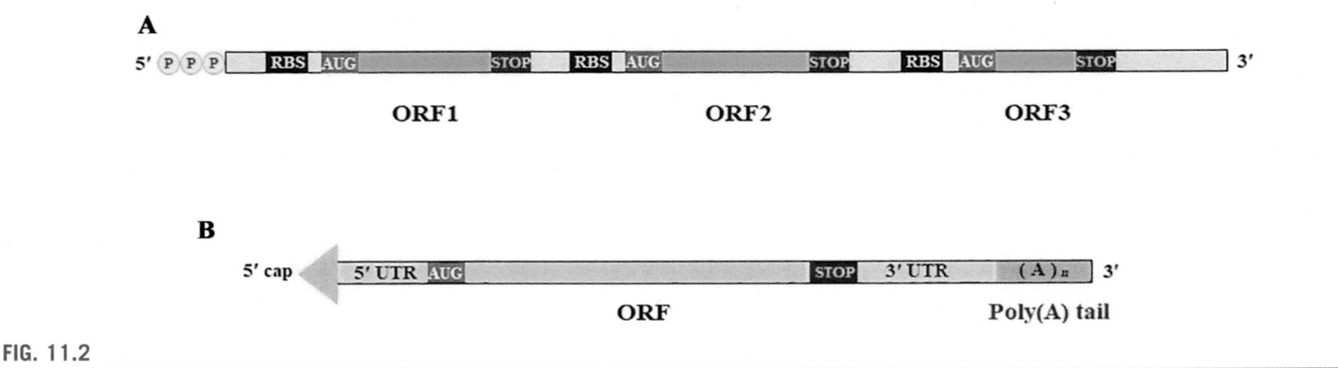

FIG. 11.2

Sequence features of messenger RNA. (A) Prokaryotic. (B) Eukaryotic.

would be $4^3 = 64$. Each of 61 codons (out of 64) specifies an amino acid. As for example, UGG specifies tryptophan. The remaining three codons, which do not specify any amino acid, are chain-terminating signals. The genetic code is, therefore, the mapping of 64 3-letter codons to 20 amino acids and a chain-terminating or "stop" signal.

Many of the 20 amino acids are specified by more than one codon. This is known as the degeneracy of the genetic code. For example, the amino acid aspartic acid is specified by GAU and GAC, while leucine is specified by as many as six codons—UUA, UUG, CUU, CUC, CUA, and CUG.

In principle, any segment of an mRNA should be readable by the translation machinery in triplets in three different frames (reading frames). Yet, only certain segment(s) of an mRNA is(are) decoded. The reading frame is determined by the presence of a "start" codon which is usually 5′-AUG-3′ (sometimes 5′-GUG-3′ or even 5′-UUG-3′). The reading frame ends with a stop codon which is either of the three—5′-UAG-3′, 5′-UGA-3′, or 5′-UAA-3′. A nonoverlapping string of codons with a start codon and a stop codon is called an open reading frame (ORF). Translation begins at the start codon, proceeds one codon at a time along the ORF, and terminates at the stop codon (Fig. 11.2).

Every mRNA molecule contains at least one ORF. mRNAs containing a single ORF are called monocistronic, while those containing multiple ORFs are polycistronic. Eukaryotic mRNAs are mostly monocistronic, whereas prokaryotic mRNAs are often polycistronic.

For translation to begin at a start codon, the ribosome must be recruited to the mRNA. Authentic start codons in prokaryotic mRNAs are mostly preceded by a short purine-rich sequence which is the ribosome-binding site (RBS) (Fig. 11.2A). This sequence element is also called Shine-Dalgarno (SD) sequence. The SD is a subset of the sequence 5′-AGGAGG-3′, typically 4–5 nt in length. In general, lengthening of the SD does not lead to an increase in translation. Nevertheless, a stronger-than-average SD is effective when the start codon is other than AUG or the initiation site is veiled by secondary structures. On the other hand, in the case of an A/U-rich initiation site, which does not form a stable secondary structure, SD interaction may not be essential for translation.

Eukaryotic mRNA, as we have seen in the previous chapter, possesses two distinct features as a result of 5′-end and 3′-end processing of the primary transcripts—an m^7G-cap at the 5′-end and a poly(A) tail at the 3′-end (Fig. 11.2B). Both features play important roles in mRNA stability and translation initiation.

11.1.3 Transfer RNA—Means of information transfer

Now that we know the format in which the mRNA stores genetic information and contains the signals for the translation machinery to recognize and start decoding the information, it can be easily appreciated that translational information transfer is much more challenging than transcription of DNA into mRNA. Transcription is guided by the H-bonding interactions between deoxyribonucleotide bases of DNA and ribonucleotide bases of RNA. In contrast, specific interactions between the triplets of ribonucleotide bases and the side chains of amino acids are almost nonexistent. To resolve the paradox, Francis Crick had proposed that the amino acid is carried to the mRNA template by an "adaptor" molecule. He further speculated that the adaptors would be containing nucleotides so that they could interact with the ribonucleotide triplets of the template in accordance with Watson-Crick base-pairing rules. Subsequently, the adaptor molecule was found to be transfer RNA (tRNA).

In all three domains of life—bacteria, archaea, and eukaryotes—tRNA genes are transcribed to synthesize precursors which are posttranscriptionally modified to form mature and functional tRNA. The posttranscriptional events include removal of 5′-leader and 3′-trailer sequences as well as modifications at individual bases such as the formation of pseudouridine (ΨU) and dihydrouridine (D). The mature tRNA molecules, which are 75–95 ribonucleotides in length, display highly conserved secondary and tertiary structures. They contain regions of self-complementarity, forming limited stretches of double helix, and can be represented schematically by a cloverleaf (Fig. 11.3).

The tRNA cloverleaf displays the following features: (a) the acceptor arm, formed by pairing between 5′- and 3′-ends of the molecule, (b) the D-arm, characterized by the presence of dihydrouridines in the loop, (c) the anticodon arm, (d) the variable loop, and (e) the TΨC-arm (T-arm) with a ΨU-loop. A set of tertiary interactions, mostly between the ΨU-loop and the D-loop, fold the tRNA into an L-shaped three-dimensional structure (Fig. 11.4).

All tRNAs carry the sequence 5′-CCA-3′ at the 3′-terminus. However, in several organisms, this sequence is not encoded by the gene, but added posttranscriptionally by CCA enzymes. The cognate amino acid is covalently linked by an ester bond to either the 2′- or 3′-OH of the end of CCA, thus forming an aminoacyl-tRNA (Fig. 11.5A, B). This is a very high-energy linkage; the hydrolysis of this ester bond is associated with a large change in favorable free energy. The energy of this bond is coupled to the formation of a peptide bond that covalently links the amino acid to a growing polypeptide chain.

The other region of the tRNA crucial for its function is located in the anticodon arm. It is essentially a set of three consecutive nucleotides that constitute the anticodon. Guided by base-pairing interaction between its anticodon and the complementary codon in the mRNA (Fig. 11.6), the tRNA carries the amino acid attached at the 3′-teminus to the growing polypeptide chain. Now taking into consideration the role of tRNA in the synthesis of a polypeptide chain, Fig. 11.1 can be revised accordingly (Fig. 11.7). As shown, the carboxyl end of the polypeptide chain (the nth amino acid) remains activated by its covalent linkage to a tRNA molecule (called the peptidyl-tRNA). In the course of addition of the $(n+1)$th amino acid to the C-terminus, the high-energy covalent linkage is disrupted and replaced by an identical linkage to the $(n+1)$th amino acid. In other words, the nth amino acid of the growing polypeptide chain carries with it the activation energy for the addition of the $(n+1)$th amino acid.

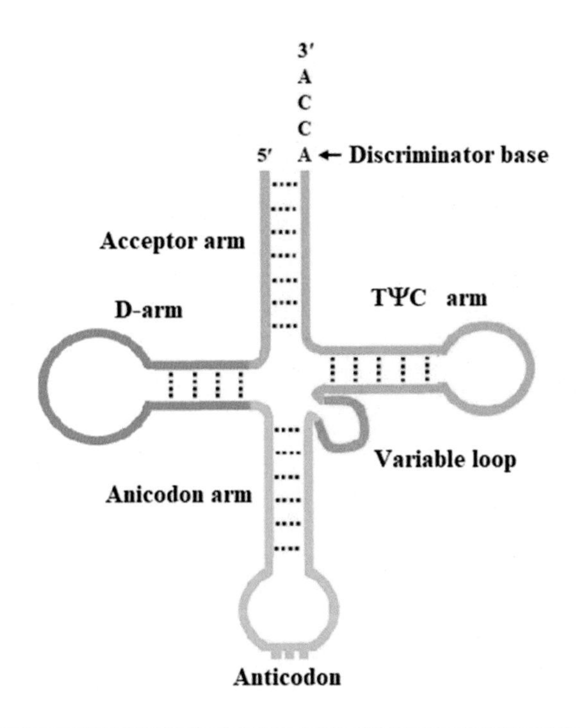

FIG. 11.3

Cloverleaf representation of tRNA.

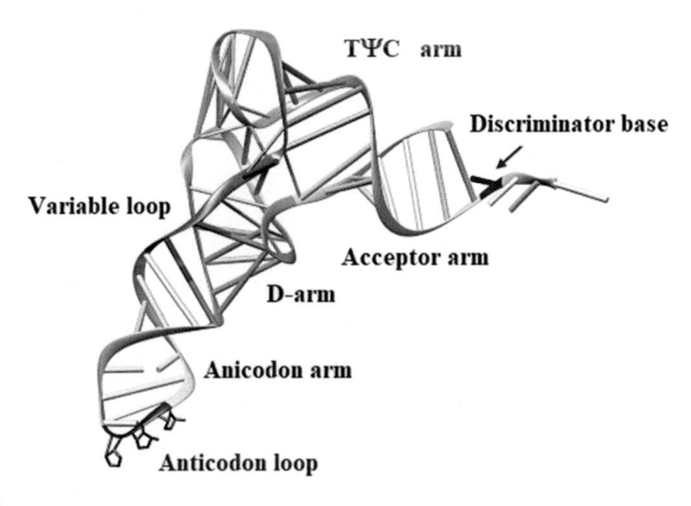

FIG. 11.4

Three-dimensional structure of tRNA.

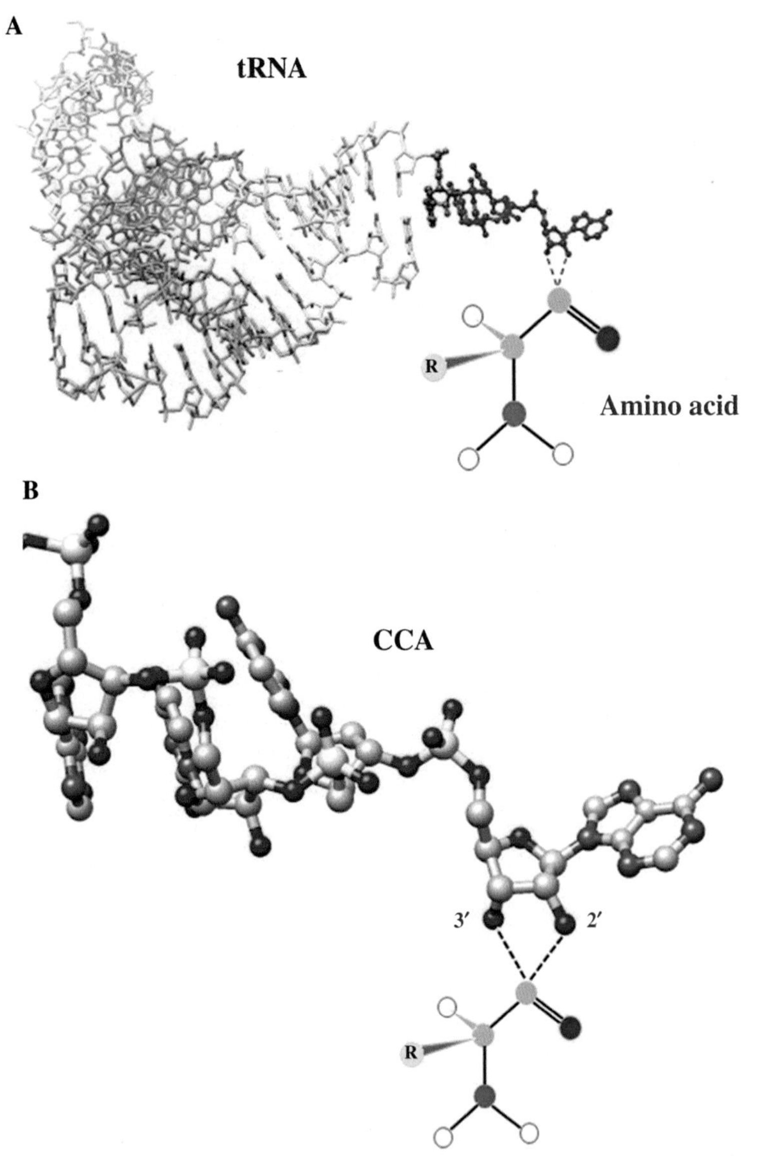

FIG. 11.5

Aminoacyl-tRNA linkage. (A) Entire tRNA molecule; (B) CCA-end.

We have seen earlier that 61 codons specify 20 amino acids. In other words, the genetic code is redundant. This implies that either some of the amino acids can be linked to more than one tRNA or some of the anticodons can base-pair with more than one codon. The bottom line is that, in any organism, the number of different tRNAs may be more that 20 but certainly far fewer than 61.

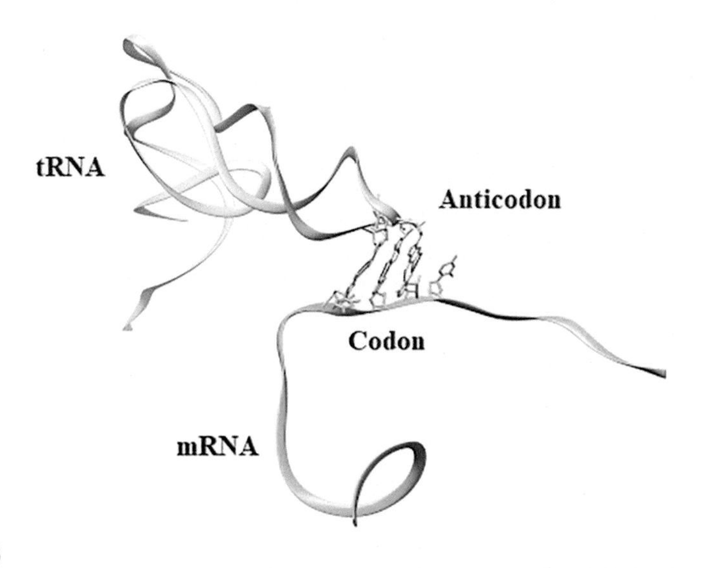

FIG. 11.6

Codon-anticodon interaction. H-bonds shown by *red* lines.

Source: Jenner, L.B., et al., 2010. Nat. Struct. Mol. Biol. 17, 556–560.

Francis Crick attempted to address the problem by proposing his wobble hypothesis. According to this hypothesis, the first two positions of the codon are uniquely identified by the tRNA by forming canonical Watson-Crick base-pairing. Ambiguity remains in the third position of the codon (first position of the anticodon) where non-canonical hydrogen bonding or "wobble" base-pairing, including the bonding of a purine with another purine, is possible. Thus, uridine (U) at the first position of the anticodon would base-pair with guanosine (G) at the third position of the codon (Fig. 11.8).

Posttranscriptional modifications of tRNA also play important roles in the decoding process. Several modifications are found in the tRNA anticodon loop, particularly at nucleotide positions 34 and 37. For example, insosine (I) at position 34 (the first position of the anticodon or the wobble position) can base-pair with U, C, and A. In general, modifications at the wobble position modulate codon recognition by restricting, extending, or altering the decoding capabilities of the tRNAs. It has been shown by X-ray crystallographic analysis that a 5-methylaminomethyl-2-thiouridine (S, mnm^5s^2U) modification at the anticodon position 34 of $tRNA^{Lys}_{UUU}$ forms an unusual base-pair with guanosine at the wobble position and can thereby recognize both AAA and AAG codons; yet, the modified tRNA discriminates against near-cognate codons AUA and UAA.

11.1.4 Aminoacyl-tRNA synthetase—Activation of tRNA

Thus, we have seen that each of the tRNA molecules that participate in the synthesis of a polypeptides chain is covalently linked to an amino acid and said to be charged or activated. The enzyme that catalyzes the activation process (aminoacylation) is

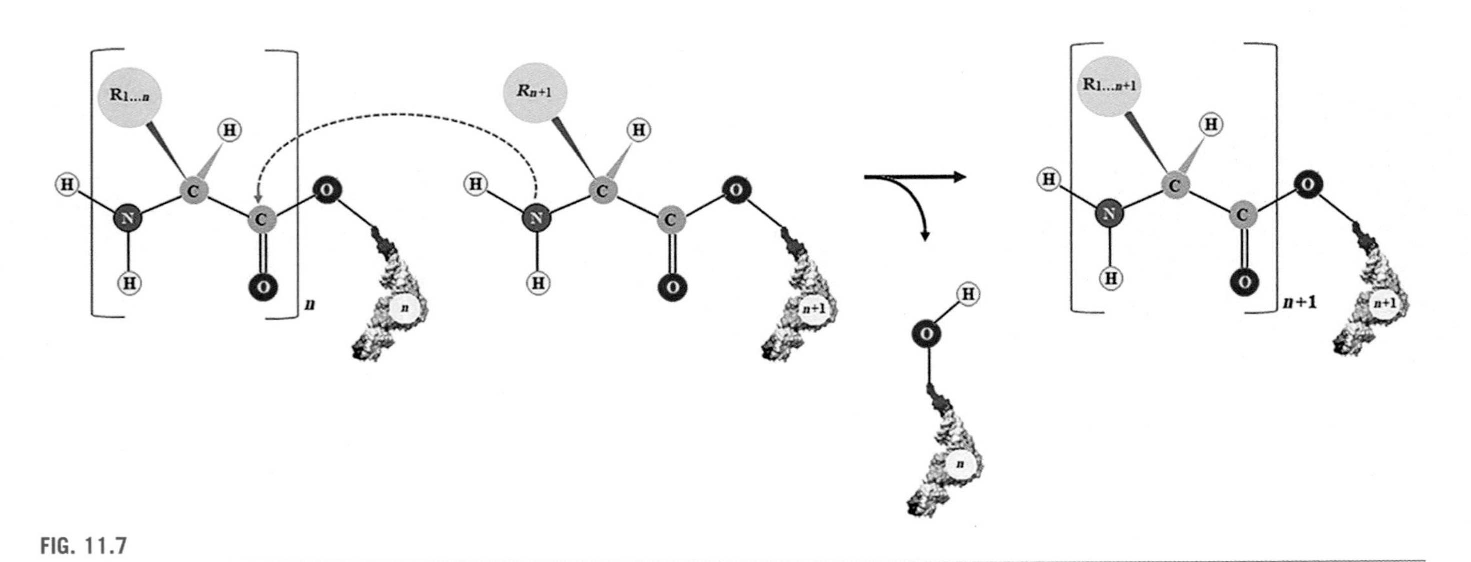

FIG. 11.7

Role of tRNA in amino acid incorporation to growing polypeptide chain.

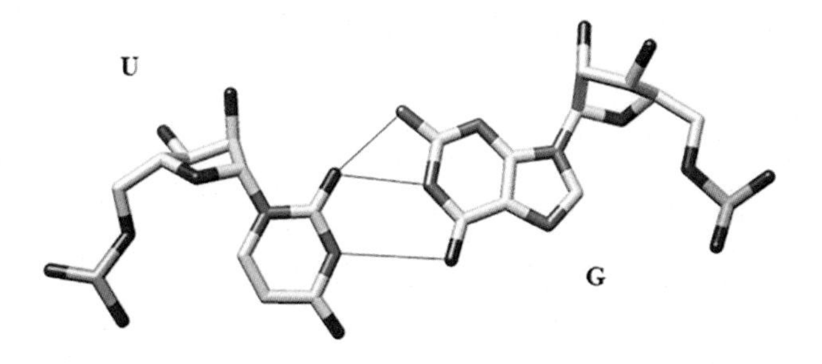

FIG. 11.8

U/G wobble base-pair. H-bonds shown by *red* lines.

aminoacyl-tRNA synthetase (aaRS). For each of the 20 amino acids, there is a single dedicated synthetase enzyme. So, there are 20 different aminoacyl-tRNA synthetases in the cell. Since most amino acids are specified by more than one codon, it is usual for a specific synthetase to recognize more than one tRNA. As for example, a specific synthetase attaches serine to all tRNAs (isoacceptors) that recognize the codons for serine.

Nevertheless, all aminoacyl-tRNA synthetases are united by a common aminoacylation reaction that occurs in two steps (Fig. 11.9). In step one, which is known as adenylylation, the synthetase activates the amino acid using ATP and Mg^{2+}, producing a high-energy intermediate, aminoacyl-adenylate (aa-AMP), and at the same time, releases pyrophosphate. Like polynucleotide synthesis (Chapter 8), the adenylylation reaction is also driven by the subsequent hydrolysis of pyrophosphate by pyrophosphatase. The aa-AMP remains bound to the enzyme. In step two, the aminoacyl moiety is transferred from the aa-AMP to the 3′-terminal adenosine of the tRNA and AMP is released. The amino acid is now linked to either 2′- or 3′-O of adenosine sugar.

The family of aminoacyl-tRNA synthetases has been divided into two unrelated classes (I and II) based on their active site architecture and consensus sequence elements. The class I active site is characterized by a Rossman dinucleotide-binding fold and two consensus sequence motifs, KMSKS (Lys-Met-Ser-Lys-Ser) and HIGH (His-Ile-Gly-His). The class I enzymes are monomeric. In contrast, the class II synthetases are usually dimeric or tetrameric and their active site consists of a novel antiparallel β-fold. Besides, the class II enzymes contain three consensus sequence motifs which are different from those in the class I.

The two classes also differ in their mode of binding the acceptor stem of tRNA and ATP. The class I synthetases bind ATP in an extended conformation, whereas the class II enzymes bind ATP in a bent conformation. Further, the class I aaRSs bind the tRNA acceptor stem on the minor groove side; in contrast, the class II enzymes bind the acceptor stem on the major groove side.

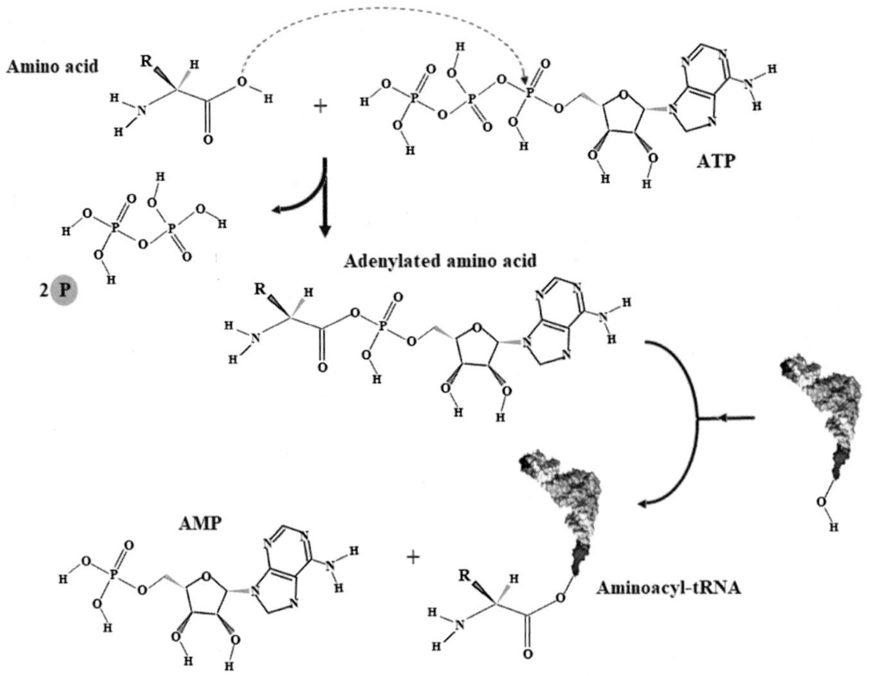

FIG. 11.9

Two-step aminoacylation.

Mechanism of aminoacylation

Here, we shall focus on a class I enzyme, namely, glutamyl-tRNA synthetase (GluRS), and examine the structural detail to understand the mechanism of the aminoacylation reaction. GluRS is one of the aaRSs which catalyze the first step of aminoacylation (amino acid activation) only in the presence of their cognate tRNAs. Nevertheless, GluRS does bind both ATP and glutamate in the absence of tRNA. Fig. 11.10 shows the crystal structure of a *Thermus thermophilus* GluRS·ATP·Glu complex.

GluRS contains at its catalytic site both the KMSKS and HIGH motifs (in GluRS, they are KISKR and HVGT, respectively). For binding, the ATP-Mg^{2+} substrate interacts with these motifs and some additional residues (Fig. 11.11). The adenine base is held in a hydrophobic pocket formed by H15, Y20, L235, L236, and I244. N1 and N6 of adenine form H-bonds with L236 and I244, whereas the phosphate group H-bonds with K243, S245, and R247. The Mg^{2+} ion coordinates the α-, β-, and γ-phosphate oxygen atoms of ATP and three water molecules to stabilize the ATP phosphate conformation. The ATP also interacts with E208 and W209.

The glutamate binds GluRS close to the ATP-binding site (Fig. 11.12). The α-ammonium group of glutamine H-bonds with the main chain of A7 and side chains

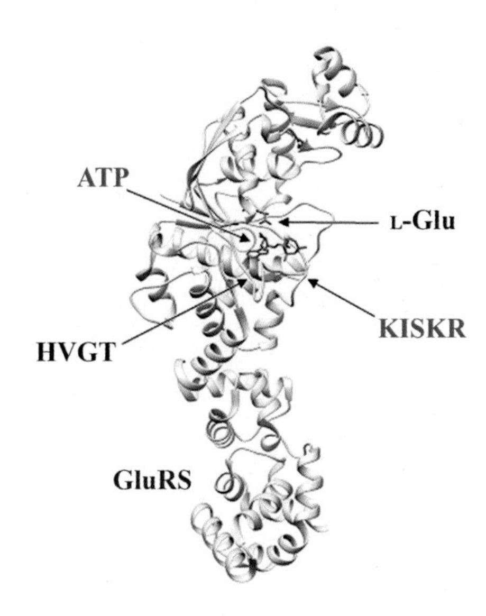

FIG. 11.10

Thermus thermophilus GluRS·ATP·Glu complex.

Source: Sekine, S., et al., 2003. EMBO J. 22, 676.

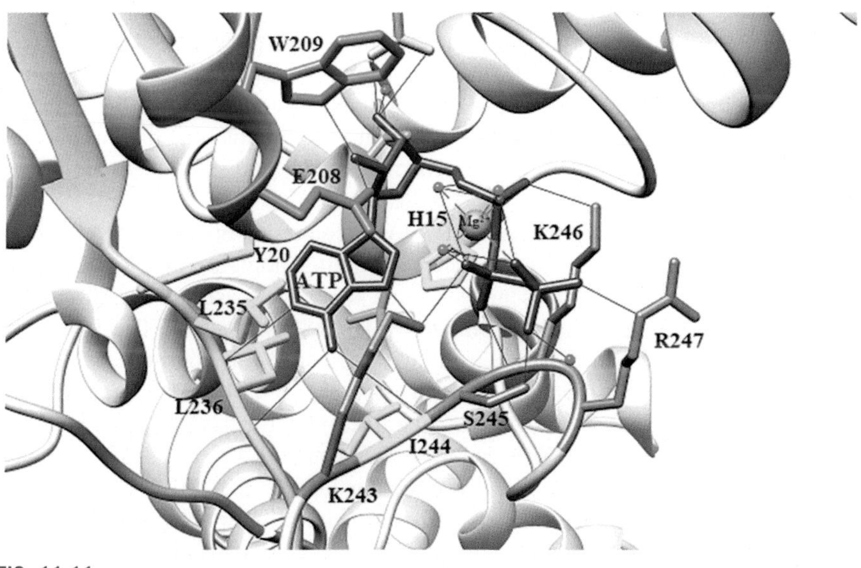

FIG. 11.11

GluRS residues interacting with ATP-Mg^{2+} substrate. H-bonds are shown by *blue* lines.

Source: Sekine, S., et al., 2003. EMBO J. 22, 676.

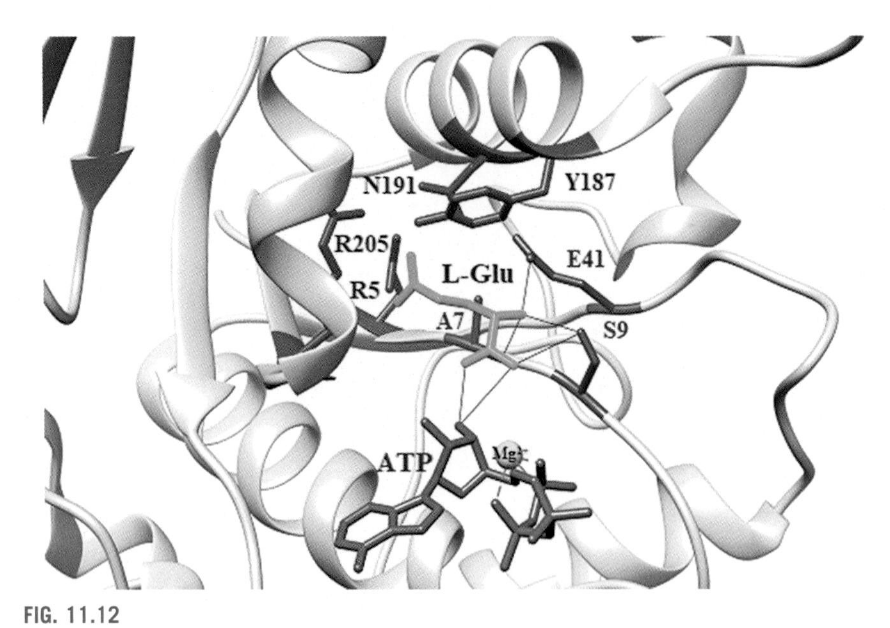

FIG. 11.12

GluRS residues interacting with glutamine. H-bonds shown by *red* lines.

Source: Sekine, S., et al., 2003. EMBO J. 22, 676.

of S9 and E41. Four residues, R5, Y187, N191, and R205, form a pocket complementary to the γ-carboxyl group of glutamate and determine the amino acid specificity. However, the α-phosphate of ATP and α-carboxyl of glutamate are too far apart to react. Therefore, despite both the ATP and glutamate binding reasonably tightly to their respective sites in GluRS, the arrangement is nonproductive. tRNA-binding is essential for accurate positioning of the two molecules and subsequent amino acid activation.

A major site of interaction in GluRS for accommodating the tRNA acceptor arm is the connective peptide or acceptor-binding domain (Fig. 11.13). Here, the $tRNA^{Glu}$ C74, which does not stack on the other bases in the 3′-terminal region, is trapped in a pocket formed by GluRS residues, E107, R147, and S181. H-bonds are formed between the C74 base and the main chains of E107 and S181 and also between the C74 phosphate and R147. The 3′-end is bent towards the active site of the enzyme. The discriminator base A73 moves into a stacking arrangement with C75 and A76 bases and W209 and Y187 aromatic rings. The close proximity of the discriminator base to the active site facilitates specific protein-RNA interactions. Additional specificity for tRNA-binding is provided by the anticodon-binding domain of some aaRSs which include GluRS. $tRNA^{Glu}$-binding brings about significant conformational changes in GluRS, particularly at the substrate-binding catalytic site (Fig. 11.14).

As $tRNA^{Glu}$ binds GluRS, the two molecules combine to form a cleft that is geometrically and electrostatically complementary to the glutamate substrate. While the

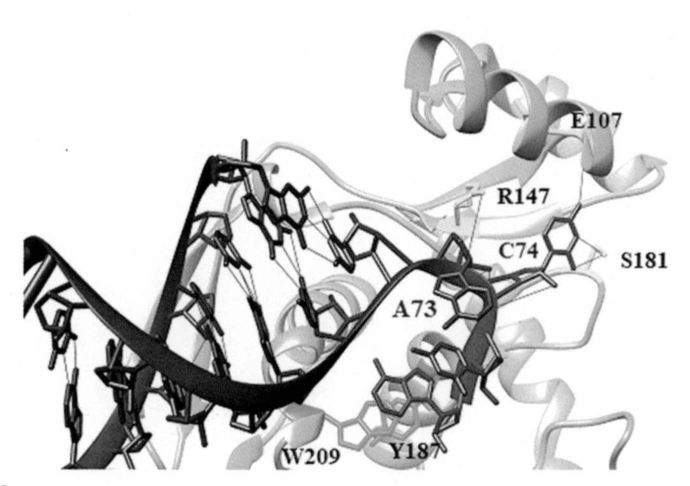

FIG. 11.13

GluRS connective peptide domain interaction with tRNA acceptor arm. GluRS—*light green*; tRNA—*brown*. H-bonds shown in *blue*.

Source: Sekine, S., et al., 2003. EMBO J. 22, 676.

FIG. 11.14

Conformational changes in GluRS upon tRNA-binding. Coloring scheme: (a) GluRS without tRNA—*gray*; GluRS with tRNA—*light blue*.

Source: Sekine, S., et al., 2003. EMBO J. 22, 676.

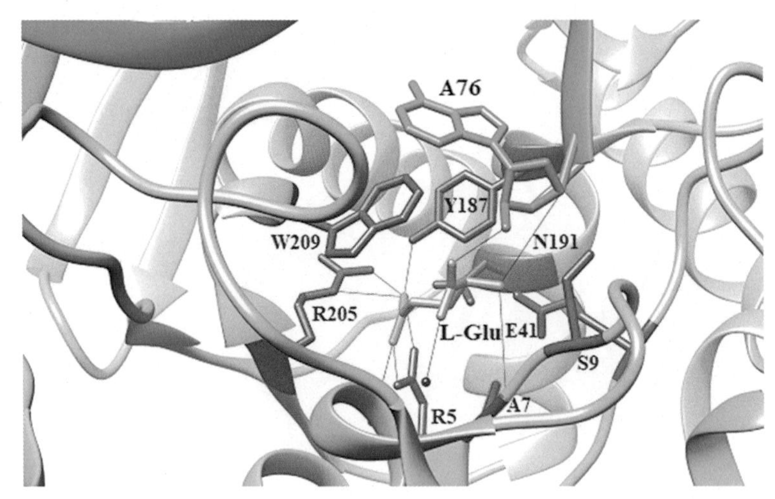

FIG. 11.15

Specific binding of glutamate to GluRS in presence of tRNA. Coloring scheme: GluRS—*light green*; tRNA—*dodger blue*. H-bonds shown by *red* lines.

Source: Sekine, S., et al., Structure 14, 1791.

glutamate-binding site is not very well-defined in the absence of the tRNA, the binding site becomes highly specific in its presence (Fig. 11.15). Direct contacts (that includes several H-bonds) with seven GluRS residues, R5, A7, E41, Y187, N191, R205, and W209, and A76 of $tRNA^{Glu}$ increase the glutamate-binding affinity of the enzyme. The glutamate is pulled to the bottom of the cleft leaving the opening unobstructed for ATP to bind and react with the amino acid.

In the GluRS·tRNA·ATP complex (Fig. 11.16), the adenine ring of the ATP is held in the same hydrophobic pocket as mentioned earlier. However, the ATP ribose moves deeper into the active site cleft. In presence of the tRNA, some of the earlier contacts, existing in its absence, are replaced by newer contacts. The 2′-hydroxyl of the ATP ribose interacts with A206, while the 3′-hydroxyl interacts with R5 through a water molecule. The I21 side chain makes van der Waals contacts with both the sugar and base of ATP. Interestingly, the 3′-terminal adenosine of $tRNA^{Glu}$ H-bonds with all the three phosphates of ATP. Thus, in the presence of $tRNA^{Glu}$, the ATP molecule undergoes a significant change in its orientation that brings it closer to the glutamate (Fig. 11.17).

Editing by aminoacyl-tRNA synthetase

High accuracy of the aminoacylation reaction is essential to the fidelity of amino acid incorporation. The aaRS selects the correct amino acid by a "double sieve mechanism." The first sieve (pretransfer editing), which discriminates between cognate and noncognate amino acids, is based on the affinity of an amino acid for the active site. Nevertheless, it is difficult to distinguish between two closely related amino

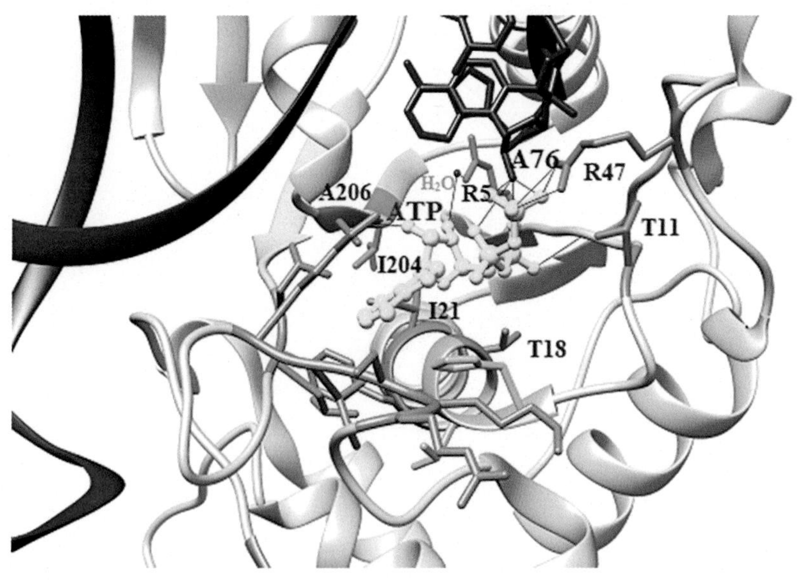

FIG. 11.16

GluRS·tRNA·ATP complex. Coloring scheme: tRNA—*firebrick*; GluRS—*light blue*; ATP—*yellow*. H-bonds are shown in *red*.

Source: Sekine, S., et al., 2003. EMBO J. 22, 676.

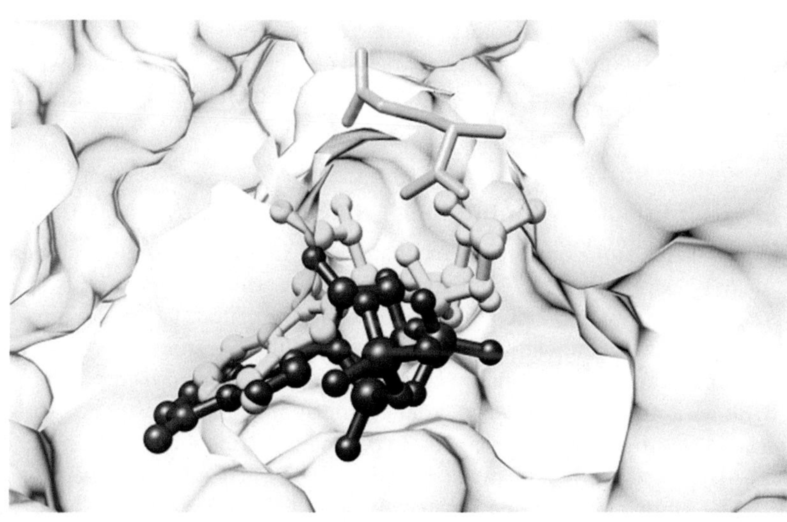

FIG. 11.17

Orientations of ATP in GluRS in absence *(firebrick)* and presence *(yellow)* of tRNA. Glutamine shown in *light green*.

Source: Sekine, S., et al., 2003. EMBO J. 22, 676; Sekine, S., et al., 2006. Structure 14, 1791.

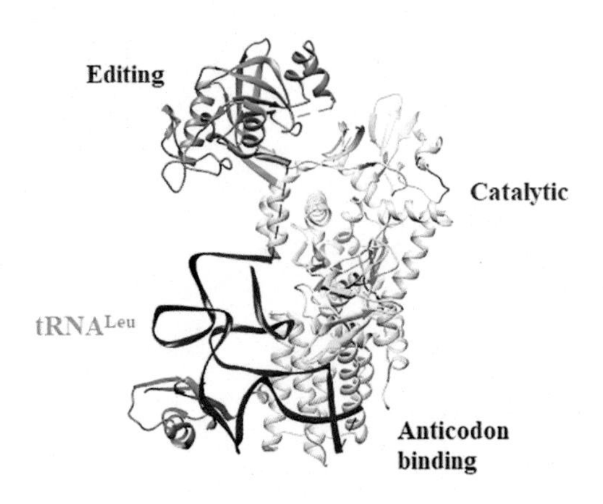

FIG. 11.18

E. coli LeuRS-tRNA^{Leu} in editing state.

Source: Palencia, A., et al., 2012. Nat. Struct. Mol. Biol. 19, 677.

acids, such as those that are isosteric (valine vs threonine) or differing by only a methyl group (isoleucine vs valine). Some of the aaRSs, therefore, depend on a post-transfer editing mechanism with the help of a second, "finer" sieve.

The crystal structure of a functional aminoacylation complex of *Escherichia coli* LeuRS has provided a mechanistic basis for posttransfer editing (Fig. 11.18). When the tRNA is charged (or mischarged) in the synthetic or catalytic site of LeuRS, it is translocated to the editing site, located at a distance of ~35 Å in an independently folded domain. This sieve excludes the correctly charged cognate amino acid, most likely based on size or hydrophilicity, while giving access to the noncognate amino acid. The latter is then hydrolyzed and released from the enzyme.

11.1.5 The ribosome—Multicompartmental reaction vessel

Once the tRNA is aminoacylated, it can participate in the amino acid incorporation reaction as shown in Fig. 11.7. Initiation of a polypeptide chain synthesis and successive addition of amino acids, by peptide bond formation, to the growing chain occur in a multicompartment reaction vessel called the ribosome. It is a huge ribonucleoprotein complex that enhances the rate of the amino acid incorporation reaction by at least 10^5-fold. The basic architecture of an active ribosome is shown in Fig. 11.19. All ribosomes consist of two subunits (ribonucleoprotein subcomplexes)—large and small. The intact ribosome, individual subunits, and their macromolecular components are designated by their "S" or Svedberg values, which refer to their rate of sedimentation when subjected to a centrifugal force.

The small subunit of bacterial ribosome, known as the 30S subunit, is composed of a 16S ribosomal RNA (rRNA) and 21 ribosomal proteins (r-proteins). The large (50S) subunit contains 5S and 23S rRNAs and 33 r-proteins. The 16S and 23S rRNAs

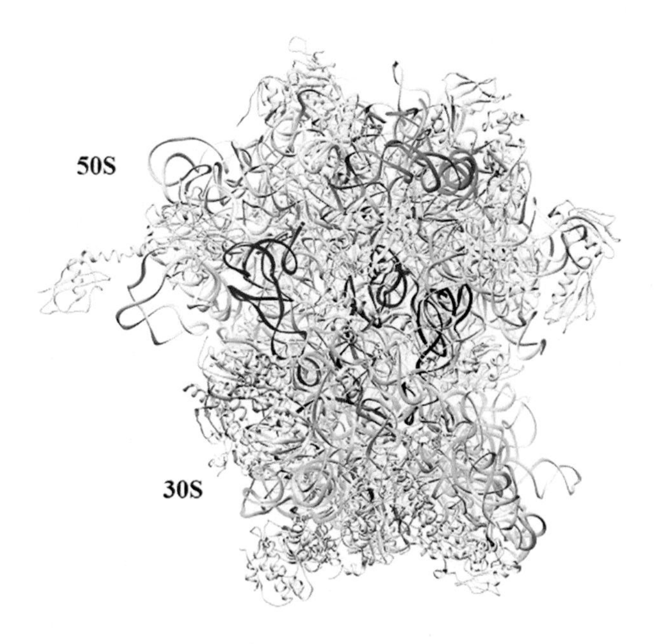

FIG. 11.19

Structure of *Thermus thermophilus* 70S ribosome. Coloring scheme: 23S rRNA *(light blue)*; 16S rRNA *(light green)*; 5S rRNA *(dodger blue)*; mRNA *(red)*; A-site tRNA *(brown)*; P-site tRNA *(dark magenta)*; E-site tRNA *(dark blue)*; 50S ribosomal proteins *(yellow)*; 30S ribosomal proteins *(pink)*.

Source: Voorhees, R.M., et al., 2009. Nat. Struct. Mol. Biol. 16, 528.

have been partitioned into domains based on secondary structures and molecular interaction between secondary structure elements. The 16S rRNA has been divided into four domains, while the 23S rRNA into seven domains (including the 5S rRNA as domain VII). Together, the two subunits form a 2.5-megadalton 70S ribosome.

Eukaryotic and prokaryotic ribosomes are similar in their core (Fig. 11.20). Nevertheless, the 3.3-megadalton eukaryotic (yeast) 80S ribosome is 30% larger than its bacterial counterpart. The large 60S subunit of the yeast ribosome contains three rRNAs (25S, 5.8S, and 5S) and 46 r-proteins, while the small 40S subunit consists of one rRNA (18S) and 33 r-proteins. This is no surprise since the eukaryotic ribosome is subjected to more complex regulation in its assembly and the initiation and termination processes.

The two subunits of the ribosome are not permanently bound to each other; they come together only when a polypeptide chain is synthesized. The association of the large and small subunits, as we shall see, is linked to the initiation of protein synthesis and remains as the synthesis continues (elongation). Almost immediately after the end or termination of synthesis, the two subunits dissociate to repeat the process. This is known as the ribosome cycle. The association and dissociation are regulated by different protein factors.

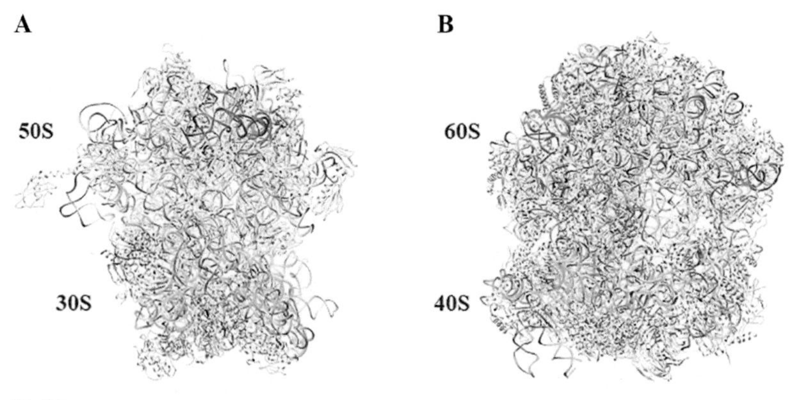

FIG. 11.20

Core structures of prokaryotic and eukaryotic ribosomes. (A) *T. thermophilus* 70S ribosome: 23S rRNA *(light blue)*, 16S rRNA *(light green)*, 5S rRNA *(dodger blue)*, 50S ribosomal proteins *(yellow)*, 30S ribosomal proteins *(pink)*. (B) *Saccharomyces* 80S ribosome: 25S rRNA *(light blue)*, 18S rRNA *(light green)*, 5S rRNA *(dodger blue)*, 5.8S rRNA *(cyan)*, 60S ribosomal proteins *(yellow)*, 40S ribosomal proteins *(pink)*.

Source: Voorhees, R.M., et al., 2009. Nat. Struct. Mol. Biol. 16, 528; Ben-Shem, A., et al., 2011. Science 334, 1524.

The ribosome catalyzes the peptide-bond formation, as shown in Fig. 11.7, by bringing together the 3′ ends of the two tRNAs—aminoacyl-tRNA and peptidyl-tRNA—into proximity. As the reaction involves a transfer of the growing polypeptide chain from the peptidyl-tRNA to the aminoacyl-tRNA, it is called the peptidyl transferase reaction and takes place at a site or "center" in the ribosome known as the peptidyl transferase center (PTC). The PTC is located in the large subunit (50S in bacteria).

As required, the two tRNAs have their respective binding sites in the ribosome—the A site and the P site. The 3′ ends of the A- and P-site tRNAs are in proximity at the PTC. Additionally, there is an E (or exit) site to which the deacylated P-site tRNA moves before it is ejected from the ribosome. It is to be noted that the binding sites are formed at the interface between the two subunits so that each of the bound tRNAs can span the distance between the PTC in the large subunit and the decoding center in the small subunit.

Two other structural features of the ribosome are indispensable for well-organized and regulated polypeptide synthesis. They are related to the movement of the mRNA and nascent polypeptide chain. The genetic information carried by the mRNA is presented to the ribosome at the decoding center, whereas the nascent polypeptide chain needs to escape from the PTC as synthesis continues. The two centers of action are incidentally buried within the intact ribosome. Nevertheless, two "tunnels" in the ribosome have addressed the problem.

The RNA tunnel is present in the small subunit. Through this tunnel, the mRNA enters and exits the decoding center in a single-stranded form. Besides its interaction with tRNAs, the mRNA is also involved in other interactions in the tunnel which ensures high fidelity reading of the encoded message (Fig. 11.21).

The other ribosomal tunnel is in the large subunit (Fig. 11.22). It facilitates the exit of the newly synthesized polypeptide chain from the ribosome. There is growing evidence that the nascent chain interacts with the exit tunnel to regulate translation and influence initial folding events.

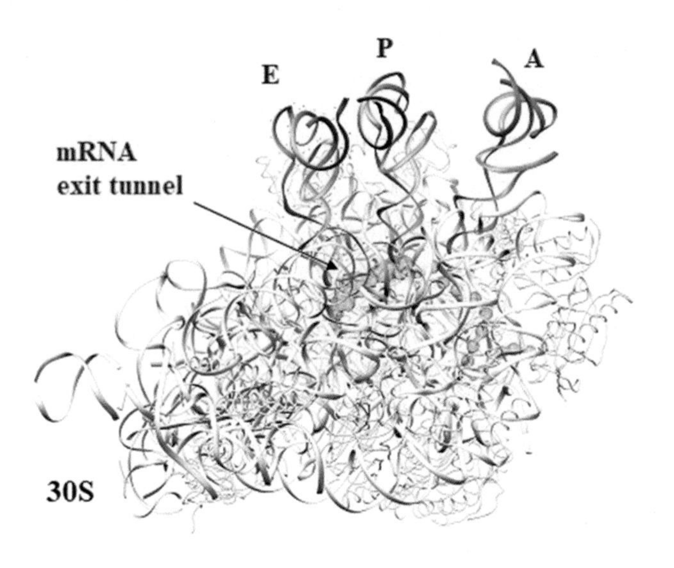

FIG. 11.21

Path of mRNA through 30S subunit. A-, P-, and E-site tRNAs are marked.

Source: Yusupova, G.Z., et al., 2001. Cell 106, 233.

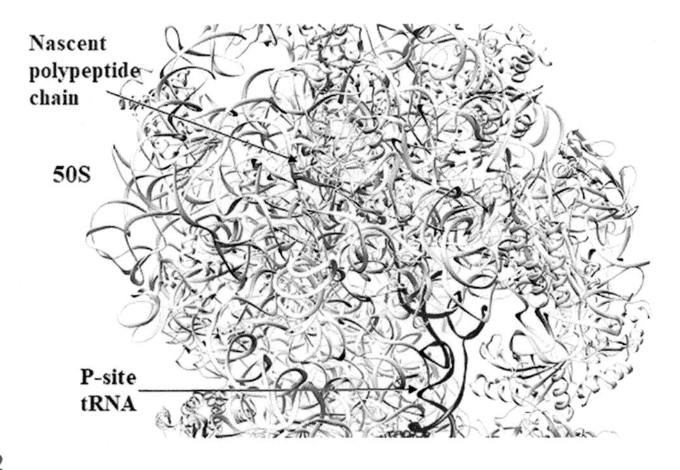

FIG. 11.22

Exit tunnel for nascent polypeptide chain in 50S ribosomal subunit.

Source: Su, T., et al., 2017. eLife 6, e25642. doi: 10.7554/eLife.25642.

11.2 Initiation of translation

11.2.1 Prokaryotic

Initiation of translation requires the correct positioning of an initiator tRNA, which in bacteria is an aminoacylated and formylated initiator tRNA (fMet-tRNA^{fMet}), in the ribosome. This occurs in sequential steps and necessitates the involvement of translation initiation factors (Fig. 11.23). Nevertheless, the initiation pathway in bacteria is not always linear as the order of binding of the initiation factors has been found to be flexible.

In bacteria, there are three initiation factors IF1-3. These factors bind to the 30S subunit only when the latter is split from the 50S at the end of the previous ribosomal cycle. We shall see later in this chapter that the dissociation of the 30S from the 50S is executed by ribosome recycling factor RRF and elongation factor G (EF-G) after translation termination. The binding of IF3 ejects the leftover mRNA and deacylated tRNA from the previous round of synthesis.

IF3 binds the 30S ribosomal subunit near the P-site. Its CTD is positioned next to IF1 and a linker extends the NTD away from 16S rRNA (Fig. 11.24). The residues involved in binding 16S rRNA are mostly hydrophilic and basic. However, IF3 can adopt multiple conformation during initiation. Its NTD changes position near

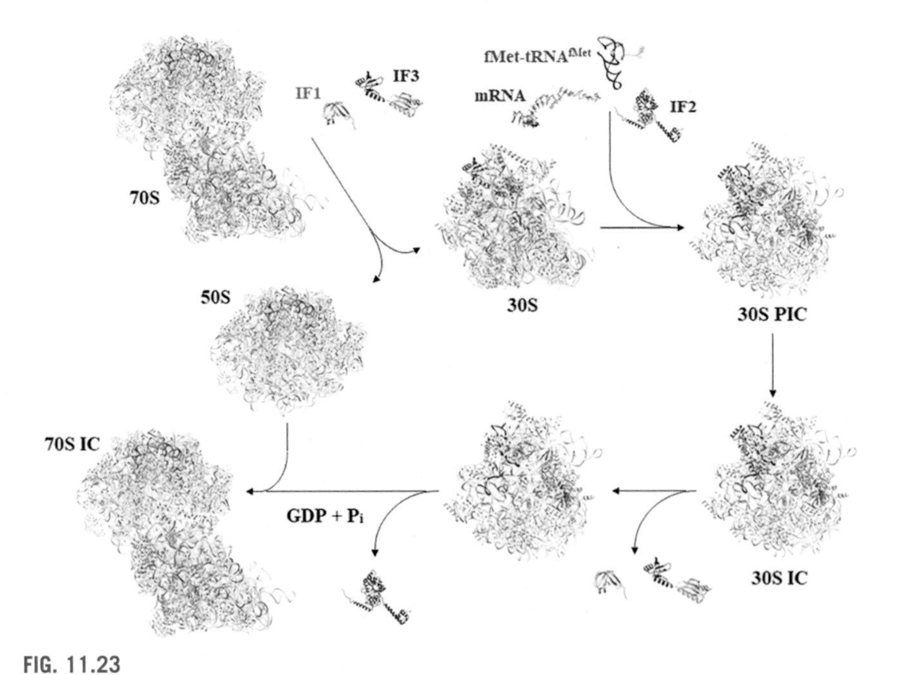

FIG. 11.23

Bacterial translation initiation pathway. IFs—initiation factors; PIC—preinitiation complex; IC—initiation complex.

Source: Hussain, T., et al., 2016. Cell 167, 133; Zhang, X., et al., 2014. Nucleic Acids Res. 42, 13430.

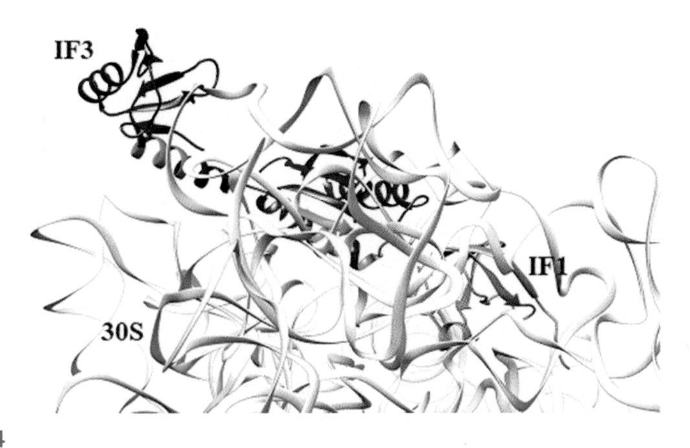

FIG. 11.24

Location of IF3 in 30S ribosomal subunit. Hydrophilic and basic residues of IF3 involved in binding 16S rRNA are shown in *dodger blue*.

Source: Hussain, T., et al., 2016. Cell 167, 133.

the P-site to accommodate the initiator tRNA. It can discriminate against elongator tRNAs by influencing codon:anticodon interaction.

IF1 locates itself at the base of the A-site in the small ribosomal subunit. This way, it blocks the A-site and directs the initiator tRNA to the P-site. IF1 also stimulates the activity of IF3 and promotes the dissociation of the ribosomal subunits. In the 30S subunit, IF1 interacts with ribosomal protein S12 and 16S RNA (Fig. 11.25). Bases A1492 and A1493 of 16S rRNA flip out of the helix and the noncanonical base-pairing between A1413 and G1487 is disrupted. A conformational change of the subunit takes place.

As the subunits dissociate, IF2, mRNA, and initiator tRNA (fMet-tRNA^{fMet}) join the 30S, possibly in random order. The association of the mRNA with the small

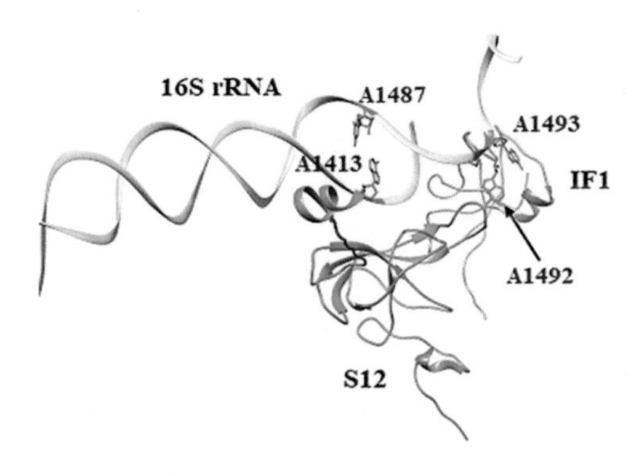

FIG. 11.25

Interactions of IF1 in the 30S ribosomal subunit.

Source: Carter, et al., 2001. Science 291, 498.

ribosomal subunit is facilitated by the interaction of its Shine-Dalgarno (SD) sequence with the anti-SD sequence of the 16S rRNA. In this preinitiation complex (30S-PIC), fMet-$tRNA^{fMet}$ is bound to the 30S in a codon-independent manner (Fig. 11.26). Aided by the initiation factors, the initiation codon is positioned in the P-site of the ribosome.

The initiator tRNA, fMet-$tRNA^{fMet}$, is produced by aminoacylation of $tRNA^{fMet}$ with methionine followed by the addition of a formyl group to the α-NH_2 group by a methionyl-tRNA transformylase (MTF). To be recognized by MTF, $tRNA^{fMet}$ contains specific structural features present in the acceptor end and anticodon stem loop distinct from those in the elongator $tRNA^{Met}$. The formyl group prevents fMet-$tRNA^{fMet}$ from interacting with elongation factor EF-Tu and rather facilitates its recognition and binding by IF2.

IF2 has been divided into several structural domains and evidence indicates that its function depends upon allosteric communications between the domains. It is a GTPase; it can bind a GTP, GDP, or ppGpp. The bound ligand influences the conformation of the protein. IF2-GTP has several-fold higher binding affinity for the 30S compared to apo IF2, IF2-ppGpp, and IF2-GDP.

The interaction between IF2 and the initiator tRNA is strongest in presence of the 30S ribosomal subunit and not affected by GTP or GDP. The interaction enables IF2

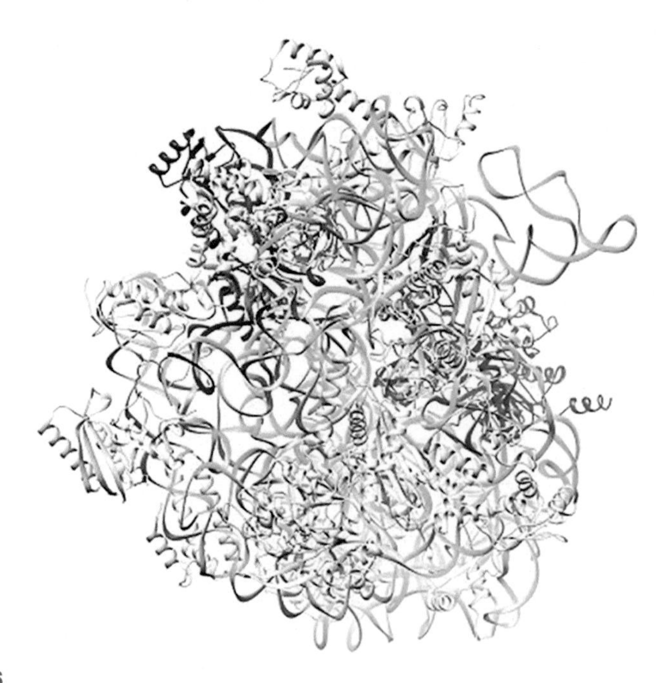

FIG. 11.26

Bacterial 30S-PIC. Coloring scheme: 16S rRNA—*light green*; mRNA—*orange*; IF1—*coral*; IF2—*blue*; IF3—*brick red*; fMet-$tRNA^{fMet}$—*dark magenta*.

Source: Hussain, T., et al., 2016. Cell 167, 133.

to correctly position the initiator tRNA in the 30S subunit and lead to the formation of a functional 30S initiation complex (30S IC) (Fig. 11.23). IF1 and IF3 are ejected probably at this stage and the 50S subunit joins the functional 30S IC to form the 70S initiation complex (70S IC) through inter-subunit salt bridges and interaction between IF2 and ribosomal protein L12. As the two subunits associate with each other, IF2 makes contacts with the GTPase activation center, GTP is hydrolyzed, and IF2 leaves the ribosome. The 70S complex now containing an fMet-tRNA^{fMet} at the P-site is ready to accept a charged tRNA into the vacant A-site and form the first peptide bond (Fig. 11.23).

11.2.2 Eukaryotic

Initiation of translation in eukaryotes is similar to that in prokaryotes in many aspects. Like bacteria, eukaryotic cells, too, assemble the mRNA and initiator tRNA in the small ribosomal unit (40S in case of eukaryotes) with the help of auxiliary factors, place the initiator tRNA at the start codon, and recruit the large subunit (60S in case of eukaryotes) to form a functional ribosome. Yet, the initiation mechanisms in eukaryotes are associated with a greater number and complexity of the participating factors. As for example, in bacteria, there are three translation initiation factors, IF1, IF2, and IF3; in yeast, 11 initiation factors consisting of 24 independent gene products are involved in the translation initiation process. Further, as we shall see here, there are significant differences in some of the individual steps of the initiation pathway.

Fig. 11.27 shows the pathway for yeast translation initiation. We can compare it with Fig. 11.23 to understand the basic similarities and differences between the prokaryotic and eukaryotic translation initiation mechanisms. Like bacteria, eukaryotic protein synthesis begins with the dissociation of the ribosomal subunits and assembly of the preinitiation complex (PIC). However, significant differences are manifested between the two systems in the mechanism of formation of the PIC.

In eukaryotic cells, association of the initiator methionyl-transfer RNA (Met-tRNA^{Met}) with the small ribosomal subunit always precedes that of the mRNA. So, at first, the initiator tRNA associates with GTP-bound eukaryotic initiation factor 2 (eIF2) to form a ternary complex (TC). Eukaryotic initiation factors (eIFs) 1, 1A, and 3 bind to the 40S subunit, followed by the TC and eIF5 to form a 43S PIC (Fig. 11.27). The eIF4 family of initiation factors then prepares the mRNA for binding the 43S PIC to form a 48S PIC.

Presentation of the mRNA to the small ribosomal subunit in eukaryotic cells is distinct from that in bacteria. We have seen in the previous chapter how in eukaryotic cells RNA polymerase II-generated transcripts are suitably processed to form translatable mRNA. Specifically, the modifications at the two ends, 5′ cap and 3′ poly(A) tail, are important for translation initiation. The eukaryotic initiation factor eIF4F complex, which comprises cap-binding protein eIF4E, a scaffolding protein eIF4G, and an ATP-dependent RNA helicase eIF4A, binds the 5′ cap of the mRNA, while poly(A)-binding protein (PABP) binds the 3′ poly(A) tail. Together, these

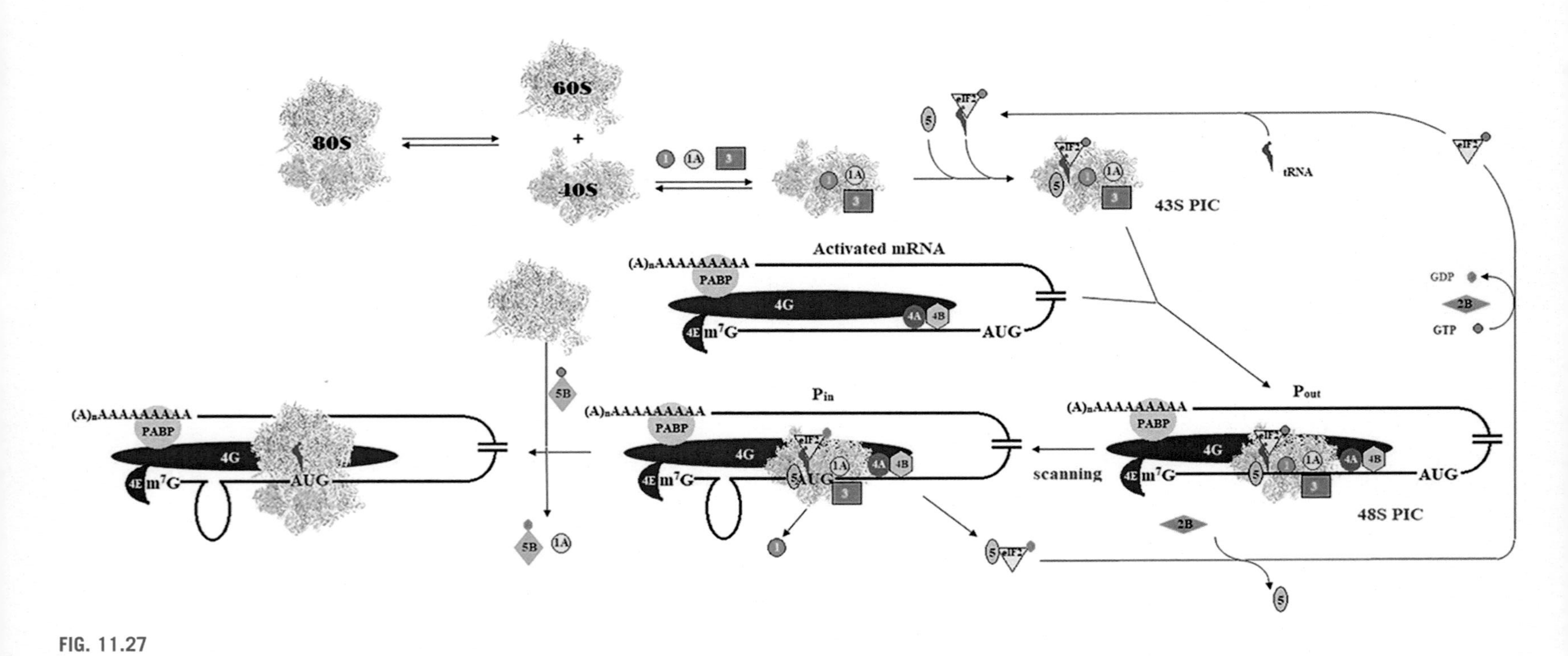

FIG. 11.27

Eukaryotic translation initiation pathway. The numbers represent the eIFs.

Source: Ben-Shem, A., et al., 2011. Science 334, 1524.

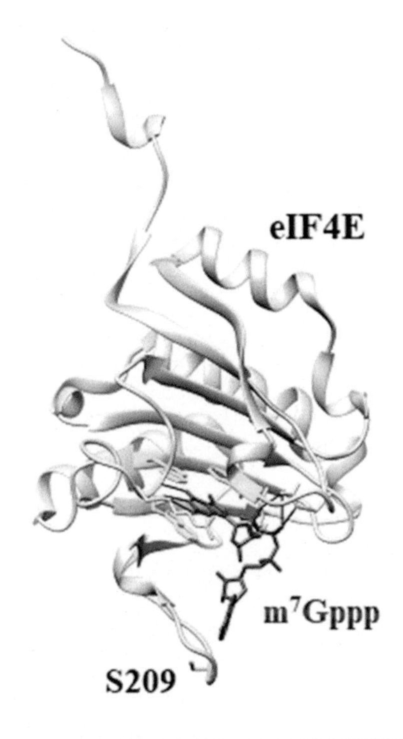

FIG. 11.28

eIF4E complexed with m^7GpppA. Hydrophobic residues are shown in *yellow*; hydrophilic residues are shown in *light blue*.

Source: Tomoo, K., et al., 2002. Biochem J. 362, 539.

interactions form a stable "closed loop" mRNP complex, which associates with the 43S PIC to form a 48S PIC (Fig. 11.27).

eIF4E recognizes the cap structure (7-methylguanosine triphosphate) of mRNA. The crystal structure of eIF4E complexed with a cap analog m^7GpppA (Fig. 11.28) shows that the interactions are mediated by hydrophobic residues, W56, W102, and W166, and hydrophilic residues, E103, R112, R157, and K 162, of the protein. The m^7G moiety of the cap is sandwiched between the aromatic residues W56 and W102. The stacking interactions are stabilized by several H-bonds. The flexible C-terminal loop (residues 203–212) may also have a role in cap-binding as phosphorylation of S209 in the loop alters the electrostatic environment in the binding site.

Poly(A)-binding protein (PABP) interacts with 3′ poly(A) tail of mRNA through its RNA recognition motifs (RRMs). We have seen in Chapter 10 that an RRM motif folds in a typical $\beta_1\alpha_1\beta_2\beta_3\alpha_2\beta_4$ conformation. PABP contains four RRMs. The crystal structure of a PABP fragment containing RRM1-2 complexed with polyadenylate RNA (Fig. 11.29) shows that RRMs form a long, narrow RNA-binding trough with the antiparallel β sheets acting as a base of the trough. The poly(A) RNA adopts an extended conformation to place itself into the trough. Adenine bases are recognized

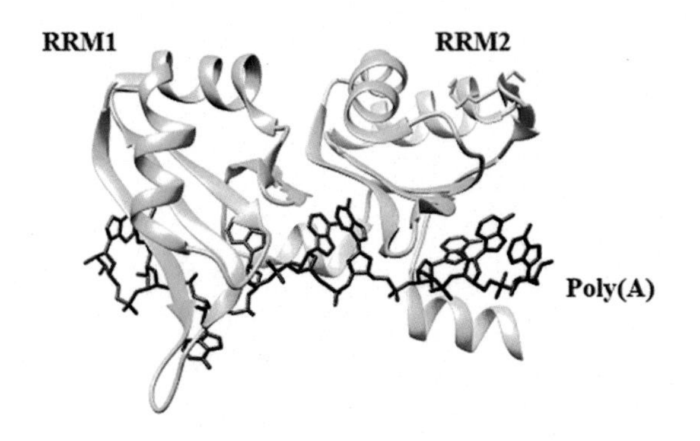

FIG. 11.29

Poly(A) RNA complexed with PABP RRM1-2.

Source: Deo, R.C., et al., 1999. Cell 98, 835.

by conserved residues in the RRMs through van der Waals contacts, H-bonds, and stacking interactions.

eIF4G is a multifactor-binding protein, as it is expected to be, that brings together the complexes at the 5′- and 3′-ends of the mRNA. It has separate binding sites for eIF4E as well as PABP. eIF4G interacts with eIF4E using its conserved motif, known as canonical eIF4E-binding motif (4E-BM), with the consensus sequence Yx(4)Lϕ (where x is any amino acid and ϕ is a hydrophobic residue). This motif binds to a conserved hydrophobic patch on the dorsal surface of eIF4E opposite the cap-binding pocket and, upon binding, adopts an α-helical conformation (Fig. 11.30).

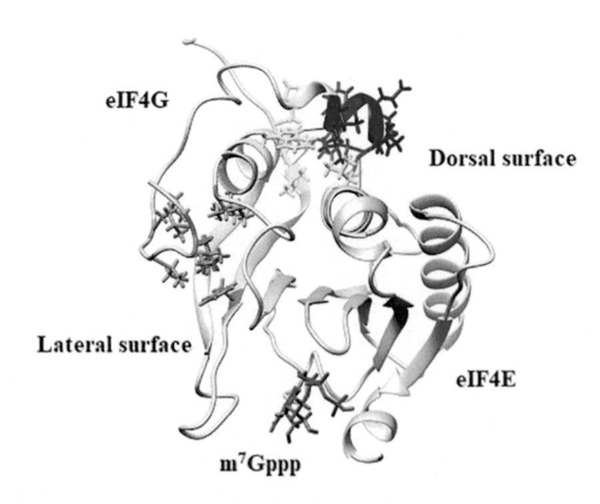

FIG. 11.30

eIF4G bound to eIF4E. eIF4G residues (shown in *dark blue*) interacting with conserved residues *(yellow)* on eIF4E dorsal surface. Hydrophobic residues of eIF4G *(light blue)* interacting with hydrophobic residues on eIF4E lateral surface *(pink)*.

Source: Gruner, S., et al., 2016. Mol. Cell 64, 467.

The hydrophobic residues of the canonical 4E-BM (human L617 and L618) interact with conserved residues, V69, W73, and L135, located on the dorsal surface of eIF4E, while the tyrosine Y612 forms an H-bond with a conserved H37-P38-L39 motif of eIF4E. In addition, hydrophobic residues (L633, L636, V639, and V640) in a noncanonical loop of human eIF4G interact with hydrophobic residues, F47, I63, L75, and I79, present on the lateral surface of eIF4E.

At the same time, another domain of eIF4G can interact with the RRM2 domain of PABP (Fig. 11.31A). The interactions between the two are both hydrophobic and polar in nature. eIF4G residue I180 penetrates the hydrophobic core of PABP. Further, hydrophobic contacts are formed between I182 and I192 of eIF4G and M158, M161, and L163 of PABP (Fig. 11.31B). Polar contacts (backbone H-bonds) form between eIF4G residues 181–183 on one side and RRM2 residues 162–164 on the other (Fig. 11.31C). The side chains of eIF4G R181 and R183 also interact with PABP N164 and D165 by making salt bridges.

Attachment of the 43S PIC to the 5′-end of the mRNA forms the 48S preinitiation complex (48S PIC). Initially, the 48S PIC is in an "open" conformation stabilized by eukaryotic initiation factors, eIF1 and eIF1A. In this conformation, the TC is in a metastable state, P_{out}, where Met-tRNA^{Met} is not fully base-paired within the P site. The conformation promotes scanning in which the initiator tRNA samples successive triplets entering the P site for a complementary codon. Probably before AUG

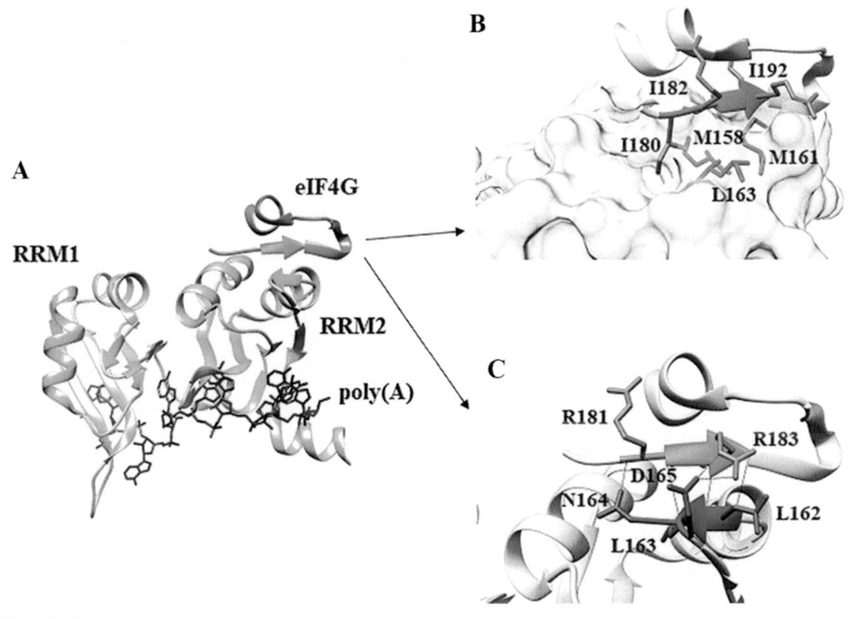

FIG. 11.31

eIF4G interacting with PABP. (A) eIF4G bound to PABP RRM2. (B) Hydrophobic contacts. (C) Polar contacts. H-bond shown by *red* lines.

Source: Safaee, N., et al., 2012. Mol. Cell 48, 375.

recognition, eIF5 stimulates hydrolysis of TC-bound GTP, but the release of P_i is prevented by the presence of eIF1 in the complex.

Start codon recognition by the TC induces a rearrangement of the factors in the PIC, which shifts to a scanning-arrested "closed" conformation. eIF1 is displaced from its location near the P site and eventually ejected from the complex. P_i is released from eIF2·GDP followed by the departure of eIF2·GDP and eIF5 from the 48S PIC. The Met-tRNA^{Met} is now tightly bound to the PIC having its anticodon base-paired with the codon in the P site. This is the P_{in} state of the initiator tRNA. Finally, eIF5B·GTP binds the 48S complex and facilitates 60S subunit joining—an elongation-competent 80S initiation complex is formed.

11.3 Translation elongation

Each elongation cycle is in fact the pathway as depicted In Fig. 11.7. In bacteria, at the beginning of an elongation cycle, the 70S ribosome contains in the P-site a peptidyl t-RNA attached to a nascent polypeptide chain of n amino acid residues ($n=1$, 2, …) and an empty A-site. The cycle consists of three sequential steps. In the first step (decoding), the next aminoacyl-tRNA is carried in a ternary complex (TC) of aminoacyl-tRNA·GTP·EF-Tu and delivered to the A-site. In the following step (peptide-bond formation), the polypeptide chain is extended by one amino acid. In the third step (translocation), catalyzed by EF-G, the tRNAs and mRNA move with respect to the ribosome.

Thus, we see that two proteins, known as elongation factors, EF-Tu and EF-G, are involved in the elongation process. EF-Tu is a GTPase; it has a nucleotide-binding site which can bind either GTP or GDP. Accordingly, it can switch between two conformational states—active (GTP) and inactive (GDP). The active state has a much higher binding affinity for aminoacyl-tRNA than the inactive state.

Initially, the TC binds very rapidly to the ribosome (Fig. 11.32). This binding is mRNA-independent and aided by ribosomal protein L7/L12. The aminoacyl-tRNA that the TC carries to the ribosome is now bound in the A-site in what is known as the A/T state. The A/T conformation, which allows simultaneous interactions with the mRNA codon and EF-Tu, requires the aminoacyl-tRNA to be distorted in two regions—the anticodon stem and the D-stem. As a result, the tRNA is in a position to sample the codon-anticodon interaction.

11.3.1 Decoding

During decoding, the ribosome selects the correct tRNA as dictated by the mRNA codon. However, level of accuracy in tRNA selection cannot be explained just based on free energy difference between a correctly paired codon-anticodon and a mismatched one. An additional and effective mechanism of discrimination between the two, which ensures a high fidelity of codon recognition, involves three conserved bases of the 16S rRNA—G530, A1492, and A1493 (Fig. 11.33). When tRNA is not

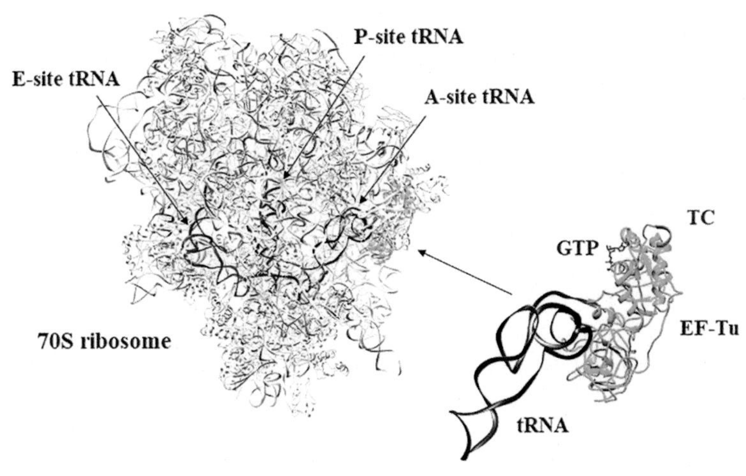

FIG. 11.32

Binding of aminoacyl-tRNA·GTP·EF-Tu TC to 70S ribosome.

Source: Schmeing, T.M., et al., 2009. Science 326, 688.

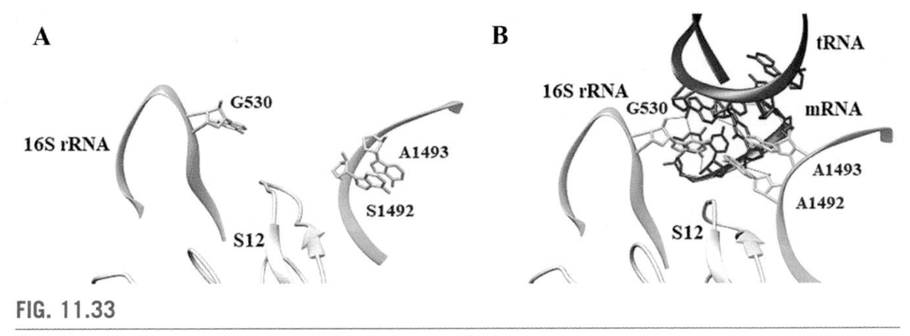

FIG. 11.33

Monitoring of decoding by 16S rRNA bases. (A) Apo-ribosome not bound to tRNA. (B) A cognate tRNA bound to A-site. H-bonds shown by *blue* lines.

Source: Ogle, J.M., et al., 2001. Science 292, 897; Wimberly, B.T., et al., 2000. Nature 407, 327.

bound to the ribosome (apo-ribosome), the two adjacent adenine bases are stacked (Fig. 11.33A). Binding of a correct tRNA causes all the three conserved bases to change their conformation and interact with the first two base-pairs of the codon-anticodon mini-helix by forming H-bonds (Fig. 11.33B). This "monitoring" of base-pair geometry by the 16S rRNA excludes from its purview the wobble position, the degeneracy of the genetic code thus remaining unaffected. The additional energy of interaction is utilized, in part, to increase the relative affinity for the correct tRNA as also to effect conformational changes in the ribosome and the TC for subsequent steps.

The correct codon-anticodon base-pairing transmits a signal to the GTPase center of EF-Tu. The GTPase activity of EF-Tu is ribosome-dependent. The ribosome stimulates the GTPase activity in two ways—by repositioning the catalytic residues of

EF-Tu and directly involving some of its own components for catalysis. Hydrolysis of GTP and release of P_i direct EF-Tu to its inactive (GDP) state, causing its release from the ribosome. The strained aminoacyl-tRNA is now held in position only through interactions with the decoding center. Possibly after going through a process termed as "kinetic proofreading," it is either accommodated into the peptidyl transferase center or ejected from the ribosome.

11.3.2 Peptidyl transfer

Peptidyl transfer reaction or peptide bond formation is the most crucial step in protein synthesis. It involves the nucleophilic attack of the α-amino group of the aminoacyl-tRNA on the aminoacyl ester of the peptidyl-tRNA. The rate of this aminolysis reaction in the ribosome is $\sim 10^7$-fold higher than the spontaneous rate in solution. The peptidyl transferase center (PTC) of the ribosome is located in the 50S subunit and consists exclusively of RNA. This implies that the ribosome is a ribozyme.

Let us try to understand the basic thermodynamic principle of the ribosome-mediated peptidyl transfer reaction. It is known that in cases where polar forces such as H-bonds and electrostatic interactions are involved in stabilizing the transition state, the concerned enzymes catalyze the reactions by lowering the heat of activation. However, in the present case, the enthalpy of activation for aminolysis was experimentally found to be less favorable on the ribosome than in solution. This result led to the interpretation that the ribosome catalyzes the peptide bond formation essentially by positioning the substrates in and excluding water from the active center, thus lowering the entropy of activation. In other words, the catalytic effect of the ribosome is primarily entropic in origin.

It is also known that, besides correct positioning of the substrates, another strategy enzymes employ is stabilization of the transition state. Extensive experimental studies have shown that the ribosome does not provide any ionizing group that can assist catalysis. Instead, as crystal structures have shown, the A76 2′OH of the peptidyl tRNA is a functional group appropriately positioned to play an important role in what is known as a "substrate-assisted catalysis."

Considering the involvement of peptidyl tRNA A76 in the peptide bond formation, it was initially proposed that protons shuttle between the attacking nucleophile, the 2′OH group, and 3′OH group of A76. However, based on further experimental evidence and with the inclusion of a water molecule that is found in an appropriate position in the crystal structure of 50S subunit, it has been hypothesized that peptide bond formation is catalyzed by a concerted eight-membered proton shuttle mechanism. In this, protons originate from the α-amine, a crystalline water molecule, and the 3′OH of A76 of the peptidyl tRNA (Fig. 11.34).

At the completion of the peptide bond formation, the A-site is occupied by the newly elongated peptidyl-tRNA and the P-site is left with a deacylated tRNA. Next, the A-site must be vacated for a new round of elongation to occur. This is carried out

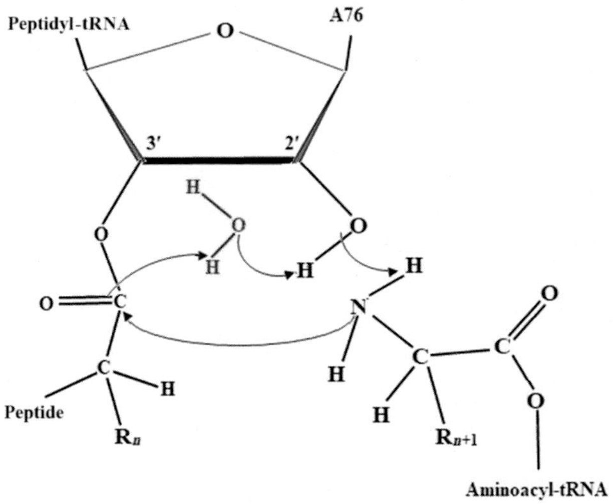

FIG. 11.34

Proton shuttle mechanism of peptidyl transfer reaction.

in the ribosome by a process called translocation, which is divided into a few intermediate steps as described below.

11.3.3 Translocation

Immediately after peptidyl transfer, the tRNAs are in a "canonical A/A and P/P pretranslocation state." In the first step of translocation, the acceptor ends of A- and P-site tRNAs move, respectively, to the P- and E-sites of the 50S subunit, while their anticodon ends, and the mRNA remains anchored in their original positions in the 30S subunit. Thus, A/P and P/E hybrid states are formed. It is to be noted that the E-site can specifically accommodate deacylated tRNA as it is too small for any aminoacyl-tRNA. Hence, the hybrid state can form only after peptidyl transfer.

Translocation is complete when the anticodon stem loops (ASLs) of the tRNAs, along with their associated mRNA codons, have moved, respectively, from the 30S A and P sites to the P and E sites to restore the canonical state of the ribosome. In bacteria, the entire process of translocation is catalyzed by the GTPase elongation factor EF-G. It is made up of four distinct domains and structurally similar to the aminoacyl-tRNA·EF-Tu·GTP ternary complex (TC); its GTPase domain resembles that of EF-Tu and domain IV mimics the tRNA anticodon stem loop (ASL).

Binding of the GTP-form of EF-G to the ribosome leads to a rotation of the 30S subunit with respect to the 50S. This rotational movement has been termed as "ratcheting." It is likely that the entire translocation process, from hybrid state formation to restoration of the canonical state of the ribosome, is coupled to this ratcheting. GTP hydrolysis by EF-G accelerates translocation, either directly or indirectly.

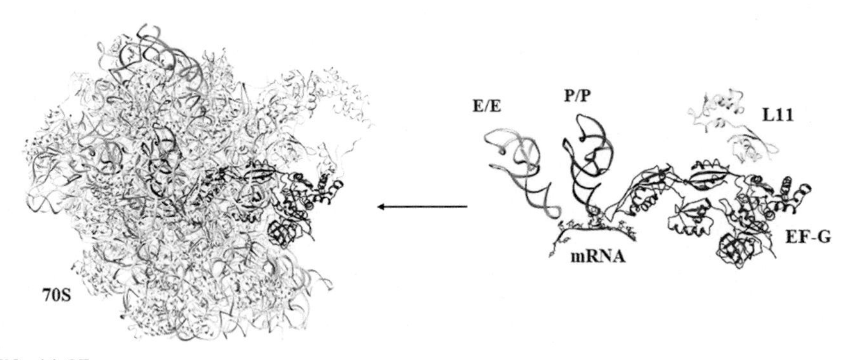

FIG. 11.35

Posttranslocation 70S ribosome-EF-G complex.

Source: Gao, Y.G., 2009. Science 326, 694.

Crystal structures have shown that the ribosomal protein L11 interacts with EF-G. L11 is thought to be involved in the GTPase activation in the ribosome. Also, EF-G domain IV appears to be interacting with the mRNA and peptidyl-tRNA in the decoding center. However, EF-G does not interact with the A-site codon (Fig. 11.35). In the *T. thermophilus* EF-G-ribosome complex, two bases of 16S rRNA, C1397 and A1503, are found to be intercalated between bases at two positions of the mRNA. It is likely that the two bases act as molecular pawls to fix the position of the mRNA ratchet.

11.4 Translation termination

Thus, we see that after the translocation phase of each elongation cycle, when the next mRNA codon has moved into the A site, it is ready to accept its cognate aminoacyl-tRNA. The cycle is repeated until an mRNA stop codon moves into the A site and signals the end of the coding sequence. The newly synthesized polypeptide chain must now be cleaved from the P-site tRNA and released from the ribosome. This process of translation termination is carried out by proteins called class I release factors. There are two class I release factors in bacteria—RF1 and RF2. Among the three stop codons, UAA is recognized by both factors, whereas UAG and UGA are recognized by RF1 and RF2, respectively. In eukaryotes, all three stop codons are recognized by a single factor eRF1, which shares little sequence homology with either RF1 or RF2. Here, we shall discuss some structural-functional aspects of RF2 catalyzed translation termination (Fig. 11.36).

RF2 consists of four domains (Fig. 11.36A)—domain 1 interacts with the large ribosomal protein L11 while domains 2 and 4 form a compact superdomain which is involved in stop-codon recognition in the 30S subunit. A tripeptide motif SPF confers specificity to RF2 for UGA. A network of H-bonds between mRNA nucleotides and reading head of domain 2 is involved in the recognition of the termination signal;

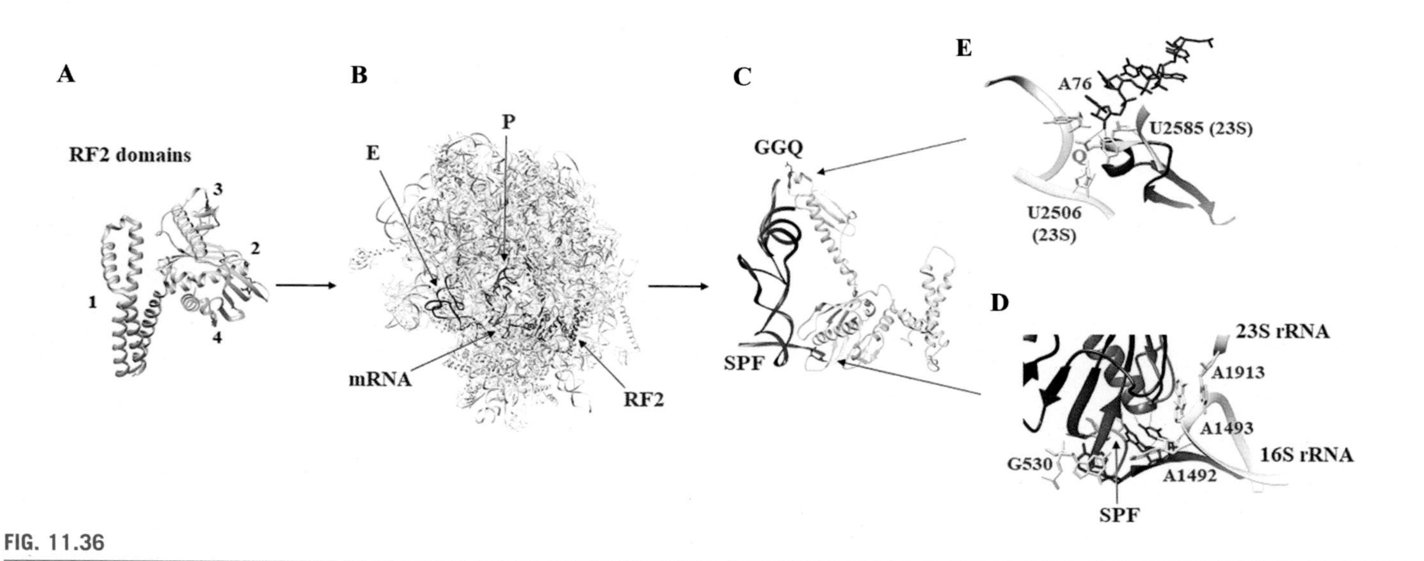

FIG. 11.36

Interactions between RF2 and ribosomal components for translation termination. H-bonds shown by *blue* lines.

Source: Weixbaumer, A., et al., 2008. Science 232, 953.

however, the SPF motif makes limited contacts with the stop codon. A GGQ motif is possibly involved in peptide hydrolysis by RF2.

The binding of RF2 does not lead to large-scale conformational changes in the ribosome. Nevertheless, some rearrangements of conserved bases in the decoding center do occur. These rearrangements are distinct from those observed during decoding of tRNA (described earlier). In this case, A1493 of 16S rRNA stacks on A1913 of 23S rRNA (Fig. 11.36D).

On the other hand, crystal structures of free RF2 and ribosome-bound RF2 have been found to be remarkably different. In the compact free form, the GGQ motif is about 23 Å from the SPF motif. When bound to the ribosome, RF2 extends itself to insert its domain 3 containing the GGQ motif into the PTC (Fig. 11.36C).

Recognition of the stop codon by RF2 leads to the hydrolysis of the ester bond between the peptidyl moiety and the terminal nucleotide A76 peptidyl-tRNA. The release factor directly participates in this catalysis. As the GGQ motif is inserted into the PTC, it contacts the nucleotides of 23S rRNA and P-site tRNA. The glutamine in GGQ makes H-bonds with A76 of the deacylated P-site tRNA. 23S rRNA nucleotides, U2506 and U2585 (Fig. 11.36E), are retracted from the A-site binding pocket to expose the ester bond to attack by a nucleophile. With all this conformational change, it is likely that the glutamine coordinates a water molecule for the nucleophilic attack. One possibility is that the water molecule enters through an opening in the PTC created by the GGQ motif. Alternatively, GGQ may carry with it a bound water molecule into the PTC.

After the newly synthesized peptide is released, the class I factors dissociate from the ribosome. The dissociation is dependent upon a class II release factor RF3, although there is also a suggestion that RF2 can dissociate spontaneously. RF3 is a translational GTPase that binds to the ribosome-RF1/RF2 complex in its GDP form and exchanges its bound GDP for a GTP. The GTP form of RF3 induces conformational changes in the ribosome, leading to destabilization of the binding of class I release factors and their subsequent exit from the ribosome. GTP hydrolysis by RF3 results in its own dissociation from the ribosome, leaving the mRNA and a deacylated tRNA in the P site (posttermination complex).

GTP hydrolysis-dependent dissociation of the posttermination ribosome into its large (50S) and small (30S) subunits, called "ribosome recycling," is induced by a combined action of ribosome recycling factor (RRF) and EF-G. IF3 then binds to the 30S subunit to prevent its reassociation with the 50S subunit. It has also been proposed that IF3 accelerates dissociation of the deacylated tRNA and mRNA from the 30S subunit. The separated ribosomal subunits are now available for a new round of translation initiation.

From existing knowledge, one could infer that the basic mechanisms of translation elongation and peptide bond formation are conserved between prokaryotes and eukaryotes. The same cannot be said about translation termination. In any case, several mechanistic questions on the termination processes in both systems remain unanswered. It can be expected that, with the resolution of additional molecular structures, mechanistic similarities and differences between prokaryotic and eukaryotic translation will be more evident.

Sample questions

1. How does a tRNA molecule function as an interface between an mRNA codon and an amino acid to be incorporated into a growing polypeptide chain?
2. Glutamyl-tRNA synthetase binds both ATP and glutamine even in the absence of tRNA, yet it catalyzes aminoacylation reaction only in the presence of its cognate tRNA—explain the structural basis.
3. What is the structural basis for posttransfer editing by *E. coli* leucyl-tRNA synthetase?
4. What are the respective functions of the bacterial initiation factors in the formation of translation initiation complex?
5. How does eukaryotic initiation factor eIF4G play a coordinating role to prepare the mRNA for binding to the 43S preinitiation complex?
6. What is the energetic basis for peptide bond formation?
7. What is the role of 16S rRNA in the decoding process?
8. Explain the structural basis of translation termination by the release factor RF2.
9. How does the start codon recognition by the eukaryotic translation initiation complex differ from that in bacteria?

References and further reading

Deo, R., et al., 1999. Recognition of polyadenylate RNA by the poly(A)-binding protein. Cell 98, 835–845.

Dever, T.E., Kinzy, T.G., Pavitt, G.D., 2016. Mechanism and regulation of protein synthesis in *Saccharomyces cerevisiae*. Genetics 203, 65–107.

Eliseev, B., et al., 2018. Structure of a human cap-dependent 48S translation pre-initiation complex. Nucleic Acids Res. 46 (5), 2678–2689.

Gualerzi, C.O., Pon, C.L., 2015. Initiation of mRNA translation in bacteria: structural and dynamic aspects. Cell. Mol. Life Sci. 72, 4341–4367.

Hussain, T., et al., 2014. Structural changes enable start code recognition by eukaryotic translation initiation complex. Cell 159, 597–607.

Hussain, T., et al., 2016. Large-scale movements of IF3 and tRNA during bacterial translation initiation. Cell 167, 133–144.

Ibba, M., Söll, D., 2000. Aminoacyl-tRNA synthesis. Annu. Rev. Biochem. 69, 617–650.

Korostelev, A.A., 2011. Structural aspects of translation termination on the ribosome. RNA 17, 1409–1421.

Kozak, M., 2005. Regulation of translation via mRNA structure in prokaryotes and eukaryotes. Gene 361, 13–37.

Ling, Y., et al., 2014. tRNA synthetase: tRNA aminoacylation and beyond. Wiley Interdiscip. Rev. RNA 5 (4), 461–480.

Lorenz, C., Lünse, C.E., Möri, M., 2017. tRNA modifications: impact on structure and thermal adaptation. Biomolecules 7, 35. https://doi.org/10.3390/biom702035.

Macé, K., et al., 2018. The structure of an elongation factor G-ribosome complex captured in the absence of inhibitors. Nucleic Acids Res. 46 (6), 3211–3217.

Marcotrigiano, J., et al., 1999. Cap-dependent translation initiation in eukaryotes is regulated by a molecular mimic of eIF4G. Mol. Cell 3, 707–716.

Martin Schmeing, T., Ramakrishnan, V., 2009. What recent ribosome structures have revealed about the mechanism of translation. Nature 461, 1234–1242.

Martin Schmeing, T., et al., 2009. The crystal structure of the ribosome bound to EF-Tu and aminoacyl-tRNA. Science 326, 688–694.

Moras, D., 2010. Proofreading in translation: dynamics of the double sieve model. Proc. Natl. Acad. Sci. U. S. A. 107, 21949–21950.

Palencia, A., et al., 2013. Structural dynamics of the aminoacylation and proof-reading functional cycle of bacterial leucyl-tRNA synthetase. Nat. Struct. Mol. Biol. 19 (7), 677–684.

Rodina, M.V., 2013. The ribosome as a versatile catalyst: reactions at the peptidyl transferase center. Curr. Opt. Struct. Biol. 23, 595–602.

Rozov, A., et al., 2016. Novel base pairing interactions at the tRNA wobble position crucial for accurate reading of the genetic code. Nat. Commun. 7, 10457. https://doi.org/10.1038/ncommun10457.

Safaee, N., et al., 2012. Interdomain allostery promotes assembly of the poly(A) mRNA complex with PABP and eIF4G. Mol. Cell 48, 375–386.

Sekine, S., et al., 2006. Structural bases of transfer RNA-dependent amino acid recognition and activation by glutamyl-tRNA synthetase. Structure 14, 1791–1799.

Sekine, S.-I., et al., 2003. ATP binding by glutamyl-tRNA synthetase is switched to the productive mode by tRNA binding. EMBO J. 22 (3), 676–688.

Søgaard, B., et al., 2005. Initiation of protein synthesis in bacteria. Microbiol. Mol. Biol. Rev. 69 (1), 101–123.

Su, T., et al., 2017. The force-sensing peptide VemP employs extreme compaction and secondary structure formation to induce ribosomal stalling. eLife 6, 25642. https://doi.org/10.7554/eLIfe.25642.

Tomoo, K., et al., 2002. Crystal structure of 7-methyguanosine 5′ triphosphate (m^7GTP)- and P^1-methylguanosine -P^3-adenosine-5′,5′-triphosphate (m^7GpppA)-bound human full-length eukaryotic initiation factor 4E: biological importance of the C-terminal flexible region. Biochem. J. 362, 539–544.

Vestergaard, B., et al., 2001. Bacterial polypeptide release factor RF2 is structurally distinct from eukaryotic eRF1. Mol. Cell 8, 1375–1382.

Voorhees, R.M., Ramakrishnan, V., 2013. Structural basis of translation elongation cycle. Annu. Rev. Biochem. 82, 203–236.

Yusupova, G.Z., et al., 2001. The path of messenger RNA through the ribosome. Cell 106, 233–241.

Zhou, J., et al., 2013. Crystal structures of EF-G-ribosome complexes trapped in intermediate states of translocation. Science 340 (6140), 1236086. https://doi.org/10.1126/science.1236086.

CHAPTER

DNA damage and repair 12

The survival and evolution of a species require the maintenance of its genetic material (mostly DNA) allowing only optimal changes (mutations). A mutation is a permanent alteration in the DNA sequence. Mutation rates higher than the normal would be detrimental to the species. On the other hand, perpetuation of the genetic material with absolute fidelity would preempt all chances of evolution.

12.1 Structural chemistry of DNA damage

12.1.1 Spontaneous reactions

Two factors have the potential to cause mutation in the genetic material: (a) inaccuracy in DNA replication and (b) damage, or chemical alteration, to the DNA. Replication error is the result of tautomerization of DNA bases, as for example, conversion of cytosine from its more predominant keto form to rare imino form (Fig. 12.1). Such errors, as we have seen in Chapter 8, are partly rectified by the proofreading mechanism of the DNA polymerase itself. However, some errors, which evade proofreading, require postreplication repair mechanisms to be corrected.

Damage to DNA can be caused by chemical reactions either occurring spontaneously or driven by environmental factors such as radiation and some chemical agents. Spontaneous hydrolysis of the phosphodiester linkage is extremely slow (almost insignificant) under normal physiological conditions. Considerably more frequent are the hydrolytic damages involving DNA bases. Spontaneous hydrolysis deaminates cytosine to uracil (Fig. 12.2A). Deamination of adenine and guanine are much slower than cytosine deamination. However, deamination of 5-methylcytosine (5-mC) is 2–3 times faster than that of unmodified cytosine, 5-mC being converted to a thymine (Fig. 12.2B).

Spontaneous hydrolysis of the glycosidic bond crops away nucleobases from the sugar-phosphate backbone. The reaction is called depurination as it is much more efficient for purines than pyrimidines.

12.1.2 Chemical damage

Alkylation

Besides spontaneous hydrolysis, DNA is vulnerable to damage by electrophiles and free radicals. Alkylating agents such as ethylmethanesulfonate (EMS) are potent electrophiles. We have seen in Chapter 4 that electrophiles look for centers of

Fundamentals of Molecular Structural Biology. https://doi.org/10.1016/B978-0-12-814855-6.00012-2

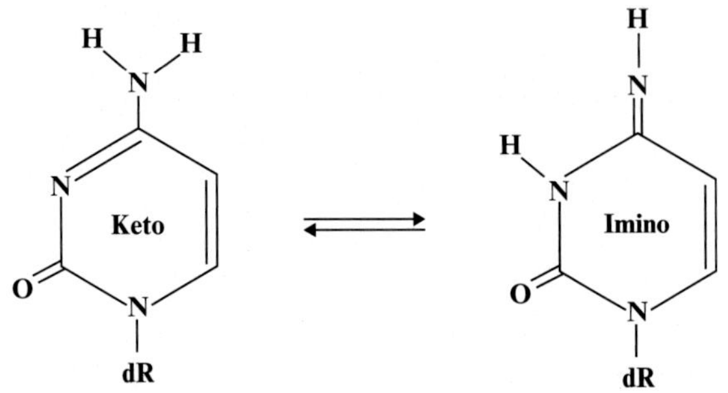

FIG. 12.1

Tautomerization of cytosine.

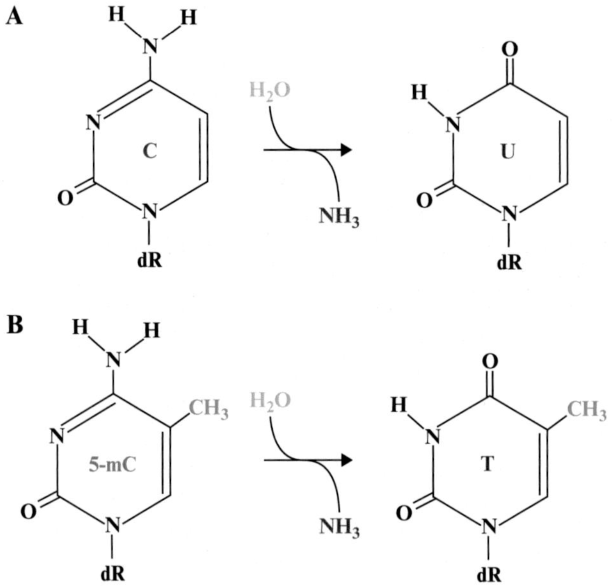

FIG. 12.2

Spontaneous deamination of (A) cytosine and (B) methylcytosine.

negative charge or nucleophiles in other molecules and interact with them. All the heteroatoms (such as nitrogen, oxygen, phosphorous, etc.) in DNA are potential nucleophiles. N7-position in guanine is the most nucleophilic site in the DNA bases and, therefore, most reactive with alkylating agents. Nevertheless, alkylating agents like EMS show strong preference for oxygen nucleophiles in DNA. EMS transfers an ethyl group to the O6 oxygen of guanine to form O6-ethylguranine (Fig. 12.3).

All the damages described so far lead to "simple" point mutation (replacement of one base with another) at the site of the damage. The replacements are of two kinds:

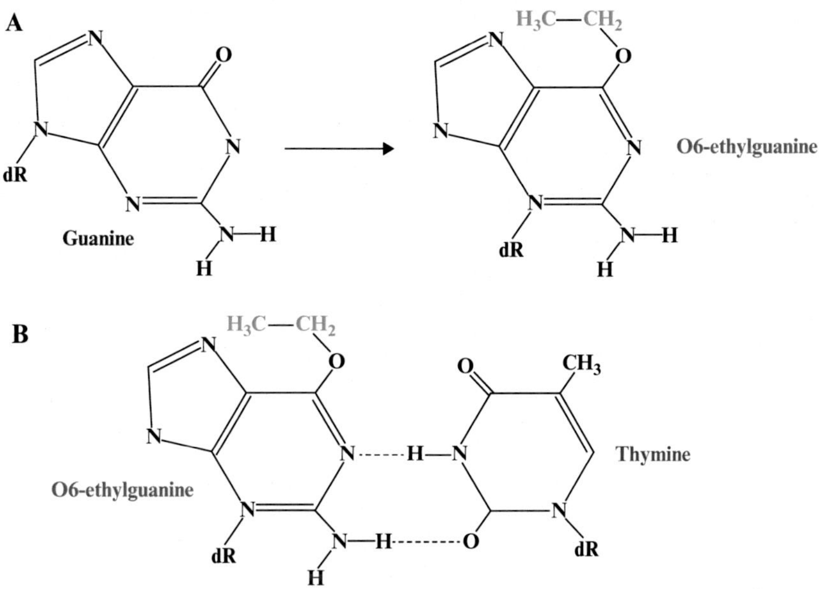

FIG. 12.3

(A) Alkylation of guanine by EMS and (B) base-pairing of O6-ethylguanine with thymine.

(a) transition, where a purine is substituted for another purine or a pyrimidine for another pyrimidine, and (b) transversion, which is either pyrimidine-to-purine or purine-to-pyrimidine substitution. Spontaneous hydrolysis to uracil causes a G→A transition in the strand opposite to the damage. Similarly, the O6-ethylguanine, which is created by the action of EMS, base-pairs with thymine instead of cytosine, thus resulting in a C→T transition.

Aromatic amines

Actions of aromatic amines (arylamines) and reactive oxygen species at certain positions in the DNA bases can form chemically stable adducts. Such adducts then undergo conformational rearrangements which may involve rotation about one or more chemical bonds, but no bond breakage. The N-glycosyl bond in DNA is a major site of the conformational rearrangements. We have seen in Chapter 5 that the N-glycosyl bond in B-form DNA is held in the *anti*-conformation with the nucleobase away from the deoxyribose sugar. This allows the canonical Watson-Crick (WC) base-pairing interactions of the complementary nucleobases. The bulky adduct forces the N-glycosyl bond into the *syn*-conformation and hinder WC H-bonding interactions.

>An arylamine whose action on DNA has been well-documented is *N*-acetyl-2-aminofluorene. 2-Aminofluorene (AF) is acetylated in the cell to form *N*-acetyl-2-aminofluorene which, in turn, reacts with DNA to produce a 2-amino-*N*-(deoxyguanosin-8-yl)fluorine (C8-G AF) adduct (Fig. 12.4). When present in the sequence

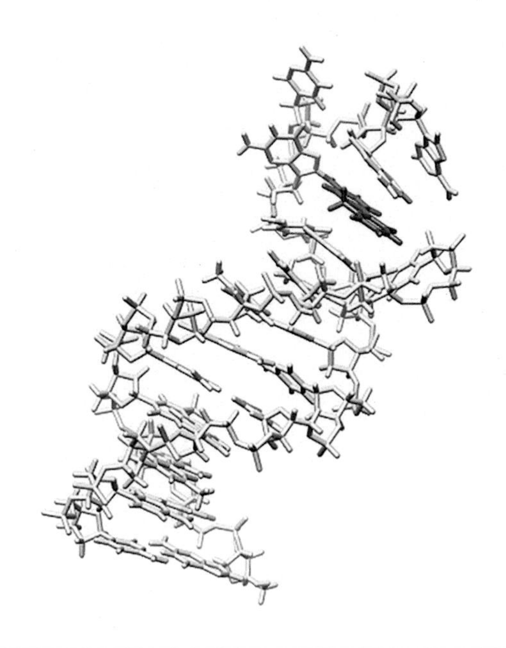

FIG. 12.4

C8-G AF adduct (colored *orange-red*) in DNA.

Source: Gu, Z., et al., 1999. Biochemistry 38, 10855.

5′-AXG-3′ or 5′-CXC-3′, the damaged nucleotide (X) can adopt either of the two conformations—(a) external- (*anti*) and (b) inserted- (*syn*) with respect to the N-glycosyl bond. In the external-AF conformation, X(*anti*):C(*anti*) WC base-pairing is allowed. In contrast, in the inserted-AF conformation, the AF moiety is within the DNA helix and WC base-pairing at X(*syn*):C(*anti*) is disrupted leading to a G→T transversion (Fig. 12.4).

Oxidative damage

Similar adducts are formed by oxidative damages to DNA. Such damages are mostly unavoidable as a number of reactive intermediates are formed due to existence of life in an aerobic environment. The reduction of O_2, which provides energy to the living cell, also produces reactive oxidants such as superoxide anions (O_2^-), OH^-, and H_2O_2. Deoxyguanosine is oxidized by these reactive intermediates at C8 position to form 7,8-dihydro-8-oxoguanine (8-oxo-G) (Fig. 12.5). Like C8-G AF, 8-oxo-G can adopt two different conformations—8-oxo-G(*anti*):C(*anti*) can form WC H-bonds, whereas 8-oxo-G(*syn*):A(*anti*) mispairs leading to G→T transversion.

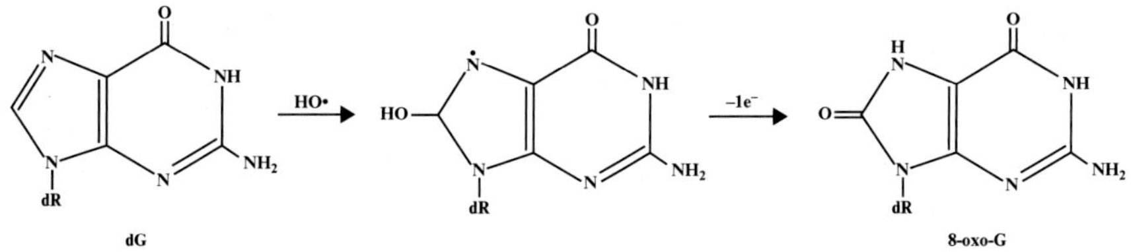

FIG. 12.5

Formation of 8-oxo-G by oxidation of deoxyguanosine.

12.1.3 Radiation damage

Radiation is also another prominent cause of damage to DNA. Ultraviolet (UV) radiation of wavelength 200–300 nm is absorbed by DNA bases and consequently, in certain cases, may lead to the formation of cyclobutane pyrimidine dimer (CPD)—thymine dimer (T<>T) in particular. T<>T is the result of photochemical fusion between two neighboring thymine (T) bases through C5-C5 and C6-C6 linkages (Fig. 12.6). The fusion reduces the aromacity of the dimerized thymines and thereby weakens their H-bonding with complementary bases. The lesion also perturbs the local structure of the DNA, generating a kink in the helical axis.

Ionizing radiation such as X-rays and γ-rays can not only cause extensive DNA base damage, but even produce single-strand breaks (SSBs) in the DNA. This is accomplished either directly or by high energy radiolysis of H_2O, generating reactive products such as OH^- and H_2O_2, which can attack the sugar-phosphate backbone. High doses of such radiation can frequently produce two SSBs in complementary DNA strands within one helical turn, thus causing double-strand breaks (DSBs). Homologous recombination (HR) and nonhomologous end joining (NHEJ) are the two pathways for DSB repair (DBR). HR, which is relatively more accurate of the two processes, will be discussed in the next chapter.

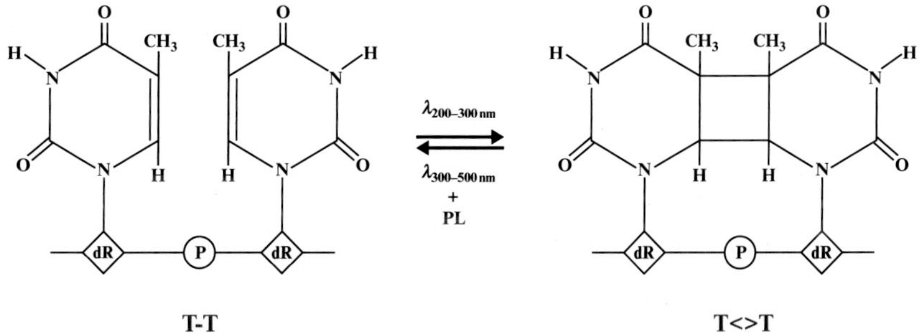

FIG. 12.6

Formation of cyclobutane thymine dimer (T<>T) by UV radiation and reversal to normal thymines (T-T) by near-UV and visible *(blue)* light and photolyase (PL).

12.2 Mismatch repair

Replication errors that escape the proofreading mechanism of bacterial DNA polymerase are mostly rectified by the mismatch repair (MMR) system. In *Escherichia coli*, MMR requires the coordinated action of more than 10 proteins, of which MutS and MutL are the key players. The process involves the following steps. (a) The mismatch is recognized by MutS which, then, binds ATP. The protein undergoes a conformational change and forms a sliding clamp on the DNA. (b) The MutS sliding clamp recruits MutL which, in turn, activates the endonuclease MutH. (c) The lack of methylation at d(GATC)'s of the newly synthesized DNA enables MutH distinguish between the parental and daughter strands and generate a nick in the daughter strand. (d) UvrD, which is also activated by MutL, then unwinds the DNA starting from the nick, while a specific set of exonucleases hydrolyze the released ssDNA. (e) The gap thus created in the DNA is filled up by the DNA polymerase III complex and, eventually, DNA ligase seals the nick (Fig. 12.7).

MutS binds the mismatch containing DNA as a dimer to form the initial recognition complex (IRC). In the DNA-bound complex, the homodimer forms two channels and is arranged asymmetrically (Fig. 12.8). Only one subunit makes specific contacts with the mismatch, whereas second subunit contacts only the sugar-phosphate backbone.

Each subunit of the dimer consists of a number of domains (Fig. 12.8). Two domains are responsible for binding of the DNA—the clamp domain, which holds the duplex, and the lever domain, whose helices connect the clamp domain to the core of the protein. The N-terminal mismatch-binding domain of one subunit is responsible for the mismatch recognition.

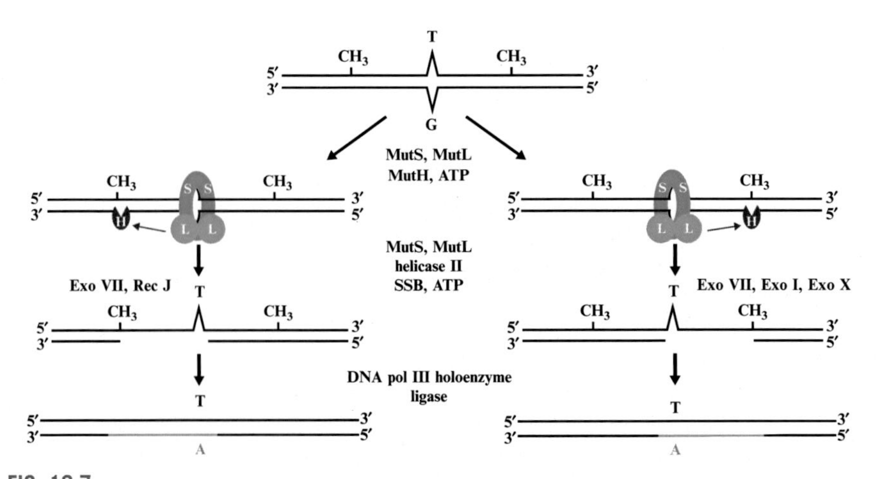

FIG. 12.7

Flowchart of MMR.

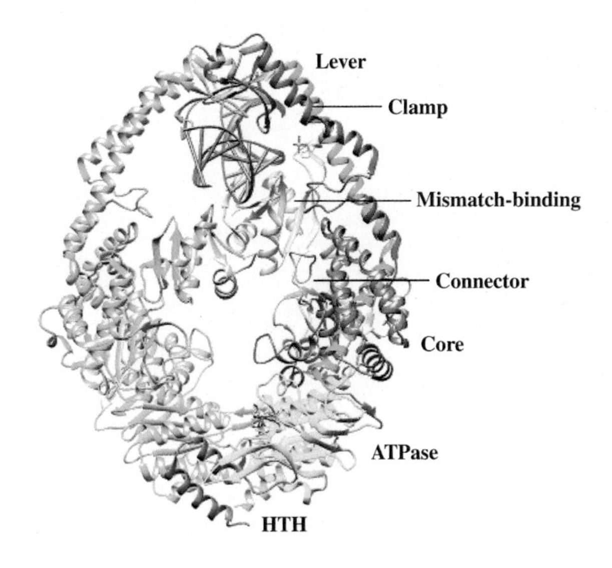

FIG. 12.8

MutS-DNA complex. The DNA is shown in *orange*. One of the two subunits of MutS dimer is shown in *light green*, while the domains in the other subunits are labeled and colored differently.

Source: Lamers, M.H., et al., 2000. Nature 407, 711.

Adjacent to the mismatch-binding domain is the connector domain which, together with the core and ATPase domains, is involved in the interaction with MutL. Further, a helix-turn-helix (HTH) domain, located towards the C-terminal end, is required for the stability of the dimer.

MutS belongs to the ABC (ATP-binding cassette) family of ATPases. Like the other members of the family, its activity is regulated by ATP-binding. The asymmetry of MutS is also evident in regards to ATP- (or ADP) binding. Only the subunit in contact with the mismatch binds an ATP/ADP molecule at the ATP-binding site. The adenine base is held in position by H760 and F596 on two sides. Further, I597 forms two H-bonds with the nucleotide and the β-phosphate of ATP/ADP, together with S621 and four water molecules, coordinates a Mg^{2+} ion (Fig. 12.9).

ATP binding to MutS causes significant conformational changes in its ATPase domain and these changes, in turn, induce rearrangements in the DNA-binding domain. This signal transduction is mediated by the α-helices connecting the two domains and the flexible loops of the ATPase domain.

As already mentioned, the binding of MutS with DNA is asymmetric—several residues of each subunit form multiple contacts with the DNA, yet they are all different. The majority of the contacts are hydrophilic—the amino acids interacting with the sugar-phosphate backbone—and, therefore, independent of the nucleotide sequence.

The only specific contact made with the mismatch is by a phenylalanine residue (F36) of a conserved N-terminal motif GxFxE motif. F36 inserts into the DNA

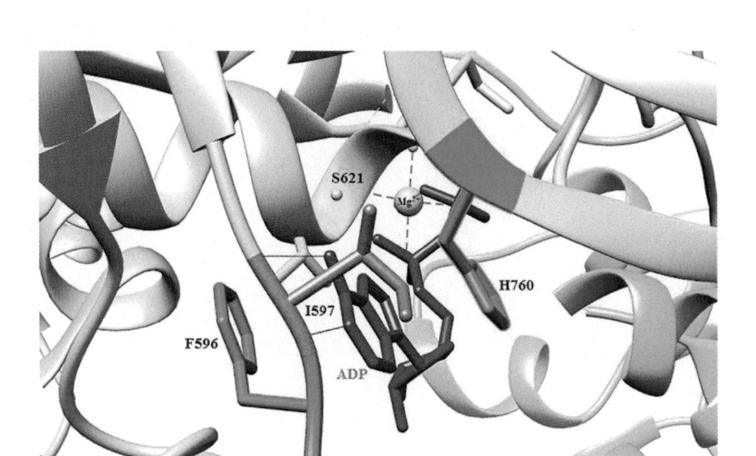

FIG. 12.9

Nucleotide-binding by MutS. The two subunits are shown in *light blue* and *light green*, respectively. The residues of only one subunit *(light blue)* are involved in nucleotide-binding. H-bonds are shown by *red* lines.

Source: Lamers, M.H., et al., 2000. Nature 407, 711.

duplex and stacks onto one of the mispaired bases. E38 forms an H-bond with N7 or N3 of the mispaired purine or pyrimidine, respectively, and stabilizes the stacking interaction. A 60° bend in the DNA is thereby induced and the minor groove widens to the dimension close to that of a major groove (Fig. 12.10).

It is to be noted that the mismatch, which is the recognition site for MutS, is hundreds of base-pairs away from the site of subsequent hydrolysis of the daughter strand of DNA. A molecular coordination of all the proteins involved in MMR is, therefore, necessary in both time and space. MutL protein is the "molecular coordinator." It coordinates the early stage of the process involving MutS with the subsequent stages, which engage MutH, UvrD, DNA pol III, polymerase processivity factors, and exonucleases.

Like MutS, MutL also functions as a dimer. Although the structure of a full-length protein has not been determined as yet, separate structures of the N-terminal and C-terminal domains are available (Fig. 12.11). It is presumed that the two domains are interconnected by an unstructured linker (Fig. 12.11). The C-terminal domains are responsible for the formation of the primary dimerization interface.

ATP-binding and hydrolysis are essential for the function of MutL. The N-terminal domains contain ATP-binding sites. The protein belongs to the GHKL family of ATPases. The members of the family are known to form homodimers

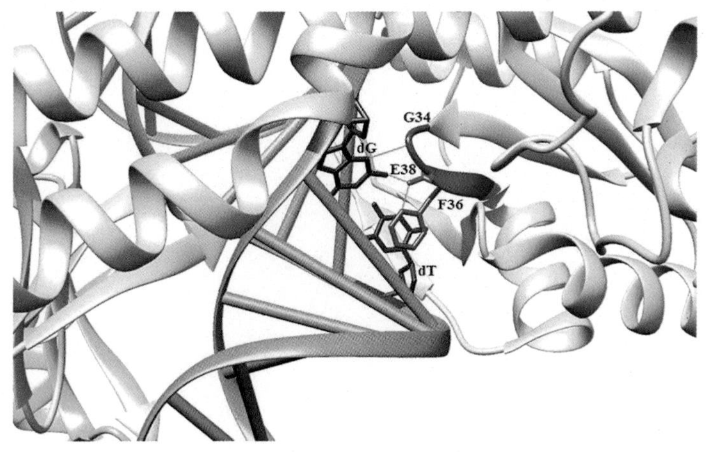

FIG. 12.10

Specific contacts to mismatch by MutS. The mismatched bases are shown in *brick red*, while the rest of the DNA is colored *orange*. H-bonds are shown by *green* lines.

Source: Lamers, M.H., et al., 2000. Nature 407, 711.

or heterodimers through their C-terminal regions. As ATP-binding causes dimerization of the N-terminal domains as well, MutL encircles the duplex DNA. Two loops located near the N-terminus of MutL interact with MutS and a groove formed along the surface of the N-terminal domain binds MutH.

The cleavage of the newly synthesized DNA strand is executed by an endonuclease. In most of the organisms, the activity resides in the C-terminal domain of MutL. Only in some Gram-negative bacteria, where GATC sites remain temporarily hemimethylated after replication, a site-specific endonuclease, MutH, is responsible for the cleavage.

In order to perform its assigned role, MutH is required to (i) interact with the right substrate, that is, the hemimethylated GATC, (ii) bind metal ion for its catalytic activity, and (iii) be activated by MutL for efficient cleavage.

MutH functions as a monomer. The crystal structure of *Haemophilus influenzae* MutH (which has $\sim$60% sequence similarity with *E. coli* MutH) has been determined (Fig. 12.12). It shows that the enzyme poses like a "clamp" consisting of two "arms," N-arm and C-arm, housing a DNA-binding pocket. The catalytic center is located in the N-arm, while the amino acid residues involved in DNA-binding are found in the C-arm.

MutH binds hemimethylated DNA, specifically to 5′-Gm^{6}ATC-3′/3′-CTAG-5′, with significantly higher affinity than fully methylated DNA. Though the specific interactions are restricted to the GATC sequence, nonspecific contacts are found on either side. A loop (residues 184–190) specifically interacts with GATC in the major groove—R184, K186, and G187 are H-bonded with a G/C base-pair

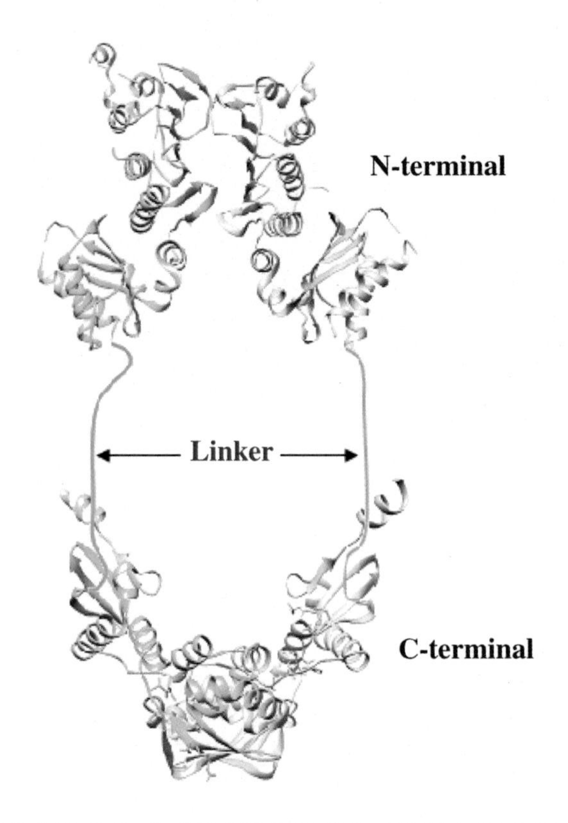

FIG. 12.11

E. coli MutL structural model. The two subunits are shown in *light blue* and *light green*, respectively.

Source: Guarne, A., et al., 2004. EMBO J. 23, 4134; Ban, C., Yang, W., 1998. Cell 95, 541.

(Fig. 12.13). F94 is packed against K186 which, in turn, is H-bonded with the guanine adjacent to the active site (Fig. 12.13). The OH group of Y212 is H-bonded with N6 of the unmodified adenine—a methyl group attached to a modified adenine would hinder the interaction.

The hydrolytic (nicking) activity of MutH on a newly synthesized DNA strand requires Mg^{2+} ions. The protein contains a characteristic motif D-$(x)_n$-E-x-K (also known as DEK motif), which is essential for the metal ion coordination and cleavage (Fig. 12.14). The lysine residue (K79) coordinates a nucleophilic water molecule.

As mentioned earlier, for efficient catalytic (cleavage) activity, MutH needs to be activated by MutL. This can be explained in terms of MutH binding to DNA. The DNA-binding channel in the apo-form of MutH is not wide enough to bind DNA. It is understood that as MutL interacts with the solvent-exposed helix (αE; Fig. 12.12)

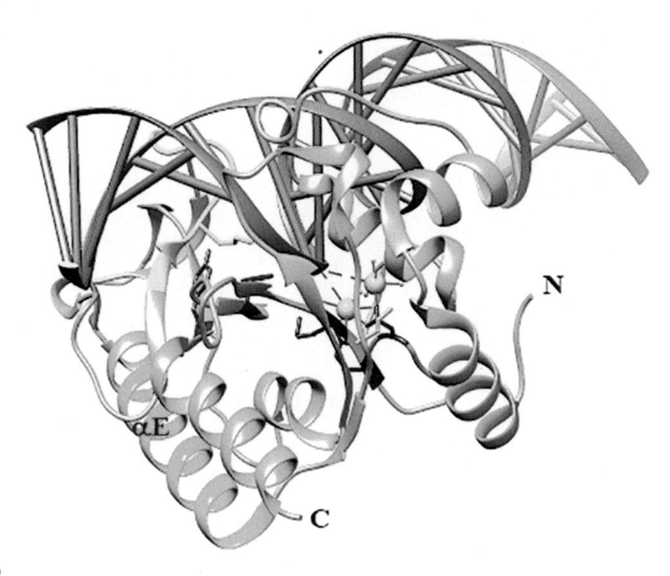

FIG. 12.12

MutH-DNA complex. The GATC sequence is shown in *dodger blue*; the rest of the DNA is colored in *orange*. αE helix of MutH is shown in *cyan*; the rest of MutH is colored in *light green*. MutH residues (colored *yellow* and *purple*) interacting with GATC are labeled in Fig. 12.13.

Source: Lee, J.Y., et al., 2005. Mol. Cell 20, 155.

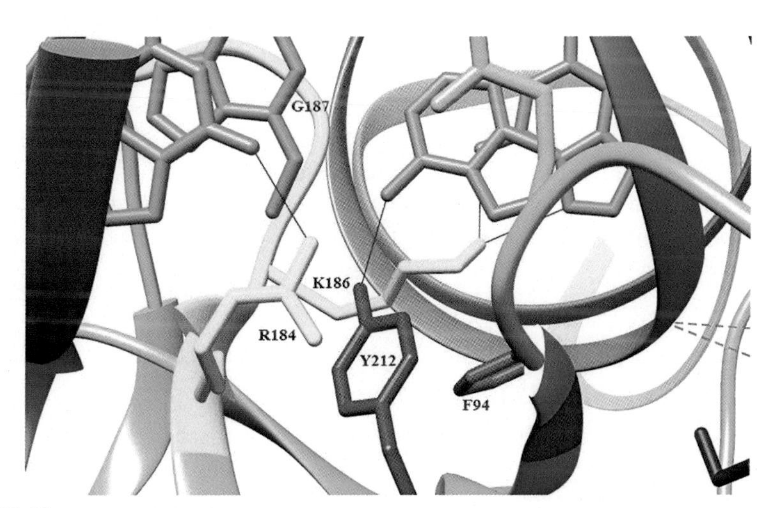

FIG. 12.13

Specific interactions of MutH with GATC (colored in *dodger blue*). H-bonds are shown by *red* lines.

Source: Lee, J.Y., et al., 2005. Mol. Cell 20, 155.

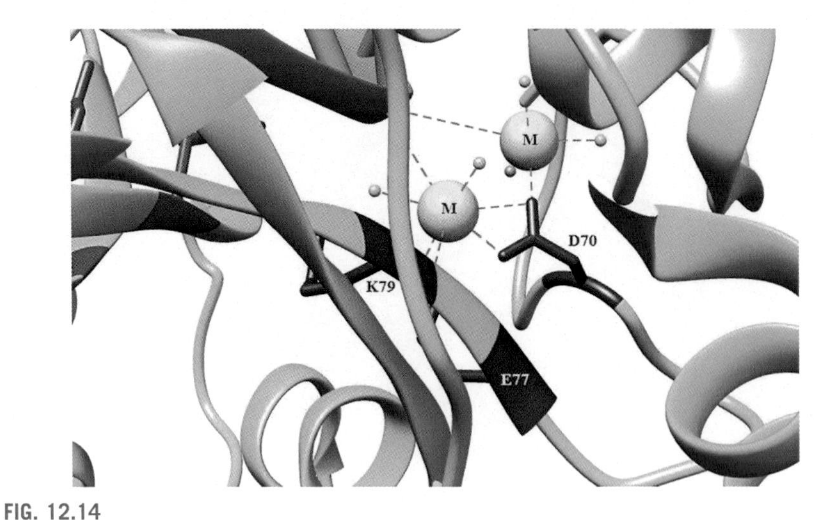

FIG. 12.14

MutH DEK motif *(brown)* coordinating metal ions (M).

Source: Lee, J.Y., et al., 2005. Mol. Cell 20, 155.

of MutH, the C-arm is set in motion to widen the DNA-binding channel. As a result, the rate of MutH binding to DNA is enhanced.

MMR pathway exists in eukaryotic cells as well. Initial steps of the eukaryotic MMR pathway, executed by MutS and MutL homologs, are essentially same as those in *E. coli*. Nonetheless, the eukaryotic homologs are compositionally different from and their actions somewhat more complex than their *E. coli* counterparts. In humans, MutSα, a heterodimer of two MSH (MutS homolog) proteins, MSH2 and MSH6, recognizes the base-base mismatch. On the other hand, MutLα, formed by human MLH1 and PMS2, is the "molecular coordinator" in MMR. One prominent difference between *E. coli* and human MMR systems is that, in human cells, the incision in the newly synthesized DNA strand is carried out by the MutLα complex. Proliferating cell nuclear antigen (PCNA) is responsible for activating MutLα endonuclease activity and strand discrimination.

12.3 Direct reversal of damage

Direct reversal of damage can be one of the DNA repair mechanisms, particularly if the damage is the outcome of a simple chemical reaction. Indeed, in a number of organisms, including *E. coli*, the UV-induced damage in the form of a T<>T is directly reversed (repaired) by an enzyme called photolyase (PL). The enzyme can absorb photons of near-UV and visible (blue) light (wavelength 300–500 nm) and split the dimer into two normal thymines (Fig. 12.6). Since the repair process is driven by light energy, it is also called photoreactivation, while the enzyme can be considered as a "photon-powered nanomachine."

In general, photolyases contain two kinds of chromophoric cofactors. One cofactor, a reduced flavin-adenine diucleotide ($FADH^-$), directly interacts with the substrate (CPD) and performs a catalytic role. The other is a light-harvesting cofactor which can be a 5,10-methenyltetrahydrofolic acid (MTHF) or an 8-hydroxy-5-dezaflavin (8-HDF). The light-harvesting cofactor acts as a panel (like a solar panel) to harvest light energy which is then transferred to $FADH^-$. In *E. coli* photolyase, $FADH^-$ is bound at the C-terminal region, while MTHF is associated with the N-terminal half (Fig. 12.15). The three-dimensional structure of photolyase shows that $FADH^-$ is deeply buried in the middle of a helical domain.

Photolyase can recognize the CPD-induced perturbation in the local structure of the DNA and bind to the damage site. Electrostatic interactions between a positively charged groove (Fig. 12.16) on the enzyme surface and the negatively charged DNA phosphodiester backbone steer the T<>T from within the helix into the active site of the enzyme. The T<>T is now able to make van der Waals contacts with the isoalloxazine ring of $FADH^-$ and possibly some protein residues such as W277, M345, and W384. The T<>T-binding site is surrounded by positively charged residues, R226, R342, R397, and K407 (Fig. 12.17).

However, the binding alone does not trigger any catalytic reaction; MTHF has to absorb a photon and transfer the excitation energy to $FADH^-$. The excited flavin, $FADH^{-*}$, is then able to repair the T<>T by a redox reaction.

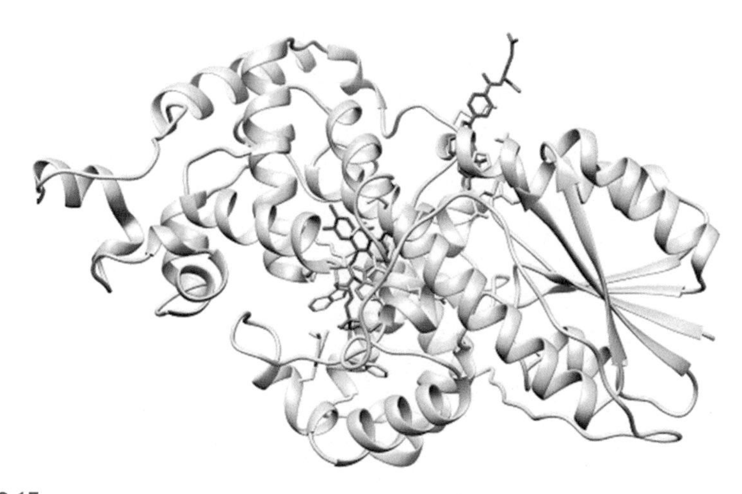

FIG. 12.15

E. coli photolyase containing two cofactors: $FADH^-$ *(orange)* and MTHF *(purple)*.

Source: Park, H.W., et al., 1995. Science 268, 1866.

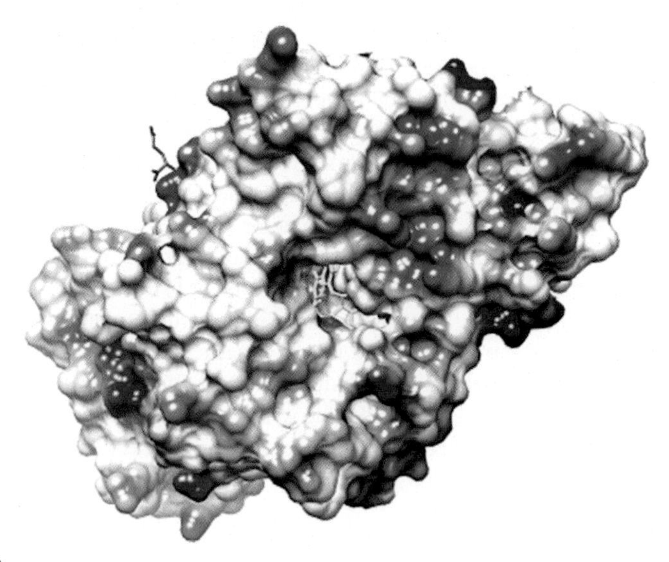

FIG. 12.16

Electrostatic surface of *E. coli* photolyase showing FADH$^-$ *(yellow)* in a groove: *blue*—positive; *white*—neutral; *red*—negative.

Source: Park, H.W., et al., 1995. Science 268, 1866.

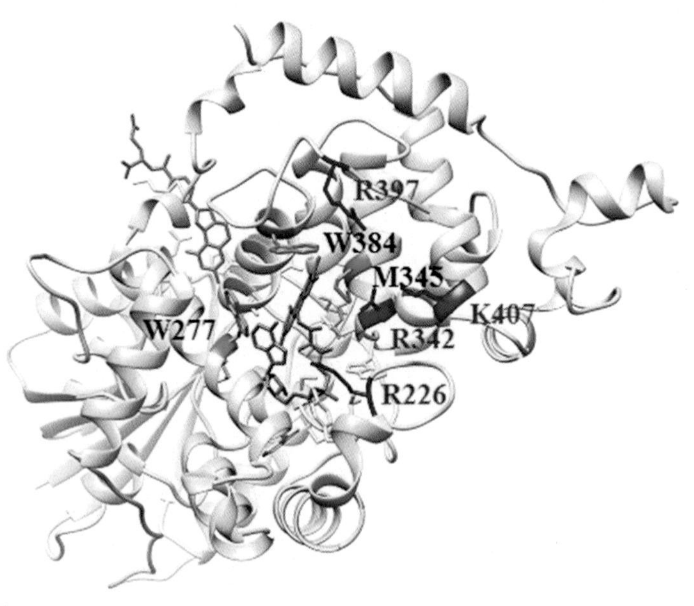

FIG. 12.17

Photolyase residues *(green)* making contacts with T<>T and positive residues *(blue)* that promote interaction with DNA phosphodiester backbone. FADH$^-$—*orange*; MTHF—*purple*.

Source: Park, H.W., et al., 1995. Science 268, 1866.

12.4 Base excision repair

Some of the DNA damages described above, such as the oxidation of guanine, affect individual bases, but do not significantly disfigure the double-helical structure. These lesions are mostly repaired by the base excision repair (BER) mechanism.

BER involves essentially two consecutive steps. In the first step, an enzyme called glycosylase cleaves the glycosidic bond between the damaged base and deoxyribose—the process being known as deglycosylation (Fig. 12.18). Evidently, BER requires that the glycosylase should recognize and remove the damaged base.

The second step is catalyzed by another enzyme, an apurinic/apyrimidinic (AP) endonuclease. The enzyme recognizes an abasic product of the glycosylase action and cleaves 5′ to the lesion. Thus, a normal 3′-OH is generated for subsequent DNA resynthesis (Fig. 12.18).

Every organism, bacterial or eukaryotic, contains multiple glycosylases. However, each lesion is recognized by only one or a subset of the glycosylases in a cell. The structures of different glycosylases may or may not be related, but there is noteworthy functional commonness among them. Nucleotide flipping, DNA kinking, and discontinuous base-stacking are the common features. The active site of every glycosylase can accommodate one and only one flipped-out extrahelical base. The damaged base is recognized on the basis of its shape, H-bonding potential, and charge distribution which are different from an undamaged base.

Let us take the example of the damage to DNA due to the formation of 8-oxo-G. In *E. coli*, the glycosylase MutM (Fpg) catalyzes the excision of 8-oxo-G. In eukaryotes, the excision is carried out by the functional analog 8-oxoguanine-glycosylase (OGG1).

A significant feature of hOGG1 (human) is a helix-hairpin-helix (HhH) motif. With the help of this HhH architecture, hOGG1 produces a kink in the duplex DNA, disrupts the 8-oxo-G:C base-pair, and ejects 8-oxo-G out of the helix into a binding pocket in the enzyme's active site (Fig. 12.19A). The carbonyl oxygen of G42 makes an H-bond with N7 of 8-oxo-G. This is perhaps the only contact specific to 8-oxo-G and responsible for discrimination between 8-oxo-G and normal G by the enzyme.

Additional interactions between 8-oxo-G and residues of hOGG1 also contribute to the recognition process. F319 and C253 stack against the opposite π-faces of 8-oxo-G. The NH_2 group of Q315 interacts with O6 of 8-oxo-G and its side chains carbonyl forms two H-bonds with 8-oxo-G (Fig. 12.19B).

E. coli Endo IV and human APE1 belong to two different classes of AP endonucleases. There is very little structural resemblance between the two, yet there are remarkable similarities in their mode of action. Here we shall restrict our discussion to APE1.

In mammalian cells, most of the AP site cleavage is undertaken by APE1. The enzyme searches for AP sites and subsequently catalyzes the incision of the DNA phosphodiester bond 5′ to the AP site, producing a single-stranded nick with 3′-hydroxyl and 5′-deoxyribose phosphate (dRP) termini.

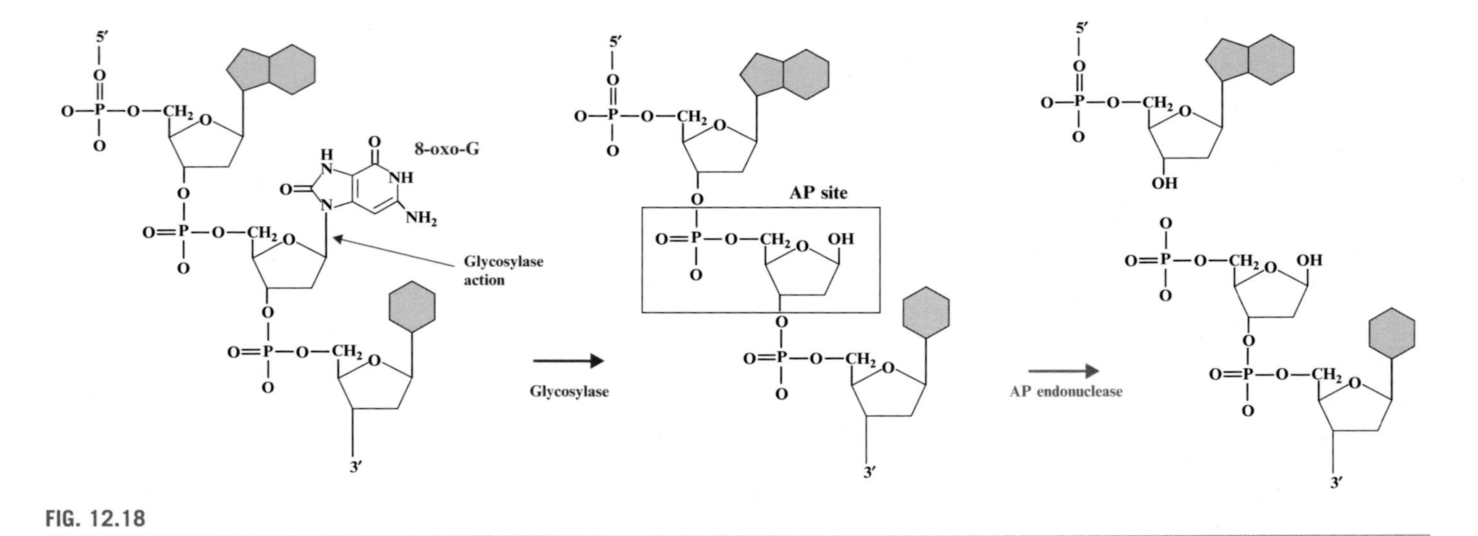

FIG. 12.18

BER pathway. Damaged base (8-oxo-G) labeled; normal bases shown in *light blue*.

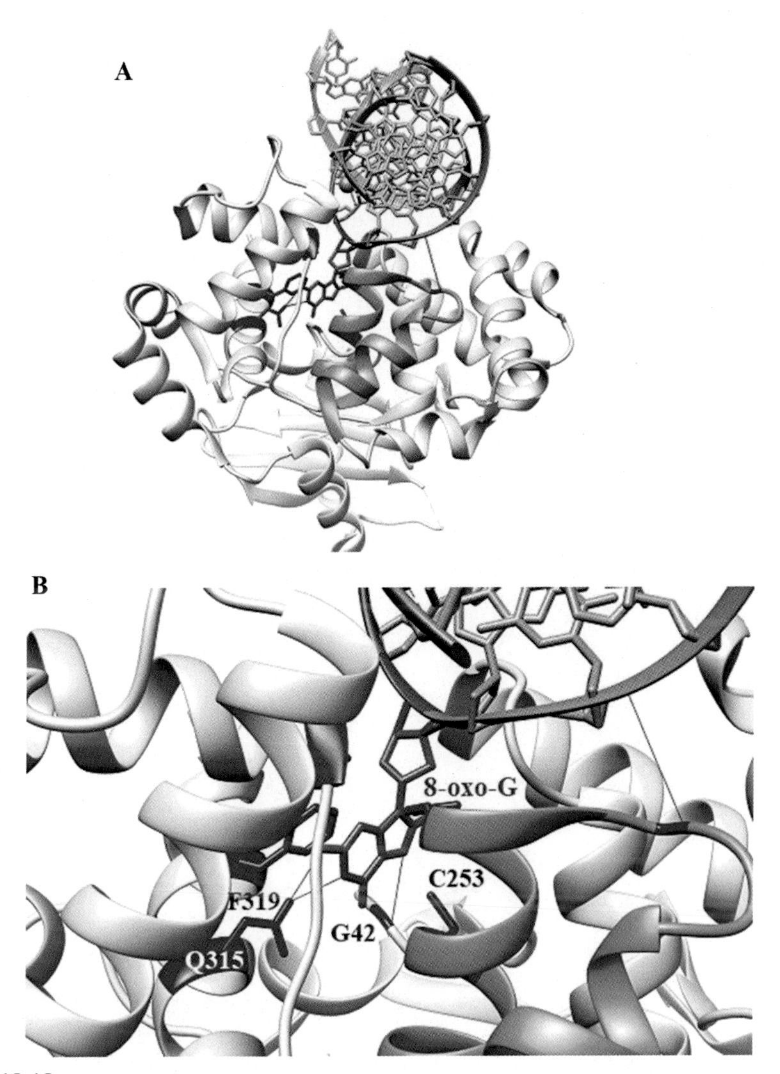

FIG. 12.19

hOGG1—8-oxo-G-containing DNA complex. (A) Overall structure. 8-Oxo-G is shown in *purple*, the rest of DNA in *orange*. HhH motif is colored in *dodger blue*, while the rest of hOGG1 colored in *light blue*. (B) Contact interface. Interacting residues of hOGG1 are colored *deep blue*; H-bonds shown by *red* lines.

Source: Brunner, S.D., Norman, D.P., Verdine, G.L., 2000. Nature 403, 859.

In and around the AP site, APE1 makes several contacts with the DNA by binding to both major and minor grooves. R73, A74, and K78 make contacts with phosphates on the 5′ side of the AP site (Fig. 12.20A). G127 and Y128 widen the minor groove, R177 penetrates the major groove to form an H-bond with a phosphate at the AP site, while M270 moves into the minor groove and displaces the base opposite to the AP site (Fig. 12.20A). These interactions enable APE1 to kink the DNA by ~35° and flip the AP-site out of the double helix and put it into the active site binding pocket. The active site, formed by a set of amino acid residues E96, Y171, N174, D210, N212, D308, and H309, carries out the incision at the AP-site (Fig. 12.20B).

Subsequently, DNA polymerase β adds a dNMP to the 3′ end of the incision and removes the 5′-dRP residue; the remaining nick is sealed by DNA ligase III in conjunction with XRCC1 protein.

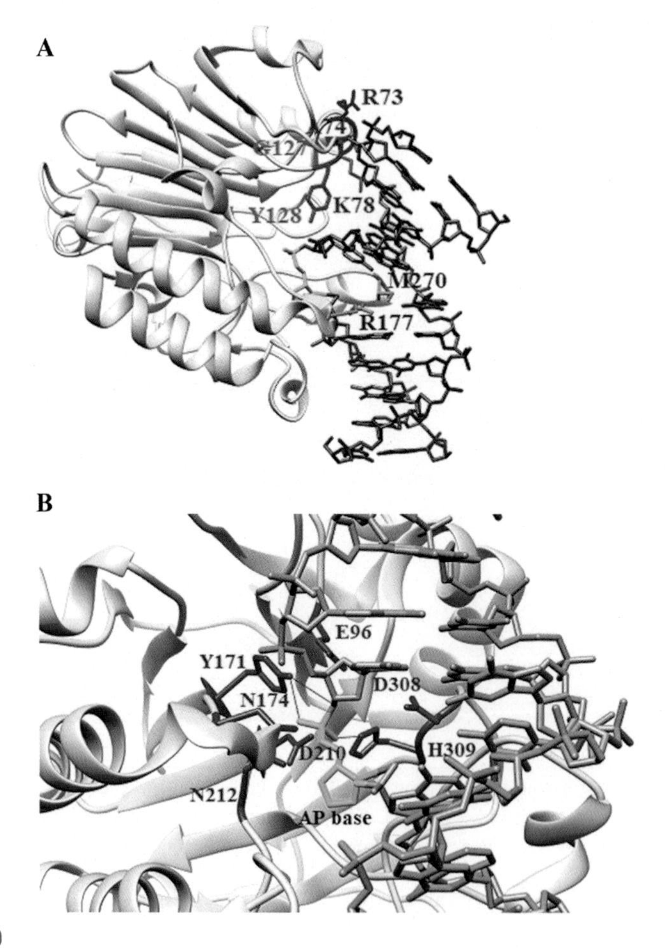

FIG. 12.20

APE1—AP site-containing DNA complex. (A) APE1 residues involved in modification of the DNA structure. (B) Active site residues responsible for incision at the AP-site. H-bonds shown by *red* lines.

Source: Mol, C.D., et al., 2000. Nature 403, 451.

12.5 Nucleotide excision repair

Nucleotide excision repair (NER) remedies bulky DNA lesions by recognizing local distortions in the double-helical shape of the DNA caused by the lesions. The mechanisms of NER in prokaryotes and eukaryotes are largely similar. They involve the following steps:

(a) Damage detection—the damage is detected by the joint action of proteins UvrA and UvrB (that is, UvrAB). In case a transcribing RNA polymerase (RNAP) halts at a damaged site before UvrAB detects the latter, the RNAP has to be displaced by a transcriptional-repair coupling factor (TRCF or Mfd) before UvrAB is recruited.

(b) Once the damage has been identified, there is a conformational change in the UvrAB complex such that the DNA is partially unwound near the lesion and wrapped around UvrB. Thus, the damage is "handed over" to UvrB and the formation of a tight UvrB-DNA scaffold facilitates the dissociation of UvrA—the preincision complex is established.

(c) Next, the UvrB-DNA complex recruits UvrC that contains two nuclease domains—one cleaves the phosphodiester bond eight nucleotides 5′ to the damaged site, while the other cleaves the phosphodiester bond four to five nucleotides 3′ to the damaged site.

(d) Following the cleavages, the postincision complex is removed by the dual action of UvrD and DNA polymerase I (PolI). UvrD excises the damage-containing oligonucleotide, while PolI fills up the gap thus created. In the final step, DNA ligase seals the newly generated repair patch.

UvrA plays a lead role in damage recognition. It binds the damaged DNA without the help of any other NER component and subsequently recruits UvrB to form the preincision complex.

UvrA binds DNA as a dimer. Besides, it possesses an ATPase/GTPase activity which is essential for the recognition of the lesion. Hence, we need to look at the structural features of UvrA which enable it to interact with DNA as well as UvrB. Fig. 12.21 shows the three-dimensional structure of UvrA from *Bacillus stearothermophilus*.

UvrA is a member of the ATP-binding cassette (ABC) superfamily of ATPases. Like other ABC ATPases, UvrA possesses a nucleotide-binding domain which consists of Walker A/P loop, Q-loop, ABC signature (LSGGN), Walker B, D-loop, and H-loop.

Each UvrA monomer carries two ABC ATPase structural modules. The dimeric form, therefore, possesses four nucleotide-binding sites formed in an intramolecular manner.

The interface of the UvrA monomeric subunits extends over a large surface area involving hydrophobic and H-bonding interactions. However, the interface does not contain bound nucleotides. The nucleotides are rather located within the monomers. Therefore, UvrA can dimerize even in the absence of a nucleotide.

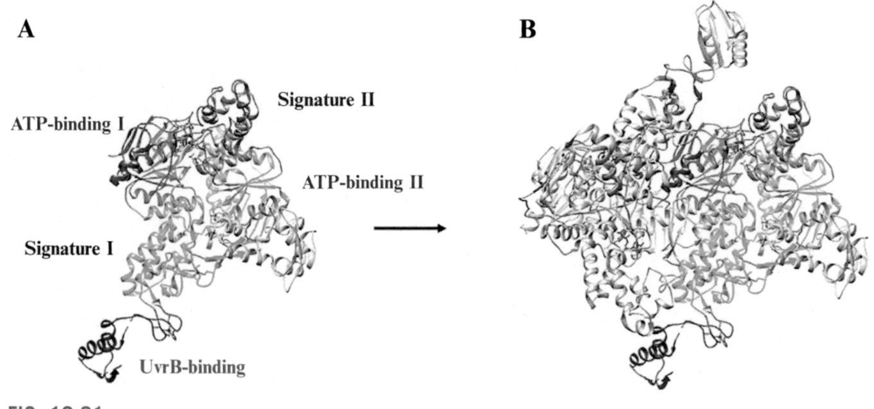

FIG. 12.21

Structure of *Bacillus stearothermophilus* UvrA: (A) monomer—domains labeled; (B) dimer—domains of only one subunit colored differently.

Source: Prakotiprapha, D., et al., 2008. Mol. Cell 29, 122.

Now, the question is how UvrA recognizes a lesion in the DNA. The question has been answered to some extent by the structural characterization of UvrA from *Thermotoga maritama*. It has been observed that the modified bases in the lesion are slightly pushed out of the double helix causing a stretch in the surrounding bases. The DNA in this region is bent by ~15° and unwound by ~20°. The weakened base-pairing leads to a general destabilization of the double helix that forms the basis of damage recognition.

The conformation of the UvrA dimer also changes in the course of binding of distorted DNA, which binds in the cleft formed by the dimer. UvrA interacts only with the DNA backbone in a sequence-independent manner. The most important contacts between UvrA and the DNA involve residues within or close to the signature domain II (G670, T679, Y680, R688, K704, S705, S708, and N710). H-bonding plays a major role in some of these contacts (Fig. 12.22).

Thus, it appears that an indirect readout mechanism enables UvrA to recognize lesions that may differ in both structure and chemical composition. The signature domain in each monomer grabs the DNA at a distance away from the lesion. In fact, UvrA neither interacts directly with the lesion nor distinguishes between the damaged and undamaged strands. The task is left to another protein, UvrB, to verify the presence of a lesion in the locally destabilized region.

UvrB plays a central role in the NER pathway. It interacts with not only UvrA for damage verification, but also UvrC, UvrD, and DNA polymerase I. The structure of *Bacillus caldotenax* UvrB is shown in Fig. 12.23. It consists of four domains—1a, 1b, 2, and 3. An important feature in the structure is a β-hairpin that bridges domains 1a and 1b.

UvrB from *Bacillus subtilis* has been crystallized as a dimer. The domain structure of each of its subunits is similar to that of *B. caldotenax* UvrB. The dimeric structure is shown in Fig. 12.24. The dimeric interface consists of the β-hairpin, parts of domain 2, and a helix (S481-K495) present at the outer edge of domain 3.

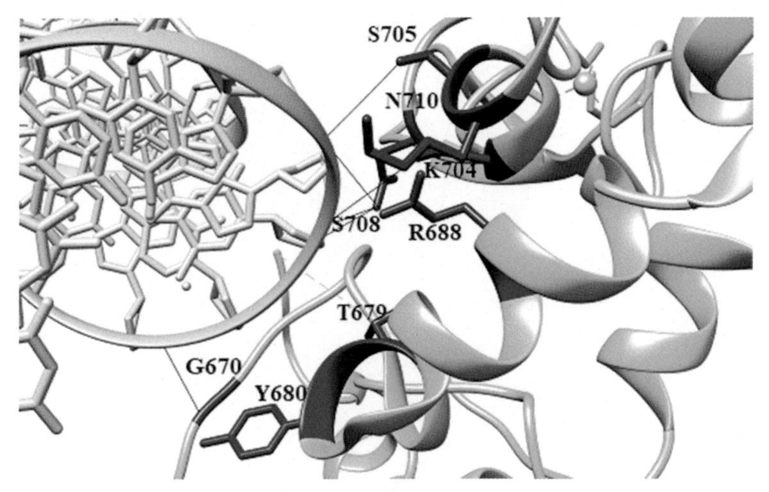

FIG. 12.22

DNA-UvrA contacts: DNA—*light blue*; UvrA—*light green*; *red* lines represent H-bonds.

Source: Prakotiprapha, D., et al., 2008. Mol. Cell 29, 122.

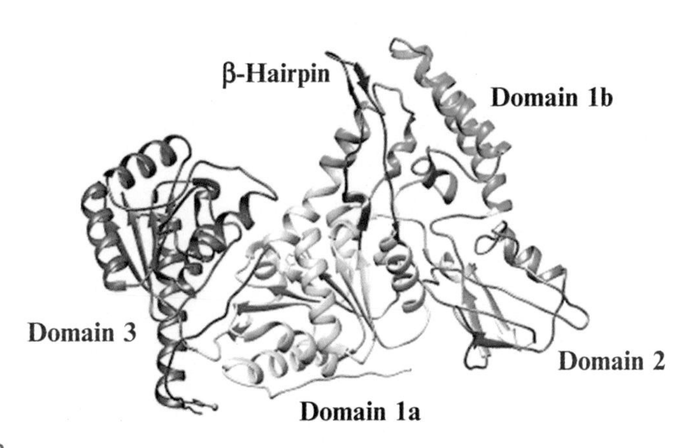

FIG. 12.23

Structure of *B. caldotenax* UvrB.

Source: Theis, K., et al., 1999. EMBO J. 18, 6899.

The contacts forming the interface are mostly electrostatic and van der Waals in nature. However, the interface in the UvrB dimer is much smaller than that in the UvrA dimer.

The β-hairpin plays a prominent role in the damage verification process. The structure of UvrB-DNA complex (Fig. 12.25) shows that one DNA strand threads behind the hairpin. One of the bases is flipped into a hydrophobic pocket. Tyrosine residues of the β-hairpin interact with the damage.

UvrB associates with UvrA to form a complex for damage recognition and verification. The complex contains two molecules of each protein. Domain 2 of UvrB interacts with the UvrB-binding domain of UvrA. The interaction interface

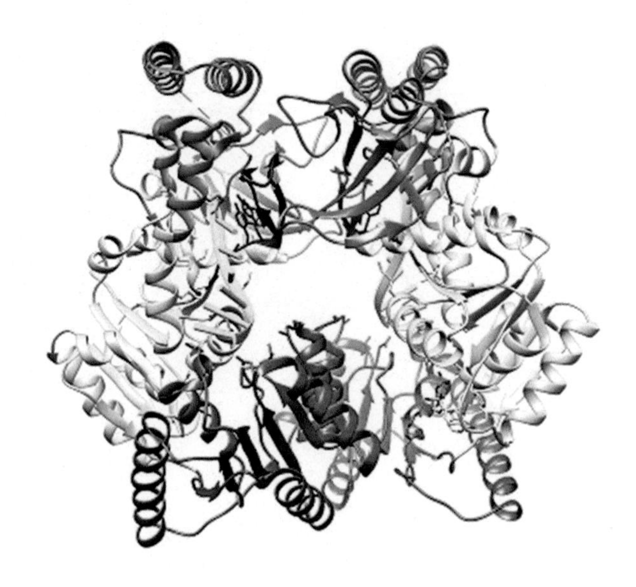

FIG. 12.24

Structure of *B. subtilis* UvrB dimer. Coloring scheme: UvrB domain 1a—*light blue*; domain 1b—*forest green*; domain 2—*medium purple*; domain 3—*orange red*; and β hairpin—*firebrick*. The DNA is colored *golden*.

Source: Webster, M.P., et al., 2012. Nucleic Acid Res. 40, 8743.

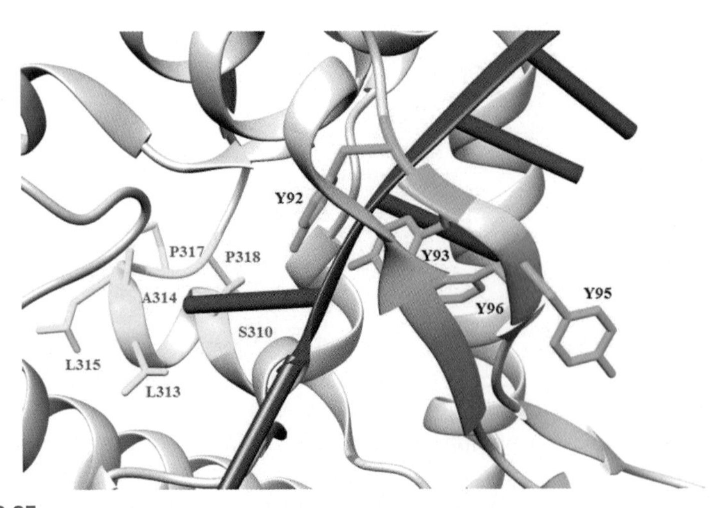

FIG. 12.25

UvrB-DNA interaction. Coloring scheme: UvrB—*grey*; β-hairpin—*dodger blue*; DNA—*purple*; tyrosine—*sea green*; hydrophobic residues—*yellow*.

Source: Truglio, J.J., et al., 2006. Nat. Struct. Mol. Biol. 13, 360.

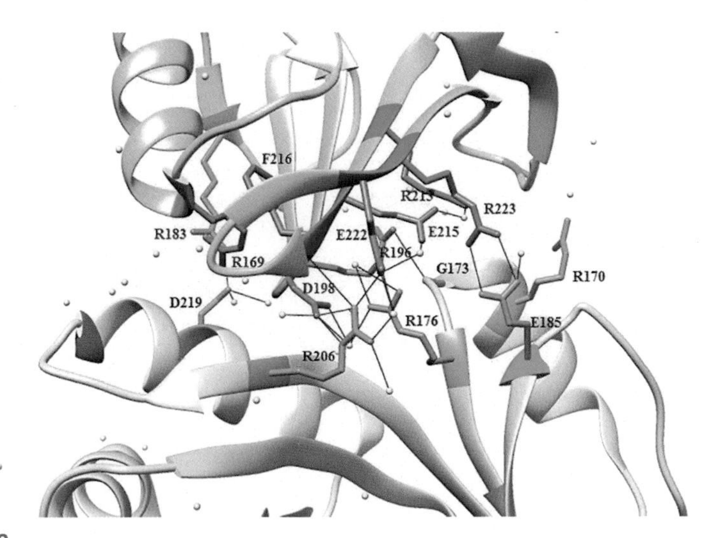

FIG. 12.26

UvrA-UvrB interaction. Coloring scheme: interacting residues in UvrA—*dodger blue*; rest of UvrA—*light blue*; interacting residues in UvrB—*forest green*; rest of UvrB—*light green*; water molecules—*light blue* dots; H-bonds—*red* lines.

Source: Pakotiprapha, D., et al., 2009. J. Biol. Chem. 284, 12837.

is shown in Fig. 12.26. It is a polar interface populated with a number of H-bonds (including water-mediated ones) and electrostatic interactions.

Once the damage is recognized and verified, UvrA dissociates from the complex leaving UvrB bound to the DNA. Next, UvrB recruits UvrC. UvrBC domain of UvrC interacts with the C-terminal domain of UvrB to form a complex that carries out the incision of the damaged strand.

The interaction between the respective domains of UvrB and UvrC can be understood on the basis of a high degree of sequence homology between *E. coli* UvrB residues 634–668 (call it UvrB′) and *E. coli* UvrC residues 205–239 (UvrC′). UvrB′ adopts a helix-loop-helix fold and dimerize. The dimerization involves hydrophobic residues M643, A647, L650, F652, A655, and A656 and polar residues E653 and R659. The hydrophobic residues pack together to form a core, while E653 of one monomer subunit forms a salt bridge with R659 of the other. With UvrC′ also showing the propensity to form a helix-loop-helix structure, similar interactions between the homologous domains of UvrB and UvrC are possible (Fig. 12.27).

The catalytic residues for the 3′ incision activity are present in the UvrC N-terminal domain. UvrC follows a one-metal mechanism for the cleavage of a phosphodiester bond. In *T. maritama* UvrC, a single divalent cation (Mg^{2+}; Fig. 12.28) is coordinated by E76 and five water molecules in an octahedral arrangement. The water molecules coordinating the metal also form additional contacts with other protein residues—hydroxyl of Y29, amide of K32; additionally, a hydrogen bond is formed between a water molecule and a carboxylate of E76 (Fig. 12.28).

On the other hand, the catalytic domain responsible for 5′ incision resides in the C-terminal half of UvrC. The core of this 5′ endonuclease domain shares significant

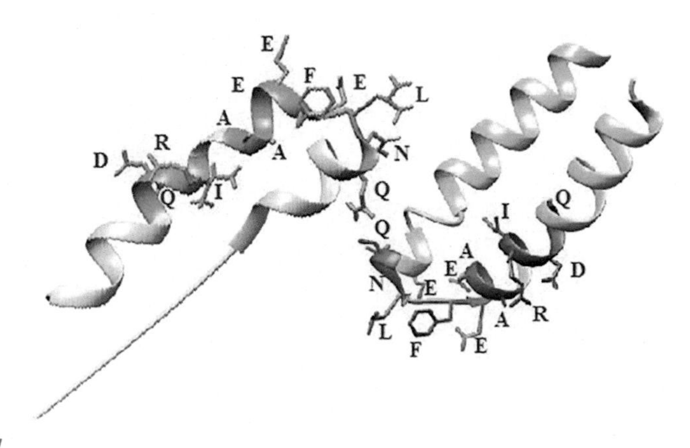

FIG. 12.27

Homologous domains of UvrB *(light green)* and UvrC *(light blue)*. Homologous residues are labeled and those in UvrB and UvrC are colored *deep green* and *dodger blue*, respectively.

Source: Karakas, E., et al., 2007. EMBO J. 26, 613.

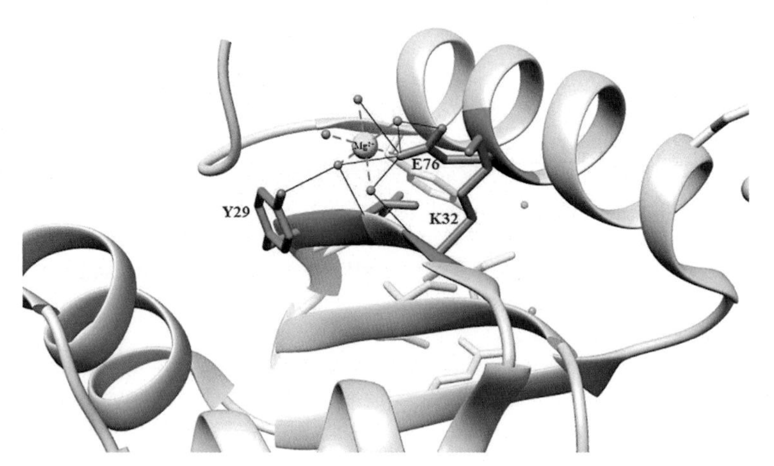

FIG. 12.28

Coordination of Mg^{2+} ion by UvrC. The metal ion is shown as a *golden sphere*. Hydrogen bonds with Y29, K32, and E76 are shown as *red* lines.

Source: Truglio, J.J., et al., 2005. EMBO J. 24, 885.

structural homology with the RNaseH enzyme family. However, unlike the other members of the family which contain a DDE triad at the catalytic center, the endonuclease domain of UvrC contains a DDH triad. The first two residues of the triad (D367 and D429 in *T. maritama*; D399 and D466 in *E. coli*) are essential for the 5′ incision reaction. The third residue (*T. maritama* H488) is required to be in the vicinity of the above two aspartates. Opposite to the active site is a positively charged surface which forms a protein-DNA interface that has an influence on both the 5′ and 3′ incision reactions (Fig. 12.29).

The endonuclease domain is connected to two helix-hairpin-helix (HhH) motifs, which are adjacent to each other, by a short flexible linker. The HhH motif is found in

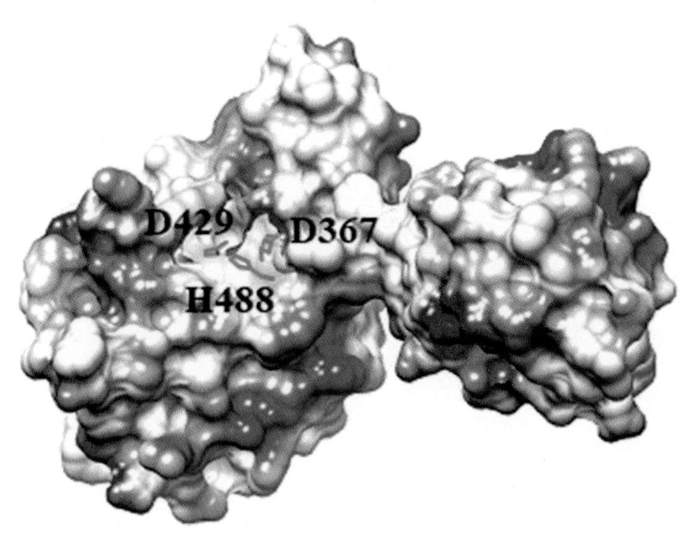

FIG. 12.29

Electrostatic surface potential of the C-terminal half of UvrC. Most negative—*blue*; neutral—*white*; most negative—*red*.

Source: Karakas, E., et al., 2007. EMBO J. 26, 613.

a number of other proteins and interacts nonspecifically with DNA by making contacts with the phosphate backbone of the minor groove. In UvrC, it is found to be essential for 5′ incision and, in certain cases, for 3′ incision (Fig. 12.30).

The next to move into action in the NER pathway is UvrD. Originally known as DNA helicase II in *E. coli*, UvrD unwinds the DNA in the 3′–5′ direction. Structurally, it has been delineated into four domains: 1A, 1B, 2A, and 2B (Fig. 12.31). It also contains a C-terminal extension.

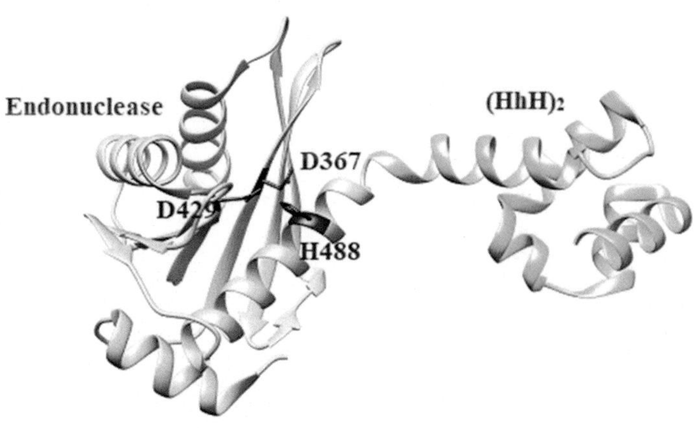

FIG. 12.30

Structure of 5′ endonuclease and $(HhH)_2$ domains of UvrC.

Source: Karakas, E., et al., 2007. EMBO J. 26, 613.

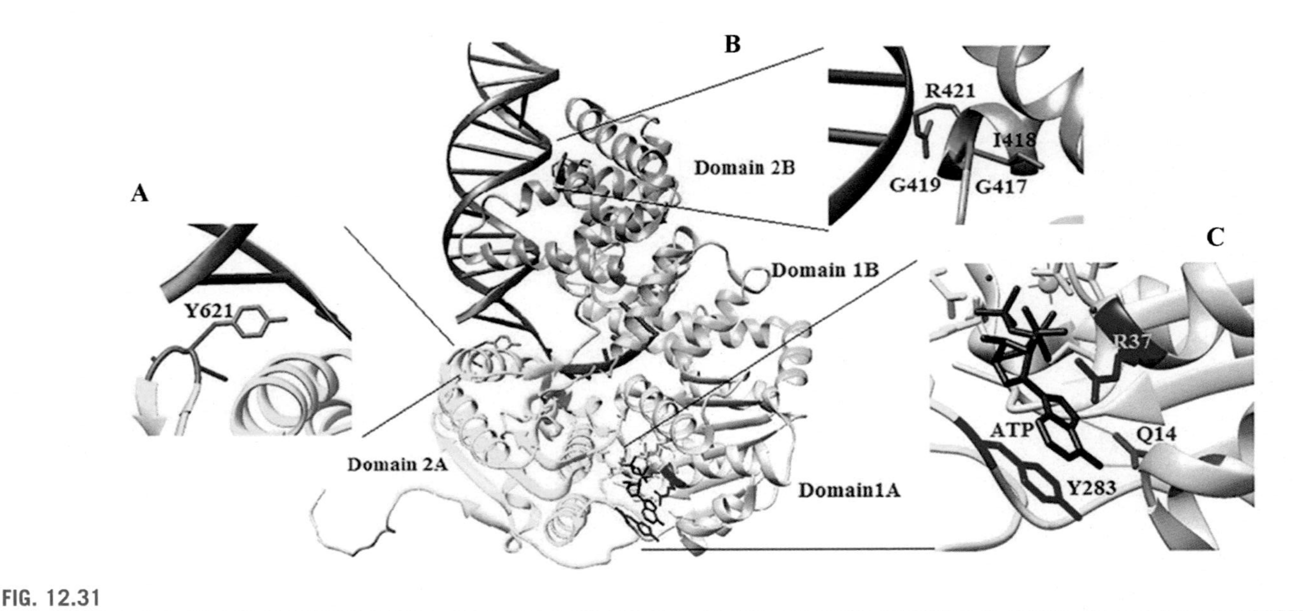

FIG. 12.31

Structural features of *E. coli* UvrD. Inset A: residues 620–623 of domain 2A essential for helix unwinding shown in *green*; inset B: GIG motif; inset C: ATP-binding site.

Source: Lee, J.Y., Wang, W., 2006. Cell 127, 1349.

UvrB interacts with UvrD to recruit the latter to the postincision NER complex. There are two regions in UvrD that interact with UvrB—domain 1A and the unstructured C-terminal extension.

Each ds-ss DNA junction binds a UvrD monomer and assumes an "L" shape with the ds and ss regions almost perpendicular to each other (Fig. 12.31).

ATP binds UvrD at the interface of domains 1A and 2A with the adenine base sandwiched between Y283 and R37, and H-bonded to Q14 (Fig. 12.31; inset C). ATP-binding leads to a significant conformational change in the protein inducing an ~20° rotation between domain 2A and rest of the protein. Subsequent ATP hydrolysis translocates the UvrD complex in a 3′–5′ direction with the ssDNA moving through a helix in domain 2B, known as the "gating helix"—1 bp translocation occurs at each step.

Residues 620–623 in domain 2A, which includes Y621 at the tip, are essential for helix unwinding (Fig. 12.31; inset A). Additionally, interactions exist between residues in the GIG motif and the adjacent R421 with the dsDNA (Fig. 12.31; inset B). The action of UvrD releases UvrC and the lesion-containing oligonucleotides, while UvrB, which remains bound to the undamaged strand, is removed by PolI.

12.6 Translesion synthesis

So far we have discussed in this chapter different cellular mechanisms of repairing a damaged DNA. Nevertheless, some lesions may still be able to evade the actions of all these repair processes and tend to stall a replication fork. This could have been deleterious to the cell had it not a set of specialized polymerases which can bypass the lesions. The "rescue," however, comes at a cost as the bypass mechanisms are error-prone—mutations in the form of base substitution and/or deletions being introduced in the process.

In Chapter 8, we have described the Y-family of DNA polymerases which carry out translesion DNA synthesis. We have seen that like DNA polymerases belonging to two other families, the overall structure of the Y-family polymerases resembles a right hand with palm, fingers, and thumb domains. Yet, there are prominent structural differences. The fingers and thumb domains of the Y family polymerases are smaller and, as a consequence, they make fewer contacts with the DNA substrate and incoming nucleotide. Evidently, they impose fewer constraints on the nascent base-pair and DNA substrate. Their active cleft is relatively more spacious and can, therefore, accommodate a variety of non-WC base-pairs. The dearth of constraints, however, results in reduced fidelity and processivity in DNA synthesis.

Translesion DNA polymerases are present across all domains of life. *E. coli* contains three such polymerases—PolII, PolIV, and PolV. However, here we are going to focus on Dbh, a Y-family member, from *Sulfolobus acidocaldarius*, which has been introduced in Chapter 8.

Dbh can efficiently bypass some damaged forms of dG such as 8-oxo-G. In fact the enzyme makes base substitution errors only at a frequency of 1 in 10^3–10^4 nucleotides replicated. However, it is responsible for a high rate of single-base deletions.

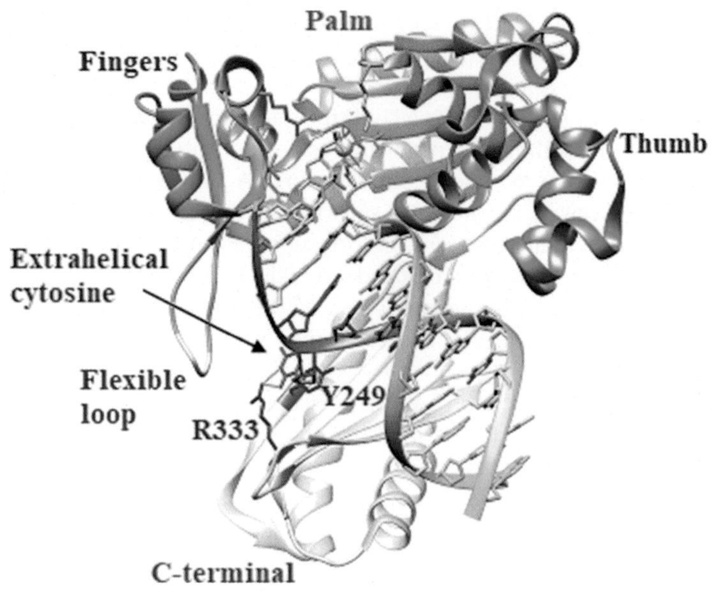

FIG. 12.32

Dbh in complex with substrate DNA containing a hotspot 5′-GCC-3′. Hotspot sequence colored *red*; rest of the DNA colored *golden*. H-bonds are shown by *blue* lines.

Source: Wilson, R.C., Pata, J.D., 2008. Mol. Cell 29, 767.

The sequence 5′-GPyPy-3′ is the "hotspot" of Dbh activity. A dC is incorporated opposite to dG (even if it is 8-oxo-G), but one of the two (identical) pyrimidines is skipped at a frequency of about 50% resulting in a –1 frameshift.

The basis of this single-base deletion has been explained with the help of crystal structures of Dbh in complex with substrate DNA containing a hotspot. What has been revealed in these structures is that a cytosine, located three bases upstream of the nascent base-pairs and immediately 3′ to the G, is unpaired and adopts an extrahelical conformation. The conformation is stabilized by two C-terminal residues of Dbh—Y249 stacks on the 3′ side of the extrahelical cytosine base, while R333 forms a bidentate H-bond with it (Fig. 12.32).

Additionally, Dbh residues (35–39) in the flexible loop may be interacting with a bulged base next to the extralelical cytosine. It is likely that all these interactions enable Dbh to generate a single-base deletion by a "template slippage" mechanism.

****

While concluding the chapter, it can be unambiguously stated that the different repair pathways described here are not completely independent of each other. Overlapping specificities do exit and often they complement each other. An example would be the Mfd- or TRCF-initiated pathway involved in the repair of 8-oxo-G. Normally, TRCF has a preference for BER; however, if the BER pathway is compromised in the cell, or its capacity is overwhelmed by the extent of damage, Mfd can recruit NER. The relative contribution of a specific pathway varies under differing contexts such as the stage of the cell cycle or other DNA transactions.

Sample questions

1. Why does the C8-G AF adduct formation in the DNA lead to the disruption of Watson-Crick base-pairing? What kind of mutation is introduced as a result?
2. What are the structural features based on which MutL protein plays the role of "molecular coordinator" in MMR?
3. How does electrostatic interactions facilitate the catalytic activity of *E. coli* photolyase?
4. How does human 8-oxoguanine DNA glycosylase (hOGG1) specifically interact with 8-oxo-G damage in the DNA?
5. Elucidate the structural features and interactions that help UvrB play a central role in the NER pathway.
6. Explain, on the basis of the structural features of a Y-family DNA polymerase, why translesion DNA synthesis is error-prone?

References and further reading

Dyrkheeva, N.S., Lebedeva, N.A., Lavrik, O.I., 2016. AP endonuclease 1 as a key enzyme in repair of apurinic/apyrimidinic sites. Biochemistry (Moscow) 81 (9), 951–967.

Goodman, M.F., Woodgate, R., 2013. Translesion DNA polymerases. Cold Spring Harb. Perspect. Biol. https://doi.org/10.1101/cshperspect.a010363.

Groothuizen, F.S., Sixma, T.K., 2016. The conserved molecular machinery in DNA mismatch repair enzyme structures. DNA Repair 38, 14–23.

Groothuizen, F.S., et al., 2015. MutS/MutL crystal structure reveals that the MutS sliding clamp loads Mutl onto DNA. eLife 4, e06744. https://doi.org/10.7554/eLife.06744.

Jaciuk, M., et al., 2011. Structure of UvrA nucleotide excision repair protein in complex with DNA. Nat. Struct. Mol. Biol. 18 (2), 191–197.

Jiricny, J., 2013. Postreplicative mismatch repair. Cold Spring Harb. Perspect. Biol. 5, a012633.

Karakas, E., et al., 2007. Structure of the C-terminal half of UvrC reveals an RNase H endonuclease domain with an argonaute-like catalytic triad. EMBO J. 26, 613–632.

Kisker, C., Kuper, J., Van Houten, B., 2013. Prokaryotic nucleotide excision repair. Cold Spring Harb. Perspect. Biol. 5, a012591.

Koval, V.V., Knorre, D.G., Fedorova, O.S., 2014. Structural features of the interaction between human 8-oxoguanine DNA glycosylase hOGG1 and DNA. Acta Nat. 6 (3), 52–65.

Krokan, H.E., Bjørås, M., 2013. Base excision repair. Cold Spring Harb. Perspect. Biol. 5, a012583.

Lee, J.Y., Yang, W., 2006. UvrD helicase unwinds DNA one base pair at a time by a two-part power stroke. Cell 127, 1349–1360.

Michael, P.S., et al., 2011. Chemistry and structural biology of DNA damage and biological consequences. Chem. Biodivers. 8, 1571–1614.

Miroshnikova, A.D., et al., 2016. Effect of mono- and divalent metal ions on DNA binding and catalysis of human apurinic/apyrimidinic endonuclease 1. Mol Biosyst. 12, 1527–1539.

Pakotiprapha, D., et al., 2008. Crystal structure of *Bacillus stearothermophilus* UvrA provides insight into ATP-modulated dimerization, UvrB interaction, and DNA binding. Mol. Cell 29, 122–133.

Pakotiprapha, D., Liu, Y., Verdine, G.L., Jeruzalmi, D., 2009. A structural model for the damage-sensing complex in bacterial nucleotide excision repair. J. Biol. Chem. 284 (19), 12837–12844.

Perevoztchikova, S.A., et al., 2013. Modern aspects of the structural and functional organization of the DNA mismatch repair system. Acta Nat. 5 (3), 17–34.

Robertson, A., Pattishall, A.R., Matson, S.W., 2006. The DNA binding activity of MutL is required for methyl-directed mismatch repair in *Escherichia coli*. J. Biol. Chem. 281 (13), 8399–8408.

Sancar, A., 2016. Mechanisms of DNA repair by photolyase and excision nuclease (Nobel lecture). Angew. Chem. Int. Ed. 55, 8502–8527.

Sohi, K., et al., 2000. Crystal structure of *Escherichia coli* UvrB C-terminal domain and a model for UvrB-UvrC interaction. FEBS Lett. 465, 161–164.

Webster, M.P.J., et al., 2012. Crystal structure of the UvrB dimer: insights into the nature and functioning of the UvrAB damage engagement and UvrB-DNA complex. Nucleic Acids Res. 40 (17), 8743–8758.

Wilson, R.C., Pata, J.D., 2008. Structural insights into the generation of single-base deletions by the Y-family DNA polymerase Dbh. Mol. Cell 29, 767–779.

Wu, J., 2017. DNA binding strength increases the processivity and activity of a Y-family DNA polymerase. Nat. Sci. Rep. 7, 4756. https://doi.org/10.1038/s41598-017-02578-3.

Yang, W., 2008. Structure and mechanism for DNA lesion recognition. Cell Res. 18, 184–197.

Zhang, M., Wang, L., Zhong, D., 2017. Photolyase: dynamics and electron-transfer mechanisms of DNA repair. Arch. Biochem. Biophys. 632, 158–174.

CHAPTER

Recombination

13

Exchange of genetic information between DNA molecules is continuously at work in nature. The phenomenon is known as genetic recombination. A very prominent way this exchange can occur is based on similarity or near-similarity of nucleotide sequences between two DNA molecules. This is called homologous recombination (HR).

Examples of homologous recombination are as common as that associated with eukaryotic meiosis or observed in bacterial conjugation. The two phenomena play a crucial role in the respective organisms in the generation of genetic diversity. Homologous recombination also plays an essential role in the maintenance of genetic stability by repairing DNA double-strand breaks (DSBs).

DNA damage, as we have seen in Chapter 12, can occur as a consequence of endogenous metabolic reactions in the cell or driven by environmental factors such as radiation. Some of the damages can take the form of single-strand gaps (SSGs) or double-strand breaks (DSBs). Both SSGs and DSBs are produced in every cell cycle as a result of DNA replication on a damaged template. One way DSBs can be repaired is by nonhomologous end joining (NHEJ). Nevertheless, being template-driven, HR is a relatively more accurate process of DNA repair.

13.1 Homologous recombination

The basic steps of recombinational DNA repair are conserved in the three domains of life (Fig. 13.1). They are as follows: (a) initiation of recombination—"resection" or enlargements of the DNA break by nucleases and DNA helicases resulting in the generation of single-stranded DNA (ssDNA); (b) homologous DNA pairing and strand exchange—search for complementarity by the newly created ssDNA leading to reciprocal exchange of DNA strands and formation of a "joint molecule" (Holliday junction, HJ); (c) branch migration—the joint molecule (single or double HJ) migrates in either direction to extend or collapse the heteroduplex DNA; and (d) separation of the joined DNA molecules by nucleolytic or topological means. Needless to say, each step involves multiple protein factors.

Fundamentals of Molecular Structural Biology. https://doi.org/10.1016/B978-0-12-814855-6.00013-4

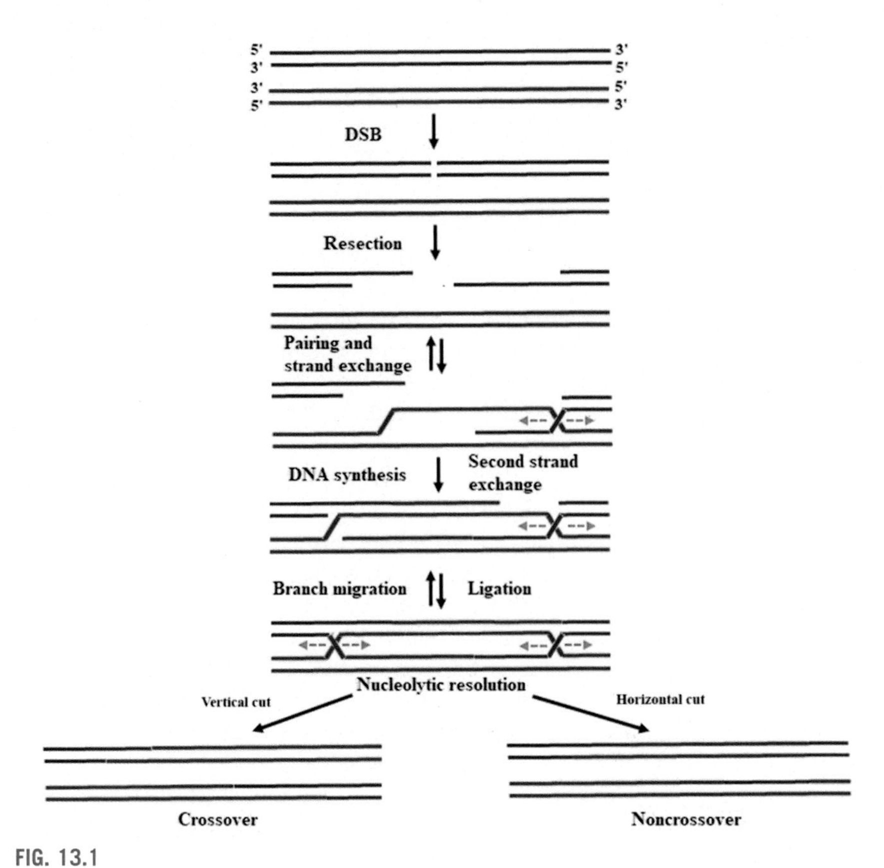

FIG. 13.1

Basic steps of recombinational DNA repair.

13.1.1 Initiation of recombination: Resection and recombinase loading

We know that DNA sequence information is stored as duplex DNA; nonetheless, the information is read in the form of ssDNA. Hence, the primary objective of the initiation process is the creation of ssDNA suitable for homologous DNA recognition. Processing of DSBs at this stage involves specific helicases and nucleases.

Bacteria

In bacteria, two enzymes are primarily involved in resection—RecBCD and RecQ. RecQ is a 3′→5′ DNA helicase that functions in the RecF pathway which is mostly responsible for SSG repair. Here, we are going to elaborate on the RecBCD pathway which is considered to be the major pathway for DNA DSB repair.

Escherichia coli RecBCD is a multifunctional heterotrimeric enzyme complex. Its action is dependent on the helicase and nuclease activities of its component subunits and a Chi (crossover hotspot instigator) sequence element 5′-GCTGGTGG-3′ in the DNA. The processing of DSBs by RecBCD is carried out in the following sequential steps (Fig. 13.2): (A) The enzyme binds tightly to the dsDNA at the break. (B) ATP-dependent DNA unwinding is initiated. The ssDNA products generated by continuous unwinding are cleaved asymmetrically, the 3′-terminated tail being cleaved more frequently than the 5′-tail. (C) DNA translocation continues till the enzyme encounters a Chi site in the 3′-termnated strand and pauses. (D) Interaction with the Chi sequence dramatically modifies the biochemical properties of the enzyme. The nuclease polarity is reversed. Cleavage on the 3′-tail ceases at or within

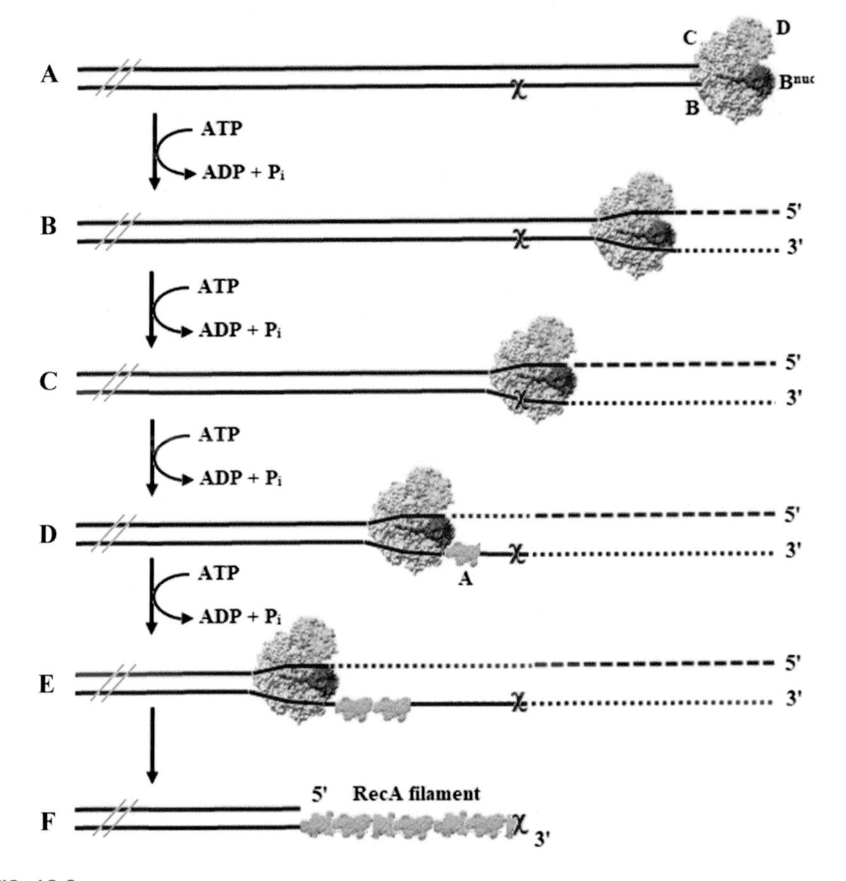

FIG. 13.2

Double-strand break processing by RecBCD enzyme. RecBCD subunits (including RecB^{nuc} domain), RecA and χ (Chi) have been labeled.

a few bases 3′ to the Chi; the 3′-tail is protected from further digestion. The rate of hydrolysis of the 5′ ssDNA tail escalates. At this stage, RecBCD facilitates the loading of the recombinase RecA protein on the 3′-tail. (E) RecBCD-mediated deposition of RecA continues and a filament grows primarily in the 5′→3′ direction. (F) RecBCD dissociates from the DNA. The RecA-coated 3′ ssDNA tail with Chi at its terminus can now initiate homologous recombination. Evidently, most of the steps mentioned above are ATP-dependent.

The multiple enzyme activities catalyzed by RecBCD can be attributed to its subunits—RecB, RecC, and RecD (Fig. 13.3). RecB has both 3′→5′ helicase and nuclease activities; RecC interacts with Chi and RecD is a 5′→3′ helicase. Recognition of ssDNA by the helicases is nonspecific, most likely guided by electrostatic and aromatic interactions.

RecB is a DNA-dependent ATPase. The main body of RecB comprising four domains—1A, 1B, 2A, and 2B—is connected to domain 3, the nuclease domain ($RecB^{nuc}$), by a long linker of about 70 amino acids (Fig. 13.3). The 3′-tail of the bound DNA spans domain 2A. The nuclease domain contains three conserved acidic residues (E1020, D1067, and D1080) and a lysine residue (K1082) in its active site (Fig. 13.4). D1080 is considered to be crucial for the nuclease activity.

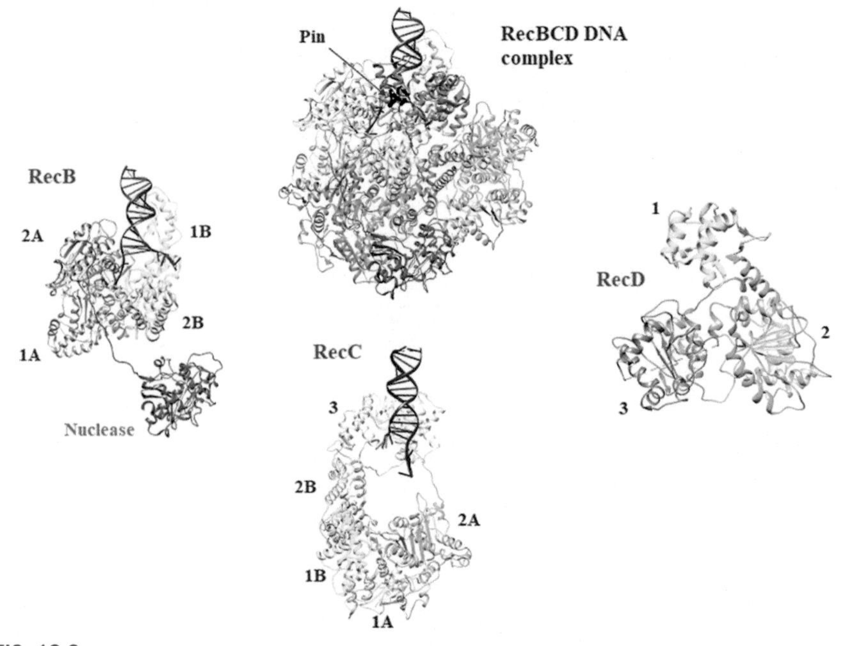

FIG. 13.3

Structure of RecBCD-DNA complex and individual subunits. Domains of each subunit, when shown separately, colored differently and labeled.

Source: Singleton, M.R., et al., 2004. Nature 432, 187.

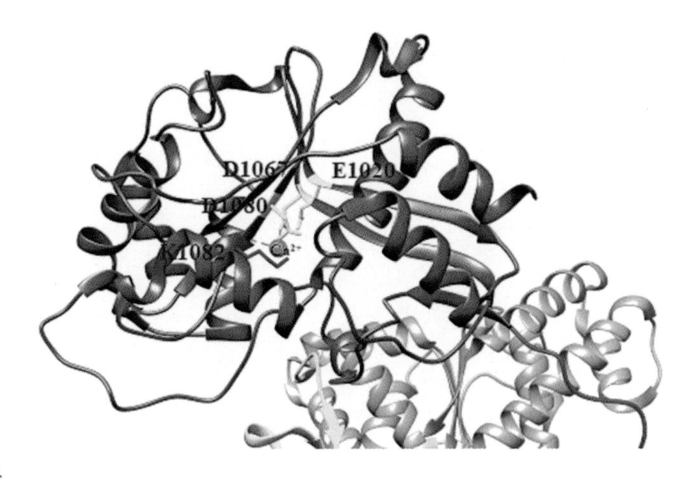

FIG. 13.4

Active site of RecB nuclease domain.

Source: Singleton, M.R., et al., 2004. Nature 432, 187.

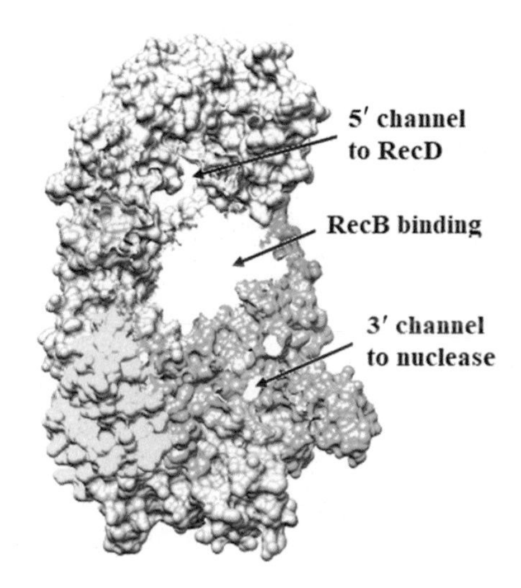

FIG. 13.5

Channels through RecC subunit.

Source: Singleton, M.R., et al., 2004. Nature 432, 187.

There are three large channels through the RecC subunit (Fig. 13.5). The largest channel accommodates the 2B domain of RecB. Through the other two channels, ssDNA tails move towards or away from the two helicase subunits. RecC contacts both strands of the DNA and split them across its "pin" to feed the 3′-tail to the RecB subunit and 5′-tail to RecD.

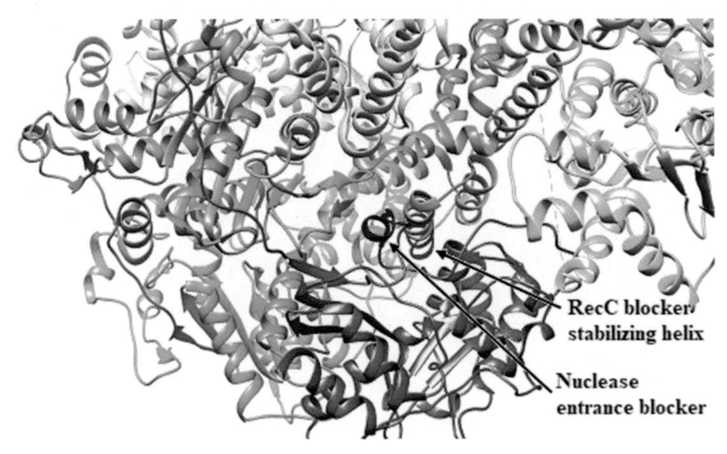

FIG. 13.6

RecB α-helix (colored red) blocking entrance to nuclease active site. RecC blocker-stabilizing helix colored *purple*.

Source: Singleton, M.R., et al., 2004. Nature 432, 187.

Like RecB, the RecD subunit also drives DNA unwinding, although in the opposite $5' \to 3'$ direction, by acting as an ssDNA motor. RecD contains three domains—1, 2, and 3—the last two being the motor domains.

There are two exit channels for the 3′-tail as it emerges from RecC. One exit channel is located on the back side of the nuclease active site. The other is blocked by an α-helix (residues 913–922) within the linker region of RecB (Fig. 13.6). This helix is stabilized by stacking interactions with another α-helix (residues 620–627) in the RecC subunit. Evidently, the nuclease activity is suppressed. Based on the three-dimensional structure of RecBCD complexed with a suitable DNA substrate, it has been hypothesized that as the 5′-tail grows beyond a certain length (~10nt), RecD induces a large conformational change in the complex affecting the RecB and RecC subunits. As a result, the blocking α-helix swings out of the way and the nuclease activity of RecB is stimulated.

The nuclease activity switches between the 5′- and 3′-tails by the intervention of Chi. Before RecBCD encounters Chi, as the 3′-tail finds its way through the RecC subunit and emerges at the nuclease active site, it is processively hydrolyzed. On the other hand, the 5′-tail, which is fed to RecD, emerges at the opposite side of the nuclease active site. In the competition for access to the site, the 5′-tail is in a conformationally less favorable disposition than the 3′-tail. Hence, the 5′-tail is digested less frequently.

Since the 3′-tail passes through the RecC subunit, the latter is able to scan the DNA for a Chi sequence. On encountering Chi, RecC binds tightly to this sequence element and prevents further digestion of the 3′-terminated strand. The 5′-tail is now cleaved more frequently as its access to the nuclease site is less hindered (Fig. 13.2).

As the ssDNA is created in the process of resection, it needs to receive the recombinase molecules for filament formation. In RecBCD-mediated recombination, the

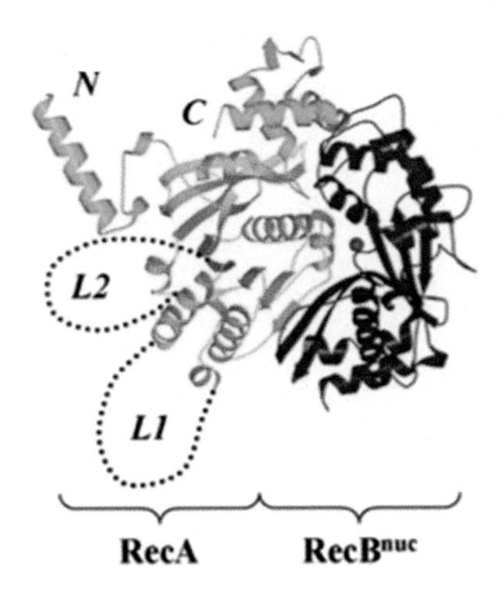

FIG. 13.7

Structure of the RecA·RecB^{nuc} complex predicted by docking algorithm.

Courtesy: Spies, M., Kowalczykowski, S.C., The RecA-binding locus of RecBCD is a general domain for recruitment of DNA strand-exchange proteins. Mol. Cell, February 17, 2006, with permission from Elsevier.

recombinase RecA faces competition from ssDNA-binding protein (SSB) for binding to the ssDNA. In *E. coli*, the problem has been addressed by the direct involvement of RecBCD in loading RecA on to the ssDNA in response to Chi recognition. The RecA-binding site resides in the nuclease domain (RecB^{nuc}) of RecBCD. Only on Chi recognition, the site is exposed for interaction and binds a RecA dimer (Fig. 13.7). There are additional proteins that facilitate the assembly of RecA on ssDNA.

Eukaryotes

In eukaryotes, the Mre11-Rad50-Nbs1 (MRN) complex (Mre11-Rad50-Xrs2 or MRX in *Saccharomyces cerevisiae*) plays a central role in signaling, processing, and repairing DNA DSBs. In fact, homologs of Mre11 and Rad50 are present in all three domains of life. The MRN complex consists of two subunits of Mre11, two subunits of Rad50, and either one or two Nbs1 subunit(s). (Fig. 13.8). The N-terminal domain of each Mre11 subunit contains phosphodiesterase motifs that are required for both $3' \rightarrow 5'$ dsDNA exonuclease and ssDNA endonuclease activities. Rad50 is an ATPase; it contains ATP-binding motifs in both N- and C-terminal regions. The two motifs are separated by an intramolecular coiled-coil. The Nbs1 subunit is specific to eukaryotes. Its core domains in the N-terminal region bind several phosphorylated proteins to regulate MRN interactions.

MRN binds DNA and functions as a sensor to DNA DSB. ATP-dependent DNA-binding is essentially mediated by Rad50 (Fig. 13.9). The DNA duplex is placed in the positively charged groove between the two coiled-coils of the Rad50 dimer. Rad50 also regulates the nuclease activity of Mre11 in an ATP-dependent manner.

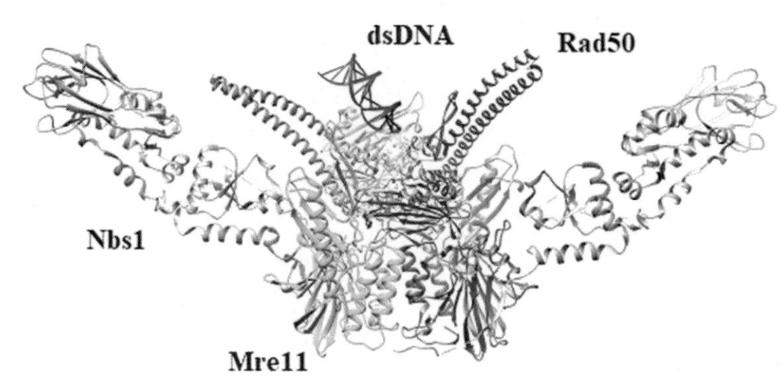

FIG. 13.8

Eukaryotic MRN complex bound to dsDNA. Two Mre11 subunits colored *dodger blue* and *cyan*; two Rad50 subunits colored *golden* and *coral*; both Nbs1 subunits colored *gray*.
Source: Seifert, F.U., et al., 2016. EMBO J. 35, 759; Schiller, C.B., et al., 2012. Nat. Struct. Mol. Biol. 19, 693.

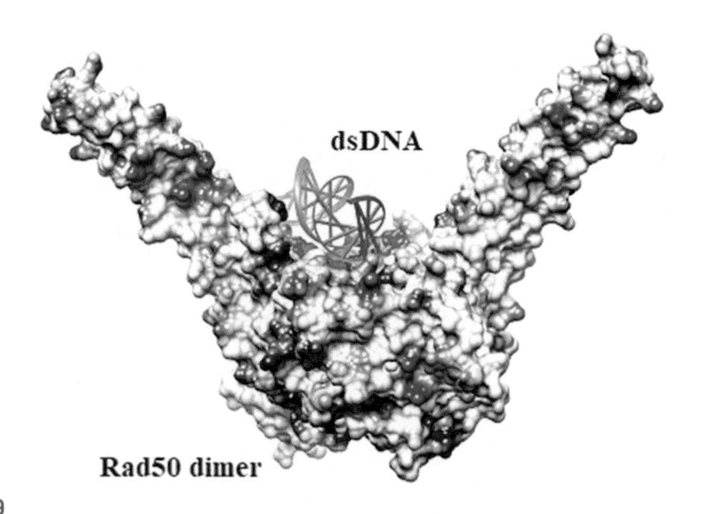

FIG. 13.9

Electrostatic surface potential of Rad50 dimer. Coloring: *blue*—positive; *white*—neutral; *red*—negative.

Source: Seifert, F.U., et al., 2016. EMBO J.

In the presence of ATP, Rad50 dimer binds Mre11 nuclease active site and blocks its access to DNA. In the absence of ATP, Mre11 binds the Rad50 ATPase domain near the base of the coiled-coils, making the nuclease site accessible to DNA for resection.

MRN function in DSB resection is aided by CtIP (Sae2 in *S. cerevisiae*). Initially, Mre11 makes an endonucleolytic cleavage in both 5′-terminated strands several base pairs away from the DSB. Subsequently, in conjunction with phosphorylated CtIP, Mre11 exonuclease activity digests DNA in the 3′→5′ direction from the nick

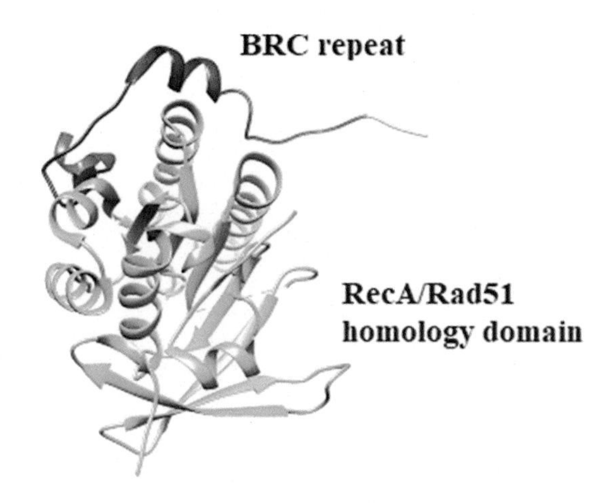

FIG. 13.10

RecA-homology domain of Rad51 bound to a BRC repeat of BRCA2.

Source: Pellegrini, L., et al., 2002. Nature 420, 287.

towards the DSB end. At the same time, $5' \rightarrow 3'$ exonucleases, such as EXO1, extend the resection from the nick away from the DSB. Altogether, 3′ ssDNA overhangs, which in many cases exceed 1000 nt in length, are generated. The 3′-tail is now committed to HR for which the recombinase protein Rad51 must be assembled on it.

In human cells, one of the several proteins that influence Rad51-mediated nucleoprotein filament formation is the tumor suppressor protein BRCA2. One of the key motifs in BRCA2 is a 35 aa sequence. This motif, repeated eight times in human BRCA2, is referred to as the BRC repeat. Through these repeats, BRCA2 interacts with the RecA-homology domain of Rad51 and loads the latter on to Replication protein A (RPA)-ssDNA produced by DSB resection (Fig. 13.10).

13.1.2 Homologous DNA pairing and strand exchange

Bacteria

As we have seen that at the end of DSB resection in bacteria by RecBCD, a RecA filament is formed on the resultant ssDNA. RecA is a recombinase that lies at the core of ATP-dependent homologous recombination in bacteria. ATP-bound filament formation is a primary requirement for the recombinase activity of RecA. As the filament appears before DNA pairing or synapsis, it is also called a "presynaptic complex."

RecA, a 38 kDa protein, is the founding member of a family of "strand-exchange proteins." It forms a right-handed nucleoprotein filament on both ssDNA and dsDNA. The binding is cooperative and faster on ssDNA than on dsDNA. Crystal structures of ssDNA-bound presynaptic and dsDNA-bound postsynaptic RecA filaments have provided us some understanding of RecA-mediated homology search

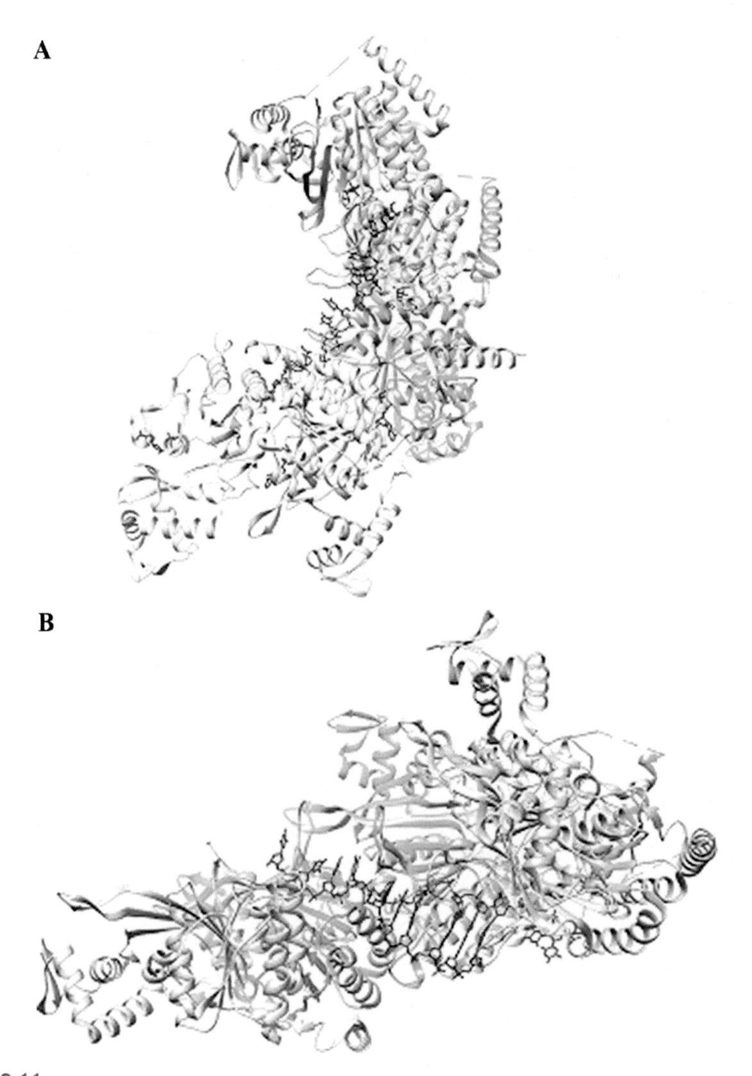

FIG. 13.11

RecA monomers forming (A) presynaptic (with ssDNA) and (B) postsynaptic (with dsDNA) nucleoprotein filaments. Each monomeric subunit has been colored differently. The DNA has been colored *brown*; H-bonds have been shown by *dark blue* lines.

Source: Chen, Z., Yang, H., Pavletich, N.P., 2008. Nature 453, 489.

and recognition (Fig. 13.11). It has been observed that, within the nucleoprotein filament, the ssDNA is extended to about 150% relative to a dsDNA molecule of the same length. This stretching is essential for subsequent DNA-strand exchange. However, the stretching is not uniform at the nucleotide level. Distinct sets of nucleotide triplets are seen in the filament. Each triplet maintains more or less normal B-form

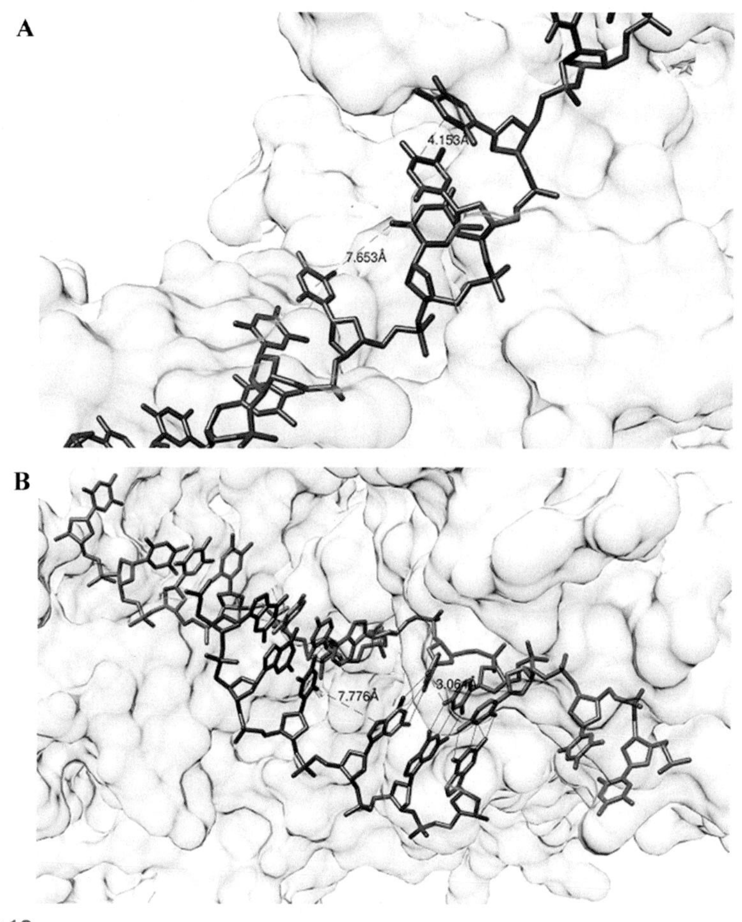

FIG. 13.12

(A) ssDNA and (B) dsDNA stretched within RecA filaments into nucleotide triplets. RecA has been represented as surface and H-bonds have been shown by *dark blue* lines. Internucleotide distance within the triplets and separation between the triplets have been labeled.

Source: Chen, Z., Yang, H., Pavletich, N.P., 2008. Nature 453, 489.

dimensions which are ~3.2–3.5 Å for dsDNA and ~3.5–4.2 Å for ssDNA, whereas between the triplets there is a separation of 7–8 Å (Fig. 13.12A).

The presynaptic complex is the active form of RecA that is responsible for DNA homology searches followed by pairing and exchange of DNA strands. Synapsis is a bimolecular collision process between the RecA-based nucleoprotein filament and the homologous DNA molecule. The process is dependent on ATP-binding, and not ATP hydrolysis. Homology is searched by sampling many transient and weak contacts. The B-form triplets function as "units" or "genetic words" for the homology search (Fig. 13.12B).

When a sufficient amount of homology is found (>8 bases), the free energy of ATP-binding to stretch and unwind the ssDNA within the filament is utilized to bind the incoming dsDNA through base-pairing interactions. The dsDNA binds to a secondary site in the RecA filament. RecA then facilitates the formation of Watson-Crick H-bonds between the DNA strand at the primary binding site and its complement in the duplex DNA bound at the secondary site. Thus, a three-stranded structure or "joint molecule" is created.

It is to be noted that each RecA protomer interacts with an entire triplet which is closest to itself. Additionally, it also interacts with a nucleotide of the preceding triplet and one of the following triplets. So, effectively, each nucleotide triplet is contacted by three protomers.

There are three domains in RecA—N-terminal, central, and C-terminal. The central contains the catalytic center for ATP hydrolysis. The central domain also contains two flexible loops—L1 (residues G157–M164) and L2 (residues E194 to T209) (Fig. 13.13A). An aspartate residue (D161) is located at the tip of L1. It prevents dsDNA, and not ssDNA, from binding to the primary site (Fig. 13.13B). Thus, D161 in the flexible loop L1 determines the preference for ssDNA-binding to the primary site and influences the DNA-binding order (Fig. 13.13B, C).

Inside the cell, RecA faces competition from ssDNA-binding protein (SSB) for binding ssDNA. The competition serves a dual regulatory purpose. On the one hand, it suppresses potentially detrimental recombination events and, on the other hand, SSB denatures local secondary structures that inhibit filament formation.

Eukaryotes

The eukaryotic counterpart of RecA is the Rad51 recombinase. As expected, several features and behaviors are shared between the two; yet certain differences do exist. Like RecA, Rad51 also engages both ssDNA and dsDNA in clusters of nucleotide triplets. Here again, two flexible loops, L1 and L2, are involved in DNA interactions in both presynaptic and postsynaptic complexes.

In the presynaptic complex, Rad51 interacts with each nucleotide triplet of the ssDNA on the phosphate backbone, while the bases point outwards (Fig. 13.14A). These interactions stretch the ssDNA in the Rad51 filament to ~1.5 times the length of B-form DNA (Fig. 13.14B).

In the postsynaptic complex, the complementary strand forms Watson-Crick base pairs with the invading ssDNA, also in triplet clusters (Fig. 13.15A). In this Rad51-organized complex, a specific Loop1 residue (R235), which is not found in RecA, intrudes into the space between adjacent nucleotide triplets and interacts with the phosphate backbone of the complementary strand (Fig. 13.15B). It is likely that the residue helps stabilize the base-pairing. In the event, the pairing between the invading and complementary strands is restricted to less than eight nucleotides; only two R235 residues can engage in stabilization of the nascent DNA joint which is, therefore, inherently labile.

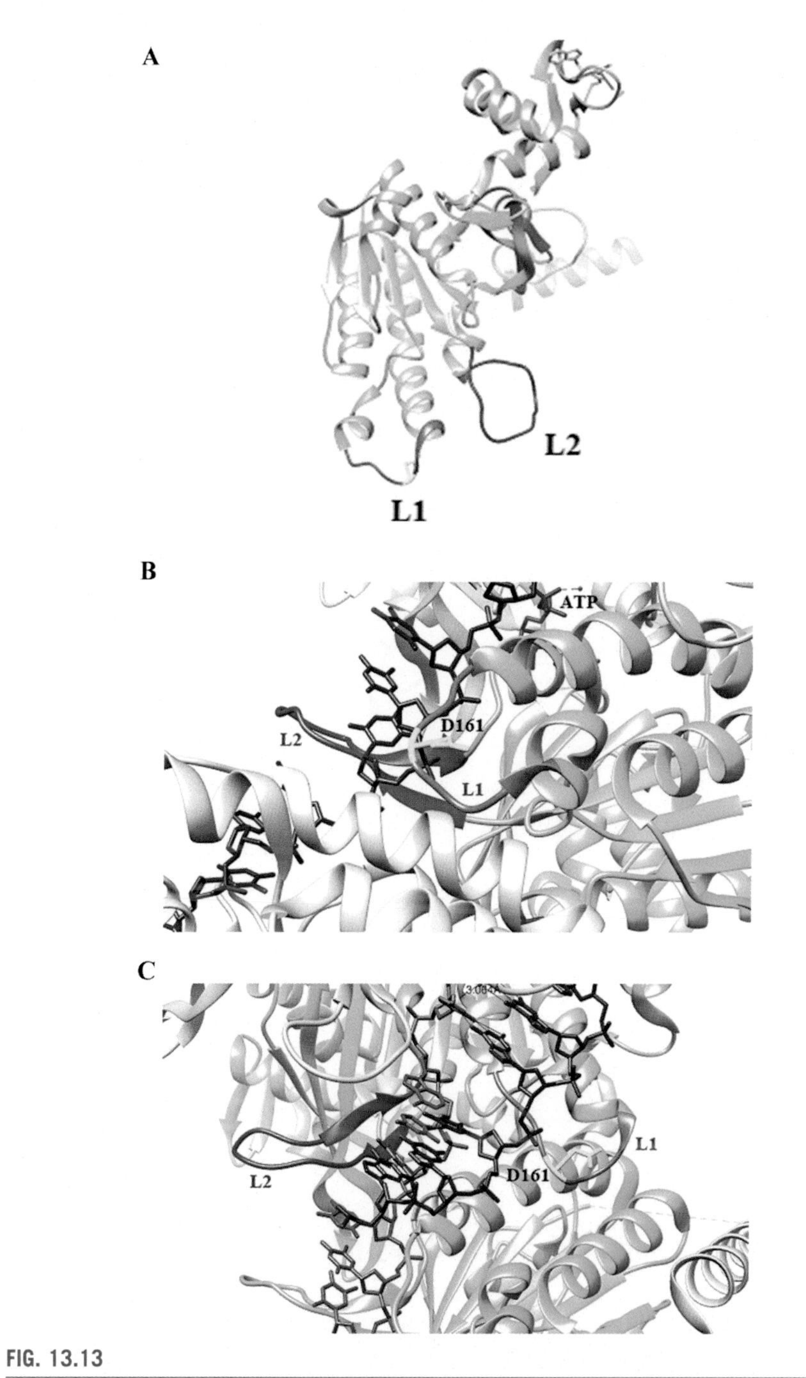

FIG. 13.13

(A) Flexible loops L1 and L2 of RecA (B) L1 with D161 at the tip present in the primary (ssDNA) binding site. (C) L2 in the secondary (dsDNA) binding site.

Source: Chen, Z., Yang, H., Pavletich, N.P., 2008. Nature 453, 489; Shinohara, T., et al., 2015. Nucleic Acids Res. 43, 973.

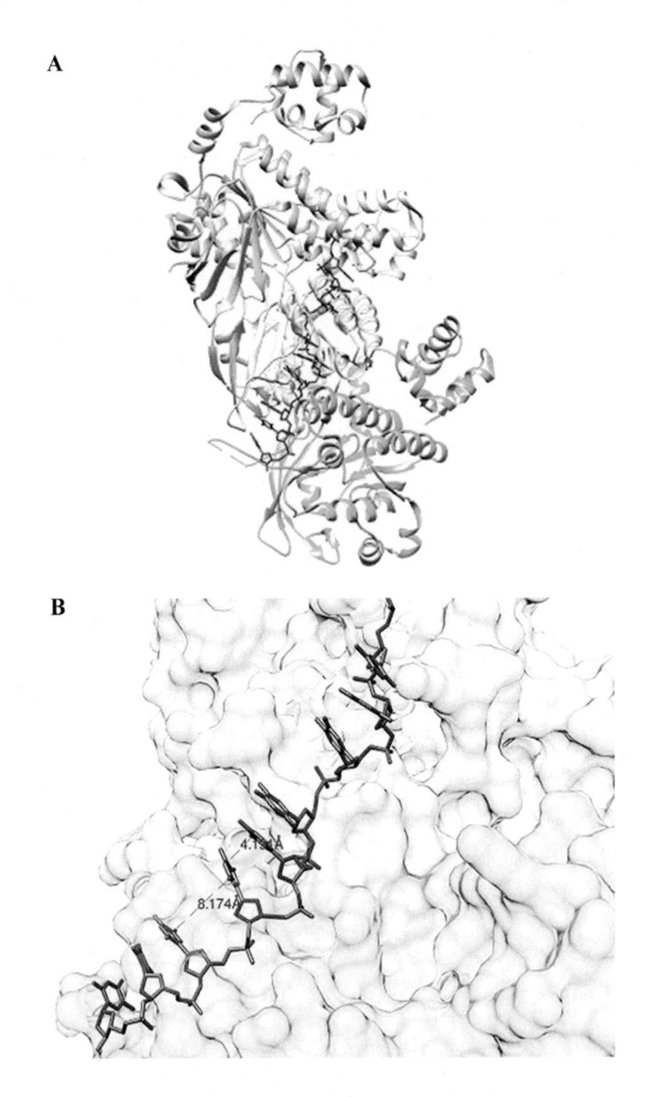

FIG. 13.14

(A) Rad51-ssDNA presynaptic complex. Three Rad51 monomers colored differently; ssDNA shown in brown. (B) ssDNA stretched on Rad51 surface; distances labeled.

Source: Xu, J., et al., 2017. Nat. Struct. Mol. Biol. 24, 40.

13.1.3 Branch migration and Holliday junction resolution

Bacteria

After synapsis, there is a reciprocal exchange of strands between two DNA molecules which are then connected by a DNA branch. This key intermediate of the recombination process is known as the Holliday Junction (HJ; also called cross-strand exchange) (Fig. 13.16). The length of the heteroduplex is controlled by

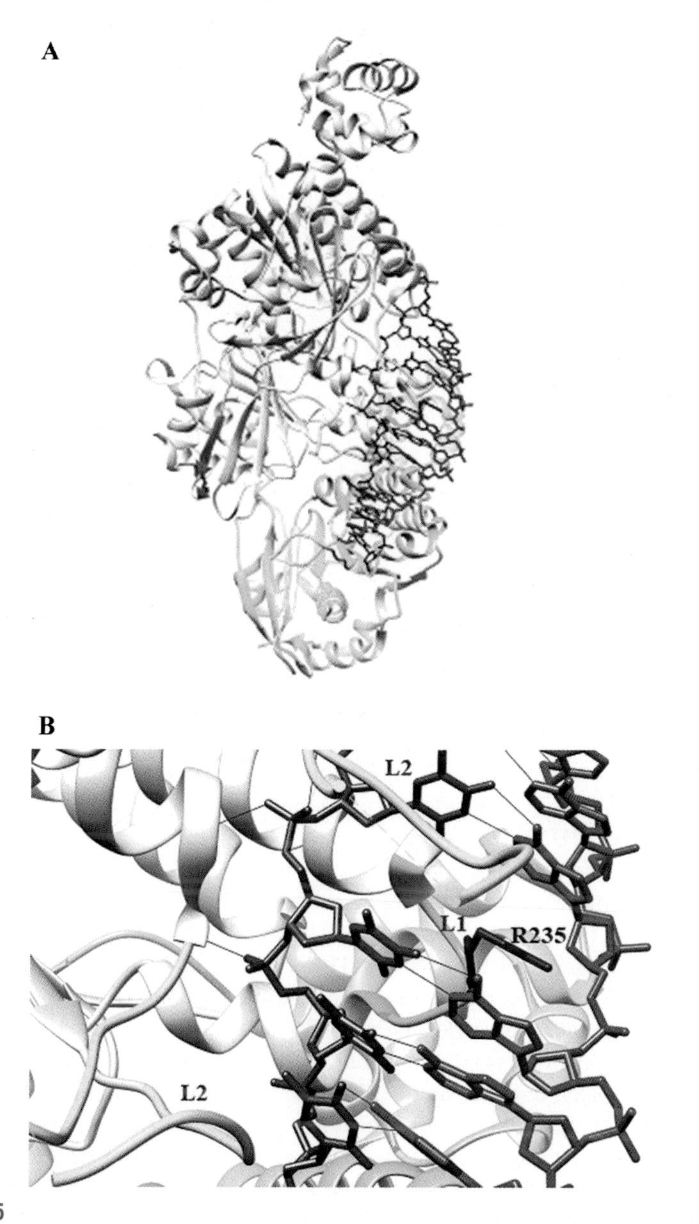

FIG. 13.15

(A) Rad51-dsDNA postsynaptic complex. (B) Close-up view of R235 interaction. All L1 loops colored *cyan*; all L2 loops colored *yellow*. H-bonds are shown by *blue* lines.

Source: Xu, J., et al., 2017. Nat. Struct. Mol. Biol. 24, 40.

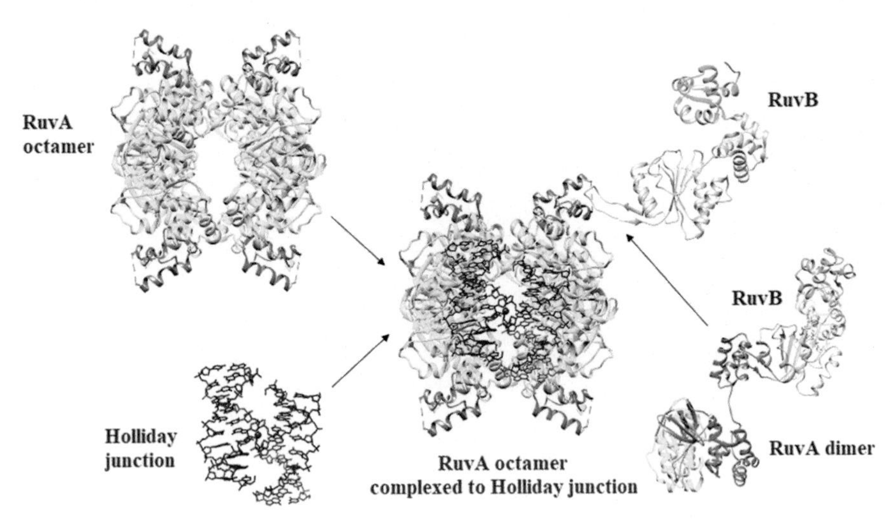

FIG. 13.16

RuvA octamer complexed to a Holliday junction (in the middle). Four RuvA domain IIIs (color: *coral*) come out in two opposite directions. A β-hairpin protruding from RuvB N domain interacts with RuvA domain III. The structure of a RuvA-RuvB complex is also shown.

Source: Roe, S.M., 1998. Mol. Cell 2, 361; Eichman, B.F., 2000. Proc. Natl. Acad. Sci. U. S. A. 97, 3971; Yamada, K., et al., 2001. Proc. Natl. Acad. Sci. U. S. A. 98, 1442; Yamada, K., et al., 2002. Mol. Cell 10, 671.

three-strand or four-strand branch migration. *In E. coli*, branch migration is carried out by the RuvAB ATP-dependent protein motor.

RuvA binds HJ as an octamer—two RuvA tetramers sandwich the HJ (Fig. 13.16). Each RuvA monomer consists of three domains—I, II, and III. Domains I and II are responsible for DNA-binding and tetramerization, while domain III interacts with RuvB (Fig. 13.16).

RuvB is a member of the AAA^+ ATPase family of proteins. It contains three domains, N, M, and C, forming a crescent-shaped structure (Fig. 13.16). Domains N and M are involved in ATP-binding, ATP hydrolysis, and hexamer formation. A hydrophobic β-hairpin protruding from RuvB N domain physically interacts with a hydrophobic patch in RuvA domain III to form the RuvAB complex (Fig. 13.16). This hydrophobic interaction is crucial for ATP-dependent branch migration.

To initiate branch migration, two RuvA tetramers bind the HJ and recruit two RuvB hexameric rings to form a tripartite structure. Four RuvA domain IIIs come out in two opposite directions and the octamer is flanked by the two RuvB rings. Together, the RuvAB complex pumps out dsDNA through the RuvA octamer core accompanied by helical rotation of the DNA. The resulting branch migration can proceed in either direction—extending the heteroduplex or disrupting the joint molecule.

The key intermediate in the genetic recombination process, the HJ, must be resolved for the process to be completed. The is achieved by the cleavage activity

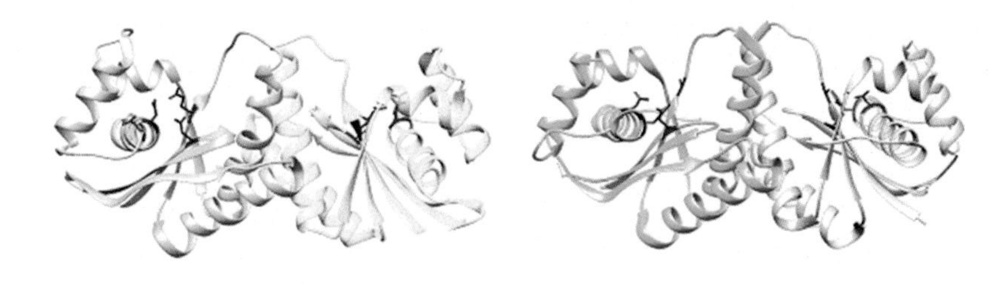

FIG. 13.17

Crystal structures of *E. coli* RuvC and *T. thermophilus* RuvC dimers. Two monomers of each enzyme have been colored differently. Catalytic residues shown in *red*.

Source: Ariyoshi, M., et al., 1994. Cell 78, 1063; Gorecka, K.M., Komorowaska, W., Nowotny, M., 2013. Nucleic Acids Res. 41, 9945.

of specific nucleases called resolvases. There exists a significant diversity among the resolvases from different organisms. In Gram-negative bacteria, HJ resolution is catalyzed by the enzyme RuvC. In fact, RuvC is considered to be a part of the RuvABC resolvasome. HJ resolution by RuvC is coupled with the RuvAB-catalyzed branch migration.

Crystal structures of RuvC from *E. coli* and RuvC-HJ complex from *Thermus thermophilus* have helped in the elucidation of the mechanism of HJ resolution (Fig. 13.17). The two proteins have 37% sequence similarity and very significant structural similarity. The structure of *T. thermophilus* RuvC-HJ complex is shown in Fig. 13.18. RuvC cleaves the HJ by introducing nicks symmetrically across the

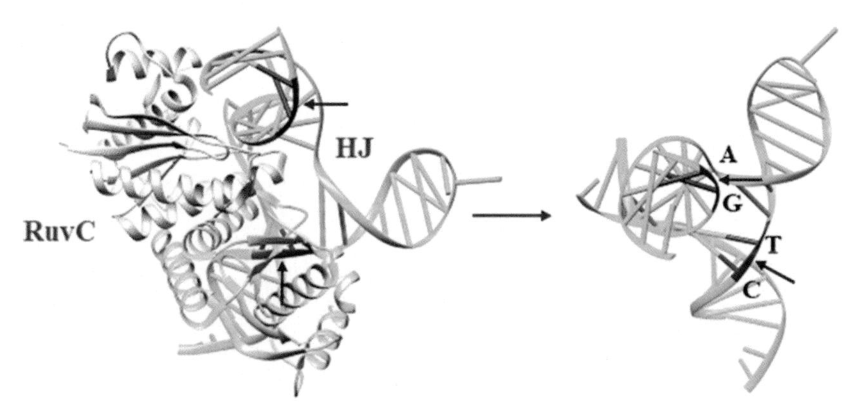

FIG. 13.18

Structure of *T. thermophilus* RuvC in complex with a synthetic HJ. Cleavage sites are indicated by *red* arrows.

Source: Gorecka, K.M., Komorowaska, W., Nowotny, M., 2013. Nucleic Acids Res. 41, 9945.

junction. The active form of the enzyme is a homodimer. It binds DNA in a structure-specific manner. The formation of RuvC-HJ complex does not depend on specific nucleotide sequences or metal ions. Each subunit of the dimer contains a large cleft, the dimension of which is sufficient to accommodate dsDNA. Basic residues lining the cleft walls mediate DNA-binding.

In contrast, divalent metal ions (Mg^{2+} or Mn^{2+}) are required for the cleavage activity of RuvC. Moreover, the cleavage is specific, the enzyme preferring the tetranucleotide consensus sequence 5′-(A/T)TT↓(G/C)-3′, where ↓ indicates the site of incision. Optimal cleavage occurs when the scissile phosphates in the DNA are located either at, or one nt away from, the point of strand exchange.

The active site in each monomer is located on the floor of the cleft and contains a tetrad of conserved residues (D7, E66, D138, and D141 in *E. coli*; D7, E71, H143, and D146 in *T. thermophilus*) (Fig. 13.17). The catalytic tetrad is possibly responsible for hydrolysis of the phosphodiester backbone. The two sites are separated by a distance of ~30 Å. The interface between the two DNA-binding clefts contains aromatic residues, the most prominent of which are F69 in *E. coli* and F73 in *T. thermophilus*, considered to be crucial for interacting with nucleotide bases close to the branch point during strand cleavage.

Eukaryotes

In eukaryotes, several DNA helicases are known to participate in branch migration. One of these helicases, RECQ1, has a substantial ability to migrate three- as well as four-stranded DNA intermediates. The structure of a RECQ1 dimer bound to two tailed DNA duplexes is shown in Fig. 13.19. Electron microscopy and biochemical analysis have shown that RECQ1 can form a flat homotetrameric complex. Based on all these results, it has been speculated that RECQ1-mediated HJ branch migration is caused by coordinated unwinding and reannealing occurring at two pairs of opposite branches. In contrast to RuvB, which is recruited to the HJ by RuvA, it appears that

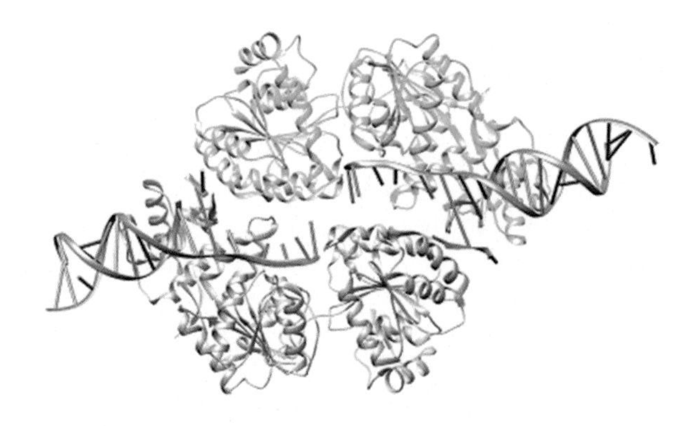

FIG. 13.19

RECQ1 dimer (two monomers shown in *gray* and *light green* bound to tailed DNA duplexes *(coral)*).

FIG. 13.20

Structure of human GEN1 bound to HJ. The two GEN1 monomers colored *gold* and *dodger blue*; the DNA strands colored *purple*, *green*, *firebrick*, and *coral*.

Source: Lee, S.H., et al., 2015. eLife 4, e12256. doi: 10.7554/eLife.12256.

RECQ1 can directly interact with the HJ. However, it remains to be established whether, inside the cell, RECQ1 acts alone or in association with other proteins.

Eukaryotic equivalents to RuvC have been identified in yeast (Yen1) and humans (GEN1). The three-dimensional structure of a GEN1-HJ complex is shown in Fig. 13.20. GEN1 and RuvC are structurally unrelated, yet there are meaningful similarities in their mechanisms of action. Like RuvC, GEN1 is a structure-selective endonuclease and functions as a dimer. It also introduces symmetrical incisions across the HJ, releasing nicked duplex products which can be subsequently ligated. However, unlike RuvC, which functions in association with RuvAB branch migration complex, GEN1 acts alone.

13.2 DNA rearrangements

As much as the genetic processes such as DNA replication, repair, and homologous recombination tend to preserve the stability and integrity of the genome of an organism, some recombinational processes are known to rearrange DNA sequences and lead to a more dynamic structure of the genome. Two types of genetic recombination are responsible for a number of DNA rearrangements—conservative site-specific recombination (CSSR) and transposition. CSSR occurs between two defined sequence elements, whereas transposition is a recombinational event between specific sequences and a nonspecific DNA target. There are multiple examples of both in living organisms. Nevertheless, here we shall restrict ourselves to an example of each kind and elaborate on the structural basis of the processes.

13.2.1 Conservative site-specific recombination

The most widely studied CSSR is the integration of the bacteriophage λ into the host bacterial chromosome. Integration occurs by a recombinational event at specific loci on the bacterial and bacteriophage DNAs called attachment (*att*) sites (bacteria: *att*B; λ: *att*P) (Fig. 13.21). The site-specific recombinase responsible for the recombination process is the bacteriophage-encoded integrase protein, Int. The action of Int is highly directional and tightly regulated and appropriately supported by an ensemble of accessory DNA-binding proteins. All these proteins interact with their respective binding sites on a 240 bp DNA locus in the λ genome.

Each of the two recombining partner DNA molecules contains a pair of inverted repeats which are the binding sites, called core-type binding sites, for Int. The repeats

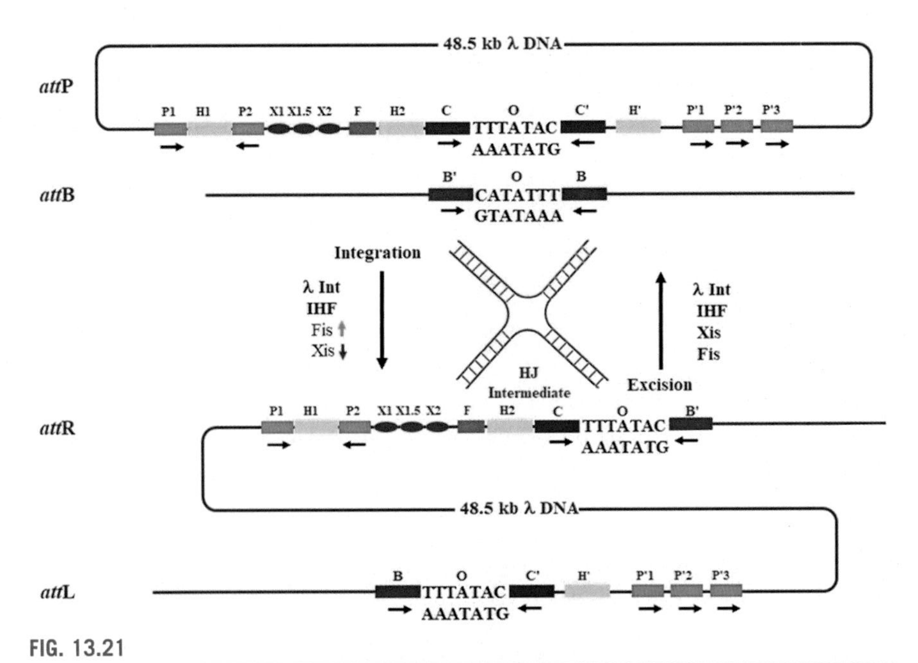

FIG. 13.21

λ integration and excision pathways. O—overlap region; C, C′ (*dark blue* boxes), B, B′ (*brown* boxes)—core-type binding sites (λ-Int-CTD); *green* boxes - λ-Int-NTD-binding sites. The binding sites for accessory proteins shown by boxes colored as follows: *gold*—IHF; *firebrick*—Xis; and *purple*—Fis. The upward *green* arrow indicates stimulation; downward *red* arrow indicates inhibition.

flank a 7-bp overlap region (O) which is identical in both DNAs. Flanking the core in the bacteriophage DNA are the "arms" which contain the binding sites for accessory proteins.

Cleavages on strands with the same polarity on the two DNA molecules followed by exchange of the cleaved strands result in a four-way DNA junction (Holliday junction, HJ). Subsequently, this intermediate structure is resolved to recombinant products by cleavages and exchange of the other two strands.

λ Int is a member of the tyrosine recombinase family. It cleaves and reseals DNA through the formation of a covalent 3′-phosphotyrosine high-energy intermediate. The enzyme consists of three functional domains—the N-terminal domain (NTD; residues 1–63) that binds to the arm sites, the core-binding domain (CBD; residues 75–175), and the catalytic domain (residues 176–336). The CBD and catalytic domains together can be said to constitute the C-terminal domain (CTD). The NTD is connected to the CTD by an α-helical coupler (residues 64–74) (Fig. 13.22).

FIG. 13.22

λ Int structure. Coloring scheme: NTD—*light blue*; α-helical coupler—*gold*; CBD—*light pink*; catalytic domain—*dodger blue*; catalytic residues—*orange red*.

Source: Biswas, T., et al., 2005. Nature 435, 1059.

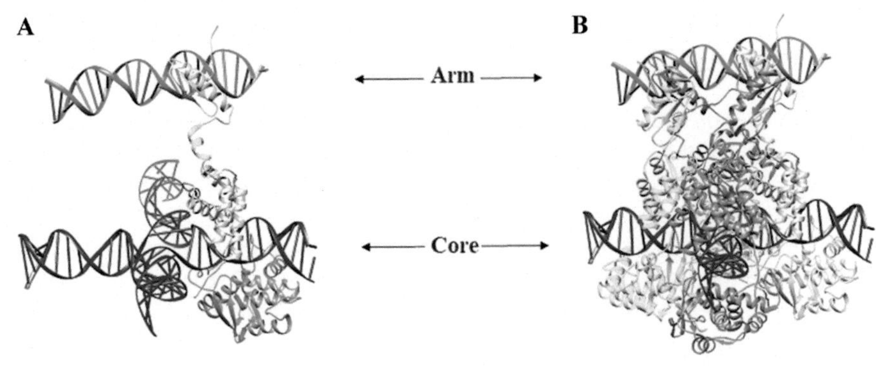

FIG. 13.23

(A) Int protomer bound to arm and core. Domains colored as in Fig. 13.22. (B) Int tetramer bound to HJ and arm. Four protomers colored differently.

Source: Biswas, T., et al., 2005. Nature 435, 1059.

Fig. 13.23A shows the structure of a λ Int protomer bound to an arm site DNA and a core site DNA. The NTD binds the arm, while the CTD binds the core. The structure of an Int tetramer bound to an HJ and arm DNA is shown in Fig. 13.23B. It is to be noted here that, in the absence of the NTD, Int can resolve the HJ, but it is unable to carry out a recombination reaction.

The structure of CTD is analogous to other tyrosine recombinases. The active site in the catalytic domain contains conserved residues, R212, K235, H308, and R311, crucial for the catalytic action of Int (Fig. 13.24). As the CTD binds the core site, a tyrosine residue (Y342) moves into the active site where it is engaged as a nucleophile with the scissile phosphate. The tyrosine cleaves and reseals the DNA with the formation of a 3′-phosphotyrosine high-energy bond (Fig. 13.24).

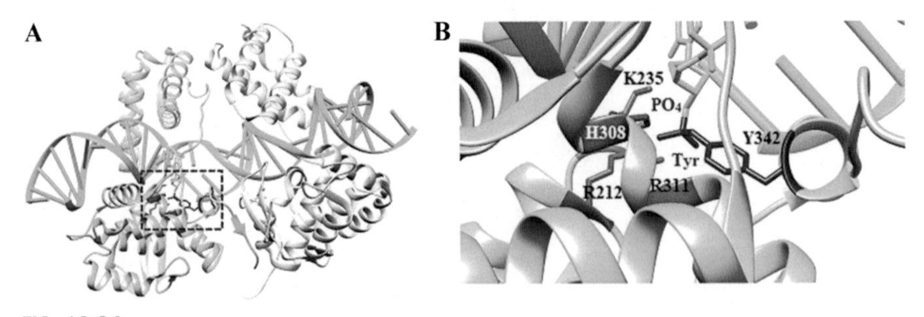

FIG. 13.24

(A) Two active Int protomers (*light blue* and *light green*) cleaving DNA *(cyan)* and forming phosphotyrosine bond. Active site marked by dotted box. (B) Active site enlarged. Catalytic residues and phosphotyrosine labeled.

Source: Biswas, T., et al., 2005. Nature 435, 1059.

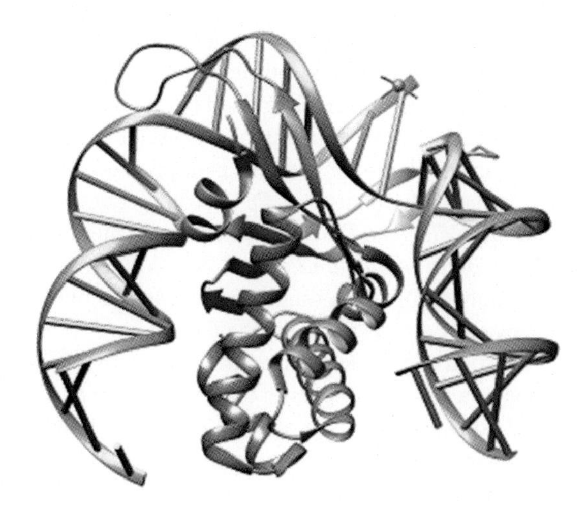

FIG. 13.25

DNA *(golden rod)* bending by IHF (α and β subunits *dodger blue* and *purple*).

Source: Swinger, K.K., Rice, P.A., 2007. J. Mol. Biol. 365, 1005.

The catalytic activity of Int is regulated by the quaternary structure of the tetramer. Only two of the subunits, out of four, are active at a time. Initially, two subunits are involved in DNA cleavage and strand exchange, while the other two remain passive. Halfway through the recombination reaction, the complex between Int and its substrates isomerizes, and the subunits switch roles. The two subunits, which were earlier inactive, are now involved in the resolution of the HJ into recombinant products.

Integration host factor (IHF) plays an important architectural role in the recombination process. IHF is a DNA-binding protein (Fig. 13.25). By binding to its cognate sites in *att*P and bending the DNA, IHF stimulates Int binding and cleavage at an otherwise low-affinity core site.

Integrative recombination between *att*P and *att*B results in a prophage flanked by recombination joints *att*L and *att*R (Fig. 13.21). This is the quiescent lysogenic state of λ. For transition to the lytic state (in which the bacteriophage multiplies), the bacteriophage DNA is excised from the host genome regenerating *att*P and *att*B. Although the excision process also involves DNA cleavage, strand exchange, and formation of HJ intermediate, it is not truly "reciprocal" to integration. The recombination machinery is configured differently during excision that requires the bacteriophage-encoded Xis protein (which, in fact, inhibits integrative recombination) and host-encoded Fis protein (Fig. 13.21).

13.2.2 Transpositional recombination

Movement of transposable elements or transposons occurs through recombination between the two ends of the element and a target DNA sequence mostly in a nonselective manner. Recombinases that carry out transposition are called transposases.

A DNA transposon contains its own transposase gene. Inverted repeat (IR) sequences present at the two ends of the element constitute the recombination sites. The terminal IRs, though not identical, are both recognized by the transposase. Transposition can occur without duplication of the element, by a "cut-and-paste" mechanism, or by a replicative mechanism in which the element is duplicated.

Tc1/mariner is a class of eukaryotic DNA transposons found in almost all animals—*Caenorbabditis elegans*, *Drosophila*, and even human. Mos1 is a mariner transposon from *Drosophila mauritiana*. The 1.3 kb Mos1 transposon possesses a 28 bp IR at each end flanked by a TA dinucleotide. Mos1 moves from one genome location to another by the cut-and-paste mechanism.

Transposition is carried out sequentially: (a) the transposase binds the IR at each end of the transposon in a sequence-specific manner, (b) the IRs form a paired-end-complex (PEC), the two DNA strands at each end are cleaved one after another, (d) the target DNA is captured (target capture complex, TCC), and (e) transposon strands are transferred to a new site (strand transfer complex, STC) (Fig. 13.26).

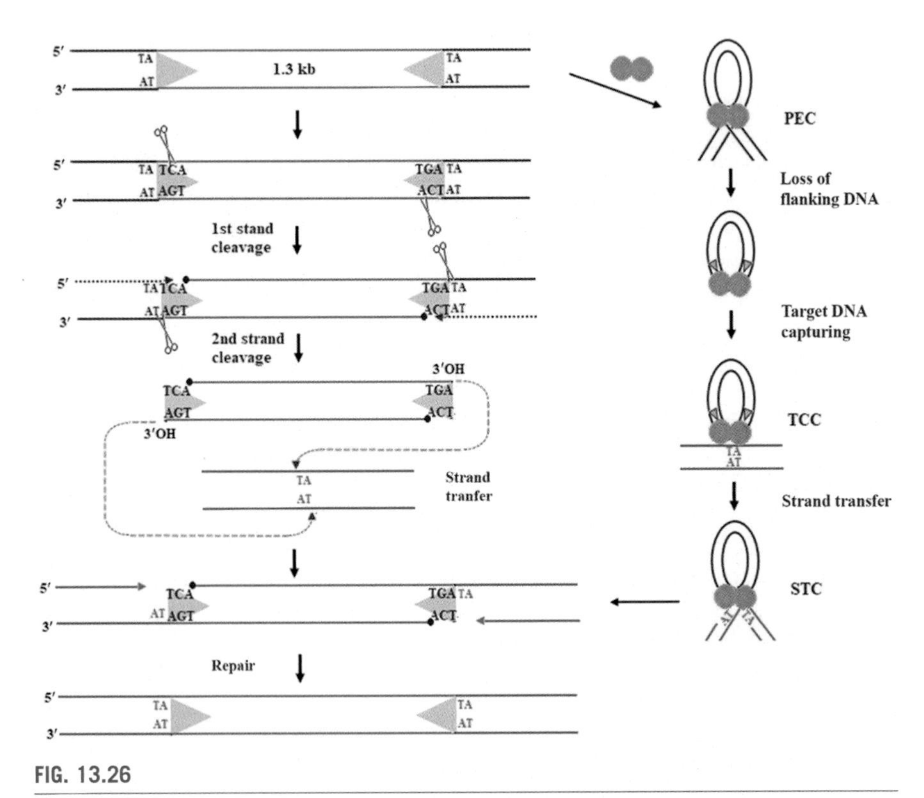

FIG. 13.26

Mos1 transposition pathway. 28 bp imperfect repeats (IRs) at two ends and Mos1 transposase shown by *gray* triangles and *orange* circles, respectively. Nucleophilic attack shown by *red* dotted lines.

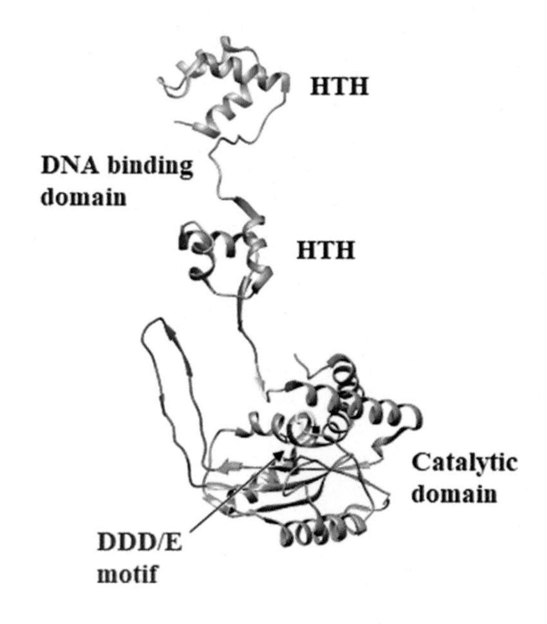

FIG. 13.27

Structure of Mos1 transposase. Coloring scheme: DNA biding domain—*dodger blue*; catalytic domain—*coral*; linker—*purple*; clamp loop—*dark blue*; DDD/E motif—*red*; WVPHEL and YSPDL—*dark blue*.

Source: Richardson, J.M., et al., 2009. Cell 138, 1096.

The Mos1 transposase contains an N-terminal DNA-binding domain and a C-terminal catalytic domain (Fig. 13.27). The DNA-binding domain contains two helix-turn-helix (HTH) motifs that bind to specific sequences in the IRs. The catalytic domain contains a conserved DDD/E motif (D156, D249, and D284). The two domains are connected by a linker (residues 113–125). A clamp loop (residues 162–189) extending from the catalytic domain is most likely involved in monomer-monomer interactions. There are two additional motifs, WVPHEL (residues 119–124) and YSPDL (residues 276–280), which may have a role in the correct positioning of DNA in the active site for cleavage (Fig. 13.27).

Mos1 functions as a homodimer that brings the transposon ends together to form the PEC (Fig. 13.28). In the Mos1 PEC crossed (*trans*) architecture, it is observed that each IR binds the DNA-binding domain of one transposase monomer while interacting with the catalytic domain of the other subunit.

Excision of the transposon is carried out by two consecutive hydrolysis reactions at each IR. The first cleavage occurs three nucleotides inside the 28 bp IR sequence on the strand which is not directly linked to the target DNA during the strand transfer reaction (hence called the nontransferred strand, NTS). This

FIG. 13.28

Structure of Mos1 paired-end complex (PEC). The two IRs are shown in *gray* and *firebrick*. All domains of the monomers colored differently.

Source: Dornan, J., Grey, H., Richardson, J.M., 2015. Nucleic Acids Res. 43, 2424.

cleavage, which produces a 5′-PO_4 on the NTS, may occur before or after the formation of PEC.

Another outcome of the first cleavage reaction is three unpaired bases on the opposite strand (transferred strand, TS). These bases interact with the transposase to position the IR end in the proximity of the catalytic residues. PEC formation is essential for the second-strand (TS) cleavage which occurs in *trans*. The TS is cleaved precisely at the IR end to generate a 3′-OH which, in turn, attacks the phosphodiester bond 5′ to a TA-dinucleotide at the target site. Integration of the cleaved transposon leads to the repair of the single-strand gap and duplication of TA dinucleotide (Fig. 13.26).

DNA transposons, virus-like retrotransposons, and retroviruses all use the cut-and-paste mechanism to insert their DNA into a new site. Yet variations exist in the individual steps. As for example, there are more than one mechanism to cleave the NTS. Besides, virus-like retrotransposons and retroviruses engage an RNA intermediate for transposition. On the other hand, although the replicative mechanism of transposition appears to be different from the cut-and-paste mechanism as regards the products of the reaction, the mechanism of recombination is quite similar.

Sample questions

1. Why is the nuclease of RecBCD initially dormant? How is this activity stimulated?
2. How does the RecBCD nuclease activity switch from the 3′-tail to the 5′-tail of the unwound DNA?
3. What is the mechanism of RecA loading on to the ssDNA by RecBCD?
4. How does Mre11-Rad50-Nbs (MRN) complex resect eukaryotic double-strand breaks?
5. What changes in the structure of DNA are brought about by RecA-based nucleoprotein filament formation?
6. How does the RecA-based nucleoprotein filament search for homology in a dsDNA?
7. What determines the preference for ssDNA-binding to the primary site of RecA?
8. What is the feature in Rad51, not found in RecA, that is involved in stabilization of the nascent DNA joint structure?
9. How does the RuvAB-dependent protein motor carry out branch migration of the Holliday junction?
10. What are the structural features of RuvC that enable it to function as a resolvase?
11. What are the similarities and differences between RuvC and GEN1 in the mechanisms of their action?
12. Illustrate the function of λ Int as a tyrosine recombinase.
13. What is the significance of paired-end-complex (PEC) in the Mos1 transposition pathway?
14. What are the two consecutive hydrolytic reactions in the excision of the Mos1 transposon from its original genomic location?

References and further reading

Bell, J.C., Kowalczykowski, S.C., 2016. RecA: regulation and mechanism of a molecular search engine. Trends Biochem. Sci. 41 (6), 491–507.

Biswas, T., et al., 2005. A structural basis for allosteric control of DNA recombination by λ integrase. Nature 435 (7045), 1059–1066.

Carreira, A., Kowalczykowski, S.C., 2011. Two classes of BRC repeats in BRCA2 promote Rad51 nucleoprotein filament function by distinct mechanisms. Proc. Natl. Acad. Sci. U. S. A. 108 (26), 10448–10453.

Cassani, C., et al., 2018. Structurally distinct Mre11 domains mediate MRX functions in resection, end-tethering and DNA damage resistance. Nucleic Acids Res. 46 (6), 2990–3008.

Dillingham, M.S., Kowalczykowski, S.C., 2008. RecBCD enzyme and the repair of double-stranded DNA breaks. Microbiol. Mol. Biol. Rev. 72 (4), 642–671.

Dornan, J., Grey, H., Richardson, J.M., 2015. Structural role of the flanking DNA in *mariner* transposon excision. Nucleic Acids Res. 43 (4), 2424–2432.

Górecka, K.M., Komorowska, W., Nowotny, M., 2013. Crystal structure of RuvC resolvase in complex with Holliday junction substrate. Nucleic Acids Res. 41 (21), 9945–9955.

Han, Y.-W., et al., 2006. Direct observation of DNA rotation during branch migration of Holliday junction DNA by *Escherichia coli* RuvA-RuvB protein complex. Proc. Natl. Acad. Sci. U. S. A. 103 (31), 11544–11548.

Kowalczykowski, S.C., 2015. An overview of the molecular mechanisms of recombinational DNA repair. Cold Spring Harb. Perspect. Biol. 7, a016410.

Krejci, L., et al., 2012. Homologous recombination and its regulation. Nucleic Acids Res. 40 (13), 5795–5818.

Landy, A., 2015. The λ integrase site-specific recombination pathway. Microbiol. Spectr. 3 (2). https://doi.org/10.1128/microbiolspec.MDNA3-0051-2014. MDNA3-0051-2014.

Laxmikanthan, G., et al., 2016. Structure of a Holliday junction complex reveals mechanisms governing a highly regulated DNA transaction. eLife 5, e14313. https://doi.org/10.7554/eLife.14313.

Pike, A.C.W., et al., 2015. Human RECQ1 helicase-driven DNA unwinding, annealing, and branch migration: insights from DNA complex structures. Proc. Natl. Acad. Sci. U. S. A. 112 (14), 4286–4291.

Punatar, R.S., et al., 2017. Resolution of single and double Holliday junction recombination intermediates by GEN1. Proc. Natl. Acad. Sci. U. S. A. 114 (3), 443–450.

Richardson, J.M., et al., 2009. Molecular architecture of the Mos1 paired-end complex: the structural basis of DNA transposition in eukaryotes. Cell 138, 1096–1108.

Schiller, C.B., et al., 2014. Structural studies of DNA end detection and resection in homologous recombination. Cold Spring Harb. Perspect. Biol. 6, a017962.

Seifert, F.U., et al., 2016. Structural mechanism of ATP-dependent DNA binding and DNA end bridging by eukaryotic Rad50. EMBO J. 35 (7), 759–772.

Shinohara, T., et al., 2015. Loop L1 governs the DNA-binding specificity and order for RecA-catalyzed reactions in homologous recombination and DNA repair. Nucleic Acids Res. 43 (2), 973–986.

Singleton, M.R., et al., 2004. Crystal structure of RecBCD enzyme reveals a machine for processing DNA breaks. Nature 432, 187–193.

Spies, M., Kowalczykowski, S.C., 2006. The RecA binding locus pf RecBCD is a general domain for recruitment of DNA strand exchange proteins. Mol. Cell 21, 573–580.

Syed, A., Tainer, J., 2018. The Mre11-Rad50-Nbs1 complex conducts the orchestration of damage signaling and outcomes to stress in DNA replication and repair. Annu. Rev. Biochem. 87, 263–294.

Tong, W., et al., 2014. Mapping the λ integrase bridges in the nucleoprotein Holliday junction intermediates of viral integrative and excisive recombination. Proc. Natl. Acad. Sci. U. S. A. 111 (34), 12366–12371.

Watt, H.D.M., Kowalczykowski, S.C., 2014. Holliday junction resolvases. Cold Spring Harb. Perspect. Biol. 6, a023192.

Wilkinson, M., Chaban, Y., Wigley, D.B., 2016. Mechanism for nuclease regulation in RecBCD. eLife 5, e18227. https://doi.org/10.7554/eLife.18227.

Xu, J., et al., 2017. Cryo-EM structures of human recombinase Rad51 filaments in the catalysis of DNA strand exchange. Nat. Struct. Mol. Biol. 24 (1), 40–46.

Yamada, K., et al., 2002. Crystal structure of the RuvA-RuvB complex: a structural basis for the Holliday junction migrating motor machinery. Mol. Cell 10, 671–681.

CHAPTER

Membrane structure and function

14

In order to sustain life, cells need to carry out a wide variety of chemical reactions. Earlier, we have discussed the synthesis of some of the most important biomolecules—DNA, RNA, and protein. However, we have to understand that the cell is able to synthesize its molecular constituents as well as carry out a number of other metabolic processes, efficiently and in an organized manner, since it is partially secluded from its more random environment by a biomembrane. The hydrophobic core of the membrane is formed by lipid assemblies.

14.1 Composition of the membrane

In a living cell, lipids perform three general functions: (a) storage of energy as triglycerol ester and steryl esters, (b) as first and second messengers in signal transduction and molecular recognition processes, and (c) formation of the matrix of cellular membranes. It is the third function that we are concerned with in this chapter.

Some bacterial cells (Gram-positive) contain just one membrane, whereas others (Gram-negative) are surrounded by two membranes—inner and outer. Eukaryotic cells are enclosed in a single plasma membrane. Some of the internal organelles of eukaryotic cells such as the nucleus, mitochondria, and chloroplasts are surrounded by double lipid bilayers, while endoplasmic reticula, Golgi apparatus, and lysosomes each contains a single lipid bilayer. Nevertheless, in all cases, lipids form the common core. Besides, the cell membranes also contain proteins and carbohydrates of different types and in varying amounts.

The major kinds of lipid found in prokaryotic and eukaryotic cell membranes are the glycerol-based phospholipids. The prokaryotic lipids are mainly phosphatidylethanolamine, phosphatidylglycerol, and cardiolipin. On the other hand, more than 50% of the phospholipids of eukaryotic membranes are phosphatidylcholine which is not usually present in prokaryotes. In addition, eukaryotic membranes also contain phosphatidylserine, phosphatidylinositol, sphingomyelin, and glycosphingolipid. Such is the lipid diversity in biomembranes (Fig. 14.1).

It is also interesting that the bilayer exhibits some degree of lipid asymmetry. As for example, in a normal red blood cell, all the phosphatidylserines and 80% of the phosphatidylethanolamines are located in the inner monolayer, whereas the outer leaflet consists mostly of phosphatidylcholines, sphingomyelins, and glycolipids.

Fundamentals of Molecular Structural Biology. https://doi.org/10.1016/B978-0-12-814855-6.00014-6

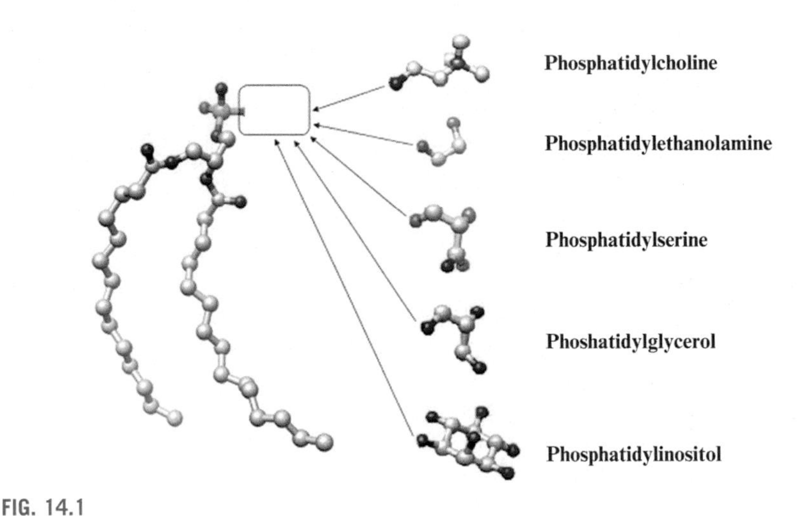

FIG. 14.1

The structure of phospholipid molecule found in the cell membrane.

14.2 Membrane structure

It needs to be emphasized here that, in spite of the diversity and asymmetry, a unifying characteristic of the lipids dictates the basic bilayer organization of the membranes, that is, their amphipathic nature. The phospholipid "head" is electrostatically charged. It is, therefore, hydrophilic in character and able to interact with polar water molecules. In contrast, the nonpolar fatty acid "tails" are hydrophobic—they tend to be sequestered from water or other hydrophilic materials.

The formation and stabilization of biomembranes are thermodynamically based on the tendency of hydrophobic structures to self-associate (entropically driven; Chapter 4) and hydrophilic structures to interact with the water molecules. In the presence of water, phospholipid molecules attain the lowest-energy organizational state by forming a bilayer in which the fatty acid tails of the two layers avoid water by facing each other, while the polar regions are oriented towards the aqueous phase (Fig. 14.2). The crystal structure of a lipid bilayer is shown in Fig. 14.3.

Evidently, the bilayer is a highly inhomogeneous region. Due to the inhomogeneity and amphipathicity, anisotropic stresses develop within the bilayer. These

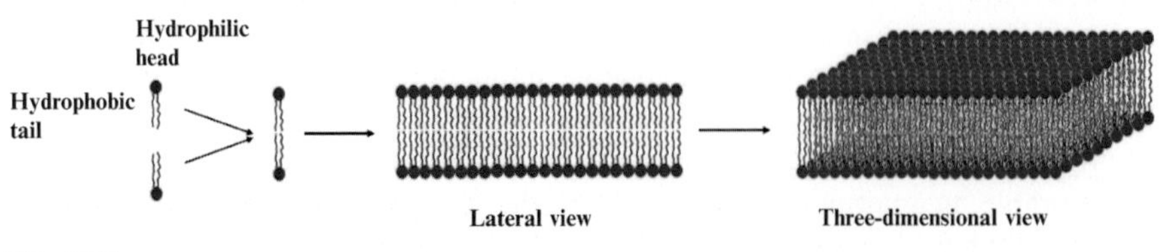

FIG. 14.2

Phospholipid molecules forming a bilayer.

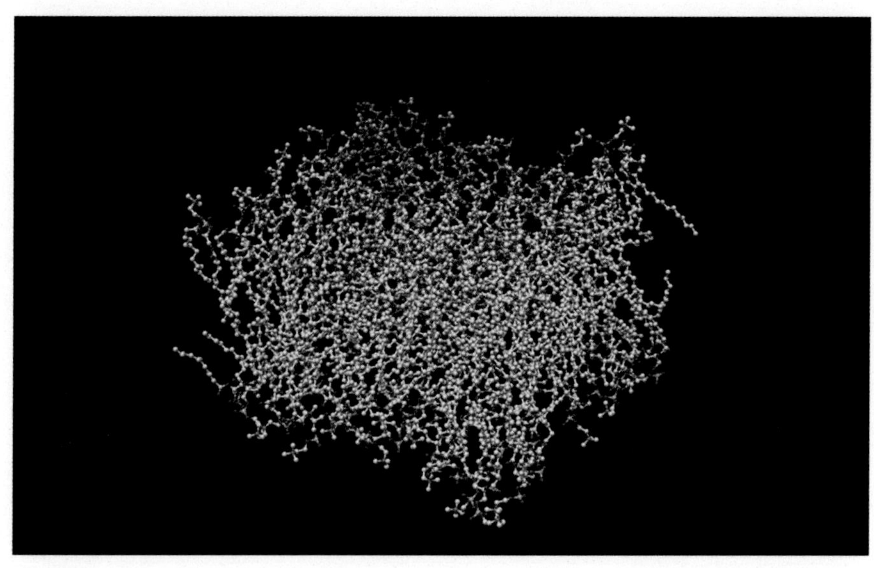

FIG. 14.3

The crystal structure of a lipid bilayer.

Source: Koppisetti, R.K., et al. 2014. Nat. Commun. 5, 5552.

stresses, which vary with the depth, are partly similar to the tension that appears at the interface of a nonpolar substance (say, oil) and water (Chapter 4). However, movement of the water molecules in an oil-water mixture is not entirely restricted and those at an interface can move to the bulk.

In contrast, the polar lipid head is stuck with its nonpolar tail embedded in the bilayer. Water molecules do not interact with the hydrocarbon tails, but strongly interact with the polar head groups. Thus, a lateral tension develops at the lipid neck. This tension is opposed by a large repulsion from the tails and a smaller repulsion between the heads.

As one can see, the environment existing for a protein molecule in the aqueous phase differs from that within a lipid bilayer. A soluble protein is bombarded from all directions, whereas a bilayer-embedded protein is subjected to anisotropic forces. The embedded protein is subjected to lateral tension at the polar-nonpolar interfaces and opposite compressional forces in the interior lipid environment (Fig. 14.4). The anisotropic forces of the lipid bilayer can reshape the embedded protein.

Biological membranes contain a wide variety of proteins. On an average, there is one protein molecule for every 25 phospholipid molecules in the plasma membrane. There are two types of membrane protein—peripheral and integral.

Peripheral membrane proteins possess very few exposed hydrophobic groups and, therefore, are not embedded in the bilayer. They have polar or charged regions that electrostatically interact with exposed segments of integral membrane proteins or polar heads of phospholipids. Peripheral proteins are found on both surfaces of the membrane.

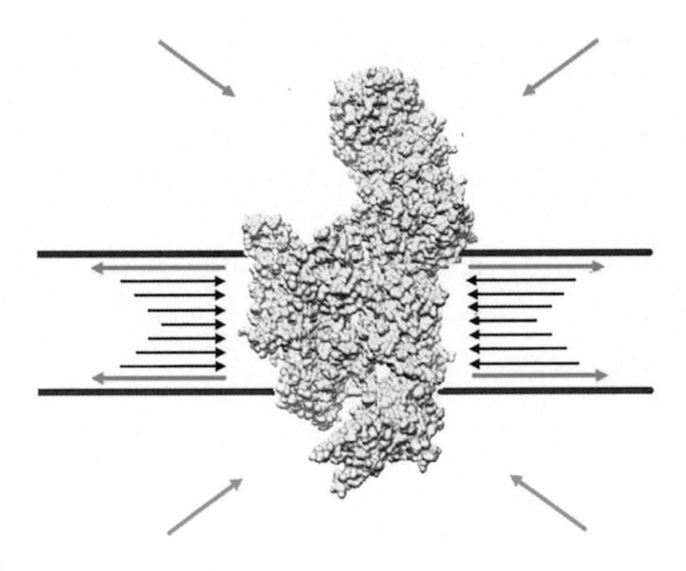

FIG. 14.4

Anisotropic forces on an embedded protein in the lipid bilayer. Besides bombardment by molecules *(light blue arrows)* in the aqueous environment on the two sides of the bilayer, the protein is subjected to lateral tension *(green arrows)* at the polar-nonpolar interfaces and opposite compressional forces *(deep blue arrows)* in the interior.

In contrast, integral membrane proteins are embedded in the bilayer. These proteins possess both hydrophobic and hydrophilic regions. The hydrophobic region traverses the hydrophobic core of the bilayer, while the hydrophilic region is found on the internal or external surface. Some integral proteins, which have two hydrophilic ends separated by an intervening hydrophobic tract, span the entire thickness of the membrane, often multiple number of times. These are known as transmembrane proteins.

Carbohydrates are another key component of the membrane. They are usually present on the outer surface of the plasma membrane and covalently bonded to lipids or proteins. A carbohydrate bonded to a lipid constitutes a glycolipid, whereas a glycoprotein is formed by a covalent bond between a carbohydrate and a protein. The bonded carbohydrate is usually an oligosaccharide containing ≤ 15 monosaccharide units.

14.3 Membrane dynamics

A lipid bilayer decorated with membrane proteins, integral and peripheral, displays a mosaic-like appearance. However, the mosaic structure is fluidic in nature, for both lipids and proteins are in constant motion. Hence, the "fluid mosaic model" of cell membranes was introduced by Singer and Nicolson in 1972 (Fig. 14.5).

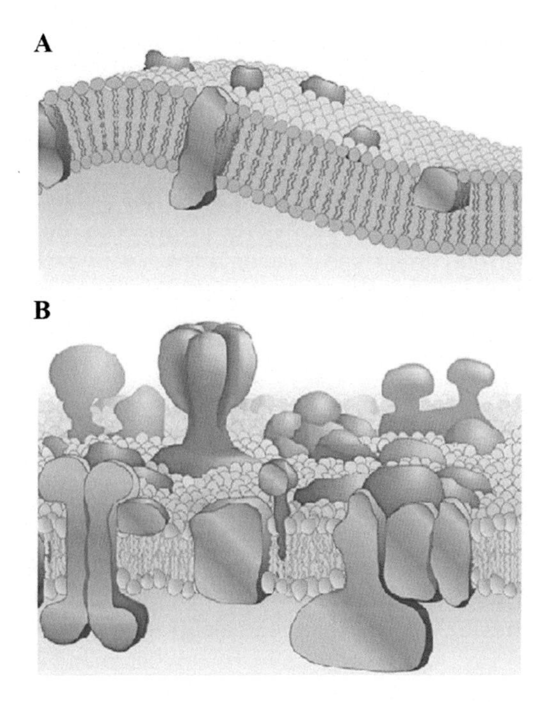

FIG. 14.5

The Singer-Nicolson model. (A) Original version. (B) Updated version.

Courtesy: Goñi, F.M., The basic structure and dynamics of cell membranes: an update of the Singer-Nicolson model. Biochim. Biophys. Acta – Biomembranes, June 2014, with permission from Elsevier.

The modes of motion of both lipids and proteins are rotational as well as translational. They rotate around their long axis which is perpendicular to the plane of the membrane. Translational diffusion occurs in two dimensions along the plane of the membrane. Diffusion barriers are almost nonexistent. However, a third kind of motion, trans-bilayer (flip-flop) diffusion, although a theoretical possibility, does not actually occur. This nonoccurrence is due to the energy barrier posed by the hydrophobic core of the bilayer to the polar groups of proteins and lipids. The asymmetry of membranes, which has been mentioned earlier, is due to the lack of trans-bilayer motion.

The basic tenets of Singer and Nicolson model remain valid even today. However, based on subsequent experimental data generated in the last 40 years, the model has been suitably updated (Fig. 14.5). One of the new concepts that have evolved concerns the density of transmembrane proteins. In the original model, the proteins are so sparsely distributed in the membrane, that is, the lipid-protein ratio is so large, that the lipids are practically unperturbed by the presence of proteins. In the modified model, the membrane is so densely populated with proteins that hardly any lipid molecule remains unperturbed.

Another conceptual change has been regarding the localization of cellular proteins with respect to the membrane. Proteins were earlier considered to be either membrane-bound or soluble. The membrane was viewed as an isolated thermodynamic system, not being involved in any exchange of matter or energy with the environment. The present concept is that different kinds of metabolic signal and other molecules including proteins "reach and leave" the membrane. In other words, several proteins spend part-time in the cytosol and part-time membrane-associated. Further, according to the current model, lateral heterogeneity exists on the membrane surface. Heterogeneous patches enriched in specific lipids and proteins are formed so that the membranes acquire characteristic functional properties.

14.4 Transport across membranes

Biological membranes regulate the internal composition of a cell by secluding it from the environment only partially. Based on their property of selective permeability, the membranes permit the passage of only certain substances, not all. Two different processes by which the permissible substances cross biological membranes are: (a) passive transport that does not require any input of metabolic energy and (b) active transport that requires the supply of metabolic energy from an external source. Passive transport of a substance, which is driven by its concentration difference on the two sides of the membrane, can be either simple diffusion through the phospholipid bilayer, or facilitated diffusion with the help of channel proteins or carrier proteins.

14.4.1 Passive transport

Ion channels are some of the well-studied channel proteins. They facilitate the movement of different ions across the membrane which is important for several biological processes. A number of ion channels have been identified, each of them being specific for a particular ion. A common structural feature in all of these channels is a hydrophilic pore that permits the movement of a particular ion through it.

Most of the ion channels function as "gates" in the membranes which can be considered as "fences." Hence, they are also called gated channels. The "gates" open and close due to changes in their three-dimensional structure brought about by chemical or electrical stimuli. Depending on the nature of the stimulus to which a gated channel responds, it can be categorized as either a ligand-gated or a voltage-gated channel.

Ligand-gated channel

A ligand-gated channel opens in response to a stimulus such as the binding of a ligand. A change in the conformation of the channel protein causes the opening of a pore lined with polar amino acids. Charged or polar substances can then pass through the open channel.

There are some ligand-gated channels which open to chloride (Cl^-) ions in response to neurotransmitter glutamate binding (Fig. 14.6). Such a glutamate-gated

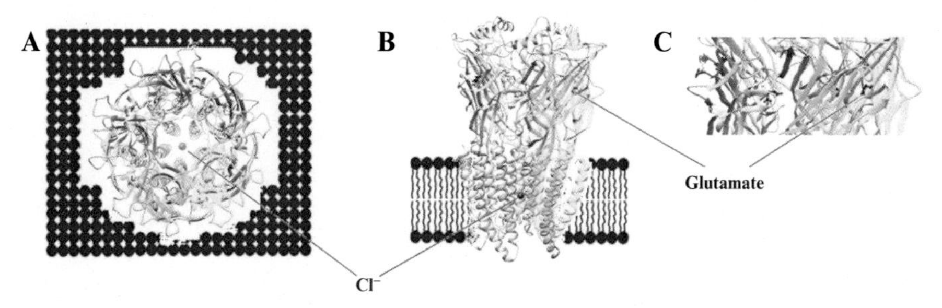

FIG. 14.6

The structure of a glutamate-gated Cl^- channel (GluCl). (A) Downward view from outside the cell. (B) Lateral view parallel to the lipid bilayer. (C) Enlarged view of glutamate-binding sites. The subunits of the homopentamer have been colored differently.

Source: Hibbs, R.E., Gouaux, E., 2011. Nature 474, 54.

channel, GluCl, has been isolated from *Caenorhabditis elegans* and its three-dimensional structure determined. GluCl is a homopentamer containing a central pore for the passage of Cl^- ions. There are glutamate-binding sites at subunit interfaces.

Voltage-gated channel

A voltage-gated channel is stimulated by a change in the transmembrane electric voltage generated by ion gradients across the cell membrane. It is to be noted that, although there is no difference in the total ionic concentration between the two sides of the membrane bilayer, the concentration of a specific ionic species (say, K^+ or Na^+) is different. Thus, an electrochemical gradient exists, and the ions diffuse down the electrochemical gradient. This results in a membrane potential of about −100mV.

Voltage-gated Ca^{2+} channels are transmembrane proteins that conduct Ca^{2+} into cells in response to the changes in membrane potential and thereby initiate a number of physiological events. In particular, the calcium channel $Ca_v1.1$ is responsible for excitation-contraction of skeletal muscles in mammals. The structure of a membrane-embedded $Ca_v1.1$ is shown in Fig. 14.7. The principal or pore-forming subunit $\alpha1$ of $Ca_v1.1$ consists of four voltage-sensing transmembrane (TM) domains (I–IV), each of which is constituted of six transmembrane (6-TM) α helices (designated S1–S6). The $\alpha1$ subunit is associated with an extracellular $\alpha2\delta$ subunit, an intracellular β subunit, and a 4-TM γ subunit. The S4 helix of each $\alpha1$ domain rotates under the influence of an electric field and serves as a voltage sensor.

Carrier protein-facilitated diffusion

The other kind of facilitated diffusion involves the actual binding of the transported molecule to a transmembrane carrier protein. Several polar molecules, such as sugars and amino acids, are transported into the cell by this mechanism. A prominent example of such a carrier protein is the glucose transporter.

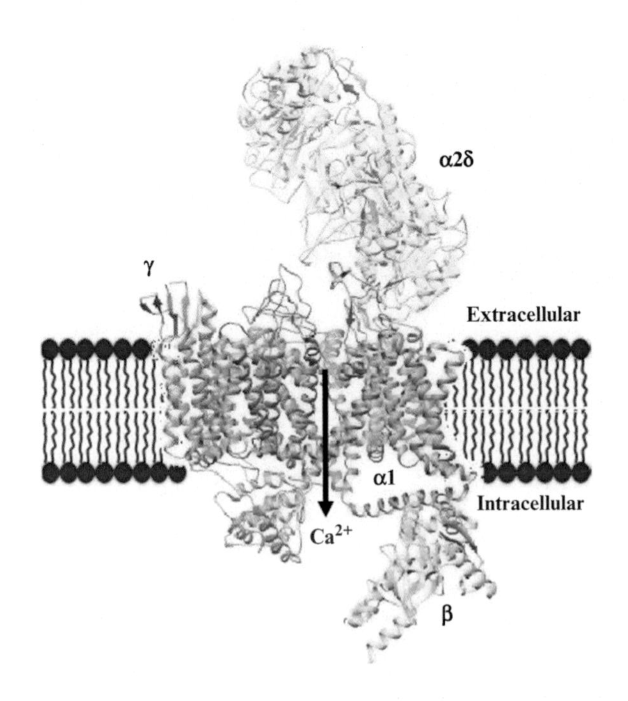

FIG. 14.7

The structure of a voltage-gated Ca^{2+} channel, $Ca_v1.1$.

Source: Wu, J.P., et al., 2016. Nature 537, 191.

In human cells, glucose is transported across the membrane by glucose transporter proteins, GLUT1-4. The three-dimensional structure of GLUT3 is shown in Fig. 14.8. Glucose binds to a site on one side of GLUT3 through H-bonds, changes its shape, and exits on the other side of the membrane. With the help of the carrier protein, glucose is able to enter the cell much faster than it would by simple diffusion through the bilayer and fulfills the energy requirement of the cell.

14.4.2 Active transport

Often, there exists a difference in the concentration of a specific ion or small molecule between the inside and outside of a cell. This concentration imbalance, which in some cases is necessary for cellular functions, is maintained by a plasma membrane protein. The concerned protein moves the ion or molecule against its concentration gradient in a process called active transport. Evidently, the process requires expenditure of energy which in many cases comes from the hydrolysis of adenosine triphosphate (ATP).

A very widely studied active transport protein is Na^+/K^+-ATPase, which is an integral membrane glycoprotein also known as the sodium-potassium (Na^+-K^+) pump (Fig. 14.9). In many cases, the concentration of K^+ is higher inside the cell

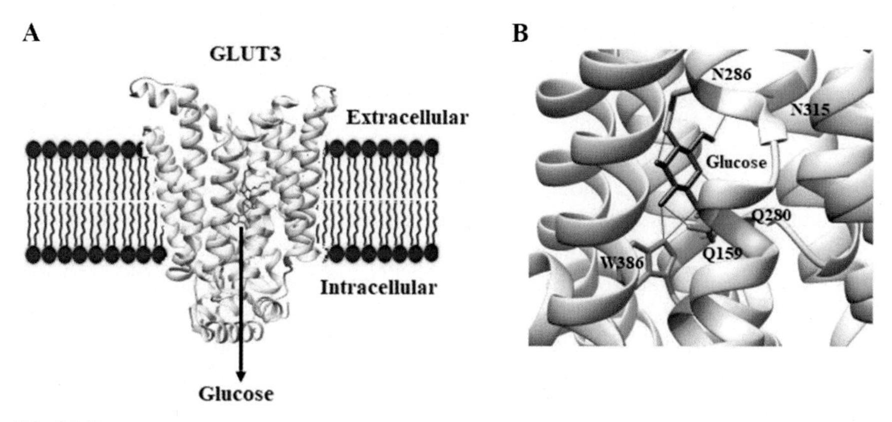

FIG. 14.8

(A) The structure of a glucose transporter protein GLUT3. (B) Enlarged view of the glucose-binding sites. H-bonds are indicated by *red* lines.

Source: Deng, D., et al., 2015. Nature 526, 391.

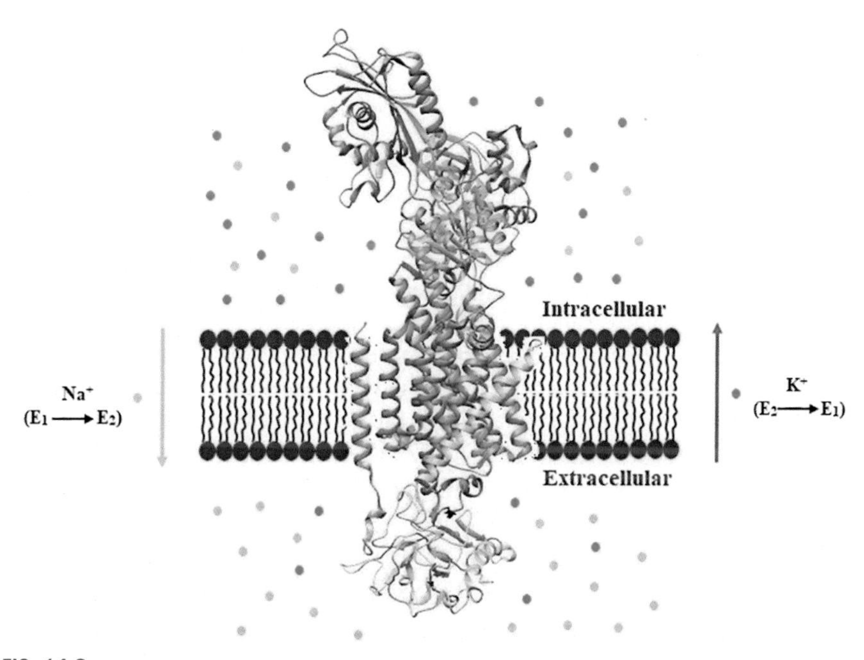

FIG. 14.9

The structure of Na^+/K^+-ATPase. The subunits have been colored differently.

Source: Shnoda, T., et al., 2009. Nature 459, 446.

than outside, while it is the opposite for Na^+. The Na^+/K^+-ATPase hydrolyzes an ATP molecule to adenosine diphosphate (ADP) and the released energy is utilized to drive two K^+ into and three Na^+ out of the cell.

The Na^+/K^+-ATPase contains a catalytic α subunit, a heavily glycosylated β subunit, and a tissue-specific regulatory subunit called the FXYD protein. It exists in two conformational states, E_1 and E_2. The two states differ in their interactions with Na^+, K^+, and ATP. At first, Na^+ and ATP bind the E_1 conformation with high affinity. Hydrolysis of ATP phosphorylates an aspartate residue of the pump protein and changes its conformation to E_2. At this stage, three Na^+ ions are released outside the cell, while the enzyme remains phosphorylated. The affinity for K^+ is increased, as a result of which K^+ ions bind the pump. The binding of K^+ induces a spontaneous dephosphorylation of the E_2 conformation, two K^+ ions are released to the cell's interior, and the pump returns to its original E_1 conformation.

14.4.3 Transport in vesicles

In general, macromolecules, such as proteins, polysaccharides, and nucleic acids, are too large to pass through biological membranes. Yet, sometimes it may be necessary for cells to pick up from or release to the external environment many of these molecules. Eukaryotic cells have specialized mechanisms to transport large molecules across the membrane—endocytosis (into the cell) and exocytosis (out of the cells).

Endocytosis

Endocytosis is the process in which the plasma membrane captures small molecules, macromolecules, and even small cells in a pocket that turns into a vesicle, and eventually the vesicle separates from the membrane to carry its contents to the cell's interior. There are different types of endocytosis. Here, we are going to discuss only an example of receptor-mediated endocytosis.

Receptors are protein molecules that receive chemical signals from outside the cell and trigger specific cellular responses. In receptor-mediated endocytosis, an integral membrane protein, located at a specific site on the extracellular surface of the plasma membrane, binds a ligand and prompts its uptake into the cell. The concerned site on the membrane is a "coated pit"—its cytoplasmic side is coated with another protein (mostly clathrin).

Most mammalian cells take up cholesterol from the extracellular environment by receptor-mediated endocytosis. Cholesterol is packaged into low-density lipoproteins (LDLs) and circulates in the blood. Cells produce LDL receptors (LDLRs) which populate the clathrin-coated pits. Here, LDLRs bind LDLs and initiate the process of endocytosis.

LDLR is synthesized as an 860 aa polypeptide. After removal of the 21 aa signal peptide, the 839 aa LDRL is inserted into endoplasmic reticulum (ER) membrane. In the ER, the protein is subjected to folding and glycosylation. The folded protein moves to the Golgi apparatus where its N-linked sugars are modified and O-linked sugars are elongated. Subsequently, the mature LDRL is transferred to the clathrin-coated pits of the membrane.

The mature LDLR contains three extracellular domains. The first domain contains seven contiguous cysteine-rich repeats, each ~40 aa in length, referred to as LDL receptor type A (LA) repeats. This is immediately followed by 400-residue region with 35% homology to epidermal growth factor precursor. The third is a 58 aa domain, rich in serine and threonine residues, involved in O-linked glycosylation. Next are the 22 aa hydrophobic transmembrane segment and a 50 aa C-terminal cytoplasmic tail.

A single copy of 4536 aa-long apolipoprotein B (ApoB) is the only protein component of LDL. The extracellular domains of LDLR bind LDL via ApoB. At least some of the LA repeats are involved in ligand binding which requires calcium ions. The structure of a ligand-binding LA repeat is shown in Fig. 14.10.

The cytoplasmic tail of LDLR has an important role in its interaction with clathrin-coated pits. An FxNPxY motif in the cytoplasmic tail is required for internalization of the receptor into the pits (Fig. 14.11). The motif is connected to the endocytosis machinery by an adaptor protein ARH1. The adaptor has a phosphotyrosine-binding (PTB) domain that binds unphosphorylated NPxY sequence and a clathrin-binding consensus motif LLDLE.

The 22-residue transmembrane segment of LDLR is enriched in aliphatic and other hydrophobic residues. This stabilizes the structure in the nonpolar interior of the lipid bilayer. Any change (by a mutation) that causes LDLR to be less efficiently inserted into the ER membrane invites metalloproteinase cleavage of the receptor protein.

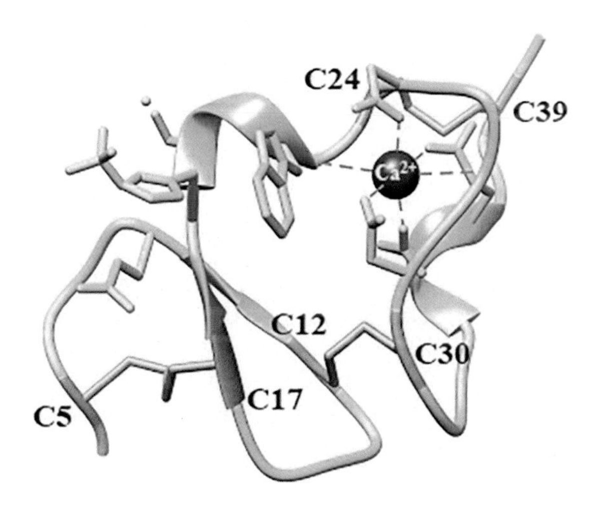

FIG. 14.10

The structure of an LDL-binding LA repeat of LDRL. The calcium ion and cysteine residues have been labeled.

Source: Faas, D., et al., 1997. Nature 388, 691.

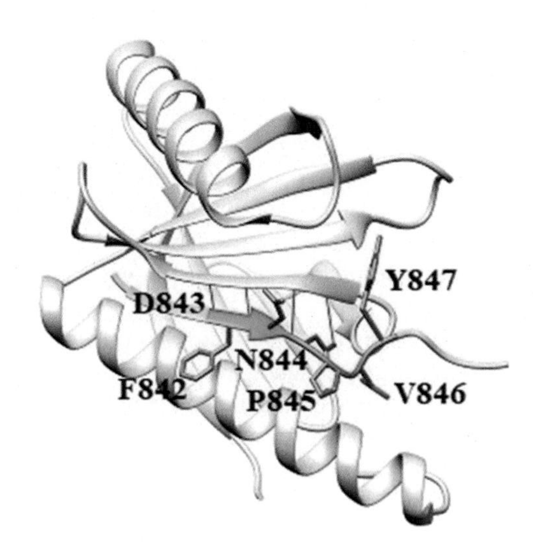

FIG. 14.11

FxNPxY motif in an LDLR tail fragment *(green)* bound to ARH1 PTB domain *(pink)*.

Source: Dvir, H., et al., 2012. Proc. Natl. Acad. Sci. U. S. A. 109, 6916.

Exocytosis

In exocytosis, materials are packaged in intracellular vesicles in the cytoplasm and released (secreted) from the cell. The common mechanism of exocytosis is that the vesicle membrane fuses with the plasma membrane, an opening to the outside of the cell is created, and through this opening contents of the vesicle are released into the extracellular space.

An important example of exocytosis is the release of chemical agents, known as neurotransmitters, by synaptic vesicles of presynaptic nerve cells (neurons). Each presynaptic neuron contains hundreds of synaptic vesicles which are filled with neurotransmitters with the help of a proton pump. Subsequently, the neurotransmitter-filled vesicles move to the "active zone" of the plasma membrane and responding to a Ca^{2+} signal completes the process of exocytosis.

Intracellular membrane vesicle fusion is mediated by soluble N-ethylmaleimide-sensitive factor attachment protein receptor (SNARE) proteins. The SNARE complex consists of v-SNARE Synaptobrevin/VAMP2 and t-SNAREs Syntaxin and SNAP25. Both VAMP2 and Syntaxin contain a C-terminal transmembrane (TM) domain which is essential for the fusion process. The TM domains of VAMP2 and Syntaxin, respectively, enter into the vesicle and target membranes and pull them together by a zippering mechanism (Fig. 14.12).

14.4.4 Signal transduction

Cells have the mechanisms to process information or "signals" from their environment. The signal may be in the form of a molecule or a physical stimulus such as light or heat. It initiates a sequence of molecular events, or a signal transduction pathway,

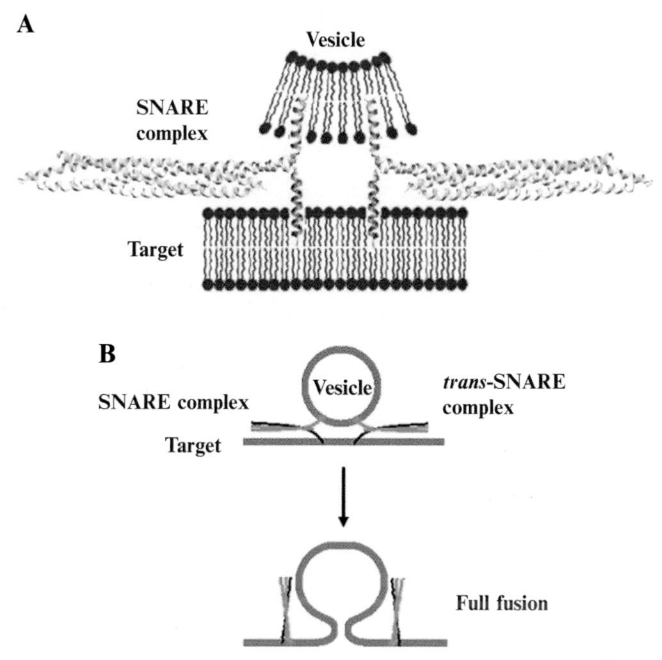

FIG. 14.12

(A) Two SNARE complexes docking a vesicle to a target membrane. (B) Schematic of SNARE-mediated membrane fusion.

Source: Ernst, J.A., Brunger, A.T., 2003. J. Biol. Chem. 278, 8630.

in the cell. The pathway may culminate in the activation of an enzyme or an alteration in the expression of a gene. However, a cell must have an appropriate "receptor" to effectively respond to a signal.

In many cases, membrane proteins function as receptors. The signal molecule (as a ligand) binds to a specific extracellular site of its corresponding receptor protein and causes a conformational change in the latter. Consequent allosteric regulation by the ligand exposes the active site in the catalytic domain of the protein which is located on the cytoplasmic side of the membrane. Thus, the signal is "transduced" through the transmembrane domain of the receptor.

The epidermal growth factor receptor (EGFR) has been considered as a "prototypical" receptor tyrosine kinase (RTK). As the name indicates, the signal transduction pathway is initiated by binding of epidermal growth factor (EGF) (ligand) to EGFR, leading to the phosphorylation of the amino acid tyrosine.

Human EGF is a single-chain polypeptide containing 53 amino acid residues. It binds the N-terminal extracellular region of EGFR (Fig. 14.13). This region contains four domains (I–IV). Domains I, II, and III form a C-shaped structure and the EGF molecule is held between domains I and III. The membrane is spanned by a short (~23 aa) transmembrane (TM) domain. Inside the cell is the tyrosine kinase domain

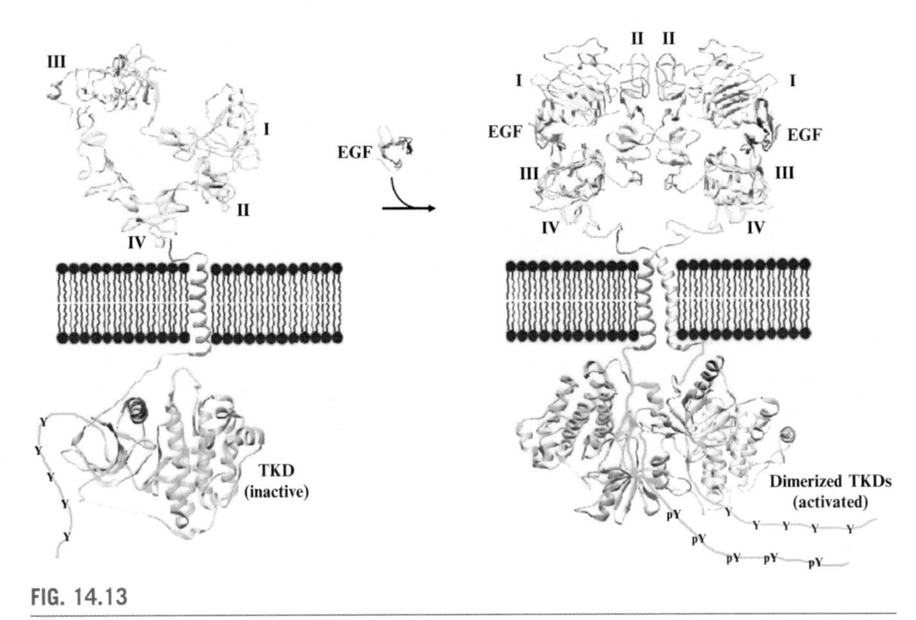

FIG. 14.13

EGF-induced dimerization and activation of EGFR. On the left, monomeric EGFR without bound EGF. On the right, dimerized EGFR with EGF bound to each monomeric subunit.

Source: Ogiso, H., et al., 2002. Cell 110, 775; Bocharov, E.V., et al., 2017. Biochemistry 56, 1697; Jura, N., et al., 2009. Cell 137, 1293.

(TKD) followed by an unstructured C-terminal tail that contains at least five tyrosine autophosphorylation sites.

EGFR uses a growth factor-mediated receptor dimerization mechanism for transmission of signals across the plasma membrane. Binding of the ligand (EGF) induces a major conformational change in the receptor, exposing a dimerization "arm" from an earlier tethered configuration, and leads to the formation of an asymmetric dimer. In the dimerized TKDs, one of the two kinase units induces kinase activity in the other kinase domain through allosteric interactions (Fig. 14.3).

The TM domain of EFGR contributes directly to dimerization and signal transduction. It has been experimentally observed that EGFR fragments consisting of the extracellular and TM domains exhibit several-fold higher affinity for dimerization than those containing only the extracellular region.

****

As a sequel to the discussion in this chapter, a couple of basic points on the nature of the membrane can be reemphasized. First, the membrane is a dynamic and heterogeneous structure with activities not limited to its function just as a boundary. Second, it is neither an autonomous cellular structure nor an isolated thermodynamic system; it continuously exchanges matter and energy with the cell's interior as well as the outside environment. Thus, it plays a crucial role in the establishment of cell-environment synergy.

Sample questions

1. What is the thermodynamic basis of lipid bilayer formation?
2. What is the nature of stresses a protein molecule is subjected to within the bilayer?
3. What are the modes of motion of lipid and protein molecules in the cell membrane?
4. What are the conceptual changes in the updated version of the fluid mosaic model in respect to the original version?
5. How is a transmembrane electric voltage generated across the cell membrane?
6. How is an active transport protein able to move an ion against its concentration gradient?
7. What is the basis of interaction of the low-density lipoprotein receptor (LDLR) with a clathrin-coated pit to trigger exocytosis?
8. How do SNAREs carry out intracellular membrane fusion?
9. How is the kinase domain of epidermal growth factor receptor (EGFR) activated by epidermal growth factor (EGF)?

References and further reading

Anishkin, A., et al., 2014. Feeling the hidden mechanical forces in lipid bilayer is an original sense. Proc. Natl. Acad. Sci. U. S. A. 111 (22), 7898–7905.

Bezanilla, F., 2008. How membrane proteins sense voltage. Nat. Rev. Mol. Cell Biol. 9, 323–332.

Brunger, A.T., 2006. Structure and function of SNARE and SNARE-interacting proteins. Q. Rev. Biophys. 38, 1–47.

Cantor, R.S., 1997. Lateral pressures in cell membranes: a mechanism for modulation of protein functions. J. Phys. Chem. 101, 1723–1725.

Cattrail, W.A., Wisedchaisri, G., Zheng, N., 2017. The chemical basis for electrical signaling. Nat. Chem. Biol. 13 (5), 455–463.

Cullis, P.R., Fenske, D.B., Hope, M.J., 1996. Physical properties and functional roles of lipids in membranes. In: Vance, D.E., Vance, J.E. (Eds.), Biochemistry of Lipids. Elsevier Science, B.V., pp. 1–33.

Deng, D., et al., 2015. Molecular basis of ligand recognition and transport by glucose transporters. Nature 526, 391–396.

Goñi, F.M., 2014. The basic structure and dynamics of cell membranes: an update of the Singer-Nicolson model. Biochim. Biophys. Acta 1838, 1467–1476.

Hibbs, R.E., Gouaux, E., 2011. Principles of activation and permeation in an ion-selective Cys-loop receptor. Nature 474, 54–60.

Jeon, H., Blacklow, S.C., 2005. Structure and physiologic function of the low-density lipoprotein receptor. Annu. Rev. Biochem. 74, 535–562.

Lemmon, M.A., Schlessinger, J., Ferguson, K.M., 2014. The EGFR family: not so prototypical receptor tyrosine kinases. Cold Spring Harb. Perspect. Biol. a020768, 6.

Li, F., et al., 2014. A half-zippered SNARE complex represents a functional intermediate in membrane fusion. J. Am. Chem. Soc. 136, 3456–3464.

Lu, C., et al., 2010. Structural evidence for loose linkage between ligand binding and kinase activation in the epidermal growth factor receptor. Mol. Cell. Biol. 30, 5432–5443.

Mishra, S.K., Watkins, S.C., Traub, L.M., 2002. The autosomal recessive hypercholesterolemia (ARH) protein interfaces directory with the clathrin-coat machinery. Proc. Natl. Acad. Sci. U. S. A. 99, 16099–16104.

Nicolson, G.L., 2014. The fluid-mosaic model of membrane structure: still relevant to the understanding the structure, function and dynamics of biological membranes after more than 40 years. Biochim. Biophys. Acta 1838, 1451–1466.

Ogiso, H., et al., 2002. Crystal structure of the complex of human epidermal growth factor and receptor extracellular domains. Cell 110, 775–787.

Rudenko, G., et al., 2002. Structure of the LDL receptor extracellular domain at endosomal pH. Science 298, 2353–2358.

Shinoda, T., et al., 2009. Crystal structure of the sodium-potassium pump at 2.4 Å resolution. Nature 459, 446–451.

Strøm, T.B., et al., 2014. Mutation G805R in the transmembrane domain of the LDL receptor gene causes familial cholesterolemia by inducing ectodomain cleavage of the LDL receptor in the endoplasmic reticulum. FEBS Open Bio 4, 321–327.

Südof, T.C., et al., 1985. The LDL receptor gene: a mosaic of exons shared with different proteins. Science 228, 815–822.

Toyoshima, C., Kanai, R., Cornelius, F., 2011. First crystal structures of Na^+, K^+-ATPase: new light on the oldest ion pump. Structure 19, 1732–1738.

Wu, Z., et al., 2017. Regulation of exocytic fusion pores by SNARE protein transmembrane domains. Front. Mol. Neurosci. 10, 315. https://doi.org/10.3380/fnmol.2017.00315.

CHAPTER

Fundamentals of structural genomics

15

In Chapter 1, we have seen how classical genetics evolved into the science of genomics. It has also been mentioned there that the initiation of this new field was prompted by genome sequencing of various organisms—from bacteria to human. Evidently, in genomics, that is, the study of genomes in their entirety, the first and the most crucial step is to obtain a complete genome sequence.

15.1 Genome sequencing

In true sense, genome sequencing had begun in 1977 when the 5380 nt-long, single-stranded genome of bacteriophage ϕX174 was completely sequenced by a so-called "plus and minus" method. This method was soon superseded by two others—Maxam and Gilbert chemical degradation and Sanger chain-termination. By the 1980s, the chain-termination method became the most preferred approach to DNA sequencing and remained so for over two decades.

15.1.1 Sanger-automated sequencing

In the meanwhile, improvements and breakthroughs in associated technologies, such as chemical, electrophoretic, and optical, converted Sanger sequencing from a manual to a semiautomated to an automated procedure and the term 'high throughput' was attributed to it. Using the automated Sanger sequencing procedure, the first genome of a free-living organism, the bacterium *Haemophilus influenzae* ($\sim 1.8 \times 10^6$ bp), was completely sequenced.

Yet, the progress in genomics demanded even faster and lower cost sequencing means. To satisfy the demand, alternative sequencing strategies, collectively termed as "next generation sequencing (NGS)", have emerged. The new technologies have redefined "high throughput sequencing" as they outperform Sanger sequencing by a factor of 100–1000 in the throughput rate. Here, we are going to briefly discuss two NGS technologies that have been widely used for genome sequencing—pyrosequencing and reversible terminator chemistry-based sequencing.

In spite of their superior performance, the two NGS technologies share the common principle of DNA synthesis with Sanger sequencing. Each of them requires a primer-template (p/t) junction, dNTPs, and a polymerase enzyme that extends the primer strand by incorporating the nucleotides in accordance with the rule of

Fundamentals of Molecular Structural Biology. https://doi.org/10.1016/B978-0-12-814855-6.00015-8

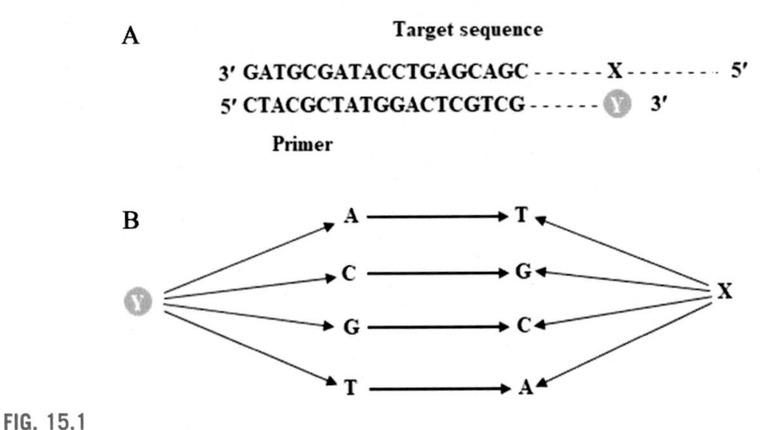

FIG. 15.1

(A) Primer-template configuration for incorporation of nucleotide Y. (B) Identification of Y leads to identification of the nucleotide X.

complementarity with the template DNA (Chapter 8). Fig. 15.1 shows the common configuration to determine the sequence of a template strand (target). Let us consider a nucleotide X at the *n*th position ($n = 1, 2, 3, \ldots$) in the template strand with respect to the p/t junction. The polymerase incorporates a complementary nucleotide Y opposite to X after extension of the primer strand up to the ($n - 1$)th position. So, it is possible to identify X if the identity of Y (A, C, G, or T) is known (Fig. 15.1). Different strategies are adopted to track and identify Y.

In Sanger capillary sequencing, Y is one of the four dideoxynucleotides (ddNTPs: ddATP, ddCTP, ddGTP, and ddTTP). Incorporation of Y at the *n*th position irreversibly terminates further extension of the strand. Termination at different "*n*" produces extension products of different molecular weights which are separated by capillary electrophoresis. Each of the ddNTPs is labeled with a different fluorescent marker (fluorochrome) so that the chain-terminated polynucleotides can be distinguished by a fluorescent detector (Fig. 15.2).

15.1.2 Next-generation sequencing

In contrast to the size-dependent separation strategy of Sanger sequencing, NGS is dependent on positional separation. Millions of different template strands are held in a two-dimensional array (mostly on a glass slide) with the position of each template remaining fixed throughout the sequencing procedure. The captured image can resolve the position of each template.

In pyrosequencing, the template is sequentially exposed to dATPαS (instead of dATP which can produce a false signal), dCTP, dGTP, and dTTP. Incorporation of the correct nucleotide Y opposite to X (Fig. 15.3) releases a pyrophosphate (PP_i). The released PP_i is converted to ATP in a reaction catalyzed by ATP sulfurylase

$$PP_i + APS \rightarrow ATP + SO_4^{2-}$$

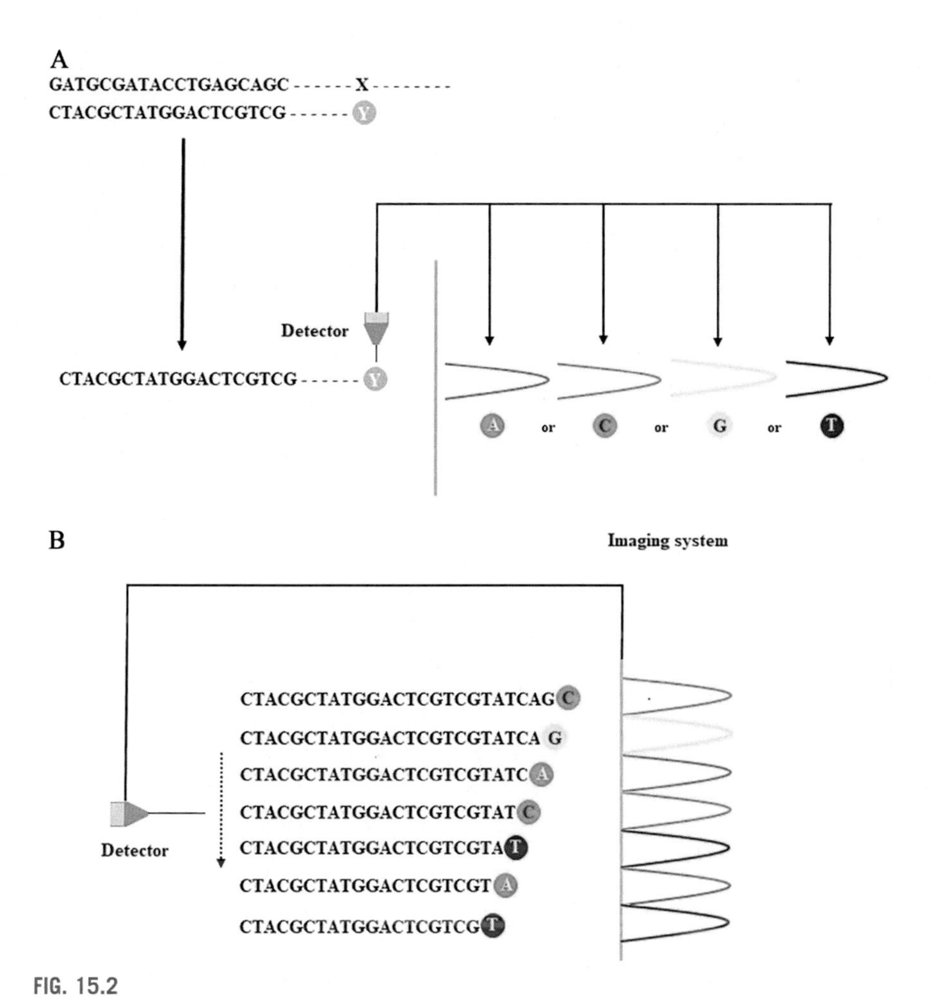

FIG. 15.2

Sanger sequencing process. (A) Incorporation of fluorescent labeled dideoxynucleotide Y. A—ddATP *(green)*, C—ddCTP *(blue)*, G—ddGTP *(yellow)*, and T—ddTTP *(red)*. (B) Separated extension products moving past the detector and the attached fluorescent read out sequentially.

where APS is adenosine 5′-phosphosulfate. ATP is a cofactor for the enzyme luciferase, oxidizing luciferin to oxyluciferin and producing light (hν).

$$\text{ATP} + \text{luciferin} + O_2 \rightarrow \text{AMP} + \text{PP}_i + \text{oxyluciferin} + CO_2 + h\nu$$

The emitted light is captured in the image (Fig. 15.3).

In reversible terminator technology, the detection of Y is based on termination of synthesis at the *n*th position similar to that in Sanger sequencing. The terminator is identified by a fluorescent label it carries. However, the terminator in this NGS technology is a reversible one. It is a nucleotide modified in two ways—it is blocked by a chemically cleavable moiety at the 3′-OH position, and it carries a chemically cleavable fluorescent label (Fig. 15.4).

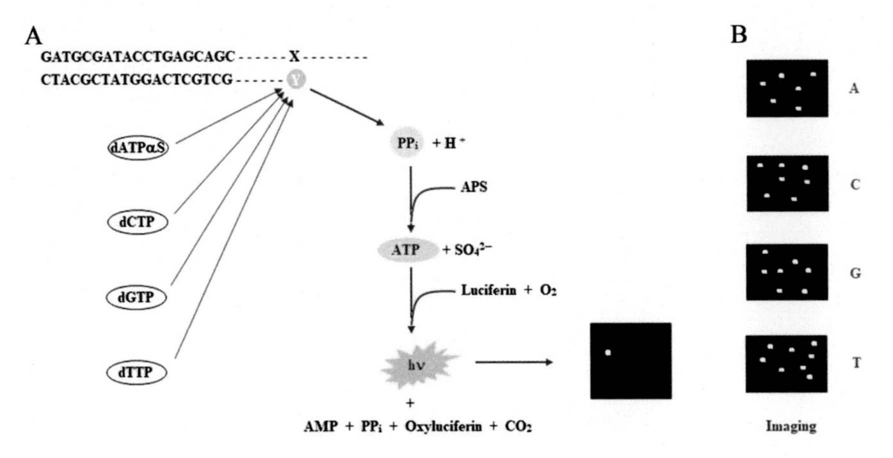

FIG. 15.3

Pyrosequencing process. (A) Templates (such as the one shown) fixed in a two-dimensional array are exposed to one of the four nucleotides (Y=dATPαS, dCTP, dGTP, or dTTP) at a time. Incorporation of Y releases a pyrophosphate which ultimately leads to the emission of a light signal. (B) Image capture from multiple templates in a four-step flow cycle.

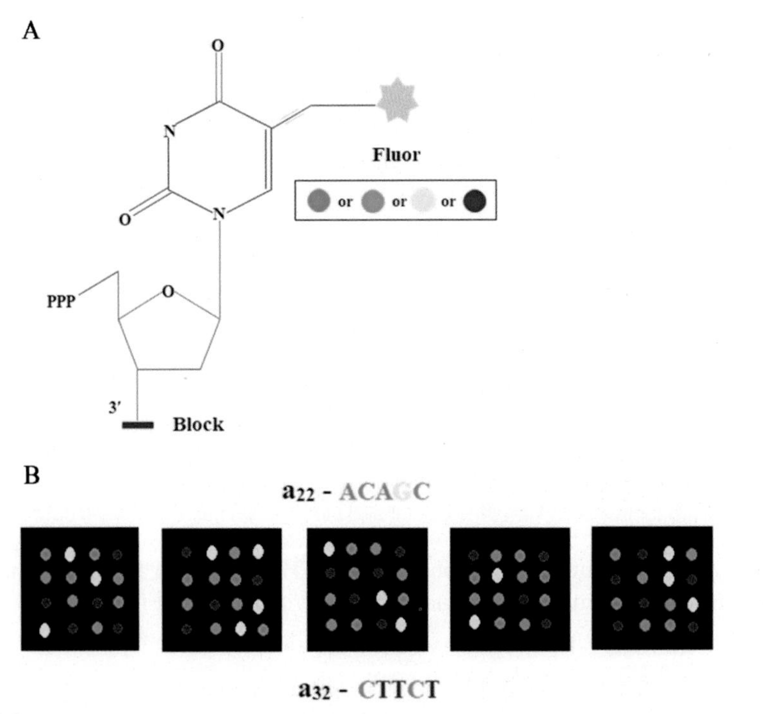

FIG. 15.4

Four-color imaging in reversible terminator technology. (A) 3′-blocked reversible terminators labeled with four different fluorochromes (A—*green*, C—*blue*, G—*yellow*, and T—*red*). (B) Sequential imaging—reads corresponding to two templates at matrix positions a_{22} and a_{32}, respectively, are shown.

All four nucleotides, A, C, G, and T, labeled differently, are added to the reaction at the same time together with the polymerase. Incorporation of the correct Y blocks further extension of the primer strand. The polymerase and unincorporated nucleotides are washed away, and Y is detected by total internal reflection fluorescence imaging using two lasers. The fluorophore and blockage are removed from Y and the p/t junction is ready for the next cycle of nucleotide incorporation. Images corresponding to five successive incorporations are shown in Fig. 15.4.

15.2 Genome annotation

In the pregenomic era, DNA sequencing, limited mostly to cloned fragments, was followed-up by an effort to identify interesting features in a DNA fragment, such as a gene or a regulatory element, and interpret their biological functions. Such annotation of a DNA sequence was even manually possible. In the postgenomic era, the identification of significant features in a genome sequence, a process called genome annotation, has assumed gigantic proportions and become totally computer-dependent. Yet, the basic molecular biological principles underlying the annotation of a DNA sequence, whether small or genomic, remain the same.

Once a genomic sequence is available, the next obvious step is to find the genes it contains, particularly the protein-coding genes. Primarily, it is the search for open reading frames (ORFs). We have seen in Chapter 11 that, in principle, a segment of an mRNA is readable by the translation machinery in three different frames. This implies that a segment of a double-stranded DNA, with any of the two strands being transcribable under appropriate conditions, has six possible reading frames, three in each direction, forward or reverse. However, an ORF is only a nonoverlapping string of codons with a start codon and a stop codon. A simple ORF scanning can locate a gene only with a limited probability. Necessary parameters and conditions are required to be associated with the scanning algorithm to make it more realistic.

ORF scanning is relatively straightforward for bacterial genomes as they contain very little noncoding sequence. Nevertheless, even here certain conditions must be imposed to obtain more meaningful results. As for example, based on the randomness of DNA sequence, a probability calculation would show that there are very few ORFs longer than 30–40 codons. In contrast, the average length of an *Escherichia coli* gene, determined from earlier experimental data, is >300 codons. Hence, setting a minimum of 100 codons as a criterion is likely to record more possible hits in the ORF search. The conserved nature of the bacterial promoter and ribosomal-binding sequence is also a helpful criterion.

For eukaryotic genomes, ORF scanning is more complex as they contain substantially more intergenic DNA and most eukaryotic genes are interrupted by introns (Chapter 10). The intron problem can be addressed by including the conserved sequences at the exon-intron boundaries (Chapter 10) in the search strategy. Additional factors that are taken into account to improve the accuracy of gene prediction are upstream regulatory sequences of the genes, cell-specific alternative splicing, RNA editing, and bias in codon usage.

15.3 Structural genomics/proteomics

Annotation of a genome, particularly by the identification of protein-coding genes, is usually followed up by characterization of the encoded proteins. Three-dimensional structures of the proteins are determined and functions assigned to them. Going further, based on structure-function correlation, the mechanisms of different biochemical processes involving the concerned proteins are understood.

Earlier, structure determination used to be undertaken only after a protein was identified and its function known. The advent of genome sequencing reversed the approach by prompting structure determination before the availability of any biological information. Moreover, since the current initiatives to understand function based on structure are usually taken at the genomic scale, the field is called structural genomics (alternatively structural proteomics). The aim of structural genomics is, therefore, to assign three-dimensional structures to the protein products of a genome (collectively called a proteome) and understand biological implications of the assigned structures. The structures can be computationally predicted and/or experimentally determined.

15.3.1 Experimental structure determination

In Chapter 6, we have discussed the three main experimental techniques that are used to derive atomic-scale structural information on biological macromolecules—X-ray crystallography, nuclear magnetic resonance (NMR) spectroscopy, and cryo-electron microscopy (cryo-EM). Of these three, X-ray crystallography has been the most widely used technique in structural genomics.

Experimental structure determination by X-ray crystallography requires the expression, purification, and crystallization of a protein. The specific gene encoding the protein is "cloned" into a "vector" by recombinant DNA technology (which has not been discussed in this book) and expressed in prokaryotic (*E. coli*) or eukaryotic (yeast, insect, etc.) cells. The protein is purified with the help of chromatographic techniques, crystallized under appropriate conditions and subjected to diffraction analysis using synchrotron X-ray sources.

The entire process from expression to structure analysis is long and time-consuming. This was acceptable so long as individual genes and their products were studied. However, in the postgenomic era, when genomic sequences are accumulating with incredible rapidity, high-throughput methods in protein structure analysis are called for. Encouragingly, worldwide structural genomic initiatives have led to remarkable advances in macromolecular crystallography. Such advances have been possible due to significant improvements in related technologies and the introduction of robotics and automation in almost all the steps in protein structure determination. Cloning of multiple genes and purification of the corresponding protein products are now carried out in parallel by highly automated procedures. Robotics are engaged in crystallization and crystal observation. However, automated protein crystal harvesting, which is expected to further improve the efficiency of the process, is yet to find wide application. More powerful X-ray sources, such as synchrotron radiation, and more sensitive X-ray detectors have substantially improved crystallographic data acquisition, while the systematic use of selenomethionine as an anomalous scatterer has made the solution of phase problem (Chapter 6) more efficient.

15.3.2 Computational structure prediction

Computational methods for protein structure prediction can be categorized into four different groups: (a) comparative modeling, (b) threading or fold recognition, (c) fragment-based recombination, and (d) physics-based ab initio modeling.

Comparative modeling

Comparative modeling is based on the premise that evolutionarily related sequences exhibit three-dimensional structural similarity. In other words, sequence similarity is an indicative of structural similarity. Two conditions are required to be fulfilled for the prediction of protein three-dimensional structure by comparative modeling—(a) the target sequence (that is, the sequence to be modeled) must have a reasonable similarity to the sequence of a known structure (to be used as the template), and (b) an accurate alignment between the target sequence and the template structure must be computable. Evidently, the accuracy of prediction by comparative modeling depends on the degree of similarity between the target and the template. Hence, the basic steps in comparative modeling are—(a) template selection. (b) target-template alignment, (c) modeling structurally conserved regions as well as side chains, and (d) model assessment.

Let us illustrate comparative modeling with a sequence from the thermophilic bacterium *Thermoanaerobacter indiensis* proteome using SWISS-MODEL ExPASy server. A search for template finds *Mycobacterium tuberculosis* thioredoxin reductase (Tdr) having 46% sequence identity with the *T. indiensis* sequence (Fig. 15.5A). *M. tuberculosis* Tdr is, therefore, used as the template (Fig. 15.5B). Modeling predicts a three-dimensional structure of the *T. indiensis* protein (Fig. 15.5C) very similar to the *M. tuberculosis* Tdr structure. Fig. 15.5D establishes the identity of the two structures by superposition.

Threading

It is interesting to note that while sequence similarity implies structural similarity, the converse is not always true. Often, proteins with similar structures are found to have little sequence similarity. Obviously, in such cases, comparative modeling may not be a reliable structure prediction approach and protein threading, also referred to as fold recognition (FR), appears to be more useful.

FR methods are prompted by two factors. First, the structure is evolutionarily more conserved than the sequence and, second, protein fold diversity is significantly more limited than sequence diversity. Threading aligns a target sequence to a library of folds. Structurally similar regions are identified in multiple templates. Hence, this

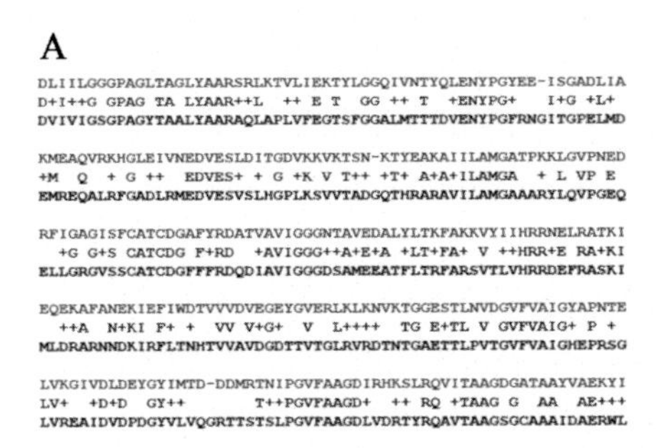

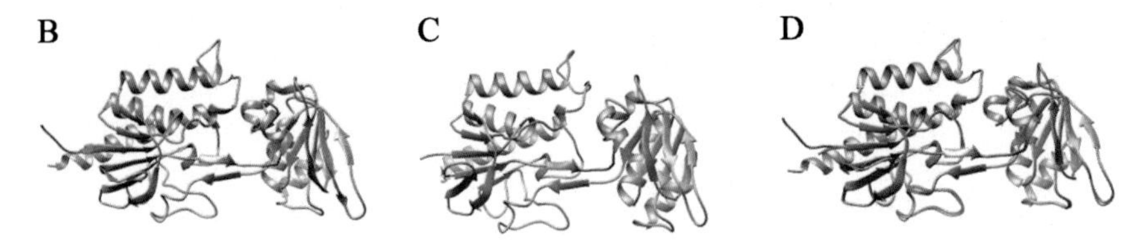

FIG. 15.5

Three-dimensional structure prediction by comparative modeling. (A) Alignment of the two sequences: query (upper) and *M. tuberculosis* Tdr (lower). (B) Three-dimensional structure of the template (PDB code: 2A87). (C) Predicted structure of *T. indiensis* protein. (D) Superposition of query and template structures.

Source (for 2A87): Akif, M., et al. 2005. Acta Crystallogr. Sect. D. 61, 1603. T. indiensis *sequence obtained from http:/www.jgi.doe.gov. Reference for comparative modeling: Guex, N., et al. given below.*

approach leads to regions of accurate structure prediction even if the target cannot be aligned with a single template as a whole.

Fragment-based recombination

The fragment-based recombination method of protein structure prediction is based on the observation that a protein backbone can be reconstructed by using short fragments (up to 10 aa) with known structures. Suitable fragments are acquired from the protein database and assembled into a structure with the help of appropriate scoring functions and optimization algorithms. Local sequence-structure correlations have been found to be significant in most cases.

Physics-based ab initio modeling

Although the knowledge-based methods, particularly comparative modeling, have been most widely used for protein structure prediction, they have some limitations. First, they are dependent on the availability of structural homology which, in many cases, may be nonexistent. Besides, these methods predict protein structures without having to acquire a basic understanding of the forces and mechanisms responsible for the structure formation.

In contrast, the first principle protein structure prediction methods are not dependent on database information. Instead, they are based on the laws of physics and, therefore, develop an in-depth understanding of the mechanisms leading to the formation of different protein structures.

The underlying principle of the physics-based methods for protein structure prediction is Anfinsen's thermodynamic hypothesis. According to Anfinsen, a protein molecule adopts a conformation which is the global minimum of its free energy. Evidently, the methods require an energy function to describe the potential energy surface of a protein molecule and a global minimization procedure. They carry out molecular dynamic (MD) simulations to analyze the movement of atoms in a protein molecule using Newton's equations of motion and suitable force fields. (The basic concepts of MD and force field will be discussed in Chapter 18.)

15.4 Function prediction

Now, we come to the ultimate step in structural genomics—prediction of protein function from structure. It is to be understood here that if structure determination efforts have been finding it difficult to cope with the speed of accumulation of genomic sequence data, assignment of a function to a structure is even more formidable. Nevertheless, once the structure of a protein is experimentally solved or computationally predicted, the aim is to assign a putative function to it. The functional assignment is carried out by comparing the sequence and three-dimensional structure of the protein of interest with those in the existing database using different computer programs. A function is 'transferred' to the 'query' protein from the closest sequence or structural match.

Sequence-based approaches are relatively simpler and, therefore, more commonly used methods of computer-aided functional annotation. These methods are primarily based on alignment of the query sequence with those of proteins with known functions and analysis of amino acid residue conservation.

Structure-based methods predict the function of a protein in one of the two ways—by speculating the overall biochemical or biological function, or by identifying specific residues that may be responsible for a certain function. For example, if the overall structure of a protein is found to be similar to those of known DNA-binding proteins, it is also likely to bind DNA. Identification of residues in the test protein similar to the DNA-binding residues in other proteins further affirms its DNA-binding property.

Functional residues are important for a protein to carry out its biological activity. Alteration (mutation) in one or more of such residues may lead to changes in its structure or, in other words, its probability distribution over the three-dimensional conformational space. A functional site contains one or more of the functional residues.

There are a number of geometry-based computer programs which examine the surface of a protein to identify and locate ligand-binding sites (active site pockets).

Besides, structure-based docking methods identify catalytic sites in a protein using small molecular probes.

Some of the function prediction methods, however, utilize both sequence and structural information of a protein. They not only analyze the surface of a protein for binding pockets, but also identify evolutionarily conserved residues to locate these pockets.

15.5 Structural genomics initiatives

Organizationally, structural genomics (SG) is the worldwide endeavor to make available the three-dimensional structures of all macromolecules of biological importance, albeit protein is the major focus. A number of SG consortia have been established in America, Europe, and Asia at the beginning of the 21st century.

Understandably, it is not feasible to match the solved protein structures with accumulating genomic sequences in sheer number. Encouragingly, unlike the sequences which are virtually unlimited in number, the 'fold space' is finite. Over 10,000 protein sequence families have been identified and, although the number is increasing, it is not expected to 'explode'.

The aim of SG initiatives, therefore, is to make available at least one representative structure of each protein family. With this goal, the SG centers have developed pipelines for high-throughput target selection, protein expression and purification, crystallization, and structure analysis by X-ray crystallography and NMR. Importantly, SG programs select proteins for structure determination based on available genomic information. In this way, known structural homologues are avoided and structural novelty is targeted.

Sample questions

1. What are the developments in chemical, electrophoretic, and optical technologies that converted Sanger sequencing into an automated procedure?
2. Why is ORF scanning of eukaryotic genomes more complex than that of prokaryotic genomes?
3. What is the basic principle shared among Sanger chain-termination, pyro-, and reverse terminator-based sequencing?
4. What are the technological improvements that made rapid advances in macromolecular crystallography possible in the postgenomic era?
5. When is threading more useful than comparative modeling for protein structure prediction?
6. What are the ways a protein can be predicted to possess a DNA-binding property?
7. What is the basis of target selection in structural genomic initiatives?

References and further reading

Brown, T. A. Genomes 4, Garland Science, 2018.

Bujnicki, J.M., 2006. Protein-structure prediction by recombination of fragments. ChemBioChem 7, 19–27.

Chen, D., 2009/2010. Structural genomics: exploring the 3D protein landscape. Biomed. Comput. Rev. 12–18.

Deller, M.C., Rupp, B., 2014. Approaches to automated protein crystal harvesting. Acta Crystallogr. F 70, 133–135.

Eswar, N., et al., 2006. Comparative protein structure modeling using Modeller. Curr. Protoc. Bioinform. https://doi.org/10.1002/0471250953.bi0506s15.

Floudas, C.A., 2007. Computational methods in protein structure prediction. Biotechnol. Bioeng. 97, 207–213.

Floudas, C.A., et al., 2006. Advances in protein structure prediction and de novo protein design: a review. Chem. Eng. Sci. 61, 966–988.

Grabowski, M., et al., 2016. The impact of structural genomics: the first quindecinnial. J. Struct. Funct. Genom. 17 (1), 1–16.

Guex, N., Peitsch, M.C., Schwede, T., 2009. Automated comparative protein structure modeling with SWISS-MODEL and Swiss-pdb viewer—a historical perspective. Electrophoresis 30, S162–S173.

Harrington, C.T., et al., 2013. Fundamentals of pyrosequencing. Arch. Pathol. Lab. Med. 137, 1296–1303.

Joachimiak, A., 2009. High-throughput crystallography or structural genomics. Curr. Opin. Struct. Biol. 19 (5), 573–584.

Kircher, M., Kelso, J., 2010. High-throughput DNA sequencing. Bioessays 32, 524–536.

Metzker, M.L., 2010. Sequencing technologies—the next generation. Nat. Rev. Genet. 11, 31–46.

Mills, C.L., Beuning, P.J., Ondrechen, M.J., 2015. Biochemical function predictions for protein structures of unknown or uncertain function. Comput. Struct. Biotechnol. J. 13, 182–191.

Ołdziej, S., et al., 2005. Physics-based protein structure prediction using a hierarchical protocol based on the UNRES force field: assessment in two blind tests. Proc. Natl. Acad. Sci. U. S. A. 102 (21), 7547–7552.

Primrose, S.B., Twyman, R.M., 2003. Principles of Genome Analysis and Genomics, 3rd ed. Blackwell Publishing.

Stewart, P.S., Mueller-Dieckmann, J., 2014. Automation in biological crystallization. Acta Crystallogr. F 70, 686–696.

Su, X.-D., et al., 2015. Protein crystallography from the perspective of technology developments. Crystallogr. Rev. 21 (1–2), 122–153.

CHAPTER

Cell signaling and systems biology 16

16.1 Cell signaling

All cells have the dynamic ability to sense and respond to environmental stimuli in order to survive and function in an appropriate manner. The response is essentially through sequences of molecular events, or signal transduction pathways, that receive and process signals originating from the external environment.

16.1.1 Pathway as signal amplifier

Most signals sensed by a cell are primarily chemical in nature (even physical signals are detected as chemical changes at the point of reception). A majority of the signal transduction processes are protein-driven, involving different proteins, or modular domains within individual proteins, along the respective pathways. Some of these proteins can modify other proteins by adding ("writing") or removing ("erasing") chemical "marks" to or from the target proteins. In addition, there are proteins that can participate in the transduction pathways as they contain domains capable of "reading" such modifications (or demodifications).

Often, more than two proteins are linked together into a pathway where one protein modifies another, which then modifies a third, and so on. In this "signal amplifier," the activity of the first modifier is amplified by the activities of the other modifiers downstream from it in the pathway.

There are several ways signaling proteins can add chemical "marks" to target proteins. For example, proteins containing kinase domains can phosphorylate specific sites on target proteins, whereas acetyltransferases can add an acetate.

16.1.2 Signaling by phosphorylation

Protein kinases catalyze the reaction shown in Fig. 16.1. A phosphoryl group (PO_3^{2-}) is transferred from ATP to the protein substrate by the formation of an ester with an amino acid residue whose side chain contains a hydroxyl group (OH^-). In mammals, phosphorylation occurs predominantly on serine (85%) and threonine (15%) residues. Nevertheless, tyrosine phosphorylation, though relatively rare, plays a very significant role in signaling.

The transferred phosphoryl group brings with it a certain amount of negative electric charge to the receiving amino acid residue. This negative charge enables the

Fundamentals of Molecular Structural Biology. https://doi.org/10.1016/B978-0-12-814855-6.00016-X

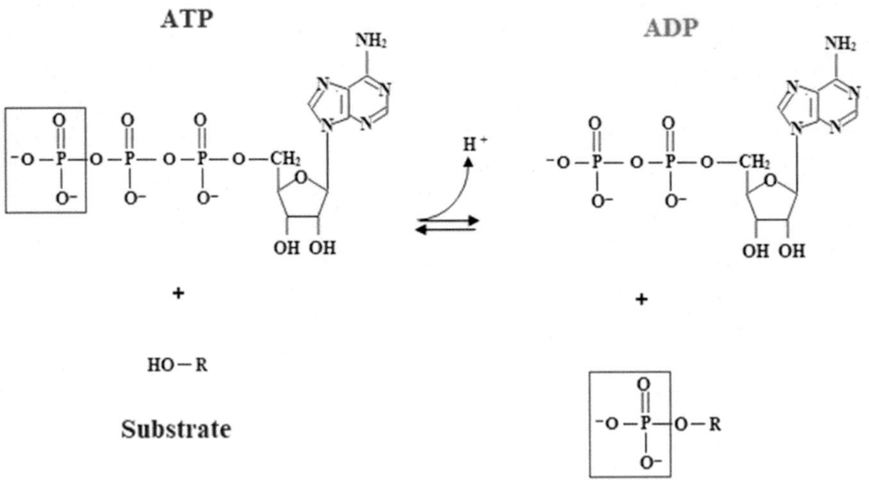

FIG. 16.1

Phosphorylation reaction.

phosphorylated residue to electrostatically interact with positively charged residues such as lysine and arginine. Often, formation of ionic interactions and hydrogen bonds leads to a significant rearrangement in the structure of the phosphorylated protein such that the binding sites for ATP and substrate are opened up, while the catalytic residues are reoriented to a conformation that facilitates the transfer of a phosphoryl group from ATP to a serine, threonine, or tyrosine residue in the substrate.

16.1.3 MAPK pathway

The Ras/Raf/MEK/ERK signaling cascades, which are dependent on phosphorylation, coordinate and regulate cell growth and differentiation in response to extracellular stimulation. They are also denoted as mitogen-activated protein kinase (MAPK) pathways. Promoted by the action of several receptor protein-tyrosine kinases, inactive Ras-GDP is converted to an active Ras-GTP which, in turn, activates the Raf kinase family. The activated Raf kinases have limited substrate specificity. They catalyze the phosphorylation, and thereby the activation, of MEK1 and MEK2.

The hydrophilic non-receptor proteins MEK1 and MEK2 are "dual-specificity" protein kinases—they mediate the phosphorylation of tyrosine and then threonine in MAP kinase ERK1 or ERK2. In contrast to the Raf kinases and MEK1/2, which have restricted substrate specificity, the activated MAP kinases catalyze numerous cytosolic and nuclear substrates.

ERK2

To understand the mechanism of activation by phosphorylation, let us look at some of the structural features of ERK2 (Fig. 16.2A). Like other protein kinases, ERK2 has a small N-terminal lobe and a large C-terminal lobe, both of which contain

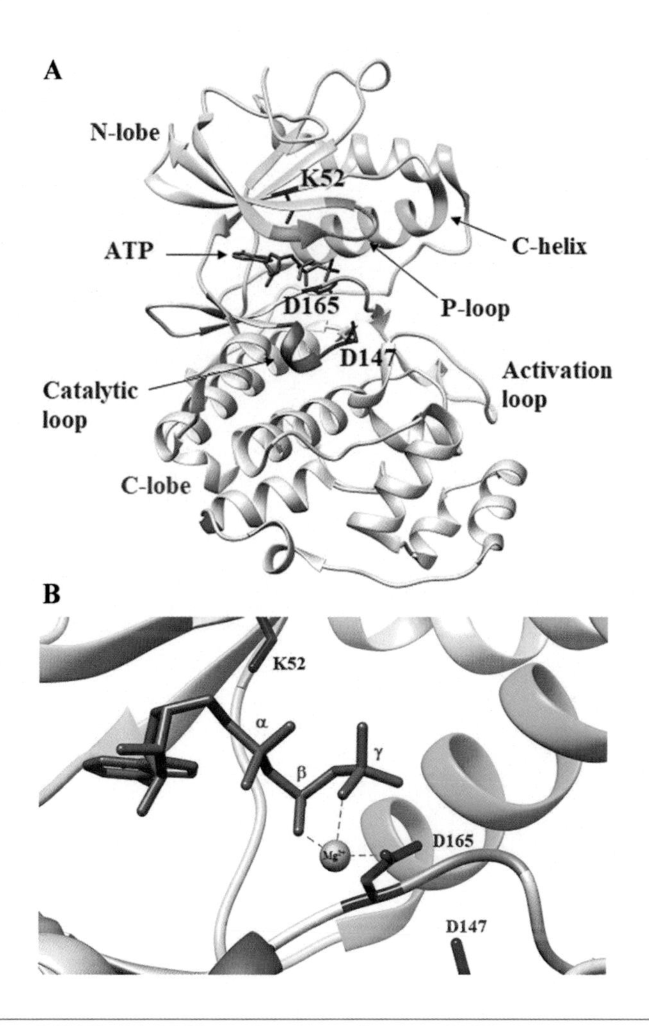

FIG. 16.2

(A) Structure of ERK2. (B) D165-mediated coordination of ATP-phosphates.

Source: Pozharski, E., Zhang, J., Shapiro, P., PDB code: 4GT3; unpublished.

several conserved α-helices and β-strands. The small lobe contains a conserved C-helix which is important for the activity of ERK2. It also contains a conserved GxGxxG ATP-phosphate-binding loop (P-loop), which helps position the β- and γ-phosphates of ATP required for catalysis. The catalytic site of ERK2, located in the cleft between the small and large lobes, is characterized by a K/D/D motif constituted of K52, D147, and D165, the last aspartate being part of a conserved DFG sequence. D165 binds Mg^{2+} ions which, in turn, coordinate the phosphates of ATP (Fig. 16.2B).

ERK2 is inactive in its nonphosphorylated state (Fig. 16.3A). Conversion to the active state requires the phosphorylation of a threonine (T183) and a tyrosine (Y185) in the sequence TEY (conserved sequence: TxY) which is part of an "activation

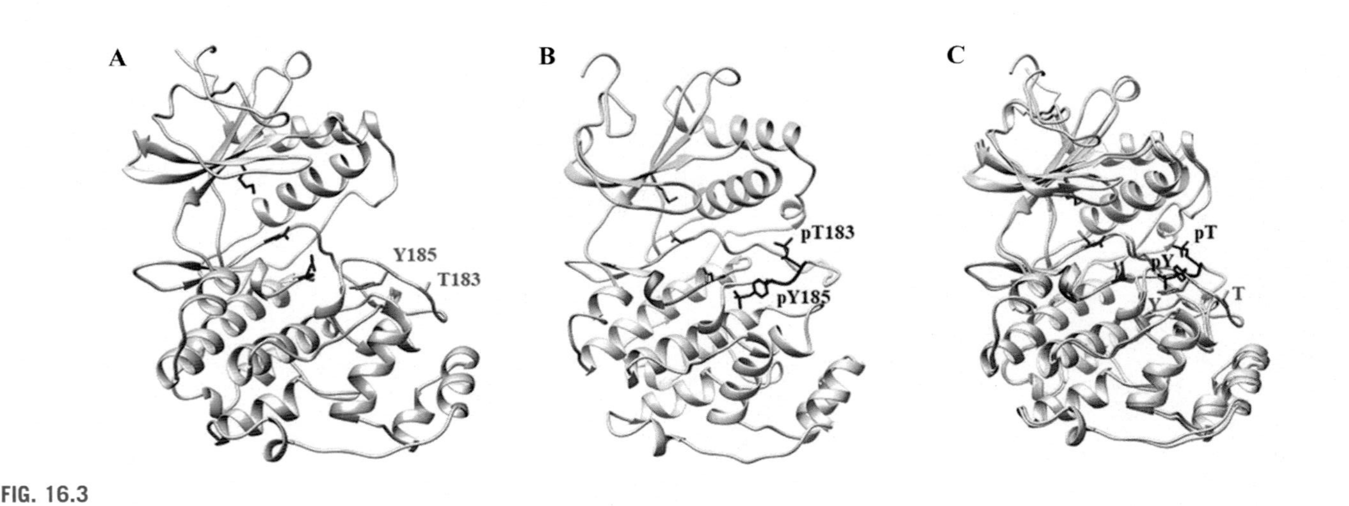

FIG. 16.3

Structure of ERK2 in nonphosphorylated (A) and phosphorylated (B) states. (C) Conformational changes revealed by superposition of the two structures.

Source: Zhang, F., et al., 1994. Nature 367, 704; Canagarajah, B.J., et al., 1997. Cell 90, 859.

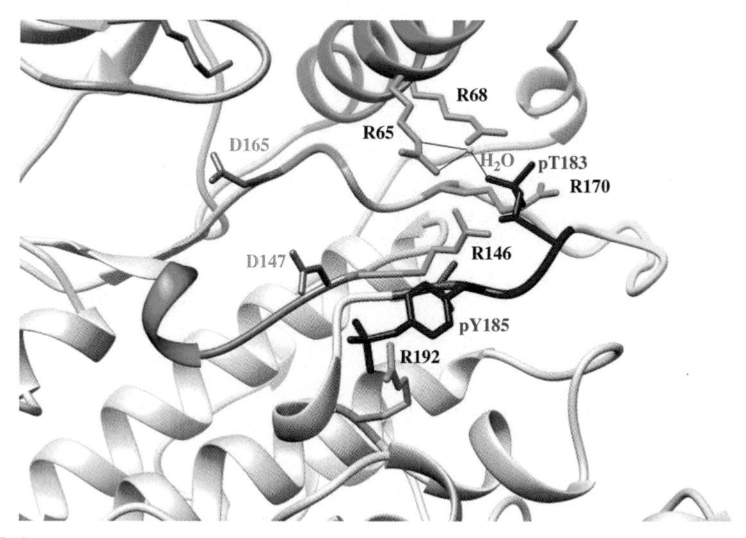

FIG. 16.4

Interaction of the phosphorylated residues of ERK2 activation loop with surrounding arginine residues.

Source: Canagarajah, B.J., et al., 1997. Cell 90, 859.

loop" (Fig. 16.3b). The two residues can interact with a network of surrounding arginine residues only when they are in a phosphorylated state (Fig. 16.4). pT183 interacts directly with R68 in C-helix, R146 in the catalytic loop, and R170 in the activation loop, and indirectly, via a water molecule, with R65 in C-helix. These are either ionic interactions or H-bonds. On the other hand, Y185 is inaccessible to solvent in the unphosphorylated ERK2 structure. pY185 moves its side chain to the surface of the kinase and, together with V186 and R192, forms a substrate recognition site.

A comparison of the three-dimensional structures of ERK2 in nonphosphorylated and phosphorylated states shows that not only the positions of the concerned residues are altered as a consequence of phosphorylation, but the entire activation loop is dragged into a new conformation. At the same time, there is a relative movement between the N-terminal and C-terminal lobes (Fig. 16.3C).

In the inactive state of ERK2, the D165 side chain faces away from the active site ("DFG-aspartate out" conformation), whereas in the active state, the D165 side chain faces towards the ATP-binding pocket ("DFG-aspartate in" conformation) and coordinates Mg^{2+}. The ability of the aspartate to bind Mg^{2+} is crucial to the kinase activity of ERK2.

MEK

MEK proteins, which phosphorylate MAPKs, are also called MAP kinase kinase (MAPKK). Like other protein kinases, MEK1/2 have an N-terminal small lobe and a C-terminal large lobe containing conserved α-helices and β-strands (Fig. 16.5).

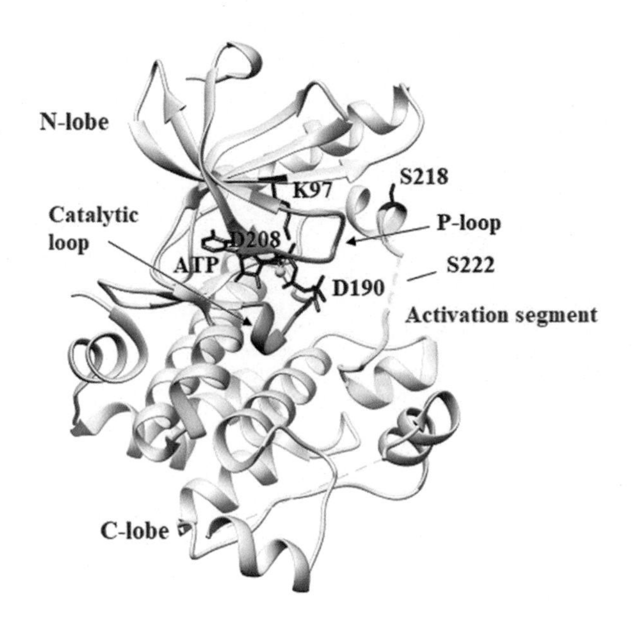

FIG. 16.5

Structural features of MEK1.

Source: Ohren, J.F., et al., 2004. Nat. Struct. Mol. Biol. 11, 1192.

The small lobe contains a conserved glycine-rich (GxGxxG) ATP-phosphate-binding loop (P-loop). The large lobe is predominantly α-helical. A catalytic loop and three catalytic residues (K97, D190, and D208 in MEK1), responsible for phosphoryl transfer from ATP to ERK substrates, are located in the cleft between the small and large lobes. MEK1 D208 binds Mg^{2+} ions which, in turn, coordinate the phosphates of ATP.

MEK1 is activated when two serine residues, S218 and S222, in the activation segment are phosphorylated by Raf kinases. The two lobes of MEK1 move relative to each other during the catalytic cycle and, thus, open and close.

For MEK1-mediated phosphorylation of ERK2, "docking" interaction between the two kinases is required. The MEK1 N-terminus contains the ERK2-docking site (KKKPTPIQL). It is likely that the side chains of two ERK2 tyrosine residues, Y316 and Y317, "dock" into a pocket formed by hydrophobic residues in the docking site of MEK1.

Activated MEK1 first phosphorylates the tyrosine in the ERK2 activation loop and dissociates from the tyrosine-phosphorylated MAPK. Then, either the same or another activated MEK reassociates with the tyrosine-phosphorylated ERK2 and catalyzes the phosphorylation of the threonine, two residues upstream from the phosphotyrosine.

MKP3

Just as the ERK kinases are positively regulated (activated) by phosphorylation, they are negatively regulated through the dephosphorylation of both threonine and tyrosine residues by a family of dual-specificity MAKP phosphatases (MKPs). One such phosphatase is MKP3 which is highly specific for ERK1/2.

Like all MKPs, MKP3 contains an N-terminal kinase-binding domain (KBD) and a C-terminal catalytic domain (CD) connected by a flexible linker (Fig. 16.6). The catalytic activity of MKP3 is controlled by its ability to recognize ERK1/2 and allosteric activation caused by binding to the MAPK substrates.

The specificity of MKP3 for ERK1/2 is based on its docking interactions with the latter. MKP3 contains a kinase interaction motif (KIM) or docking sequence (D-motif), $(R/K)_{2-3}X_{1-6}\ \Phi_AX\Phi_B$ in the KBD. In ERK2, the KIM-docking site (also called D-motif-binding site or D-site) is located in a noncatalytic region opposite to the kinase catalytic pocket (Fig. 16.7). A highly acidic patch and a hydrophobic groove together constitute the D-site. The acidic patch consists of residues E79, Y126, D160, D316, and D319. It interacts with the basic residues of the D-motif. The hydrophobic groove, formed by residues T108, L110, L113, L119, F127, and L155, accommodates $\Phi_AX\Phi_B$ (L71, P72, V73). The side of L71 plugs into the groove, while V73 makes van der Waals contacts with E107, T108, and T157.

MKPs belong to the protein-tyrosine phosphatase (PTP) superfamily. MKP3 contains the PTP signature motif, $HC(X)_5R$, also called the P-loop, in its CD (Fig. 16.6). The catalytic site of MKP3 is located in a cleft formed by the P-loop. The cleft is shallow, with a depth of 5.5 Å, yet sufficient in dimension to accommodate both phosphor-tyrosine and phosphor-threonine side chains.

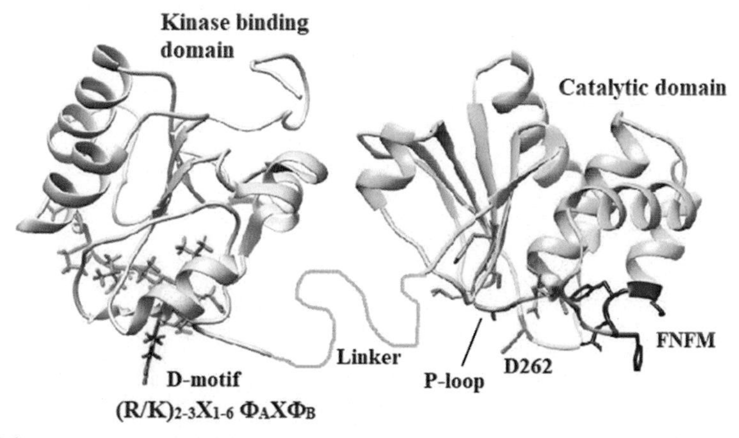

FIG. 16.6

Domain organization of MKP3. Residues in the D-motif, P-loop, and FNFM motif are highlighted.

Source: Farooq, A., et al., 2001. Mol. Cell 7, 387; Stewart, A.E., et al., 1999. Nat. Struct. Mol. Biol. 6, 174.

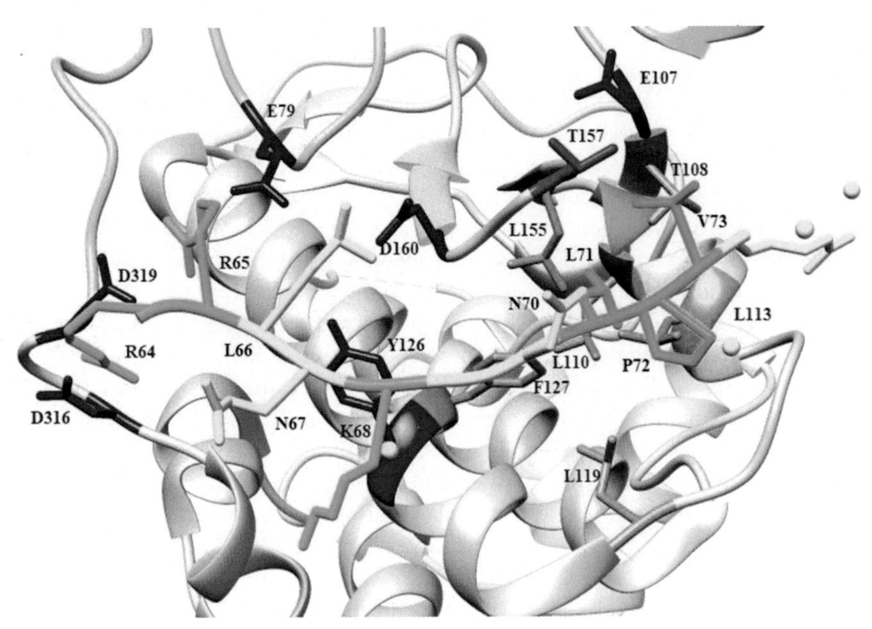

FIG. 16.7

Docking interaction between ERK2 and KIM^{MKP3}. Interacting residues are colored as follows: in ERK2, acidic residues—*red*, hydrophobic residues—*orange*; in MKP3, basic residues—cyan, $\Phi_A X \Phi_B$—lime green.

Source: Liu, S., et al., 2006. Proc. Natl. Acad. Sci. U. S. A. 103, 5326.

The MKP3 CD also contains a FNFM motif which is essential for ERK2-mediated MKP3 activation as well as MKP3-mediated ERK2 inactivation. D262 also plays a role in the ERK2-induced allosteric activation of MKP3 as the loop containing the residue is closed over the active site, thereby enhancing the catalytic efficiency of the phosphatase.

16.2 Structural basis of immune response

Antibody-antigen (Ab-Ag) interaction forms the basis of adaptive immune response of a vertebrate. When bacteria or viruses infect the human body and introduce toxic substances (antigen) into it, specialized cells (B-lymphocytes) of the body respond by producing antibody (Ab) molecules to destroy the infecting agent.

Two functions of an antibody molecule are responsible for immune response—the ability to bind specifically to its target antigen (Ag) and recruit other cells and molecules to encounter the bound Ag. Evidently, there must be sites on both the

Ab and Ag to bind to each other. The binding site on the Ag is called the epitope, while that on the Ab is called the paratope. The binding of Ag and Ab involves different types of noncovalent interactions between the epitope and the paratope. The diversity in the binding potentials of Ab molecules is indeed remarkable—they can bind to almost any "nonself" surface with high affinity and specificity notwithstanding the fact that all Abs are structurally similar. With the availability of an increasing amount of structural data, the mechanisms of Ab-Ag interaction are becoming ever more comprehensible.

As shown in Fig. 16.8, an Ab molecule consists of two identical heavy (H) chains and two identical light (L) chains. The four chains are held together by disulfide bonds to form a Y-shaped structure. Each arm of the structure is known as a Fab (fragment of antigen-binding). The Fab is composed of two variable domains (V_H and V_L) and two constant domains (C_H1 and C_L), where the subscripts H and L represent the heavy and light chains, respectively. Each H-chain contains two additional domains C_H2 and C_H3 in the F_C region.

The two variable domains, V_H and V_L, dimerize to form the F_V fragment that contains the Ag-binding site. Three hypervariable loops (H_1, H_2, and H_3) are present in V_H, while three (L_1, L_2, and L_3) are in V_L. Intervening sequences between the hypervariable regions, called Framework residues (FRs), are more or less conserved. V_H and V_L fold in such a manner as to bring the hypervariable loops together and form the Ag-binding site or paratope. The hypervariable regions are also sometimes called complementarity determining regions (CDRs).

For the Ag, there are no clear rules that would differentiate between epitopic and nonepitopic residues. In fact, under appropriate conditions, any part of the Ag surface has the potential to become an epitope. It is likely that the epitope is Ab-dependent. This is illustrated in Fig. 16.9, which shows that three epitopes on the surface of the antigen hen egg-white lysozyme bind to three different Abs.

The exact boundaries of the CDRs still remain uncertain as contacts with the Ag occur even outside the hypervariable loops. However, all Ag-binding residues are located in regions of "structural consensus" across antibodies. Sometimes, the Ag-binding residues outside the conventionally defined CDRs have a greater energetic contribution towards the stability of the Ag-Ab complex than the ones within. Fig. 16.10 shows the interaction interface of a human IL15 Ag-anti-IL15 Ab complex. Of the seven Ab L-chain residues that interact with the Ag, L46 and Y49 are outside the hypervariable region L2.

Like other ligand-binding proteins, Abs demonstrate necessary flexibility for their function. Upon Ag-binding, structural changes occur within and far from the binding site. A comparison of the free and Ag-bound structures of the antiepidermal growth factor receptor (EGFR) Ab (Fig. 16.11) shows that a significant binding-related conformational change occurs at the hypervariable region (CDR) H3. An even larger change occurs in a loop region in the C_H1 domain. It has been speculated that epitope recognition may be allosterically influenced by changes in the constant domains.

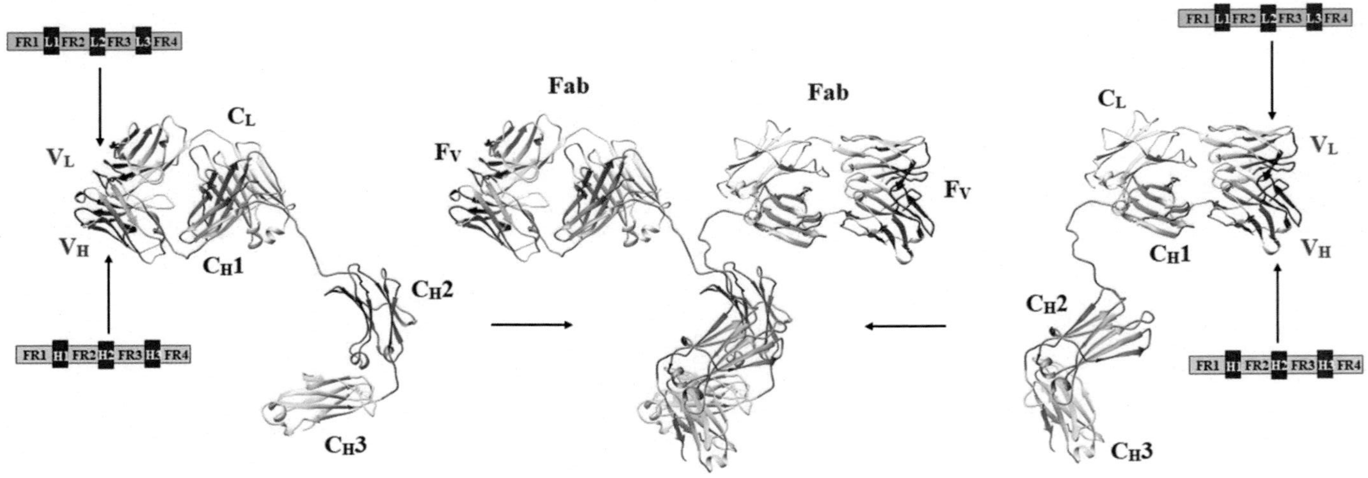

FIG. 16.8

The structure of an Ab molecule.

Source: Harris, L.J., et al., 1997. Biochemistry 36, 1581.

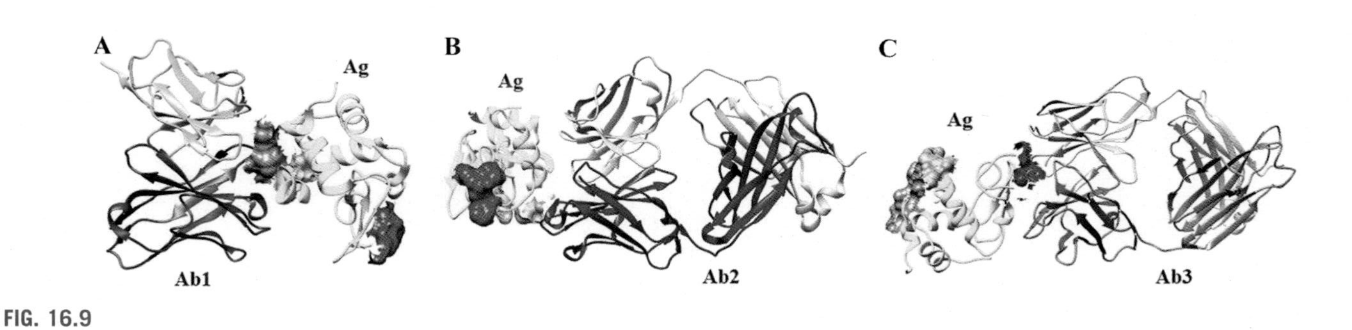

FIG. 16.9

Three Abs (Ab1, Ab2, and Ab3) bound to three different epitopes (colored *coral*, *gold*, and *purple*) in antigen (Ag) egg-white lysozyme.

Source: Chitarra, V., et al., 1993. Proc. Natl. Acad. Sci. U. S. A. 90, 7711; Acchhione, M., et al., 2009. Mol. Immunol. 47, 457; Braden, B.C., et al., 1994. J. Mol. Biol. 243, 767.

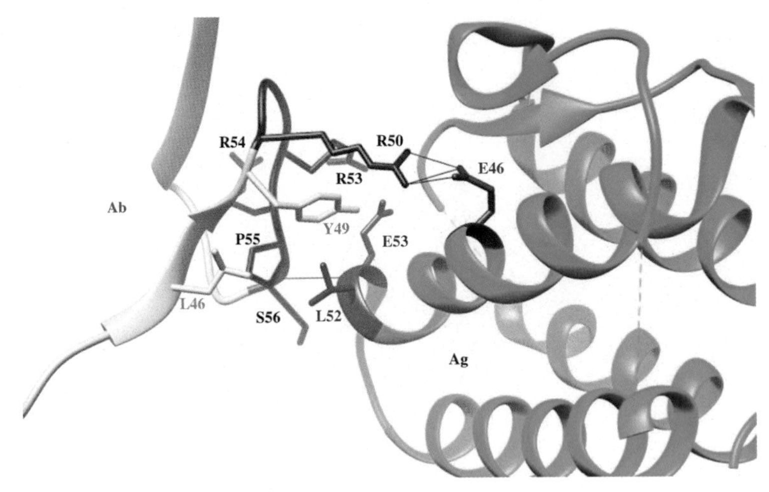

FIG. 16.10

Interaction interface of a human IL15 Ag-anti-IL15 Ab complex. Hypervariable region L2 colored *red*. Some of the H-bonds shown by *black* lines.

Source: Lowe, D.C., et al., 2011. J. Mol. Biol. 406, 160.

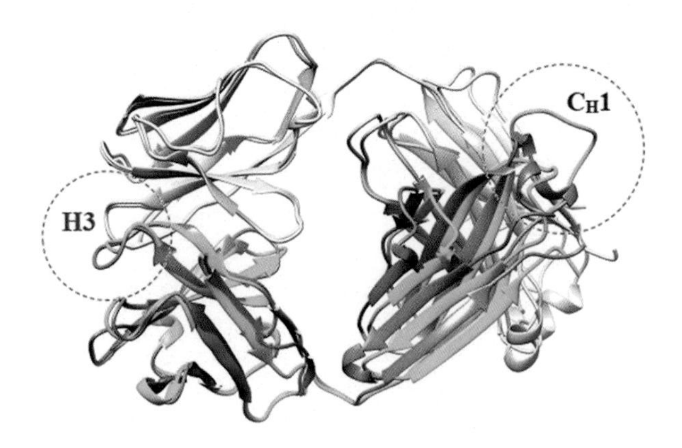

FIG. 16.11

Comparison of free *(light/hot pink)* and Ag-bound *(light blue/cyan)* structures of anti-EGFR Ab. Regions with significant changes have been indicated by dotted circles.

Source: Li, S., et al., 2005. Cancer Cell 7, 301.

16.3 Structural systems biology

A "system" is an abstract concept and not a "tangible" object. It essentially means an organization or complex consisting of interacting components and working on the basis of a set of principles of interaction. A biological system is not tangible per se, but the cells it comprises are. Similarly, a "nontangible" cellular system is made up of genes, proteins, and other small and large molecules which are, nevertheless, tangible physical objects. Hence, in order to understand the biological system, genes and other molecules have been studied, leading to the development of genetics and molecular biology. Additionally, the fact that these molecules interact on the basis of certain chemical and physical principles has led to the emergence of biochemistry and biophysics.

We have seen in earlier chapters that when the studies of individual macromolecules fell far short of explaining how a system functions as a whole, the era of "omics," pioneered by genomics, began. In the postgenomic era, we are witnessing new areas of research activity such as transcriptomics, proteomics, metabolomics, and so on. The focus has shifted to macromolecular complexes, metabolic pathways, and even the entire organism. System biology is about exploring the relationships between all the components in an organism considered together. Through detailed analyses of protein-protein, protein-metabolite, and protein-nucleic acid interactions, it aims to understand the metabolic and signaling pathways and gene-regulatory networks.

Protein-protein interactions (PPIs) play an important role in most biological processes. Among the most widely used experimental techniques to identify pairwise PPIs have been the yeast two-hybrid system and affinity purification. These efforts have been complemented by computational approaches. A commonly used method in the late 1990s had been based around "genomic context." Interactions were predicted between proteins for which there was evidence of association provided by the positioning of their corresponding genes in the known genome sequences or their expression profile. As for example, proteins whose genes are located in the same bacterial operon were considered to be functionally associated.

Besides genome information, the results of numerous experiments have led to the establishment of several metabolic and signaling pathways, each of which depicts protein-protein interactions in a series of reactions. It has been mentioned earlier in this chapter that the MAPK pathway coordinates and regulates cell growth and differentiation in response to extracellular stimulation. A couple of examples of protein-protein interaction occurring at different points along the pathway have also been discussed. The complete pathway is shown in Fig. 16.12. At each step in the pathway, signal is transmitted from a perturbed "source" node, which is a protein, to a "sink" node, which is also a protein.

The availability of genome-scale data has facilitated efforts to create PPI maps at the system-level. The initial objective is to represent the complete collection of all PPIs that take place within a cell as a complex network or an "interactome" map. In such a network map, the proteins are depicted as "nodes" and the interactions as "edges."

The interactome of a whole organism, represented as a "node-and-edge" network map, will provide a complete set of macromolecular complexes and interactions. Nevertheless, lacking in structural detail, they would be unable to elucidate the modes of interaction between the concerned proteins and, hence, the regulatory

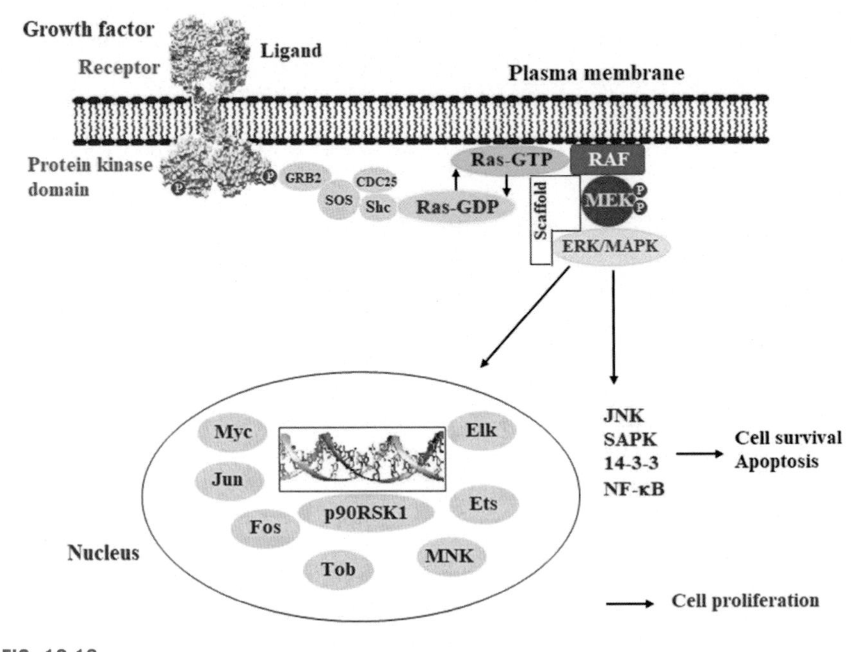

FIG. 16.12

MAPK pathway.

mechanisms. A metabolic or signaling pathway is better understood if binding sites and interactions of a protein can be predicted based on its known structure.

Classic "docking" approaches have been widely used to determine the structural basis of PPIs. These methods are based on the shape or electrostatic complementarity between protein surfaces and, for obvious reasons, require high resolution structures of the interacting proteins. There are alternative methods that are homology-based. With rapidly growing available structural data for interactions, it is becoming increasingly simple to model PPIs with newer proteins.

Fig. 16.13 shows the structural network of the ERK protein interactions. The structural network shows that seven proteins bind at different binding sites. These proteins can, therefore, interact with ERK simultaneously. The simultaneous binding of MEK1 and cPLA2 to ERK is shown in Fig. 16.13B. On the other hand, ERK, MEK2, and MP1 cannot interact with MEK1 simultaneously since they share a common binding site. It is also to be noted that PTP and Rsk2 cannot interact with ERK at the same time, despite having distinct binding sites, due to a steric clash elsewhere.

Incorporation of three-dimensional structural details into interactomes will make them ever more useful. Structural systems biology will enable the investigation and modeling of all macromolecular interactions within the cell at the atomic level.

Although still a developing field, systems biology has already made important contributions to cancer research. A network-centric approach has identified specific

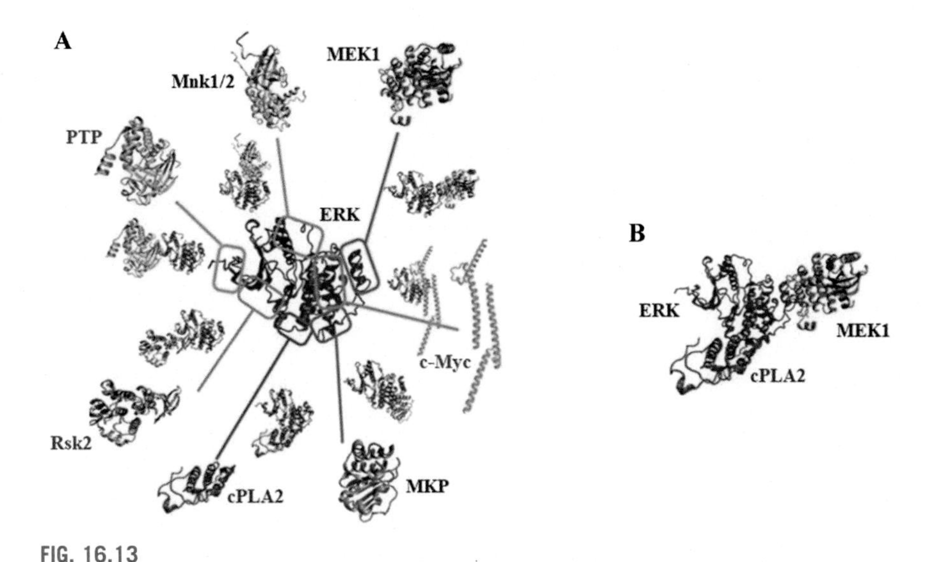

FIG. 16.13

Structural network of ERK protein interactions. (A) Proteins interacting with ERK and corresponding binding sites. (B) Simultaneous binding of MEK1 and cPLA2 to ERK.
Courtesy: Kuzu, G., et al., Constructing structural networks of signaling pathways on the proteome scale, Curr. Opin. Struct. Biol., June 2012, with permission from Elsevier.

oncogenes in certain types of cancer and led to advances in radiation therapy. Similarly, in the field of infectious disease, it is expected that structural systems biology will lead to more comprehensive models of host-pathogen interactions and, thereby, improved therapeutic drugs.

Sample questions

1. How is a signal "amplified" in a cell signaling pathway?
2. What is the effect of phosphorylation on the structure and activity of ERK2?
3. The substrate-range of ERK2 is much larger than that of MEK1—speculate on the structural basis.
4. How do ERK2 and MKP3 influence the activity of each other?
5. How is an epitopic residue differentiated from a nonepitopic residue in an antigen?
6. What are the effects of antigen binding on the structure of its antibody?
7. Under what conditions can two or more proteins bind to another protein simultaneously?
8. How can structural genomic initiatives help in the development of a more useful interactome of a biological system?

References and further reading

Aderem, A., et al., 2011. A systems biology approach to infectious disease research: innovating the pathogen-host research paradigm. mBio. https://doi.org/10.1128/mBio.00325-10.

Aloy, P., Russell, R.B., 2006. Structural systems biology: modelling protein interactions. Nat. Rev. Mol. Cell Biol. 7, 188–197.

Canagarajah, B.J., et al., 1997. Activation mechanism of the MAP kinase ERK2 by dual phosphorylation. Cell 90, 859–869.

Cusick, M.E., et al., 2005. Interactome: gateway into systems biology. Hum. Mol. Gen. 14, R171–R181.

Kunik, V., Peters, B., Ofran, Y., 2012. Structural consensus among antibodies defines the antigen binding site. PLOS Comput. Biol. 8(2). e1002388.

Kuzu, G., et al., 2012. Constructing structural networks of signaling pathways on the proteome scale. Curr. Opin. Struct. Biol. 22, 367–377.

Lee, M.J., Yaffe, M.B., 2016. Protein regulation in signal transduction. Cold Spring Harb. Perspect. Biol. 8, a005918.

Liu, C., et al., 2017. Structural and dynamic insights into the mechanism of allosteric signal transmission in ERK2-mediated MKP3 activation. Biochemistry 56, 6165–6175.

Liu, S., et al., 2006. Structural basis of docking interactions between ERK2 and MAP kinase phosphatase 3. Proc. Natl. Acad. Sci. U. S. A. 103 (14), 5326–5331.

Roskoski Jr., R., 2012. ERK1/2 MAP kinases: structure, function, and regulation. Pharmacol. Res. 66, 105–143.

Roskoski Jr., R., 2012. MEK1/2 dual-specificity protein kinases: structure and regulation. Biochem. Biophys. Res. Commun. 417, 5–10.

Sela-Culang, I., Alon, S., Ofran, Y., 2012. A systematic comparison of free and bound antibodies reveals binding-related conformational changes. J. Immunol. https://doi.org/10.4049/jimmunol.1201493.

Sela-Culang, I., Kunik, V., Ofran, Y., 2013. The structural basis of antibody-antigen recognition. Front. Immunol. https://doi.org/10.3389/fimmu.2013.00302.

Stewart, A.E., et al., 1999. Crystal structure of the MAPK phosphatase Pyst1 catalytic domain and implications for regulated activation. Nat. Struct. Biol. 6 (2), 174–181.

Vidal, M., 2009. A unifying view of 21st century systems biology. FEBS Lett. 583, 3891–3894.

CHAPTER

Macromolecular assemblies

17

Cells contain a large number of functional macromolecular assemblies, also referred to as macromolecular machines, responsible for executing vital cellular processes. In earlier chapters, we have already seen some of these assemblies such as the ribosome, spliceosome, and chaperone. Often, a macromolecular machine contains a stable core formed by the "self-assembly" of protein (and sometimes also RNA) molecules. The core defines the basic function of the complex; nevertheless, interaction of the core with peripheral protein components leads to multiple functional states of the assembly.

17.1 Molecular self-assembly

Self-assembly is a general phenomenon in nature, not restricted to biological systems. It is a process in which components spontaneously form ordered aggregates. In molecular self-assembly, molecules, or part of molecules, are the basic components. The structures generated, without human intervention, are usually in equilibrium (or sometimes in metastable states). Here, although we are not going to delve into the physics of molecular self-assembly, it may be worthwhile to consider some aspects of the phenomenon to derive at least a rudimentary understanding.

17.1.1 Principles of self-assembly

Evidently, as the term suggests, a self-assembly system involves a set of molecules, or segments of a macromolecule, that interact with each other. The molecules, or molecular segments, may be same or different; however, they must interact to form a more ordered state from a less ordered one. The interactions are mostly weak and noncovalent in nature. Such interactions have been discussed in Chapter 4. Self-assembly occurs through a balance between attractive and repulsive interactions. Occasionally, relatively weak covalent bonds (coordinate bonds) also come into play. Geometric complementarity has a crucial role in the process.

Relative weakness of the molecular interactions accords reversibility to the self-assembly process. Even after their association, the self-assembled components are able to dissociate or adjust their positions within the complex. This is possible since the strength of the interactions is comparable to the destabilizing forces (generally due to thermal motion). The process is also influenced by environmental factors.

Fundamentals of Molecular Structural Biology. https://doi.org/10.1016/B978-0-12-814855-6.00017-1

17.1.2 Biomacromolecular complexes

A large number of biomacromolecules, such as proteins and RNAs, are required to be assembled into complexes for their function. The multistep process of macromolecular assembly is analogous to the process of protein folding as depicted by Levinthal Paradox (discussed in Chapter 7). Accordingly, it is presumed that the assembly of macromolecules also proceeds through energetically favorable intermediate subcomplexes.

The intrinsic flexibility of proteins enables them to be assembled in a variety of symmetric and asymmetric structures. The flexibility facilitates binding in either of two ways. It allows structural changes that are induced upon binding. Alternatively, the intrinsic fluctuations within the unbound protein, based on its structural flexibility, help a conformational selection mechanism of binding.

Since the assembly of a macromolecular complex in a cell is a multistep process often involving a large number of components, lack of synchrony could lead to assembly aberrations or deleterious outcome for the components. However, strong regulatory mechanisms exist in the cell to ensure efficient and accurate assembly. Sometimes, the assembly process is cotranslational involving at least one protein subunit which is still in the process of being translated. On the other hand, in yeast, transcription of the rRNA genes has been found to be accompanied by cotranscriptional binding of several ribosome assembly factors.

17.1.3 Viral capsids

Viral capsids provide an interesting example of macromolecular self-assembly. Viruses are of different sizes—some of them contain even up to 10,000 protein subunits. Yet, they assemble into a complete and reproducible structure, mostly with high fidelity. There are at least two types of components in viruses—the genome, which can be DNA or RNA, and a protein shell. The nucleic acid, which is either single- or double-stranded, is protected by the protein shell, known as a capsid.

Single-stranded (ss) genome-containing viruses are usually assembled spontaneously around their nucleic acid in a single step. In case of many RNA-containing viruses, the assembly process is RNA-dependent at physiological conditions; under different ionic strengths or pH in vitro, the capsid proteins do not require the RNA to assemble into an empty shell.

The considerable stiffness and high charge density of a double-stranded (ds) genome do not permit its spontaneous encapsulation. Hence, for such a genome, packaging is a two-step process—an empty capsid is first assembled and, then, a "molecular motor," which inserts into one vertex of the capsid, hydrolyzes ATP to pump the DNA/RNA into it.

Considering the thermodynamics of capsid assembly, the process can be spontaneous only if $\Delta G < 0$ (Chapter 4, Eq. 4.26). Now, the assembly of free and disordered protein subunits into an ordered capsid structure (with or without RNA) reduces the translational and rotational entropy, that is, $\Delta S < 0$. Favorable interactions among the subunits and RNA (if present) overcome this entropy penalty.

Like most protein-protein interactions, capsid assembly involves a combination of hydrophobic, electrostatic, van der Waals, and hydrogen bonding interactions. All these interactions are short-ranged under physiological conditions, even electrostatic interactions being limited to 1 nm (10 Å). Typically, the process is primarily driven by hydrophobic interactions—electrostatics has a moderating effect, while van der Waals and hydrogen bonding interactions somewhat enforce a directional specificity.

The capsid assembly process of hepatitis B virus (HBV) has been very widely studied. HBV belongs to the *Hepadnaviridae* family of viruses. It contains a dsDNA genome, but its capsid assembles around an ssRNA pregenome. The HBV core protein (HBc) contains 183 amino acid residues. Its unstructured C-terminal tail (residues 150–183) is involved in nucleic acid binding. A truncated polypeptide (residues 1–149) is still able to assemble into capsids in vitro.

An HBc molecule has two functional interfaces—two monomers dimerize to form an intradimer interface (Fig. 17.1). Each monomer contains an amphipathic α-helical hairpin. In the course of dimerization, a four-helix bundle is formed. The bundle is stabilized by a disulfide bridge between C61 residues of the two monomers. The edges of a dimer, created by highly conserved C-terminal residues, dominate the dimer-dimer interactions (Fig. 17.2). As a result, 90 or 120 homodimers form an icosahedral capsid based on the interdimer interface (Fig. 17.3).

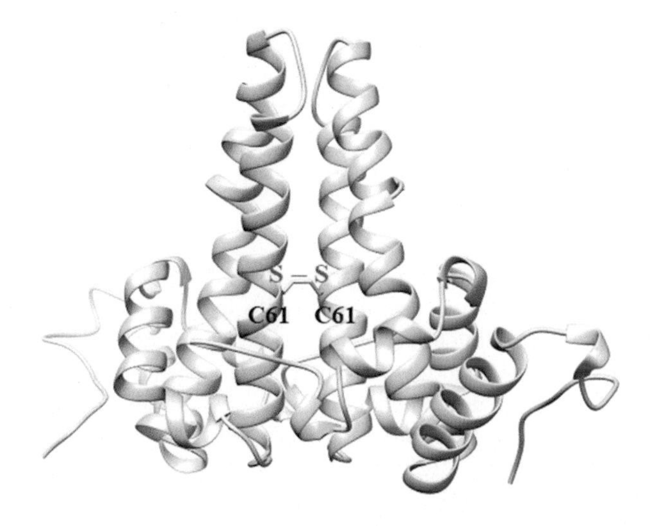

FIG. 17.1

HBV capsid dimer. The disulfide bridge (S-S) between two C61 residues is shown.
Source: Wynne, S.A., Crowther, R.A., Leslie, A.G.W., 1999. Mol. Cell 3, 771.

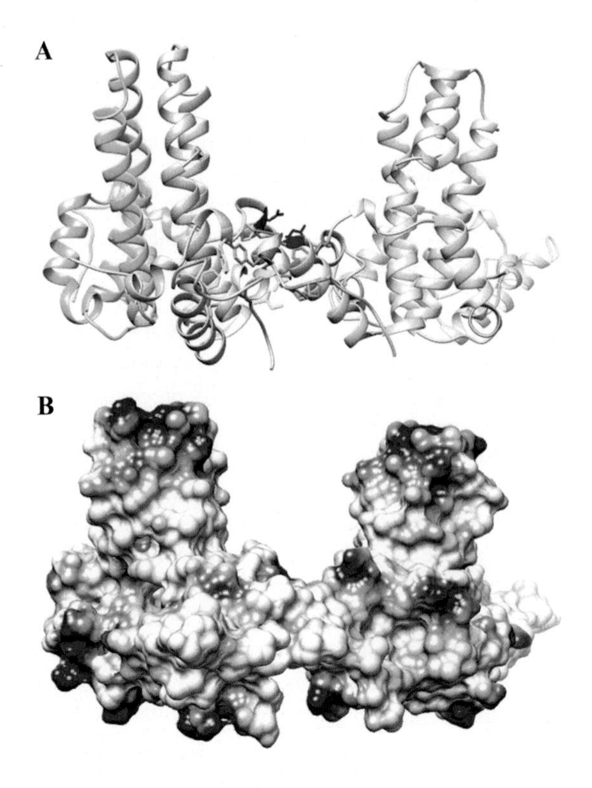

FIG. 17.2

HBV capsid dimer-dimer interactions. (A) Ribbon representation. Coloring scheme for the interacting residues: *blue*—positive; *red*—negative; hydrophobic residues on the edges of the two dimers are shown in *yellow* and *orange*, respectively. (B) Electrostatic surface: *blue*—positive; *red*—negative; *white*—neutral.

Source: Wynne, S.A., Crowther, R.A., Leslie, A.G., 1999. Mol. Cell 3, 771.

The HBV capsid assembly is driven primarily by hydrophobic effect. As much as 75% of the buried contact surface in the capsid is hydrophobic. Electrostatic interactions have some controlling effect on the capsid formation. The capsid formation can be boosted by lowering the protein charge density upto a point beyond which capsid aggregation and precipitation ensue. Based on the crystal structure, interacting residues at the interdimer interface have been identified. Hydrophobic residues F18, V120, V124, P129, Y132, P134, and P135 and charged residues E14 and R127 are present on the edge of one dimer, while hydrophobic residues F23, P25, L37, F22, and 139 and charged residues D29 and R39 are present on the edge of the other (Fig. 17.2).

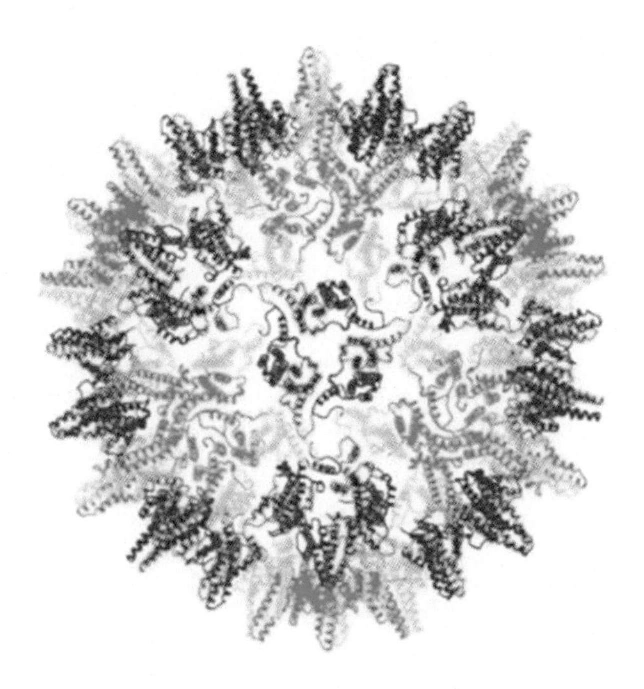

FIG. 17.3

HBV icosahedral capsid.

Courtesy: Wynne, S.A., Crowther, R.A., Leslie, A.G.W., 1999. The crystal structure of the human hepatitis B virus. Mol. Cell, with permission from Elsevier.

17.1.4 Ribosome assembly

In Chapter 11, we have discussed the structure of ribosome in relation to translation. Here, we are going to look at some aspects of ribosome assembly. For this purpose, let us select the simplest ribosomal complex—the bacterial 30S subunit (the small subunit). The subunit contains ~1540 nt 16S rRNA and about 20 proteins.

Extensive biochemical, biophysical, and structural studies have established that ribosomal assembly is effectively an RNA folding problem. Accordingly, assembly of the 30S subunit requires protein-independent and -dependent RNA folding and compaction. The rRNA secondary structures are formed mostly in a protein-independent manner, while many of the native tertiary structure contacts are created by protein-mediated events.

It has been found that the 30S ribosomal proteins recognize the shape of the folded RNA rather than specific bases. Early experiments by Nomura and coworkers had established the principle of self-assembly of the 30S subunit. It was shown that six "primary" assembly proteins bind the naked 16S rRNA. The binding of secondary assembly proteins is dependent on one or more primary assembly proteins. The tertiary assembly proteins bind after a conformational change in the 16S rRNA. Yet, in spite of these findings, the in-depth mechanisms of self-assembly remained unexplained.

X-ray crystallographic and small angle neutron scattering studies have shown that the protein-free rRNA retains some propensity of self-organization. In fact, all the expected tertiary interactions in the 5′ domain of the 16S rRNA, which forms the body of the 30S subunit, are formed unassisted by proteins. However, ribosomal proteins are needed to stabilize the structure and create binding sites for secondary assembly proteins.

Specific ribosomal proteins bind the central domain of the 16S rRNA in a cooperative manner and stabilize the tertiary interactions encoded by the rRNA. The protein binding reactions in the central domain result in a reduction of binding entropy. However, this energetic cost is compensated by a large number of favorable intermolecular contacts.

The ribosomal protein S7 induces extensive conformational changes in the 3′ domain; its binding to the 16S rRNA leads to the assembly of the small ribosomal unit head. S7 binds at the junction of the rRNA helices H28, H29, H41, and H43 (Fig. 17.4). The N-terminus of the protein is laid in the H43 groove and forms the surface of contact with the rRNA, but none of the α-helices enters the grooves of the RNA helices. Apparently, the interaction between S7 and 16S rRNA is spatial in nature.

The late stages of the assembly are chaperoned by some accessory factors and modification enzymes which ensure the fidelity of the process. These factors bind different regions of the 16S rRNA to bring about necessary conformational changes. One such protein is RimM which plays a direct role in the 3′ domain (head) assembly.

Bacterial RimM is a two-domain protein. Its C-terminal domain binds the ribosomal protein S19 (Fig. 17.5A) and the two proteins together bind a multihelix interface in the 16S rRNA (Fig. 17.5B). The binding places the N-terminal domain of RimM at the junction of several helices, such as H29, H30, H31, H32, and H42. As a result, RimM holds the 3′ domain in a conformation that enables helices 33 and 43 to fold correctly. In the absence of RimM, misfolding of these helices prevents the head domain from recruiting tertiary assembly proteins S10, S13, S14, and S19.

Another factor involved during the assembly of the small ribosomal subunit is RbfA. It binds in the vicinity of helices H28, H44, and H45. The interaction between RbfA and the 30S subunit is essentially electrostatic, but not residue-specific. The highly basic region on one side of RbfA formed by a helix-kink-helix (HKH) motif interacts with the negatively charged phosphate-oxygen backbone in the single-stranded linker region between H44 and H45 (Fig. 17.6).

The binding of RbfA to the 30S subunit induces a substantial conformational change in the 16S rRNA (Fig. 17.7). It can be seen from the figure that while the helix h18 is not much affected by the binding, the helices H44 and H45 together shift considerably. In the process, the decoding site in the 30S subunit is closed and the anti-Shine-Dalgarno sequence removed from the mRNA binding channel. It appears

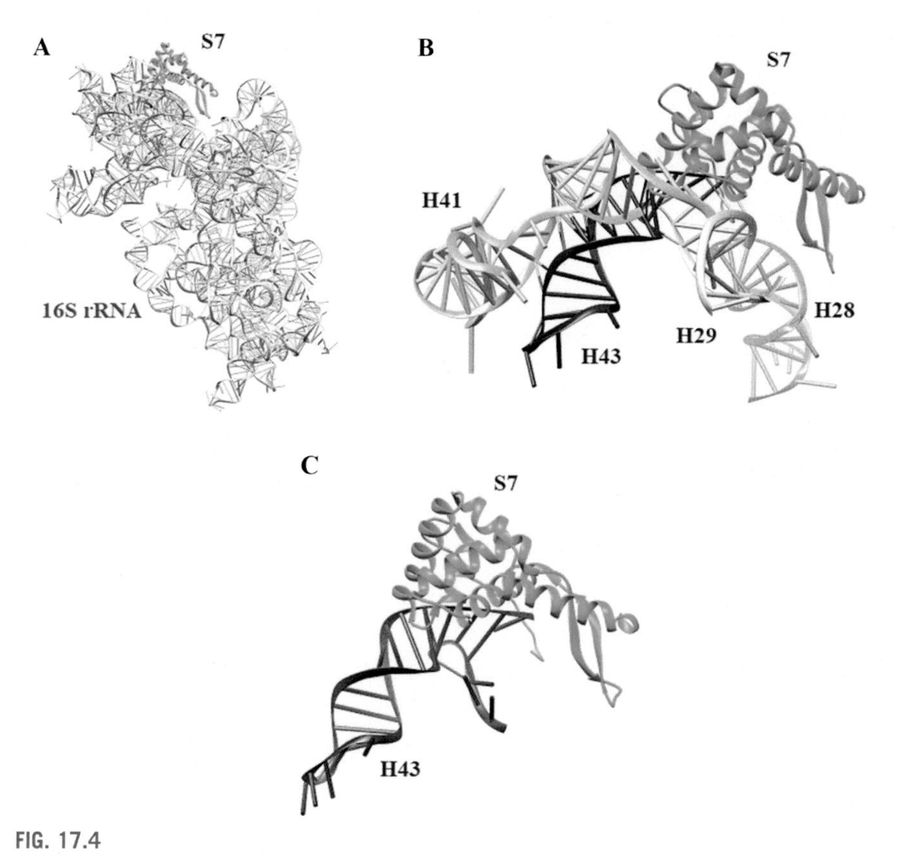

FIG. 17.4

Interaction of ribosomal protein S7 with 16S rRNA. (A) Binding to the head domain. (B) Binding at the junction of H28, H29, H41, and H43. (C) N-terminus of S7 in H43 groove.

Source: Cocozaki, A.I., et al., 2016. Proc. Natl. Acad. Sci. U. S. A. 113, 8188.

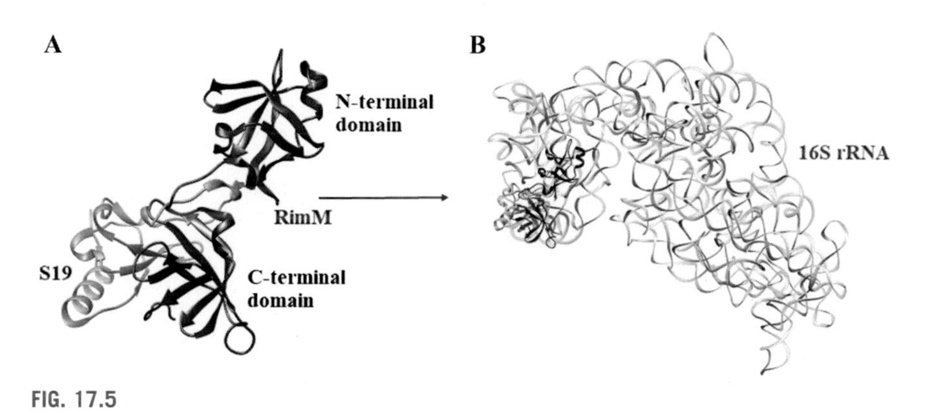

FIG. 17.5

16S rRNA—S19—RimM interaction. (A) Binding of RimM to S19. (B) RimM—S19 bound to 16S rRNA head domain.

Source: Cocozaki, A.I., et al., 2016. Proc. Natl. Acad. Sci. U. S. A. 113, 8188; Kaminishi, T., et al., https://doi.org/10.2210/pdb3A1P/pdb, unpublished.

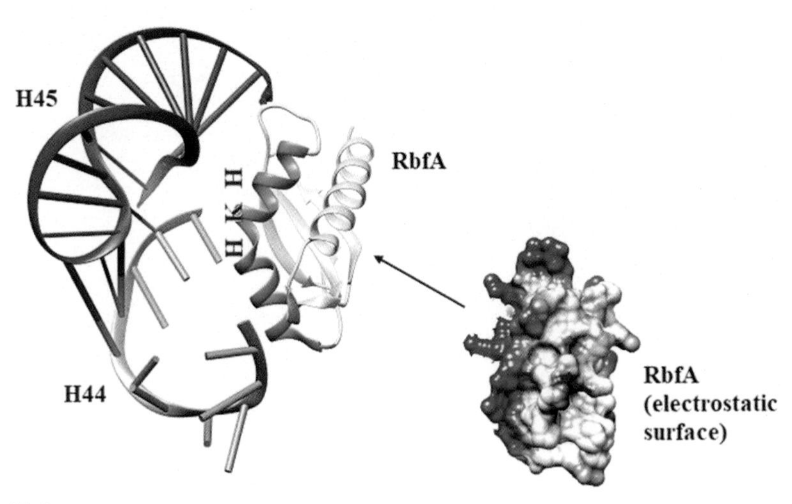

FIG. 17.6

16S rRNA RbfA interaction. Electrostatic surface of RbfA has been shown separately: *blue*—basic; *red*—acidic.

Source: Cocozaki, A.I., et al., 2016. Proc. Natl. Acad. Sci. U. S. A. 113, 8188; Datta, P.P., et al., 2007. Mol. Cell 28, 434.

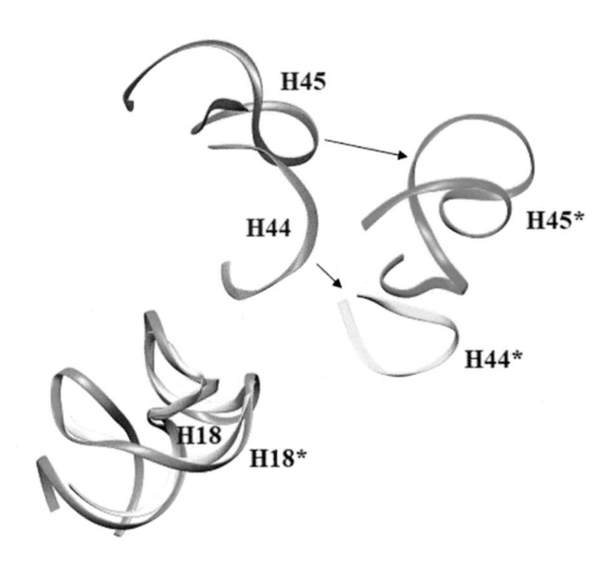

FIG. 17.7

RbfA-induced conformational changes in 30S subunit. Positions of only H18, H44, and H45 of 16S rRNA are shown. Asterisk indicates RbfA-bound 16S rRNA helices.

Source: Cocozaki, A.I., et al., 2016. Proc. Natl. Acad. Sci. U. S. A. 113, 8188; Datta, P.P., et al., 2007. Mol. Cell 28, 434.

that RbfA maintains the pre-30S complex in an inactive state to allow more time for correct assembly.

The final quality control on 30S subunit assembly is carried out by KsgA methylase. The three-dimensional structure of *E. coli* KsgA bound to the 30S ribosomal subunit has shown that the C-terminal domain of the enzyme makes extensive contacts with three helices H24, H27, and H45 of the 16S ribosomal rRNA (Fig. 17.8A). The catalytic N-terminal domain interacts with the H45 loop, which brings two adjacent adenosine residues, A1518 and A1519, close to the negatively charged active center (Fig. 17.8B). KsgA methylates the two substrate As only if they are present in a premature 30S subunit. A1518 and A1519 bases tend to interact with the minor groove of h44; KsgA can access the two residues only if H45 is swung out of position. In such event, the immature subunits are prevented from entering the translation cycle and the fidelity of translation is preserved.

17.2 Single-particle cryo-electron microscopy

In Chapter 6, we have briefly mentioned about single-particle analysis (SPA) using cryo-electron microscopy (cryo-EM). In the last decade, development of direct electron detector devices (DDDs), based on advanced complementary metal-oxide semiconductor (CMOS) technology, has led to a remarkable progress in single-particle cryo-EM. These devices have much improved sensitivity, as indicated by their detective quantum efficiency (DQE). (In terms of signal-to-noise ratio (SNR), DQE is defined as $SNR^2(k)_{out}/SNR^2(k)_{in}$, where k is the spatial frequency.)

The principle of SPA is briefly as follows. We have seen earlier (Chapter 6) that in cryo-electron tomography, multiple views of the same biological object are recorded by incrementally varying the orientation of the sample relative to the incident beam. SPA, in contrast, does not reconstruct the structure from a single biological molecule. It averages multiple views of a large number of copies of the same molecule. Hence, it is also referred to as "single-particle averaging." The analysis is based on the assumption that the three-dimensional objects, whose two-dimensional projections the DDD camera captures, are all identical. In SPA, the specimen is not subjected to multiple exposures as a "tilt-series" of projection images is not required. For this reason, the SPA specimen can tolerate a much higher electron dose than the limit for a single tomographic tilt.

In an ideal case, the SPA sample is well-distributed (and separated) identical particles vitrified in random orientations. Thousands of images of the sample are collected; each image contains up to a thousand projections of the experimental molecule in different orientations. When identical images of the same object are averaged, the SNR is found to be greatly enhanced.

The image reconstruction process repeatedly searches for the orientation of all projections. In cryo-EM, two factors lead to image distortion—electron beam-induced local motion of the specimen and defocusing of the electron lenses required for increasing the phase contrast. Motion correction is carried out during image

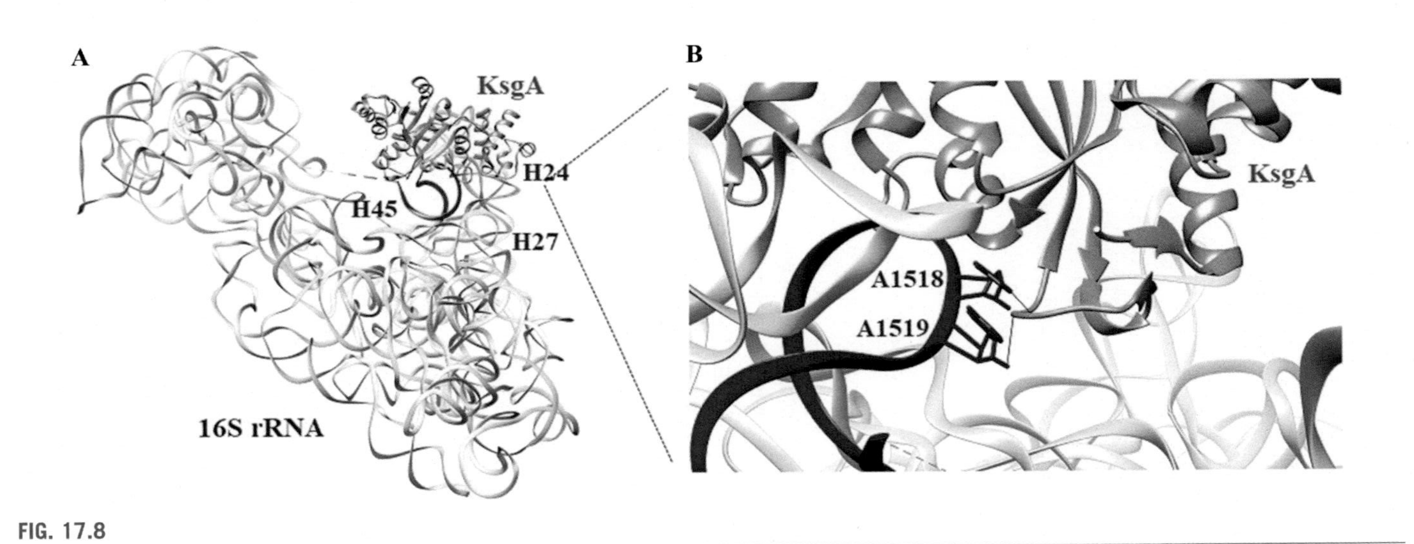

FIG. 17.8

16S rRNA—KsgA interaction. (A) KsgA C-terminal domain making contacts with 16S rRNA H24, H27, and H45. (B) KsgA N-terminal domain interacting with H45 A1518, and A1519.

Source: Boehringer, D., et al., 2012. J. Biol. Chem. 287, 10453.

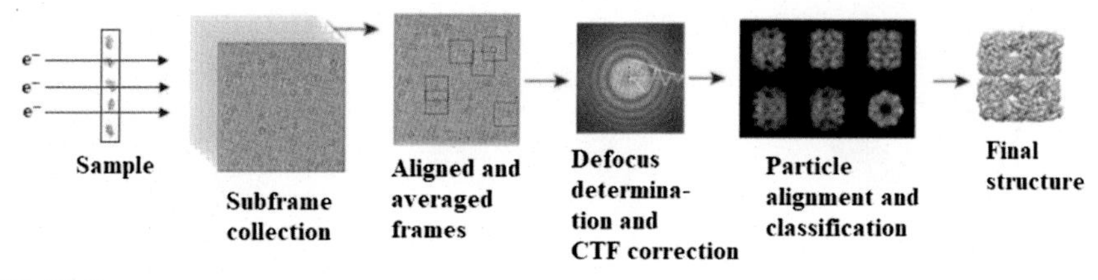

FIG. 17.9

Single-particle image reconstruction.

Courtesy: Carroni, M., Saibil, H.R., 2016. Cryo-electron microscopy to determine the structure of macromolecular complexes. Methods 95, 78–85. https://doi.org/10.1016/j.ymeth.2015.11.023 under the Creative Commons Attribution License (CC BY) http:/creativecommons.org/licenses/by/4.0/.

alignment by dedicated software packages. The changes in the image produced by defocusing are described by what is known as the contrast transfer function (CTF). Defocusing is estimated and CTF corrections are performed on motion-corrected averaged images. Aligned projections are then accumulated in a 3D-array, a 3D electron potential map is generated, and the final structure is obtained (Fig. 17.9).

At the initial stages of SPA, it used to be assumed that the sample is homogeneous, and all images are projections from identical three-dimensional objects. However, the assumption was later found out to be incorrect. As for example, most large proteins and protein complexes contain flexible domains, which can adopt more than one conformation. Even smaller proteins show conformational variability at the resolution scale of SPA. Hence, discrete classes of molecular structure exist in the ensemble and, to obtain the highest levels of resolution, 3D classification becomes necessary.

Cryo-EM has achieved the capability to computationally sort (classify) particles from structural ensembles. The process can be considered as in silico purification. With this capability of separating different conformational states of a macromolecular assembly, SPA has already been used in a number of cases to elucidate the molecular dynamics. For example, rotational movement of the yeast V-ATPase has been studied by cryo-EM.

Eukaryotic vacuolar ATPases (V-ATPases) are rotary enzymes that couple ATP hydrolysis to pump protons across membranes. The coupling occurs through the membrane-bound region of the enzyme involving the movement of a central rotor subcomplex. Near atomic resolution structures of the three rotational states of the yeast V-ATPases have been revealed by SPA (Fig. 17.10).

There has been an intimate relation between electron microscopy and ribosomes for a long time. The three-dimensional structure of a ribosome was first revealed by electron microscopy decades ago. In fact, cryo-EM has already replaced X-ray crystallography as the preferred method to study ribosomal structure and function.

The ribosome, as we have seen in Chapter 11, is a molecular machine that synthesizes proteins based on the genetic information encoded in mRNA templates in

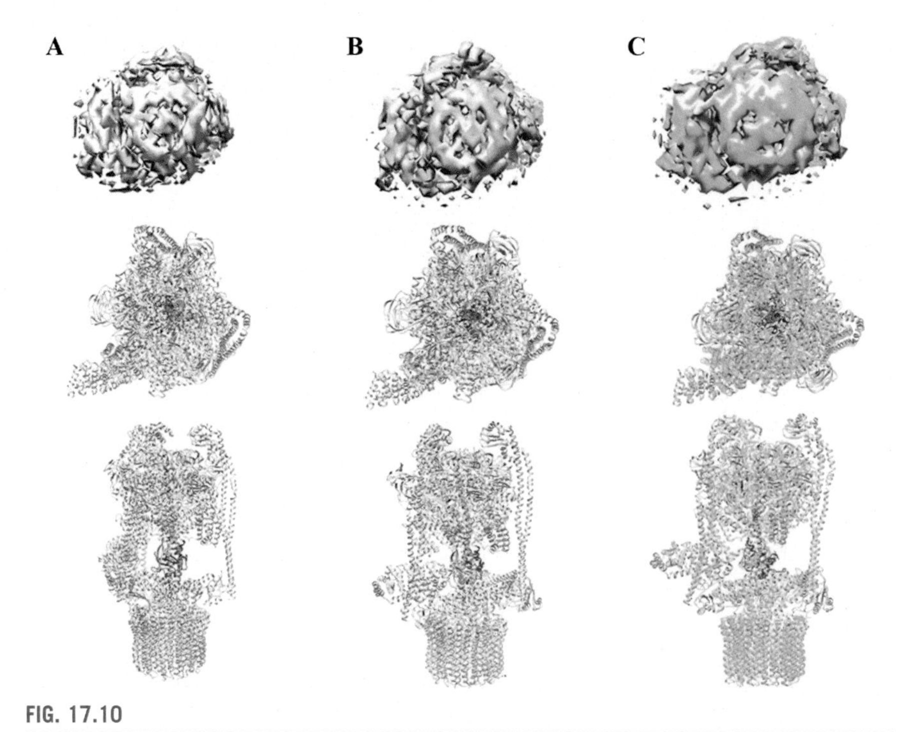

FIG. 17.10

Three rotational states of *Saccharomyces cerevisiae* V-ATPase (A, B, C). Electron micrographs are shown in the top panel. Top views (middle panel) and side views (bottom panel) of the crystal structures have been obtained by fitting models into density maps. Two subunits of the central rotor have been colored pink and blue, respectively, to visualize the movement.

Source: Zhao, J., Benlekbir, S., Rubinstein, J.L., 2015. Nature 521, 241.

the form of a linear sequence of four ribonucleotide bases. The process (translation) involves several distinct functional states of the ribosome. Many of the functional intermediates have been structurally analyzed by X-ray crystallography and cryo-EM; nevertheless, most of these structural studies have made use of in vitro-assembled complexes arrested artificially at different stages of translation. In order to explore the in vivo scenario, actively translating ribosomes were ex-vivo-derived from human cell extracts and subjected to SPA by cryo-EM. Multiple snapshots from a single specimen identified different ribosome conformations and provided a native distribution of the states as well (Fig. 17.11).

Clearly, we have seen that, not having to crystallize macromolecules and macromolecular complexes, cryo-EM have enjoyed an advantage over other structure determination techniques right from its inception. The disadvantage it suffered as regards resolution is being done away with by rapid advances in hardware and software. Soon, SPA by cryo-EM will provide a comprehensive view of the macromolecular complexes and molecular dynamics in a cell at the atomic level.

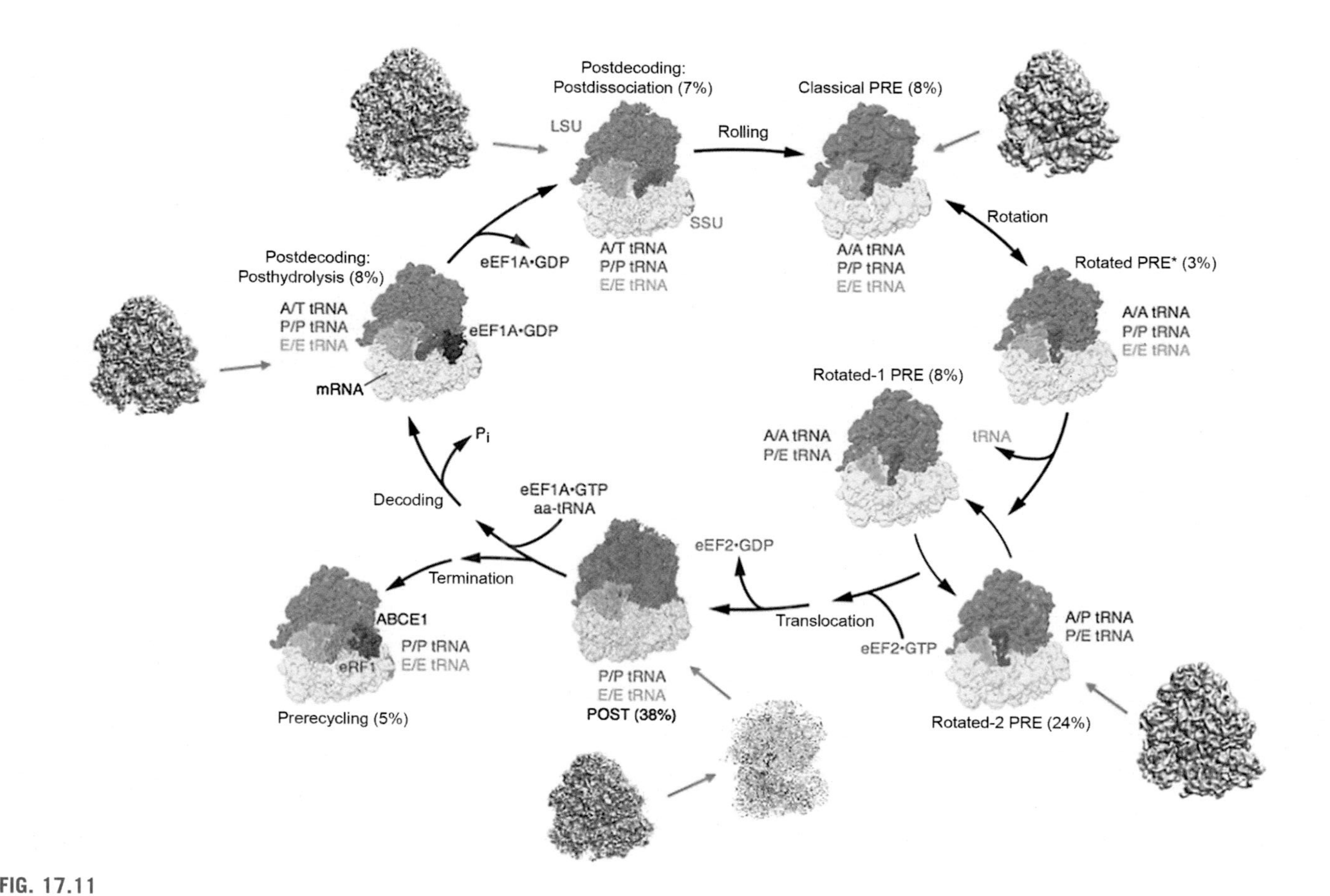

FIG. 17.11

Translation elongation cycle depicted by cryo-EM maps. Crystal structures were obtained by fitting models into density maps. Electron micrographs corresponding to some of the ribosomal states are shown alongside the crystal structures. Percentage in parentheses shows the distribution of states.

Courtesy: Brown, A., Shao, S., 2018. Ribosomes and cryo-EM: a duet. Curr. Opin. Struct. Biol., with permission from Elsevier; Behrmann, E., et al., 2015. Cell 161, 845.

Sample questions

1. What are the factors that promote self-assembly of molecules?
2. What are the ways a cell can minimize molecular assembly aberrations?
3. Does the assembly of free and disordered protein subunits into an ordered capsid structure violate the entropy principle?
4. How do hydrophobic interactions play an important role in the assembly of hepatitis B viral capsids?
5. How do ribosomal proteins bind to the 16S rRNA leading to the self-assembly of 30S subunit?
6. How does RimM help in the 16S rRNA 3′ domain assembly?
7. How does single-particle analysis (SPA) differ from cryo-electron tomography in the method of image acquisition?
8. How is the capability of cryo-EM to computationally classify particles from structural ensembles utilized in the study of macromolecular dynamics? Illustrate with an example.

References and further reading

Datta, P.P., et al., 2007. Structural aspects of RbfA action during small ribosomal subunit assembly. Mol. Cell 28, 434–445.

Davis, J.H., Williamson, J.R., 2017. Structure and dynamics of bacterial ribosome biogenesis. Philos. Trans. R. Soc. B 372, 20160181. https://doi.org/10.1098/rstb.2016.0181.

Guo, Q., et al., 2013. Dissecting the in vivo assembly of the 30S ribosomal subunit reveals the role of RimM and general features of the assembly process. Nucleic Acid Res. 41 (4), 2609–2620.

Hagan, M.F., 2014. Modeling virus capsid assembly. Adv. Chem. Phys. 155, 1–68.

McManus, J.J., et al., 2016. The physics of protein self-assembly. Curr. Opin. Colloid Interface Sci. 22, 73–79.

McMullan, G., Faruqi, A.R., Henderson, R., 2016. Direct electron detectors. Methods Enzymol. 579. https://doi.org/10.1016/bs.mie.2016.05.056.

Murata, K., Wolf, M., 2018. Cryo-electron microscopy for structural analysis of dynamic biological macromolecules. Biochim. Biophys. Acta Gen. Sub. 1862, 324–334.

Natan, E., et al., 2017. Regulation, evolution and consequences of cotranslational protein complex assembly. Curr. Opin. Struct. Biol. 42, 90–97.

Nikulin, A.D., 2018. Structural aspects of ribosomal RNA recognition by ribosomal proteins. Biochemistry (Moscow) 83, s111–s133 Suppl. 1.

Schep, D.G., Zhao, J., Rubinstein, J.L., 2016. Models for the subunits of the *Thermus thermophilus* V/A-ATPase and *Saccharomyces cerevisiae* V-ATPase enzymes by cryo-EM and evolutionary covariance. Proc. Natl. Acad. Sci. U. S. A. 113 (12), 3245–3250.

Sun, X., et al., 2018. Role of protein charge density on hepatitis B virus capsid formation. ACS Omega 3, 4384–4391.

Suzuki, S., et al., 2007. Structural characterization of the ribosome maturation protein, RimM. J. Bacteriol. 189 (17), 6397–6406.
Whitesides, G.M., Boncheva, M., 2002. Beyond molecules: self-assembly of mesoscopic and microscopic components. Proc. Natl. Acad. Sci. U. S. A. 99 (8), 4769–4774.
Woodson, S.A., 2008. RNA folding and ribosome assembly. Curr. Opin. Chem. Biol. 12 (6), 667–673.
Wynne, S.A., Crowther, R.A., Leslie, A.G.W., 1999. The crystal structure of the human hepatitis B virus capsid. Mol. Cell 3, 771–780.
Zhao, J., Benlekbir, S., Rubinstein, J.L., 2015. Electron cryomicroscopy observation of rotational states in a eukaryotic V-ATPase. Nature 521 (7551), 241–245.

CHAPTER

Computational molecular biology 18

It should not be difficult for us to appreciate that a mechanistic understanding of the biological processes discussed in different chapters would not have been possible without the availability of relevant DNA and protein structures. Interestingly, it may be noted here that, even before the first three-dimensional structure of a protein was experimentally determined, theoretical approaches had made significant contributions toward the elucidation of protein structure. Based on the properties of some small molecules, known from their crystal structures and the resonance theory of chemical bonding, Pauling, Corey, and Branson had proposed, as early as 1951, the existence of α-helices and β-sheets as the backbone of proteins. The prediction, which was later confirmed experimentally, marked the beginning of theoretical studies of biomolecules.

18.1 Modeling and simulation

The era of computational molecular biology began with the molecular dynamics (MD) simulation of a small protein, bovine pancreatic trypsin inhibitor, by McCammon and his coworkers in 1977. The simulation showed that the protein atoms undergo large-amplitude fluctuating motion on the picosecond time scale.

Computational approaches have made possible atomic resolution analysis of the dynamics of biomacromolecules, including proteins, and correlation of dynamics to function. It is true that experimentally determined structures of the same biological molecules in different environments had already demonstrated the structural flexibility of these molecules. However, early experimental efforts could recognize only the conformational end-points and not the structural details of the pathway in-between. Although in the previous chapter we have seen that an advanced experimental technique such as cryo-EM-based single-particle analysis (SPA) has been able to capture the transitional states of the macromolecular complex involved in a biological process, one should remember that the technique is extremely computation-intensive.

Modeling and simulation, utilized in almost all branches of science and technology, are a powerful means for the understanding of complex systems and processes. A simulation, though an approximate 'imitation' of the operation of a system, is particularly useful when direct experimentation with the real system is, technically or otherwise, not feasible.

Fundamentals of Molecular Structural Biology. https://doi.org/10.1016/B978-0-12-814855-6.00018-3

Evidently, "model" and "system" are the key components of simulation. Here, a system means a group or collection of interrelated elements that interact with each other to achieve a specific goal. It can be a real (existing) system, a proposed system, or a futuristic system design. Simulation, first, requires a model of the concerned system to be developed. A model is a representation and abstraction of the system. To be efficient and productive, a model should neither include unnecessary details that may increase its complexity, nor exclude the essential details, the absence of which may lead to abortive operation.

Molecular modeling and simulation have been gaining increasing importance in the studies of biological processes. It is not yet possible to elucidate the molecular mechanisms of fast processes such as enzymatic reactions or ion transport through membranes only by direct experiments. Simulations, based on the fundamental laws of physics, have already demonstrated the potential to overcome the limitations of experimentation by their ability to aptly model macromolecular movement and interaction. Simulated structures can be analyzed in atomic detail and the functions of biomolecular machines mechanistically explained. Rapid improvements in computer hardware and software, together with impressive developments in theory and algorithm, have facilitated an increasing range and depth of molecular modeling applications in biology.

18.2 Molecular dynamics

Molecular dynamics (MD), a very powerful simulation technique, has been widely used in physics, chemistry, and biological sciences. It can be considered as a 'computational microscope' that facilitates detailed atomic-scale modeling. The technique can meticulously analyze both static and dynamic properties of a molecule and correlate them with its function.

For small molecules, such as small proteins or oligonucleotides, simulations have already attained the microsecond time scale. For larger molecules (biomacromolecules), the time scale is still limited to a few hundred nanoseconds. Nevertheless, it is now possible to simulate, in atomic detail, large macromolecular complexes such as ion channels and the ribosome.

We understand that, for the functionality of a molecule, its structure and motion are both important. The motion of a protein molecule can be characterized at different levels—local, medium-scale, large-scale, and global. Localized motions, characteristic of atomic fluctuations, are responsible for ligand-docking flexibility of a protein. Medium-scale motions in the protein loops and terminal arms are often correlated with active-site conformational adaptation. Allosteric transitions of protein molecules are caused by large-scale motion of the domains, while helix-coil transition and protein-folding are examples of global motion. However, the different types of motion are interdependent and coupled to each other—the atomic-scale movement being the most crucial.

18.2.1 Basic theory

It is known that molecules are quantum mechanical entities. Hence, the dynamic behavior of a molecule is best described by the quantum mechanical equation of motion, that is, the time-dependent Schrödinger equation (Eq. 3.61). However, the equation is extremely difficult to solve for a large molecule. For this reason, MD simulations of the atomic motions of a biomolecule (say, protein) are based on classical mechanics and utilize Newton's equation of motion (Eq. 3.10)

$$\mathbf{F_i} = m_i \mathbf{a_i} = m_i \frac{d^2 \mathbf{r_i}(t)}{dt^2} \tag{18.1}$$

where $\mathbf{F_i}$ is the force acting on atom i of mass m_i. The force $\mathbf{F_i}$ is given by the negative gradient of the potential energy function $U(\mathbf{r_i})$ at position $\mathbf{r_i}$ (Eq. 3.20):

$$\mathbf{F_i} = -\nabla U(\mathbf{r_i}) \tag{18.2}$$

Evidently, the solution of Eq. (18.1) requires information on $\mathbf{F_i}$ as well as initial position, velocity, and acceleration of the atom.

A large number of biomacromolecular structures are available in the Research Collaboratory for Structural Bioinformatics (RCSB) Protein Data Bank (PDB). These structures are often used as the starting point of MD simulations. Each structure is maintained as a PDB file containing the location of the atoms in Cartesian coordinates (x, y, z) (Chapter 6). However, one must remember that these structures, a majority of which have been solved by X-ray crystallography, represent an average over all the molecules in the crystal and the entire time-course of the experiment. They may or may not represent the optimal conformation. Besides, some parts of a structure, such as loops or terminal regions of a protein, may not be resolved by crystallography.

In order to avoid unnecessary complexity in the simulation process, an atom is considered as the basic unit of a model so that only interatomic interactions are required to be parameterized. The MD atom is a 'hard ball' without directional properties and internal degrees of freedom; sometimes, it is considered to be a charged point. Depending on the conditions, an 'atom' can be anything from a simple proton to an entire amino acid. Bonding interactions and bond angles are modeled as springs (harmonic oscillators).

MD simulation requires a potential energy function that describes the interactions of all the atoms in a molecular system. The potential energy is defined by what in computational chemistry is known as molecular mechanics force field (MM FF). A set of forces act on the individual atoms depending on their positions. FF is the region where the forces act.

The potential energy, which is a function of the three-dimensional structure of the system (let us denote it as R), consists of bonding (covalent) and nonbonding (noncovalent) components.

$$U(R) = \Sigma U_{\text{bonded}}(\mathbf{r}, \theta, \phi) + \Sigma U_{\text{nonbonded}}(\mathbf{r}) \tag{18.3}$$

The bonding component consists of terms which arise out of the geometry of a chain molecule as shown in Fig. 18.1. It involves two-, three-, and four-body interactions of covalently bonded atoms and can be broken up into three parts:

$$U_{\text{bonded}}(\mathbf{r}, \theta, \phi) = U_{\text{bonds}}(\mathbf{r}) + U_{\text{angle}}(\theta) + U_{\text{tors}}(\phi) \tag{18.4}$$

where the variables (oscillatory)
$\mathbf{r}$ = bond length;
θ = valence or "bend" angle; and
ϕ = torsion or dihedral angle.

The first term, a two-body spring bond potential, represents the motion of a simple (linear) harmonic oscillator formed by a (j, k) pair of covalently bonded atoms.

$$U_{\text{bonds}}(\mathbf{r}) = k_{r,jk}\left(r_{jk} - r_{o,jk}\right)^2 \tag{18.5}$$

where $r_{jk} = |\mathbf{r_k} - \mathbf{r_j}|$ is the distance between the atoms j and k (bond length), $r_{o,jk}$ is the equilibrium distance, and $k_{r,jk}$ is the spring constant.

The angular vibrational motion occurring between a (j, k, l) triple of covalently bonded atoms is described by the three-body angular bond potential,

$$U_{\text{angle}}(\theta) = k_{\theta,jkl}\left(\theta_{jkl} - \theta_{o,jkl}\right)^2 \tag{18.6}$$

where θ_{jkl} is the angle between successive bond vectors $\mathbf{r_{jk}} = \mathbf{r_k} - \mathbf{r_j}$ and $\mathbf{r_{kl}} = \mathbf{r_l} - \mathbf{r_k}$, $\theta_{o,jkl}$ is the equilibrium angle and $k_{\theta,jkl}$ is the spring constant.

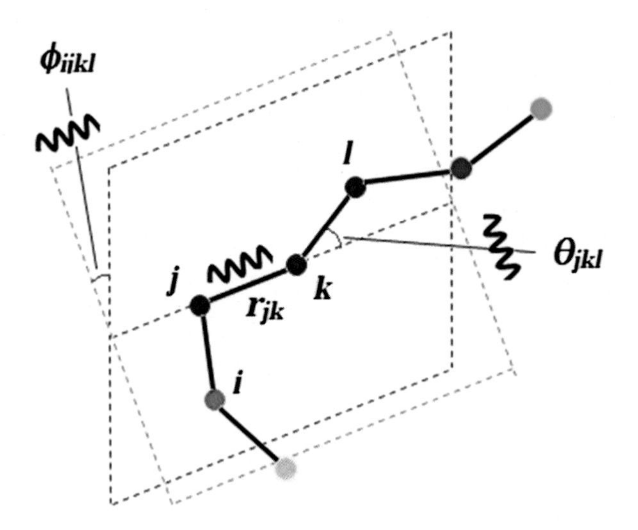

FIG. 18.1

Geometric representation of a simple chain molecule. The oscillatory nature of $\mathbf{r}$, θ, and ϕ has been indicated by springs.

The angle between the planes formed by the first three and the last three atoms of a consecutively bonded (i, j, k, l) quadruple of atoms is the torsion or dihedral angle. The four-body torsion angle potential can be expressed as

$$U_{\text{tors}}(\phi) = \sum_n k_{\varphi,n,\,ijkl}\left[1 + \cos\left(n\phi_{ijkl} - \Psi_n\right)\right] \quad (18.7)$$

where ϕ_{ijkl} is the angle between the (i, j, k) plane and the (j, k, l) plane, n is nonnegative and indicator of periodicity, Ψ is the phase angle, and $k_{\phi,n,ijkl}$ is a multiplicative constant. Eq. (18.4) can now be rewritten as

$$\begin{aligned} U_{\text{bonded}}(\mathbf{r}, \theta, \phi) &= k_{r,\,jk}\left(r_{jk} - r_{o,\,jk}\right)^2 \\ &+ k_{\theta,\,jkl}\left(\theta_{jkl} - \theta_{o,\,jkl}\right)^2 \\ &+ \sum_n k_{\varphi,n,\,ijkl}\left[1 + \cos\left(n\phi_{ijkl} - \Psi_n\right)\right] \end{aligned} \quad (18.8)$$

It is to be noted that Eq. (18.8) is somewhat oversimplification. Depending on the desired accuracy and sophistication of the simulation, cross-terms (such as the Urey-Bradley term used to describe a noncovalent spring between j and l atoms) are included in the MM FF.

The nonbonded component comprises the ion-ion interaction (el) and the Leonard-Jones (LJ) potential (Chapter 4).

$$\begin{aligned} U_{\text{nonbonded}}(\mathbf{r}) &= U_{\text{el}} + U_{\text{LJ}} \\ &= \left[\frac{q_1 q_2}{4\pi\varepsilon_0\varepsilon_r r_{ij}} + \varepsilon_{ij}\left\{\left(\frac{r_{\min,\,ij}}{r_{ij}}\right)^{12} - 2\left(\frac{r_{\min,\,ij}}{r_{ij}}\right)^6\right\}\right] \end{aligned} \quad (18.9)$$

where ε_{ij} is the minimum potential observed.

The nonbonded terms represent interactions between all (i, j)-pairs of atoms; the pairs already involved in a bonded term are usually excluded.

Eq. (18.3) then becomes

$$\begin{aligned} U(R) &= \sum_{\text{bonds}} k_{r,\,\text{jk}}\left(r_{\text{jk}} - r_{0,\,\text{jk}}\right)^2 \\ &+ \sum_{\substack{\text{bend}\\\text{angles}}} k_{\theta,\,\text{jkl}}\left(\theta_{\text{jkl}} - \theta_{0,\,\text{jkl}}\right)^2 \\ &+ \sum_{\substack{\text{torsion}\\\text{angles}}} \sum_n k_{\varphi,n,\,\text{ijkl}}\left[1 + \cos\left(n\phi_{\text{ijkl}} - \Psi_n\right)\right] \\ &+ \sum_{\substack{\text{nonbonded}\\\text{atom pairs}}} \left[\frac{q_1 q_2}{4\pi\varepsilon_0\varepsilon_r r_{\text{ij}}} + \varepsilon_{\text{ij}}\left\{\left(\frac{r_{\min,\,\text{ij}}}{r_{\text{ij}}}\right)^{12} - 2\left(\frac{r_{\min,\,\text{ij}}}{r_{\text{ij}}}\right)^6\right\}\right] \end{aligned} \quad (18.10)$$

Evidently, the energy function contains terms associated with the three-dimensional structure of a molecule as well as certain specific parameters. The values of r_{ij}, θ_{jkl}, and ϕ_{ijkl} are all obtained from experimentally determined (by X-ray crystallography, NMR, etc.) or homology-modeled structures. The parameters such as the spring constants $k_{r,jk}$, $k_{\theta,jkl}$, and $k_{\phi,n,ijkl}$ are associated, respectively, with a pair, a triple, and a quadruple of atoms. Besides, each atom is assigned a charge and LJ parameters. The precise formulation of the potential energy function, together with selection of suitable parameters, defines the force field.

The development of a force field is a nontrivial task. Experiments with small molecules (which are usually the building blocks of macromolecules of interest) and, most importantly, quantum mechanical calculations initially generate the parameters. The force field is then optimized with respect to the structural, dynamic, and thermodynamic properties of previously characterized molecules. Currently, the most widely used force fields in molecular dynamics are AMBER, CHARMM, and GROMOS.

After a molecular system has been parameterized, optimization of the potential energy function is carried out to compute its stable configuration. Such optimization is necessary for reasons mentioned earlier. Often, a multidimensional potential energy surface is used to describe the energy of a molecule in terms of its atomic positions. The stable state of the molecule corresponds to the global and local minima on this surface. In a mathematically executed energy minimization/optimization procedure, the atomic coordinates are varied to reach the lowest energy configuration (optimal structure) of the molecule.

Computation of the MD trajectory requires the solution of Newton's equation of motion (Eq. 18.1) for all the atoms in the force field at each selected timestep. The force acing on atom i is given by Eq. (18.2). Essentially, $\mathbf{r}_i(t)$ is required to be determined. A standard procedure for solving ordinary differential equations such as Eq. (18.1) is the finite-difference approach. Here, given the coordinates and velocities of the atoms at time t, those at a time $t+\Delta t$ can be obtained with a reasonable degree of accuracy. The equations are solved on a step-by-step basis using the Taylor expansion (similar to Eq. 2.67) expressed as

$$\mathbf{r}(t+\Delta t) = \mathbf{r}(t) + \dot{\mathbf{r}}\,(t)\Delta t + \frac{1}{2}\ddot{\mathbf{r}}(t)\Delta t^2 + \cdots \tag{18.11}$$

Alternatively,

$$\mathbf{r}(t+\Delta t) = \mathbf{r}\,(t) + \mathbf{v}(t)\Delta t + \frac{1}{2}a(t)\Delta t^2 + \cdots \tag{18.12}$$

where $\mathbf{v}(t)$ is the velocity and $\mathbf{a}(t)$ is the acceleration. The selection of time interval Δt depends on the properties of the molecular system and the nature of the question to be addressed by the simulation; in any case, it is significantly smaller than the simulation time scale.

Eq. (18.12) forms the basis on an integration algorithm. The initial positions $\{\mathbf{r}(0)\}$ are taken from the optimized structure of the molecule. The initial velocities $\{\mathbf{v}(0)\}$ are usually randomly assigned based on Maxwell-Boltzmann distribution of molecular speed at a temperature T (Eq. 3.47). The initial acceleration can be calculated from the force field in accordance with

$$\{\mathbf{a}(0)\} = -\left(\frac{1}{m}\right)\nabla U\{\mathbf{r}(0)\} \tag{18.13}$$

The described algorithm is very simplistic and does not produce high quality results. Other algorithms with greater accuracy, such as the Verlet algorithm, have been developed. However, in most cases the same kind of reasoning is applied.

18.2.2 Simulation of a cellular process

We know that a large number of biological processes occur in aqueous medium. Clearly, solvent molecules have an impact on molecular conformation and dynamics. Hence, solvation is necessary for accurate modeling and simulation of such a process. There are two different solvation models—explicit and implicit. In the explicit model, solvent molecules are added to the molecular system. In contrast, the implicit model represents the solvent as a dielectric continuum and computes the polarization of the continuum due to the charge distribution of the solute. The electrostatic field, thus created, forms the basis of solvent-solute interaction.

We have seen in Chapter 14 that a membrane protein functions partly in a lipid environment. Hence, simulation of the transmembrane activity of the protein requires inclusion of lipid molecules in the model.

The voltage-gated potassium channel Kv1.2 plays an important role in electrically excitable cells. It is a homotetrameric integral membrane protein that contains four voltage sensors and a pore domain, and facilitates rapid ($\sim 10^8$ ions s^{-1}) as well as selective conduction of K^+ ions across the cell membrane. Four identical segments of a highly conserved sequence TVGYG form the selectivity filter (SF) of the channel. K^+ ions and water molecules move through the SF in unison. Besides, gating of the pore involves significant changes in the protein conformation.

Mere availability of the three-dimensional structure of Kv1.2 does not reveal the status of individual ions as they permeate the channel. However, the structure provides the basis for MD simulation of the Kv1.2 pore domain that considerably elucidates the conduction mechanism.

MD simulations of the potassium channel require embedding of the protein in a lipid bilayer. Automatic insertion into the bilayer often leaves large gaps between the protein and the lipids. Filling of the gaps demands a long equilibration/optimization time. Alternatively, lipid positions are manually adjusted around the protein. A typical simulation setup (illustrated in Fig. 18.2) involves 125–150 lipids (such as dipalmitoyl phosphatidylcholine, DPPC; each DPPC molecule contains $\sim$130 atoms) and 6000–7000 water molecules.

The simulations of Kv1.2 channel uncovered two remarkable features in the mechanisms of K^+ ion conduction. The first was a validation of the Hodgkin-Keynes "knock-on" model. According to this model, the translocation of two SF-bound K^+ ions is driven by a third one, formation of the knock-on intermediate being rate-determining.

The other significant observation was that the flow of ions through Kv1.2 channel is regulated by "hydrophobic gating." This phenomenon is due to liquid-vapor (wetting-dewetting) transitions of water within the pore—dewetting causes collapse of the pore while wetting reopens it.

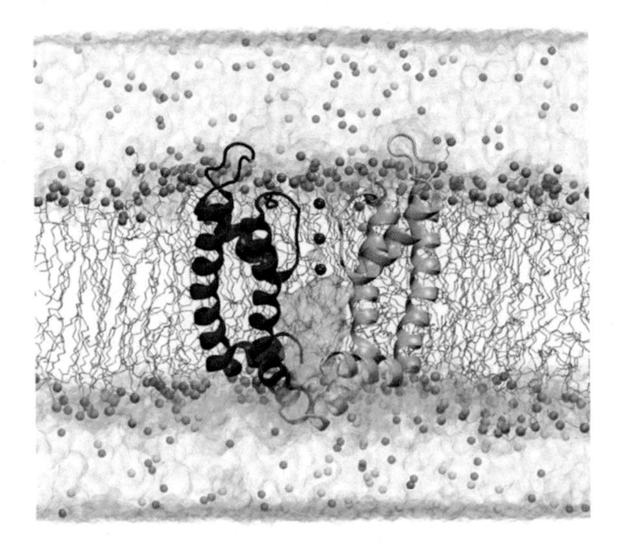

FIG. 18.2

Pore domain of Kv1.2 embedded in a lipid bilayer. Two of the four subunits of the protein shown in *dark red* and *golden color*. Three K^+ ions are in the selectivity filter.

Courtesy: Khalili-Araghi, F., Tajkhorshid, E., Schulten, K., 2006. Dynamics of K^+ ion conduction through Kv1.2. Biophys. J., with permission from Elsevier.

18.3 Drug design

Three-dimensional structures of macromolecules, especially proteins, together with advanced computational approaches, have been at the core of modern drug discovery endeavor. On the one hand, structural biology has been very effective in the mechanistic elucidation of various diseases followed by identification and characterization of corresponding biological targets. On the other hand, structure-guided experimental and computational ventures have rapidly led to the development of innumerable potent drug candidates.

18.3.1 Hemoglobin and sickle cell disease

Molecular structural biology has been intimately connected to medicine as soon as the first high-resolution macromolecular structures appeared in the 1950s and 1960s. Structures of normal and mutant hemoglobin provided a molecular-level explanation for sickle cell disease (SCD).

Hemoglobin (Hb) is a heterotetramer consisting of two α chains and two β chains. SCD is caused by a single point mutation in the β chain where a hydrophilic E6 has been exchanged for a hydrophobic V6. Under low oxygen conditions, the mutant

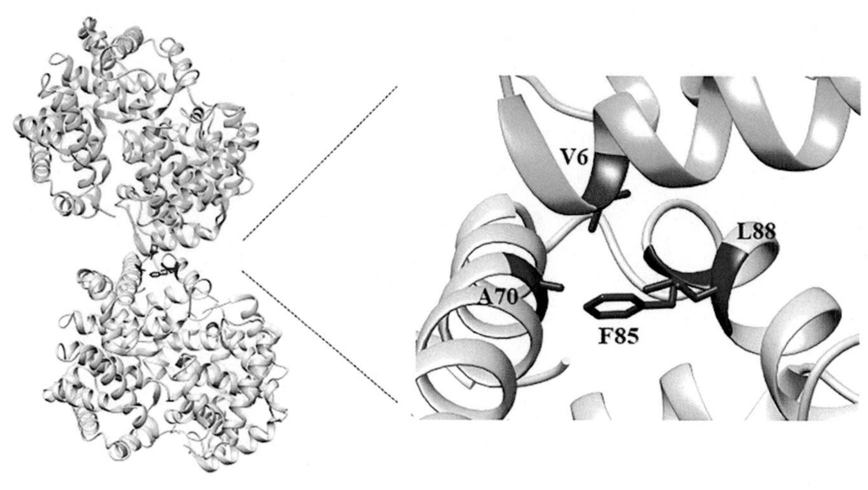

FIG. 18.3

Polymerization of HbS. Two chains (α and β), each of two adjacent tetramers, are shown in *light blue* and *light green*.

Source: Harrington, D.J., Adachi, K., Royer Jr., W.E., 1997. J. Mol. Biol. 272, 398.

hemoglobin (HbS) polymerizes as V6 from one tetramer enters a hydrophobic cavity formed by the residues A70, F85, and L88 in a β chain of the adjacent tetramer (Fig. 18.3). The red blood cell (RBC) containing these polymers loses its deformability and adopts a sickle-like shape. The sickle cells undergo hemolysis and cause anemia.

Based on the structure and polymerizing property of the HbS, it was surmised that an effective way to counter the disease would be to maintain the HbS in the oxygenated state. A structure-guided approach led to the discovery of a new potent allosteric effector GBT440. It is a small molecule that binds to the α chain by forming H-bonds with S131 (Fig. 18.4) and increases the affinity of HbS for oxygen. As a consequence, polymerization is inhibited under hypoxic conditions and sickling of RBCs prevented.

18.3.2 Enzyme inhibitors as drugs

We have seen in Chapter 1 that the first enzyme structures, such as that of lysozyme, were solved in the 1960s. The structures highlighted the selectivity of the enzymes in substrate binding and triggered the ideas about drug design. Initially, in the 1970s and 1980s, clinically important drug targets, such as human renin, were modeled on their homology with supposedly less interesting enzymes. Early inhibitors of renin were developed based on its model; later, when high-resolution crystallographic structures of renin and its complexes became available, more effective drugs were designed.

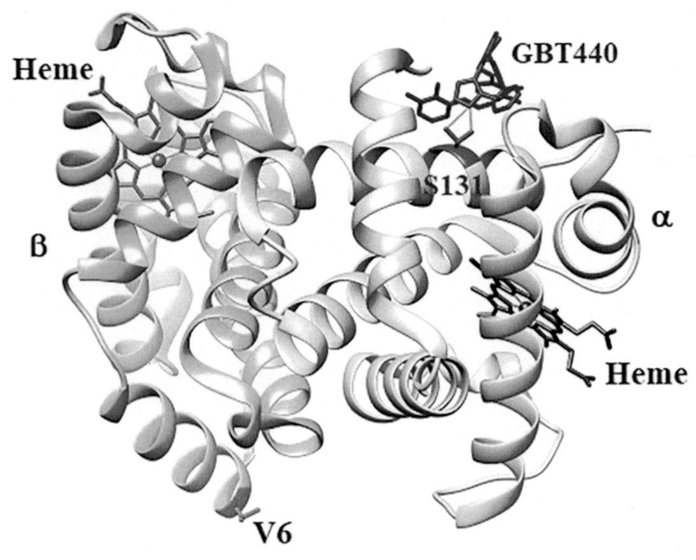

FIG. 18.4

Binding of GBT440 to HbS. H-bonds are shown by *green* lines.

Source: Oksenberg, D., et al., 2016. Br. J. Hematol. 175. 141.

The renin-angiotensin system, which controls human blood pressure, comprises renin, angiotensinogen (AGT), angiotensin converting enzyme (ACE), and angiotensin receptors. Renin is an aspartic protease that catalyzes the cleavage between L10 and V11 of AGT to release the N-terminal angiotensin I decapeptide. The latter is subsequently processed by ACE to form angiotensinogen II which raises the blood pressure by a number of actions. High substrate selectivity, activity at the rate-limiting step of the pathway and, most importantly, availability of the three-dimensional structure of renin have made it an attractive target for structure-guided drug design.

The active-site cleft in renin contains two aspartates (D32 and D215) that form a catalytic dyad. Binding of AGT to renin positions the scissile L10-V11 bond into the catalytic cleft for subsequent cleavage (Fig. 18.5). Crystal structures have shown that the inhibitors (such as aliskiren) can occupy the active sites of renin and prevent the cleavage of AGT (Fig. 18.6).

18.3.3 HIV protease as drug target

One of the most remarkable examples of structure-guided drug design effort has been the development of drugs for the treatment of human immunodeficiency virus (HIV) infection. The three-dimensional structure of HIV protease was solved in 1989. Based on its structure-activity relation, HIV protease was considered to be a good target for drug discovery.

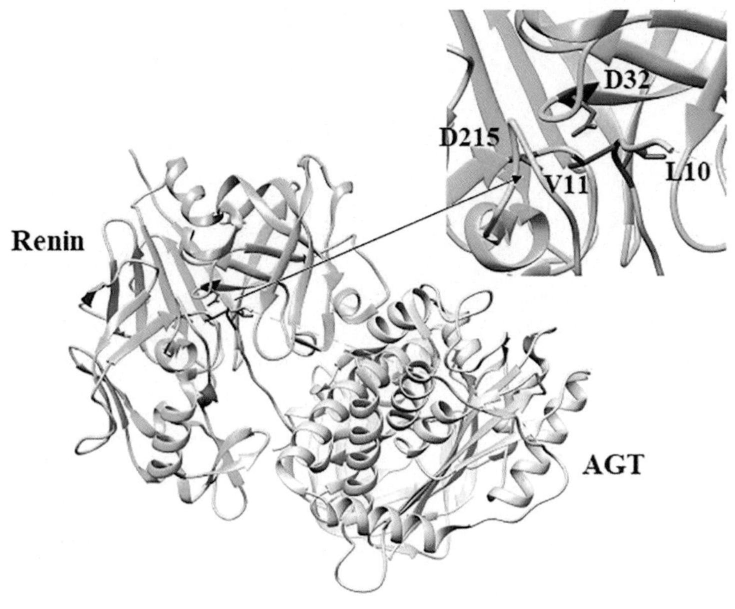

FIG. 18.5

AGT-Renin complex.

Source: Yan, Y., et al., 2018. J. Biol. Chem. doi:10.1074/jbc.RA118.006608.

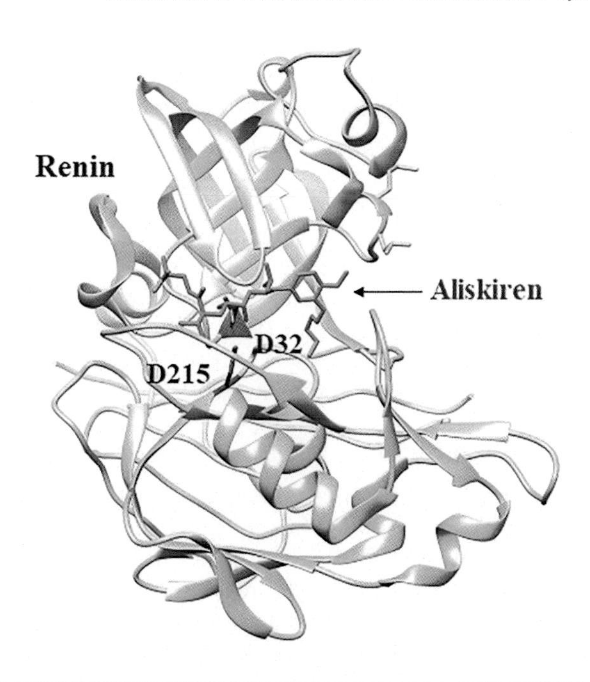

FIG. 18.6

Crystal structure of renin with inhibitor (aliskiren).

Source: Rahuel, J., et al., 2000. Chem. Biol. 7, 493.

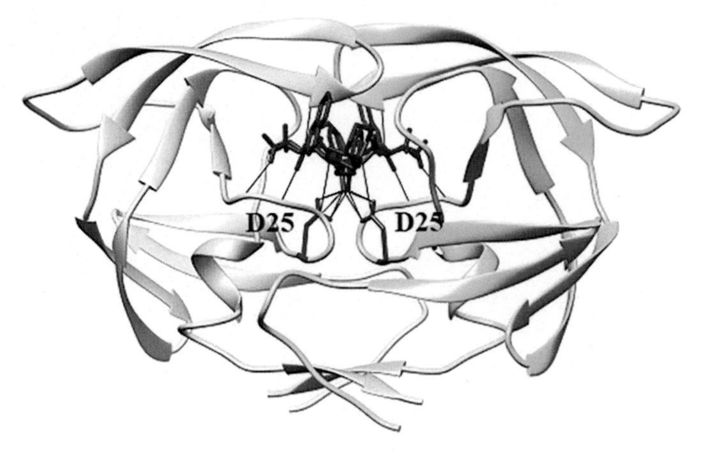

FIG. 18.7

HIV protease dimer *(light blue and light green)* complexed with saquinavir *(red)*. H-bonds are shown by *dark grey* lines.

Source: Krohn, A., et al., 1991. J. Med. Chem. 34, 3340.

The protease hydrolyzes the Gag and Gag-Pol polyproteins of the virus at different cleavage sites, producing the structural proteins (viral envelope glycoproteins), and the reverse transcriptase (RT), integrase (IN), and protease (PR) enzymes which are all packed into new virion particles. PR is a homodimer with each of the two subunits containing 99 amino acid residues (Fig. 18.7). Two aspartate residues (D25) belonging to the two subunits come together at the dimer interface and form the catalytic site of the enzyme. Two flexible glycine-rich β sheets form flaps over the top of the active site. The flaps move closer to the catalytic residues when the enzyme binds a substrate.

Based on the structure of the protease, some successful antivirals were developed. The first one to reach the market, with prompt approval from the Food and Drug Administration (FDA), was saquinavir, developed by Roche Pharmaceuticals. For its action, saquinavir binds to protease dimer forming several H-bonds with the catalytic aspartates and surrounding residues (Fig. 18.7). Binding of the normal substrate to the protease is, thus, prevented.

18.3.4 Targeting interaction networks

In the postgenomic era, it has been already recognized that a global approach is necessary to describe complex diseases. We have seen in Chapter 16 that most biological processes involve a complex network of macromolecular interactions. Likewise, the origin of diseases is most often associated, not with individual molecules, but an interaction network. Hence, a single target may not be enough to counter a disease. Sometimes, mild inhibition of multiple targets can be more effective than complete inhibition of a single one. Evidently, biomolecular interaction networks, protein-

protein interaction (PPI) networks in particular, are of paramount importance in rational drug design.

We may recall that, in a PPI network map (Fig. 18.8), the proteins are depicted as "nodes" and the interactions as "edges." Some of the proteins, behaving as "hubs," are associated with a large number of interacting partners. A clinically important protein, human p53, which is found mutated in different types of cancer, is an example of a hub protein (Fig. 18.9).

Initially, drug discovery efforts have targeted single or multiple nodes (proteins). However, such a target may occupy a central position in a network and, particularly if it is a hub protein, interact with a variety of other macromolecules. In such case, an inhibitor can block multiple pathways indiscriminately. An alternative is to target the relevant edges, that is, design specific inhibitors of PPIs. Targeting the PPI interface, a drug can modulate the interactions—destabilize or stabilize them, depending upon which state of the interactions is responsible for the concerned disease. Destabilization can be caused by competitive binding of the inhibitor at the PPI interface (orthosteric disruption) or binding of the small molecule at a site remote to the interface (allosteric disruption). A specific modulator can also increase the PPI affinity by binding to a site at the interface (orthosteric stabilization).

One of the approaches for the development of small molecule PPI inhibitors has been the identification of interfacial "hot-spots." Energetic analysis of protein binding interfaces has shown that mutations of some interfacial residues have little effect on the binding affinity, whereas hot-spot mutants have major destabilizing effects. It is assumed that, for disruption of the PPI, an inhibitor needs to interact only with a hot-spot and not the entire PPI interface. Experimental determination of hot-spots is tedious and time-consuming; however, computational methods have been developed based on the structure of the interacting partners.

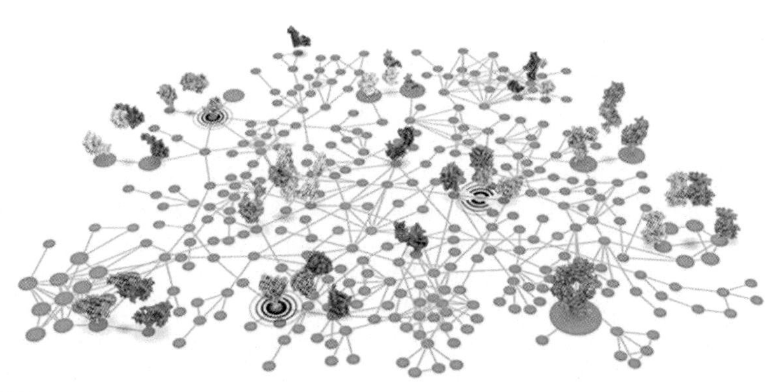

FIG. 18.8

A typical PPI network map. The thin *blue* lines indicate interactions between the proteins.

Courtesy: Duran-Frigola, M., Mosca, R., Aloy, P., 2013. Structural systems pharmacology: the role of 3D structures in next-generation drug development. Chem. Biol., with permission from Elsevier.

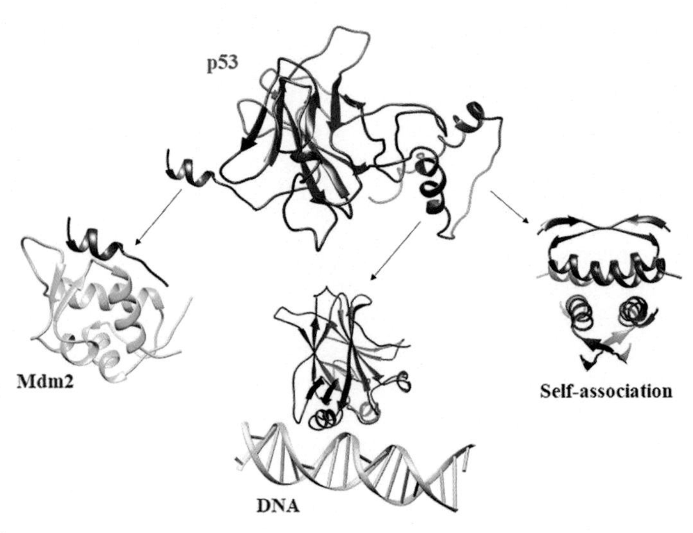

FIG. 18.9

Some interaction partners of p53.

Source: Kussie, P.H., et al., 1996. Science 274, 948; Cho, Y., et al., 1994. Science 265, 346; Lee, W., et al., 1994. Nat. Struct. Mol. Biol. 1, 877.

18.3.5 Computational drug screening

Once a drug target is identified (be it around a catalytic or a regulatory site), the next logical step is to "explore the chemical space" by screening libraries of chemical compounds for possible binding. A very powerful approach, which exploits the architecture of the target and screen chemical libraries, is the fragment-based drug discovery (FBDD).

Fragments here are small organic molecules whose size is restricted to avoid their complexity. In FBDD, a library of several thousand compounds is screened against the target of interest. Although the reduced size compromises the selectivity of the compounds, initial hits can be identified. Subsequently, they are subjected to chemical elaboration for the optimization of their binding affinity and drug-like properties.

The fragments are screened by high-throughput docking by carrying out an exhaustive search in discrete space defined by geometry, hydrogen bonds, and hydrophobic contacts. Originally, docking was based on the traditional "lock and key" concept of recognition of a ligand on a protein surface, which was considered to be "frozen and motionless." Later, crystallographic studies have revealed the role of protein flexibility in ligand binding. However, it needs to be emphasized that molecular recognition and drug binding are dynamic processes involving not only conformational changes, but "jiggling" of both target and ligand. When a small molecule, such as a drug, approaches a target (for example, a receptor), it confronts not a single, frozen structure but a macromolecule in incessant motion. Cases of a ligand fitting into a fairly static binding pocket by the "lock and key" mechanism are rare.

We have already seen that MD simulations provide insight into protein dynamics beyond that can be obtained by X-ray crystallography or even cryo-EM SPA. Hence, high-throughput docking is usually followed by MD simulations using implicit-solvent force field. For ligand optimization, simulations in explicit solvent evaluate the kinetics and thermodynamics of binding.

MD simulations can even identify what are known as cryptic binding sites on the target protein. These sites, some of which are later found to be druggable, are not detectable in ligand-free structures. A cryptic site forms a pocket only in a ligand-bound structure. These sites are mostly located away from the main functional site of the protein.

Needless to say, the chapter does not offer an exhaustive coverage of the field of computational molecular biology. Nevertheless, some fundamentals have been explained and the dynamic nature of biomolecules and biomolecular processes has been highlighted. A brief discussion on structure-based drug discovery has underlined the enormous power of computation in addressing the biomedical needs of human society. With further advances in computer hardware and algorithm design, molecular structural biology will be able to tackle problems of even greater complexity and play a pivotal role in many other fields of human interest, including environment and industry.

Sample questions

1. How does simulation overcome the limitations of experimentation in the study of biological processes?
2. What is a molecular mechanics force field? How is a force field defined for a molecular system?
3. How are MD trajectories computed for the atoms in a molecule?
4. What are the two remarkable features of the potassium channel uncovered by MD simulations?
5. What are the structural properties of HbS and renin based on which some effective drugs against these targets have been designed?
6. How has the development of antiviral saquinavir been guided by the structure of an HIV-coded enzyme?
7. What is the basis of selection of PPIs as drug targets?
8. What is the role of molecular dynamics in drug screening?

References and further reading

Allen, M.P., Attig, N., et al., 2004. Introduction to molecular dynamics simulation. In: Computational Soft Matter: From Synthetic Polymers to Proteins, NIC Series. vol. 23, pp. 1–28 ISBN 3-00-012641-4.

Bakail, M., Ochsenbein, F., 2016. Targeting protein-protein interactions, a wide open field for drug design. C. R. Chim. 19, 19–27.

Becker, O.M., MacKerell Jr., A.D., Roux, B., Watanabe, M. (Eds.), 2001. Computational Biochemistry and Biophysics. Marcel Dekker, Inc.

Duran-Frigola, M., Mosca, R., Aloy, P., 2013. Structural systems pharmacology: the role of 3D structures in next-generation drug development. Chem. Biol. 20, 674–684.

Durrant, J.D., McCammon, J.A., 2011. Molecular dynamics simulations and drug discovery. BMC Biol. 2011 (9), 71. http:/www.biomedcentral.com/1741-7007/9/71.

Fischer, G., Rossmann, M., Hyvönen, M., 2015. Alternative modulation of protein-protein interactions by small molecules. Curr. Opin. Biotechnol. 35, 78–85.

Furini, S., Domene, C., 2013. K^+ and Na^+ conduction in selective and nonselective ion channels via molecular dynamics simulations. Biophys. J. 105, 1737–1745.

Ghatge, M.S., et al., 2016. Crystal structure of carbonmonoxy sickle hemoglobin in R-state conformation. J. Struct. Biol. 194 (3), 446–450.

Ghosh, A.K., Osswald, H.L., Prato, G., 2016. Recent progress in the development of HIV-1 protease inhibitors for the treatment of HIV/AIDS. J. Med. Chem. 59 (11), 5172–5208.

Imaeda, Y., et al., 2016. Structure-based design of a new series of N-(piperidin-3-yl) pyrimidine-5-carboxamides as renin inhibitors. Bioorg. Med. Chem. 24, 5771–5780.

Jensen, M.Ø., et al., 2010. Principles of conduction and hydrophobic gating in K^+ channels. Proc. Natl. Acad. Sci. U. S. A.. https://doi.org/10.1073/pnas.0911691107.

Khalili-Araghi, F., Tajkorshid, E., Schulten, K., 2006. Dynamics of K^+ ion conduction through Kv1.2. Biophys. J.. https://doi.org/10.1529/biophyaj.106.091926 L72-74.

Levitt, M., Sharon, R., 1988. Accurate simulation of protein dynamics in solution. Proc. Natl. Acad. Sci. U. S. A. 85, 7557–7561.

Metcalf, B., et al., 2017. Discovery of GBT440, an orally bioavailable R-state stabilizer of sickle cell hemoglobin. ACS Med. Chem. Lett. 8, 321–326.

Mulholland, A., 2008. Introduction. Biomolecular simulation. J. R. Soc. Interface 5, S169–S172.

Oksenberg, D., et al., 2016. GBT440 increases haemoglobin oxygen affinity, reduces sickling and prolongs RBC half-life in a murine model of sickle cell disease. Br. J. Haematol. 175 (1), 141–153.

Rahuel, J., et al., 2000. Structure-based drug design: the discovery of novel nonpeptide orally active inhibitors of human renin. Chem. Biol. 7, 493–504.

Shanon, R.E., 1998. Introduction to the art and science of simulation. In: Medeiros, D.J. et al., (Ed.), Proceeding of the 1998 Winter Simulation Conference, pp. 7–14.

Thomas, S.E., et al., 2017. Structural biology and the design of new therapeutics: from HIV and cancer to mycobacterial infections. A paper dedicated to John Kendrew. J. Mol. Biol. 429, 2677–2693.

Van der Kamp, M.W., et al., 2008. Biomolecular simulation and modelling: status, progress and prospects. J. R. Soc. Interface 5, S173–S190.

van Holde, K.E., Johnson, W.C., Ho, P.S., 2006. Principles of Physical Biochemistry, 2nd ed. Pearson Prentice Hall, Pearson Education Inc.

Yan, Y., et al., 2018. Structural basis for the specificity of renin-mediated angiotensinogen cleavage. J. Biol. Chem.. https://doi.org/10.1074/jbc.RA118.006608.

Index

Note: Page numbers followed by *f* indicate figures, *t* indicate tables, and *s* indicate schemes.

D

E

F

G

H

I

J

K

L

N

Q

R

S

T

Printed in the United States
By Bookmasters